THE ENCYCLOPEDIA OF
FUNGI
OF BRITAIN AND EUROPE

THE ENCYCLOPEDIA OF
FUNGI
OF BRITAIN AND EUROPE

MICHAEL JORDAN

David & Charles

DIANE
for her sharp eyes and enthusiasm through sun, rain, frost and snow

---

ACKNOWLEDGEMENTS

A great many people have contributed to the making of this book, to all of whom I owe a considerable debt of gratitude. In particular the following have given their invaluable field support in locating and identifying specimens for photography: Pat Andrews, Ted Blackwell, Alison Bolton, Roger Brown, Dave Champion, Keith and Valerie Davies, Gordon Dickson, Martin Ellis, Ernest Emmett, Reg and Lil Evans, Shelley Evans, Bill Forsyth, Alan Hills, Doug and Brenda Hunt, Richard Iliffe, Diane Jordan, Roger Kemp, John Keylock, Linda Klyne, Sid Lines, Ann Leonard, Peter Long, Geoff Miller, Mike Miner, Jonathan Revett, Maurice Rotheroe, Sara Shepley, Dave Shorten, Malcolm Storey, Joan Treece, Roy Watling, John Webster, Alastair Young. Many more offered their help generously but for one reason or another it could not be taken up.

The following photographs are by kind permission:
Boletus Satanas (main illus.) David Lester
Pleurotus dryinus Graham Mattock
Rhizina undulata John Webster
Serpula lacrymans Rentokil Laboratories
I am deeply indebted to David Pegler and Brian Spooner at the Royal Botanic Gardens, Kew, for their co-operation during three years of work overall. Both offered much time and patience in respect of identifying specimens and advising on the systematics and nomenclature.

The British Mycological Society has been a constant source of support and encouragement and without the efforts of Affiliated Groups and their members in the field, many species would have never come under the camera lens.

My enduring thanks must go to Jack Marriott. Perhaps more than any other person he has contributed to the making of this encyclopedia. His constant encouragement, expertise and guidance, permitting me to join in the activities of the Forest of Dean BMS Affiliated Group, guiding me to locations, patiently identifying field material, advising on the format of text and the layout of the book, have been invaluable.

I am also indebted to the following organisations:
The British Mycological Society; Forest Enterprises (Culbin); The National Trust (Stourhead, Wilts); the Royal Botanic Gardens, Kew, London; the Royal Society for the Protection of Birds; University of Exeter; University of Leicester.

IMPORTANT NOTE TO THE READER
Many fungi are toxic, and individual reactions to them vary widely. The rule is: do not eat it unless you are sure you have accurately identified it. Neither the author nor the publishers can accept any legal responsibility or liability for errors in identification however caused, or for individual reactions to the consumption of fungi.

CONTENTS

FOREWORD

We live in an age of increasing interest in the natural world, stimulated to a large degree by excellent television programmes. The kingdom of fungi has been one of the beneficiaries of this new awareness: whereas in the past many people, in their walks through woodland or grassland, saw only flowers and butterflies, they are now conscious of the huge variety of mushrooms and toadstools at every turn. A whole new world has opened up and what was once overlooked or trampled upon or dismissed as poisonous or destructive has become an object of curiosity.

The fascination of fungi lies in their immense diversity – of shape, colour, smell, taste and ecological preferences. On any walk in autumnal woods even the experienced amateur is likely to encounter several species not seen before. And, in this country alone, it is safe to say that several new species (to Britain or to science) are discovered every year. These discoveries can be made equally well by the alert amateur mycologist, but often only the professional can provide the incontrovertible evidence required to prove that a specimen is genuinely new.

This very diversity leads to complexity and makes identification difficult. Most people need personal guidance at least in their first attempts to identify fungi in the field. The best guide is a competent mycologist and these days many local groups exist, encouraged by the British Mycological Society, where one can receive tuition in the recognition of species. But this is not available to all, and in any case knowledge gained in the field should be reinforced by private study with a good reference book. Given time and application the day will come when the identity you propose for a fungus is approved by your mentor to your immense satisfaction! Much is expected of such a book. The foremost requirement is accuracy: the illustrations must give a faithful representation of the fungus in question and the text must include the characteristics necessary to distinguish one species from another. The written material should be easy to assimilate and specialised terms must be explained: mycology, like any other branch of science, has a jargon of its own and the reader cannot be expected to understand it immediately. The book must be a pleasure to use: this is a demanding subject and nothing, whether the quality of the illustrations, the layout or the print, should be allowed to get in the reader's way. It should offer a reasonable introduction to the kingdom of fungi, sufficient to explain how it differs in so many aspects from that of plants. The number of fungi is huge, second only in size to that of insects. This makes choosing which species to illustrate extremely difficult: a balance has to be achieved between the inclusion of showy, large species and smaller, less colourful species, between the common and the rare.

Finally the book should aim to satisfy the needs not only of the beginner but also of the more experienced individual who is beginning to use microscopic characteristics in a further refinement of the identification procedure.

In my view, Michael Jordan's *Encyclopedia of Fungi* meets all these criteria. He has exercised the utmost care in ensuring accuracy of illustration and description; names have been checked by Dr David Pegler, of the Herbarium at Kew, and are correct and up-to-date; the book is a pleasure to read and a considerable achievement. It should find a place, close at hand, on the bookshelf of every field mycologist.

JACK V.R. MARRIOTT BSc, PhD
BRITISH MYCOLOGICAL SOCIETY
NATIONAL ASSOCIATES CO-ORDINATOR

INTRODUCTION

When I first caught the mushroom-hunting 'bug' at my London University College the range of mycological field guides was very limited. I suppose I thought that virtually all the larger fungi of Europe were described within the covers of the two slim volumes which reflected the extent of popular publications back in the 1960s.

Until recent years mycological field guides have been plagued by two essential shortcomings. The texts have been prepared in a manner insufficient to permit proper determination of a specimen and the illustrations have taken the form of line drawings or paintings, the latter often crude and of little value in trying to identify a specimen. Few artists have managed to emulate the superb illustrations of fungi created by Jakob E. Lange in the *Flora Agaricina Danica*.

Today, amateur interest in mushroom hunting is growing rapidly. The British Mycological Society has opened its doors beyond the professional élite and welcomes anyone, regardless of qualification, who finds a genuine interest. The Society nevertheless recognises the urgent need to standardise the method by which fungi are identified, or determined, in the field into a logical and progressive sequence of checks and tests.

The object at the outset of this work was to prepare an encyclopedia of British and north European higher fungi. No single volume can hope to provide comprehensive coverage of a group whose numbers run into many thousands and are constantly expanding. This volume is as comprehensive as time and production costs permit and includes detailed descriptions of the macroscopic and microscopic characters of each species following the BMS approved guidelines. The colour photographs are accompanied by a text which in each case reflects the specimens collected and photographed although careful cross-reference has been made to the works listed in the bibliography on page 30, particularly with respect to microscopic dimensions and characters.

It was decided that the illustrations must meet with certain criteria by representing the fungi, *in vivo*, in the habitat in which they grow and with their subtle colours and textures depicted in natural lighting conditions. Very occasionally it has not been possible to follow these criteria wholly but, wherever feasible, the specimens illustrated have been photographed where they have been found and using the normal daylight available at the time. This is a significant departure from other mycological reference books which may either remove the specimens to a controlled studio environment or rely on electronic flash lighting.

To achieve the results contained in the following pages I calculate that I have walked some 2,000 miles during the last three years and driven considerably more. My Nikon camera has withstood the exposure of about 10,000 frames of film and many thousands of specimens have passed through my hands and those of other experts. It has been a great and rewarding challenge. It will, I hope, permit mushroom hunters a fuller and a more thorough opportunity to unravel the identities of some of the most fascinating life forms on earth.

THE BIOLOGY OF HIGHER FUNGI

BIOLOGY

Fungi were once considered as a class grouped with the Algae into the sub-kingdom of plants known as the Thallophyta but modern thinking tends to regard them as a separate kingdom distinguished from most other plants by the absence of chlorophyll in their tissues and by special characteristics in their structure and origins. Because they lack the chlorophyll which green plants utilise to derive organic foods from carbon dioxide and water, they obtain their nourishment, already prepared, from outside sources which comprise living or dead animal or plant tissues. Those which rely on living plant and animal hosts are termed parasitic fungi; those which utilise dead and dying matter, including humus in the soil, are saprophytes, and fungi are thus, in company with bacteria, amongst the earth's most efficient rubbish disposers. In certain instances the fungus arrives at a mutually beneficial association with a plant, usually through the roots of a tree, an arrangement which is termed *mycorrhizal.* In these cases the fungus is usually very precise in its habitat requirement: *Leccinum holopus,* for example, is only ever associated with birches.

The individual 'building blocks' of a fungus are the *hyphae.* These are thread-like tubes or filaments, whose walls consist of fungal cellulose, and which join and are woven together to form a cotton-wool like mass, the *mycelium.* The latter represents the vegetative part of the fungus which lives on or in the substrate and which, under certain conditions, produces more solid reproductive structures, the fruiting bodies or *sporophores.* Generally, since these reproductive parts are designed for the production and distribution of spores (the fungal equivalent of seeds), they become pushed up above the surface of the substrate and it is the sporophores which form the basis of interest for the mushroom hunter.

Many fungi are wholly microscopic in size and are not of concern in this book. Those which develop fruiting bodies large enough to interest the mushroom hunter fall, by and large, into the Class Ascomycotina and the Class Basidiomycotina which are distinguished essentially by the structure of their spore-producing organs.

The Ascomycotina generate fruiting bodies known as *apothecia* (open) when they tend to be saucer-shaped, and *perithecia* (closed) which appear as more spherical objects. In either case, the germinative hyphae of these structures, which collectively form a *hymenium,* produce elongated cylindrical flask-shaped *asci,* often separated by special sterile hyphae, the *paraphyses.* It is within the *asci* that the *ascospores* develop and from which, when mature, they are released under pressure through an apical opening, which either appears by the release of a distinct lid, the *operculum,* or (*inoperculate*) by breakdown of the ascus tip. The spores are squirted out as a result of some stimulus (often a rain drop) being applied to the hymenial surface and are then carried away on air currents until they come to rest and germinate in a suitable location.

The Basidiomycotina operate on a similar principle

but with important structural distinctions. Amongst the majority, the hymenial surfaces are studded with microscopic club-shaped structures, the *basidia*, from which project small 'fingers' or *sterigmata* (usually 4, sometimes 2 and occasionally up to 8) at the tips of which develop the *basidiospores*. The fertile surfaces of the Basidiomycotina are usually protected in some way since there is no enveloping *ascus* to shield the spores, as they grow, from the rigours of the environment. The vast majority of fungi within the group thus produce their spores on the surfaces of radiating plate-like structures, the gills or *lamellae*, which hang beneath, and are protected by, a cap or *pileus*. This, in turn, is usually raised above the surrounding substrate, by a stem or *stipe*. In others gills are replaced by tubes opening through pores, or by spines. The Gastromycetes adopt a quite different structure and frequently their hymenial tissue, the *gleba*, is contained within a protective envelope or *peridium*.

Within the 'mushroom-types' of the Basidiomycotina, the fruiting body pushes up through the surface of the substrate as a 'button', the upper portion of which grows at a faster rate than the rest and develops into the familar umbrella-shaped cap on its stem. Thus, before it expands, the delicate hymenium is contained within a cavity, the *annular* cavity, in the roof of which the gills (or tubes) develop. The cavity is sealed, until the cap stretches, by a thin wall of tissue, the veil or *velum*.

When the latter ruptures it either disperses more or less without trace or it leaves a ring, or *annulus,* on the stem. This ring feature is characteristic of certain genera including *Amanita* and *Agaricus*. In a limited number of fungi the veil consists of not one but two layers, the smaller of which - the *partial veil* - stretches across the gill cavity. The more extensive wall of tissue, the *universal veil,* not only covers the gill cavity but also extends to the base of the stem. When this veil ruptures it may leave patches on the cap, belt-like rings below the *annulus* and a bag-like *volva* surrounding the stem base. The universal veil is chiefly a feature, again, of the *Amanita* genus. The veil does not always appear as a continuous skin of tissue and particularly in the *Cortinarius* genus it forms a fragile cobweb-like *cortina* whose remnants may appear on the stem as a ring-zone.

SYSTEMATICS

The orders and families included in this book are arranged into the system endorsed by the Royal Botanic Gardens, Kew, London at the time of going to press. The arrangement and nomenclature may differ somewhat from that of the Continental European system after Professor Dr Meinhard Moser (1983) and from the nomenclature supplied in the publications of Marcel Bon.

Entries in square brackets, under the Ascomycotina, are alternatives to those which appear in the page headings but which are currently under review for adoption in the revised Kew system; the entry in rounded brackets (Boletales – Coniophoraceae) has been placed out of context but in a placing in the book where it might more conventionally be expected to appear.

The names given in quotation marks in the second column are not generally regarded as falling under the headings of class or order, but nevertheless they represent divisions, based on structure, which are often mentioned in reference books.

CLASS	ORDER	FAMILY
ASCOMYCOTINA		
'Discomycetes'		
	Pezizales	
		Ascobolaceae
		Otidiaceae
		Sarcoscyphaceae
		Pezizaceae
		Helvellaceae
		Morchellaceae
	[Tuberales]	
		Tuberaceae
	Helotiales	
	[Leotiales]	Geoglossaceae
		Orbiliaceae
		Dermateaceae
		Hyaloscyphaceae
		Leotiaceae
		Sclerotiniaceae
	Rhytismatales	
		Rhytismataceae
'Pyrenomycetes'		
	Clavicipitales	
	[Hypocreales]	Clavicipitaceae
	Hypocreales	
		Hypocreaceae
	Diatrypales	
		Diatrypaceae
	Sphaeriales	
	[Xylariales]	Xylariaceae
	[Sordariales]	
		Sordariaceae
'Plectomycetes'		
	Elaphomycetales	
		Elaphomycetaceae
BASIDIOMYCOTINA		
'Homobasidiomycetes'		
	Cantharellales	
		Cantharellaceae
		Clavariadelphaceae
		Craterellaceae
		Clavariaceae
		Typhulaceae

CLASS	ORDER	FAMILY
		Clavulinaceae
		Hydnaceae
		Sparassidiaceae
	Gomphales	
		Ramariaceae
	Hericiales	
		Clavicoronaceae
		Auriscalpiaceae
		Hericiaceae
		Lentinellaceae
	Poriales	
	(non-lamellate)	Polyporaceae
		Coriolaceae
	Ganodermatales	
		Ganodermataceae
	Fistulinales	
		Fistulinaceae
	Hymenochaetales	
		Hymenochaetaceae
	Stereales	
		Hyphodermataceae
		Peniophoraceae
		Corticiaceae
		Stereaceae
		Podoscyphaceae
		Meruliaceae
		Sistotremataceae
		Botryobasidiaceae
	Thelephorales	
		Bankeraceae
		Thelephoraceae
	Schizophyllales	
		Schizophyllaceae
	(Boletales	
		Coniophoraceae)
	Agaricales	
		Hygrophoraceae
		Tricholomataceae
		Amanitaceae
		Pluteaceae
		Entolomataceae
		Agaricaceae

CLASS	ORDER	FAMILY
		Coprinaceae
		Bolbitiaceae
		Strophariaceae
	Cortinariales	
		Crepidotaceae
		Cortinariaceae
	Boletales	
		Hygrophoropsidaceae
		Paxillaceae
		Boletaceae
		Xerocomaceae
		Strobilomycetaceae
		Gomphidiaceae
		Rhizopogonaceae
		Gyrodontaceae
	Russulales	
		Russulaceae
	Poriales (lamellate)	
		Lentinaceae
	Lycoperdales	
		Lycoperdaceae
		Geastraceae
	Tulostomatales	
		Tulostomataceae
		Battarraeaceae
	Nidulariales	
		Nidulariaceae
	Sclerodermatales	
		Sclerodermataceae
		Astraeaceae
	Phallales	
		Phallaceae
		Clathraceae
'Heterobasidiomycetes'		
	Dacrymycetales	
		Dacrymycetaceae
	Exobasidiales	
		Exobasidiaceae
	Tremellales	
		Tremellaceae
	Auriculariales	
		Auriculariaceae

General Structure of Fungi

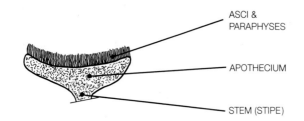

ASCI & PARAPHYSES

APOTHECIUM

STEM (STIPE)

SECTION THROUGH GENERALISED DISCOMYCETE

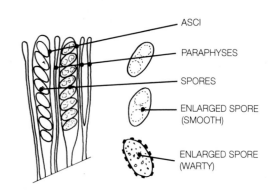

ASCI

PARAPHYSES

SPORES

ENLARGED SPORE (SMOOTH)

ENLARGED SPORE (WARTY)

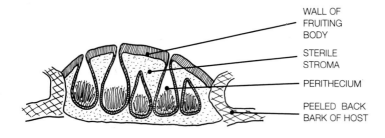

WALL OF FRUITING BODY

STERILE STROMA

PERITHECIUM

PEELED BACK BARK OF HOST

SECTION THROUGH GENERALISED PYRENOMYCETE

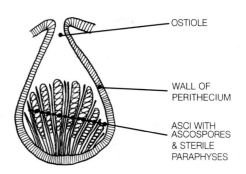

OSTIOLE

WALL OF PERITHECIUM

ASCI WITH ASCOSPORES & STERILE PARAPHYSES

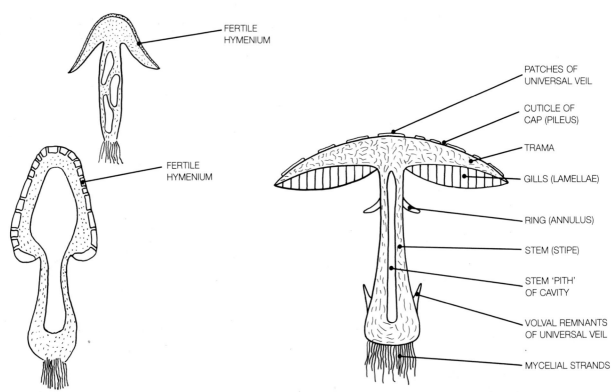

FERTILE HYMENIUM

FERTILE HYMENIUM

PATCHES OF UNIVERSAL VEIL

CUTICLE OF CAP (PILEUS)

TRAMA

GILLS (LAMELLAE)

RING (ANNULUS)

STEM (STIPE)

STEM 'PITH' OF CAVITY

VOLVAL REMNANTS OF UNIVERSAL VEIL

MYCELIAL STRANDS

SECTION THROUGH GENERALISED HELVELLA   SECTION THROUGH GENERALISED AMANITA

SHAPES OF BASIDIOMYCETE FUNGI

CAP PROFILES

HEMISPHERICAL OR 'BUN-SHAPED'   CONVEX   FLATTENED   BLUNTLY UMBONATE ('SHIELD-SHAPED')   CAMPANULATE ('BELL-SHAPED')   SHARPLY UMBONATE

UMBILICATE ('NAVEL-SHAPED')   CONICAL   ECCENTRIC   INFUNDIBULIFORM ('FUNNEL-SHAPED')

STEM PROFILES

MORE OR LESS EQUAL   CLAVATE   BULBOUS   MARGINATE BULBOUS   VENTRICOSE   ROOTING

GILL PROFILES

FREE   ADNEXED   ADNATE   EMARGINATE (NOTCHED IS SLIGHTLY MORE ABRUPT AT THE STEM)   DECURRENT

MICROSCOPIC PROFILES

SPORES

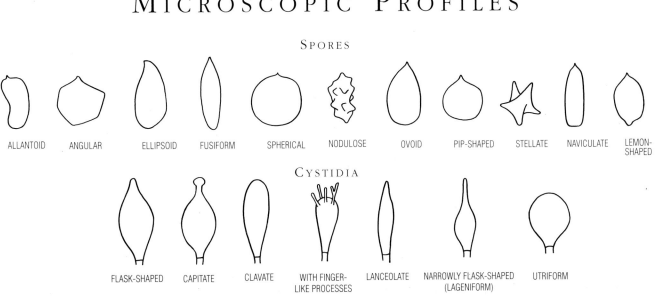

ALLANTOID   ANGULAR   ELLIPSOID   FUSIFORM   SPHERICAL   NODULOSE   OVOID   PIP-SHAPED   STELLATE   NAVICULATE   LEMON-SHAPED

CYSTIDIA

FLASK-SHAPED   CAPITATE   CLAVATE   WITH FINGER-LIKE PROCESSES   LANCEOLATE   NARROWLY FLASK-SHAPED (LAGENIFORM)   UTRIFORM

HOW TO USE THIS BOOK

DESCRIPTION OF ENTRIES

Each photograph in this book is accompanied by a descriptive text with all the information reasonably needed to identify a specimen. It is, however, important to remember that a fungal fruiting body can alter markedly in shape, size and colour according to age, moisture content and other factors. Wherever possible the photograph represents the fungus in its mature state, perhaps with immature specimens adjacent. The description again reflects the mature condition.

Each block of text is laid out in a format which most effectively permits identification in a methodical manner. This ordering has been approved, through practical and extensive field experience, by the British Mycological Society.

The entries are organised as follows:
1. A brief general description followed by the type of habitat. In some instances the latter is quite specific eg. 'with birch' or 'in sphagnum', whilst in other cases it is more general e.g. 'under broad-leaf trees'.
2. Dimensions. Macroscopic dimensions of the fruiting body are recorded in centimetres or fractions thereof. Cap diameter represents the range of measurements taken from edge to edge passing through the centre point. Stem length is taken from the base to the point of union with the cap. Stem diameter is an approximate mean measurement. It should be remembered that a fungus, like any other organism, will increase in size as it matures.

3. Fruiting body. The general description of the visible, above-ground organs in those fungi which do not possess distinct cap, gills and stem.

4. Cap. Colour, shape, surface texture and marking as well as, in certain instances, behaviour of the cuticle when peeled and presence or absence of remnants of the veil. A description of the cap flesh follows and, in some cases, microscopic details of cells in the cuticle (dermatocystidia) are described. Colour changes, where relevant, are also indicated.

5. Gills. Colour (when mature), shape, type of attachment to the stem, depth and relationship one to another. The microscopic details of spores follow, including colour, surface texture, shape, chemical reactions, contents and size. Finally the section includes details of the number of spores on each basidium, and of the gill cystidia and other microscopic structures associated with the gills etc. This section can only be utilised fully with the aid of a microscope.

6. Stem. Details are included in a similar order to those given for the cap with additional information about the presence or absence of a ring, rooting base, etc. A description of the stem flesh follows which, it should be noted, in certain instances differs markedly from that of the cap. Colour changes, where relevant, are indicated.

7. Odour and taste. Both positive and negative information is noted. Unless specifically warned against, tasting small amounts on the tip of the tongue and spitting out the remains will do no physical harm.

8. Reactions to chemical tests, where applicable, are described. The text also indicates if no distinguishing reactions take place.

9. Occurrence. This is not a critical factor in identification for which reason it is the last consideration when identifying a species. Fungi will often 'get their clocks wrong' and appear out of season. Furthermore, across a wide range of latitudes and altitudes, seasonal appearance can vary markedly and can also be affected by climatic changes, including temperature and rainfall. It should be appreciated that a species which appears commonly in one location may be extremely rare, within a similar habitat, elsewhere. Fungi are also unpredictable in their cycles of fruiting. A fungal mycelium may lie dormant for a year or more. It is thus important never to be prejudiced by remarks given under 'occurrence'.

10. The edibility of the fruiting body completes the entry under headings 'edible', 'inedible', 'poisonous' and 'lethally or dangerously poisonous'. Those identified as inedible will cause no harm if consumed but are uninteresting or unpalatable.

In a desire to conform to an internationally recognised standard when denoting the edibility or otherwise of a species, this book follows the arrangement of symbols established in *Fungi of Switzerland*, Breitenbach, J. and Kränzlin, F.

□ EDIBLE – recognised as being, to varying extent, of culinary worth when fresh and uncontaminated.

■ INEDIBLE – unpalatable but otherwise harmless if ingested.

✙ POISONOUS – causing adverse physical and/or mental symptoms, including serious illness though rarely, if ever, mortality, if ingested (this category includes certain species whose toxins are thermolabile and therefore destroyed on cooking).

✚ LETHALLY POISONOUS – almost invariably leading to death if ingested, even in small quantity. It is recommended that, where stated, no portion of the tissues of species denoted thus should be tasted (this category includes certain species whose toxins are thermolabile and may therefore be destroyed on cooking).

Note: certain individuals exhibit allergic reactions to fungi which others consume with no ill-effect. It is advisable, when eating fungi for the first time, to sample small amounts. Neither the author nor the publishers can accept any legal responsibility or liability for errors in identification however caused, or for individual reactions to the consumption of fungi.

IDENTIFICATION CHART OF COLOURS REFERRED TO IN THIS BOOK

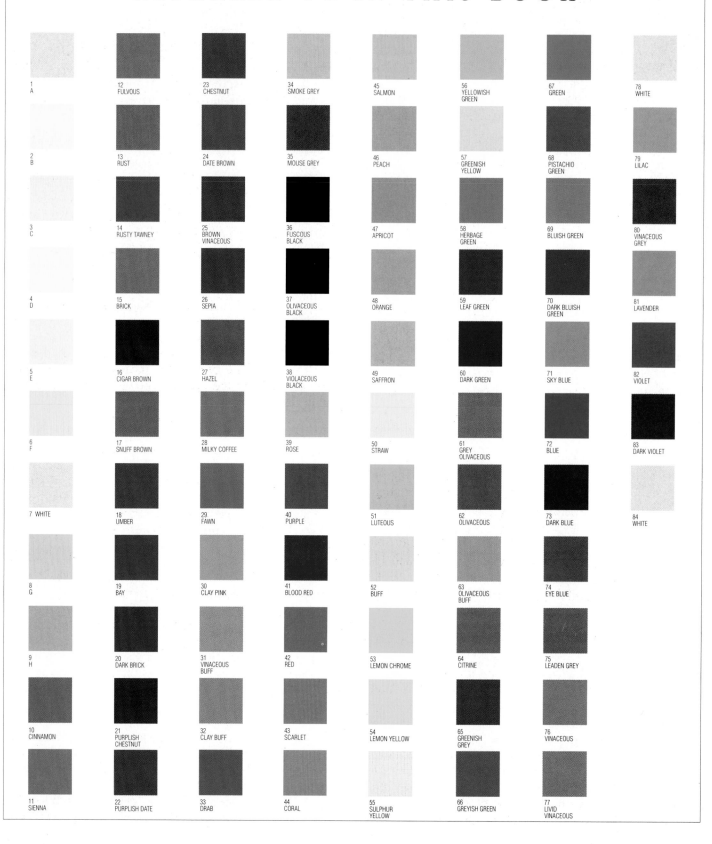

1 A	12 FULVOUS	23 CHESTNUT	34 SMOKE GREY	45 SALMON	56 YELLOWISH GREEN	67 GREEN	78 WHITE
2 B	13 RUST	24 DATE BROWN	35 MOUSE GREY	46 PEACH	57 GREENISH YELLOW	68 PISTACHIO GREEN	79 LILAC
3 C	14 RUSTY TAWNEY	25 BROWN VINACEOUS	36 FUSCOUS BLACK	47 APRICOT	58 HERBAGE GREEN	69 BLUISH GREEN	80 VINACEOUS GREY
4 D	15 BRICK	26 SEPIA	37 OLIVACEOUS BLACK	48 ORANGE	59 LEAF GREEN	70 DARK BLUISH GREEN	81 LAVENDER
5 E	16 CIGAR BROWN	27 HAZEL	38 VIOLACEOUS BLACK	49 SAFFRON	60 DARK GREEN	71 SKY BLUE	82 VIOLET
6 F	17 SNUFF BROWN	28 MILKY COFFEE	39 ROSE	50 STRAW	61 GREY OLIVACEOUS	72 BLUE	83 DARK VIOLET
7 WHITE	18 UMBER	29 FAWN	40 PURPLE	51 LUTEOUS	62 OLIVACEOUS	73 DARK BLUE	84 WHITE
8 G	19 BAY	30 CLAY PINK	41 BLOOD RED	52 BUFF	63 OLIVACEOUS BUFF	74 EYE BLUE	
9 H	20 DARK BRICK	31 VINACEOUS BUFF	42 RED	53 LEMON CHROME	64 CITRINE	75 LEADEN GREY	
10 CINNAMON	21 PURPLISH CHESTNUT	32 CLAY BUFF	43 SCARLET	54 LEMON YELLOW	65 GREENISH GREY	76 VINACEOUS	
11 SIENNA	22 PURPLISH DATE	33 DRAB	44 CORAL	55 SULPHUR YELLOW	66 GREYISH GREEN	77 LIVID VINACEOUS	

NOMENCLATURE

Very few fungi possess authentic common-or-garden names and, of those that do, most fall into the category of 'edible toadstools' – St George's Mushroom, the Cep, the Horse Mushroom, the Chanterelle and others. There is a temptation to include common English or Continental names for as many species as possible but some of those which appear in current popular books are highly questionable and so common names have been omitted unless they are well established through tradition. For the vast bulk of fungi the Latin name is the only acceptable option, a situation not quite as daunting as it may first appear – we are, after all, familiar and at ease with such everyday garden names as *Chrysanthemum, Crocus, Gypsophila* and *Buddleia*.

The Latin name involves two words (the basis of the internationally recognised binomial system) which are frequently printed in italics. (Note that in this book the principal names against the illustrations are in bold Roman type). The first is the name of the genus, e.g. *Amanita*, the second is the name of the species e.g. *muscaria*; thus one arrives at *Amanita muscaria*, the well-known red-and-white spotted 'fly agaric'.

Fungi present much greater difficulty than green plants to consign to a particular genus and, frequently, their position in classification changes as our knowledge of fungi evolves. The rate of change in nomenclature and grouping is confusing to the amateur mycologist who may feel frustrated at being constantly out-of-date but the shifting is generally a reflection of the advance in understanding of mycological structure and behaviour. It is this ongoing debate which explains the necessity to include synonyms and which, in part, accounts for the 'citation' that appears after the name of the fungus: both act as additional pointers towards accurate identification. The 'starting point' for fungal classification was, until recently, regarded as the work of Elias Fries (1794-1878) but it is now considered preferable to go back to his predecessor, Carl Linnaeus (1707-1778). In some instances the citation is simple, thus *Lactarius circellatus* Fr. means that Elias Fries originally named the species and that his naming has never changed.

When two citing authors appear separated by a colon it denotes that one author sanctioned the use of a name proposed by another. When part of the authorship appears in brackets, on the other hand, it denotes a change in the naming. *Amanita crocea* (Quél.) Kühn. & Romagn. was originally given the species name *crocea* by Quélet but the genus in which it is now included was revised by Kühner and Romagnesi, who transferred it from *Amanitopsis* to *Amanita*.

The principal titles given in this book are those endorsed at the time of going to press by the Royal Botanic Gardens, Kew but the synonyms are useful when wishing to relate a species to an entry found elsewhere under a different or obsolete title. Limitations on text space do not always permit a comprehensive listing of synonyms but should a species be identified in another volume under a name that does not appear here, the recommendation is to consult the index, beginning on page 377, in which the synonyms are listed alphabetically. In the index the species name will always precede that of the genus because names of genera tend to be less stable than those of species. Entries in roman again reflect the principal names, those in italics the more recent synonyms.

A list of names and dates of citing authors (by convention often abbreviated) is provided on pages 27-29.

KEYS

How to use the Keys

The keys are designed to assist the collector in tracing an unfamiliar specimen to its correct genus. The effectiveness of any key is limited by the amount of information it contains; while it should aim to be as comprehensive as possible, the inclusion of too much information makes the key too complicated to follow. All of the genera featured in this book, with the exception of the Myxomycetes, are included in the three keys that follow.

The first stage of identification requires you to make a preliminary choice between key A, B or C, and making the correct choice here is a skill best gained through experience. As a rough generalization, fungi without gills are to be found in keys A or B. Those which are cup-, saucer-, cushion- or antler-shaped will often be found in key A, as will the pinhead types, while bracket- or crust-like varieties and gelatinous fungi will probably be found in key B. Key C contains all of the typical toadstool-shaped specimens with either gills or pores underneath. These rules are by no means universal, but the general principle in using any key is this: if you don't reach a satisfactory determination by one route, simply retrace your steps and try another.

The keys are organised in paired groups of options, and these pairs are numbered in ascending order in the left-hand column (1a, 1b; 2a, 2b; etc.). Having first decided which key (A, B or C) to use, start at the top (pair 1) and choose the most appropriate option to describe the specimen in question. In the column to the right of the description you will find either a possible solution in the form of a name, or the number of a further option. Moving down the key to the number indicated once again offers either a possible solution or a further pair of options to choose from, and so on until the search ends with no further options offered.

The numbers in brackets which precede certain descriptions refer to the option from which the present level of the search was reached: this facilitates a reverse search step-by-step should a given line of enquiry lead up a blind alley.

The simplest way to illustrate the working of a key is to follow a hypothetical example. Let us assume that you have in your hands a specimen which might be familiar to many, a Chicken of the Woods (*Laetiporus sulphureus*), a distinctive orange-yellow lumpy bracket growing on wood.

Having selected key B, the search proceeds as follows:

1a	Encrusting or bracket-like forms on wood	**2**
2b	Fertile surface spiny or porous	**14**
14b	Hymenium consisting of tubes opening by pores	**16**
16b	Fruiting body projecting hoof- or bracket-like	**19**
19b	Comparatively large brackets, mainly over 10cm dia	**24**
24a	Bracket large, soft or corky	**25**
25a	Lumpy, soft when young (yellow or pinkish-orange)	
		Laetiporus sulphureus

Key A Ascomycotina
(Fungi reproducing by ascospores developed in asci).

1a	Mature fruiting body disc-, saucer-, cup- or top-shaped.		**2**
1b	Mature fruiting body none of these shapes.		**26**
2a	(1a) Sessile or with only rudimentary stem		**3**
2b	Possessing a stem		**20**
3a	(2a) Circumference of disc more or less regular, at most notched or fringed		**4**
3b	Circumference more typically lobed or split		**11**
4a	(3a) Very large, mostly 10-20cm dia (vernal)		*Disciotis*
4b	Smaller, generally less than 5cm dia		**5**
5a	(4b) Over 1.5cm dia		**6**
5b	Less than 1.5cm dia		**13**
6a	(5a) Thin-fleshed and brittle		**7**
6b	Thick-fleshed and rubbery		**18**
7a	(6a) Mostly cream, brown or lilac-tinged disc- and saucer-shaped forms, the margins not fringed or obviously differentiated (asci tips blued with Melzer's reagent)		**8**
7b	Not as above (asci not blued with Melzer's reagent)		**9**
8a	(7a) Mainly cream, lilac or brown		*Peziza*
8b	Blackish	(only on fire sites)	*Plicaria*
9a	(7b) Bright distinctive colours	(scarlet)	*Sarcoscypha*
		(orange)	*Aleuria*
		(yellow)	*Sowerbyella*
9b	Not bright colours		**10**
10a	(9b) Mainly pallid		*Tazzetta*
			Humaria
10b	Blackish		*Pseudoplectania*
11a	(3b) Splitting down one side		*Otidea*
11b	Bladder-like then splitting into lobes		**12**
12a	(11b) Arising above ground	(sand dunes) *Peziza ammophila*	
			Sarcosphaera
12b	Arising below ground and erupting		*Geopora*
13a	(5b) Between 0.5 and 1.5cm dia		**14**
13b	Less than 0.5cm dia		**19**
14a	(13a) On soil		**15**
14b	On other substrates		**17**
15a	(14a) On fire sites	(cream and brown)	*Trichophaea*
		(wholly brown)	*Geopyxis*
		(pinkish-orange)	*Pulvinula*
15b	Not on fire sites		**16**
16a	(15b) Margin set with bristly hairs	(orange)	*Scutellinia*
16b	Hairs inconspicuous or absent	(orange)	*Melastiza*
		(yellowish-orange)	*Octospora*
		(pinkish-orange)	*Pulvinula*
17a	(14b) On wood		*Encoelia, Scutellinia*
17b	On sycamore leaves		*Rhytisma*
18a	(6b) Fruiting body blackish		*Bulgaria*
18b	Fruiting body pink or lilac		*Neobulgaria*
19a	(13b) Brightly coloured	(yellow)	*Bisporella, Orbilia, Lachnellula*
		(green)	*Chlorociboria*
		(cream or yellow)	*Phaeohelotium*
19b	Dull coloured	(grey)	*Mollisia, Tapesia*
20a	(2b) Stemmed, gelatinous, rubbery or fleshy, on wood		**21**
20b	Stemmed, thin-fleshed, not gelatinous		**22**
21a	(20a) With very short stem, flattish	(purplish)	*Ascocoryne*
21b	With longer stem, top-shaped	(cream)	*Cudoniella*
22a	(20b) On soil		**23**
22b	On other substrates		**24**
23a	(22a) Vernal	(brown or cream)	*Helvella*
23b	Not vernal	(grey or black)	*Helvella*

24a	(22b) On other substrates (vernal)		
		(on *Ranunculus* rhizomes)	
			Sclerotinia
		(on alder catkins etc.)	*Ciboria*
24b	On wood and woody remains		**25**
25a	(24b) Outer surface smooth	(brown)	*Rutstroemia*
		(very small, white)	*Hymenoscyphus*
25b	Outer surface hairy	(very small, white or brown)	*Lachnum*
26a	(1b) Mature fruiting bodies subterranean		**27**
26b	Fruiting bodies developing above ground		**28**
27a	(26a) With distinctive 'truffle' smell		*Tuber*
27b	Without distinctive smell (false truffles)		*Elaphomyces*
28a	(26b) Fruiting bodies comparatively large, solitary or gregarious		**29**
28b	Fruiting bodies very small, typically gregarious		**38**
29a	(28a) Fruiting bodies sessile		**30**
29b	Fruiting bodies with stems		**32**
30a	(29a) Sessile, soft, brain-like		*Ascotremella*
30b	Sessile, tough or woody		**31**
31a	(30a) Cushion-shaped		*Daldinia, Rhizina*
31b	Thinly encrusting		*Ustulina*
32a	(29b) Fruiting bodies tough or woody		**33**
32b	Fruiting bodies softer		**34**
33a	(32a) Club- or top-shaped	(on wood)	*Xylaria, Podostroma*
		(on horse dung)	*Poronia*
33b	Antler-shaped or spindly	(on wood)	*Xylaria*
		(parasitic on grass heads)	*Claviceps*
34a	(32b) Fertile heads pitted or honeycombed		
		Helvella, Morchella, Verpa, Gyromitra	
34b	Fertile heads of regular outline		**35**
35a	(34a) Growing on *Elaphomyces* or buried insects		*Cordyceps*
35b	Growing on litter or soil		**36**
36a	(35b) Fertile heads thin, saddle-shaped		*Helvella*
36b	Fertile heads rubbery or gelatinous		**37**
37a	(36b) Fertile heads tongue- or fan-shaped		
		(green, black)	*Geoglossum, Trichoglossum*
		(yellow)	*Spathularia*
37b	Fertile heads swollen, knobbly		*Mitrula, Leotia*
38a	(28b) Growing on other fungi		*Hypomyces, Nectria, Peckiella*
38b	Growing on wood		**39**
39a	(38a) Microscopic spheres forming crust-like sheet		*Hypoxylon*
39b	Distinctly separated spheres		**40**
40a	(39a) Whitish, very small		*Lasiosphaeria*
40b	Brownish or reddish		*Diatrype, Hypocrea, Nectria*

Key B Basidiomycotina
(Non-gilled forms excluding boletes).

1a	Encrusting or bracket-like forms on wood	**2**
1b	Neither encrusting nor bracket-like	**32**
2a	(1a) Fertile surface without spines or pores	**3**
2b	Fertile surface spiny or porous	**14**
3a	(2a) Fruiting body encrusting or with rudimentary projection	**4**
3b	Fruiting body projecting as a bracket	**11**
4a	(3a) Very thin, like a film or powder, at most 0.5mm thick	**5**
4b	More substantial than a film or powder	**6**
5a	(4a) Smooth (white, on bark) *Hyphodontia*	
	(on bilberry) *Exobasidium*	
	(cream, pink or buff) *Cylindrobasidium, Vuilleminia*	
	(yellow, minutely downy) *Botryohypochnus*	
5b	Finely roughened (whitish, granular) *Sistotrema*	
	(grey, yellow or buff, warty) *Hyphoderma, Cerocorticium*	
6a	(4b) Waxy and soft *Phlebia, Phlebiopsis, Coniophora*	
	(puckered, appearing almost porous) *Merulius, Serpula*	
6b	Waxy and hard or brittle	**7**
7a	(6a) Hymenium distinctly bristly under lens *Hymenochaete*	
7b	Hymenium smooth under lens	**8**
8a	(7a) Whitish *Amylostereum*	
8b	More distinctively coloured	**9**
9a	(8b) Bright blue *Pulcherricium*	
9b	Other colours	**10**
	except (grey-brown) *Auricularia mesenterica*	
10a	(9b) Mainly yellowish or cream *Stereum*	
10b	Mainly grey, pink or lilac *Peniophora*	
11a	(3b) Brackets, soft and rubbery or gelatinous	**12**
11b	Brackets tough and dry	**13**
12a	(11a) Rubbery textured (white) *Meruliopsis*	
	More gelatinous (white, yellow or blackish) *Tremella, Exidia*	
12b	Greyish brown, gelatinous *Auricularia mesenterica*	
13a	(11b) Growing on soil *Thelephora, Podoscypha*	
13b	Growing on wood (yellowish) *Stereum*	
	(brownish-purple) *Chondrostereum*	
14a	(2b) Hymenium consisting of spines	**15**
14b	Hymenium consisting of tubes opening by pores	**16**
15a	(14a) Encrusting (yellowish) *Mycoacia, Cerocorticium*	
	(violet – mainly brackets, see 17a) *Trichaptum*	
15b	Bracket- or cushion-like (white) *Creolophus, Hericium*	
16a	(14b) Fruiting body encrusting or with rudimentary bracket	**17**
16b	Fruiting body projecting hoof- or bracket-like	**19**
17a	(16a) Thinly encrusting (whitish pores) *Physisporinus, Skeletocutis*	
	(violet pores) *Trichaptum*	
	(maze-like pores) *Schizopora*	
17b	Thickly encrusting or cushion-like	**18**
18a	(17b) Hard when fresh	
	(white or yellowish, pores torn or lacerate) *Antrodia*	
	(pores not lacerate) *Dichomitus*	
	(rusty-brown) *Phellinus*	
18b	Soft when fresh *Meruliopsis*	
19a	(16b) Comparatively small brackets, mainly under 10cm dia	**20**
19b	Comparatively large brackets, mainly over 10cm dia	**24**
20a	(19a) Arising in rosettes on soil *Abortiporus biennis*	
20b	Arising singly or in tiers	**21**
21a	(20a) Thickly constructed, wedge-shaped *Postia, Tyromyces*	
21b	More thinly constructed	**22**
22a	(21a) Pores coloured at maturity (grey) *Bjerkandera*	
	(brown) *Inonotus radiatus*	
22b	Pores whitish	**23**
23a	(22b) Pores rounded *Coriolus, Oxyporus*	
	(small) *Skeletocutis nivea*	
23b	Pores elongated, slot-like *Datronia*	
24a	(19b) Bracket large, soft or corky	**25**
24b	Bracket large, very hard and woody	**31**
25a	(24a) Lumpy, soft when young (yellowish or pinkish-orange)	
	Laetiporus	
	Always soft, reddish, oozing juice *Fistulina*	
25b	Leathery or corky	**26**
26a	(25a) Upper surface distinctly scaly *Polyporus squamosus*	
26b	Upper surface not scaly	**27**
27a	(26b) Pores distinctly elongated or maze-like	**28**
27b	Pores more or less rounded	**30**
28a	(27a) Bracket wedge-shaped *Daedalea quercina*	
28b	Bracket thinner	**29**
29a	(28b) Mainly on coniferous wood *Gloeophyllum*	
29b	Mainly on broad-leaf wood (whitish) *Pseudotrametes, Trametes*	
	(brownish-zoned) *Daedaleopsis, Lenzites*	
30a	(27b) Pores white at maturity *Heterobasidion*	
	(restricted to birch) *Piptoporus*	
	(oozing when young) *Ischnoderma*	
30b	Pores brown at maturity *Inonotus*	
31a	(24b) With distinctive lacquered appearance	
	Ganoderma lucidum	
31b	Not with lacquered appearance (spores brown) *Ganoderma spp.*	
	(spores whitish) *Rigidoporus, Fomes*	
32a	(1b) Fruiting body with distinct stem	**33**
32b	Fruiting body more or less sessile	**44**
33a	(32a) Trumpet-, mushroom- or somewhat disc-shaped	**34**
33b	Not trumpet-, mushroom- or disc-shaped	**37**
34a	(33a) Trumpet-shaped *Craterellus, Cantharellus*	
34b	Mushroom-like or disc-shaped	**35**
35a	(34a) Hymenium consisting of pores	
	Polyporus, Coltrichia, Phaeolus	
35b	Hymenium consisting of spines	**36**
36a	(35a) Spores print white (on soil) *Hydnum, Phellodon, Bankera*	
36b	Spore print brown (on buried pine cones) *Auriscalpium*	
	(on soil) *Hydnellum, Sarcodon*	
37a	(33b) Arising from an 'egg', slimy, foul-smelling	**38**
37b	Not arising from an 'egg', dry, not foul-smelling	**39**
38a	(37a) Conical head on tall white spongy stem *Phallus, Mutinus*	
38b	Radiating arms on short spongy stem *Clathrus archeri*	
39a	(37b) Head clearly distinguished from stem *Tulostoma, Battarrea*	
39b	Head less clearly distinguished from stem	**40**
40a	Fruiting body club-shaped, simple	**41**
40b	Fruiting body antler-like	**42**
41	(40a) On soil (white or yellowish) *Clavulinopsis, Clavariadephus*	
	Clavulina	
41b	On wood and stems (yellowish) *Macrotyphula, Calocera*	
42a	(40b) With no, or only limited, branching	**43**
42b	With dense coral-like branching *Ramaria*	
43a	(42a) On soil *Clavaria, Clavulinopsis*	
43b	On wood *Clavicorona, Clavariadelphus, Calocera*	
44a	(32b) Mature fruiting bodies sessile, rounded, lumpy, potato- or onion-shaped	**45**
44b	Mature fruiting bodies sessile but not with these shapes	**50**
45a	(44a) Small, gelatinous, on wood *Femsjonia, Dacrymyces*	
45b	Not gelatinous, on soil or wood	**46**
46a	(45b) Netted, open shapes, foul-smelling (red) *Clathrus ruber*	
46b	Closed shapes, not foul-smelling	**47**
47a	(46a) Fruiting body lumpy, of fused elements *Pizolithus*	
47b	Fruiting bodies of individual rounded appearance	**48**
48a	(47a) Outer wall splitting star-wise *Geastrum, Astraeus*	
	(not British) *Myriostoma*	
48b	Outer wall not splitting star-wise	**49**
49a	(48a) Hard, breaking unevenly at maturity	
	(yellowish) *Scleroderma*	
49b	Not hard, opening by a distinct pore *Lycoperdon, Bovista*	
50a	(44b) Fruiting body cup-shaped *Crucibulum, Cyathus*	
50b	Fruiting body not cup-shaped	**51**
51a	(50b) Lobed on soil (pinkish) *Tremiscus*	
	ear-like on wood (brown) *Auricularia auricula-judae*	
	in cauliflower-like masses on wood (cream) *Sparassis*	
51b	Gelatinous, on wood (yellowish, white, brain-like) *Tremella*	
	(blackish) *Exidia*	

Key C Basidiomycotina
(Gill-bearing and pore-bearing forms).

1a	Hymenium consisting of gills	**2**
1b	Hymenium consisting of tubes opening by pores	**57**
2a	(1a) Stem central, on soil, humus or wood	**3**
2b	Stem eccentric, lateral or absent, mainly on wood	**52**
3a	(2a) Spore powder white, whitish, cream or very pale pink	**4**
3b	Spore powder some other colour including darker pink	**33**
4a	(3a) Flesh brittle and crumbly	**5**
4b	Flesh not brittle and crumbly but more or less fibrous	**6**
5a	(4a) Gills typically with intermediates, not thick, exuding milk	*Lactarius*
5b	Gills typically without intermediates, comparatively thick, not exuding milk	*Russula*
6a	(4b) Stem with distinct ring (may be ephemeral)	**7**
6b	Stem without ring, even when young	**11**
7a	(6a) Gills free	**8**
7b	Gills not free, otherwise attached	**10**
8a	(7a) Cap with scabby patches, stem with ring (patches and ring absent in some species) and volval remnants at base	*Amanita*
8b	Cap scaly or bald, stem with ring (sometimes only when young), but without volval remnants at base	**9**
9a	(8b) Cap with scales (see also 35a)	*Lepiota, Macrolepiota*
9b	Cap fibrous or bald, flesh sometimes reddening	*Leucocoprinus* (see also 35a) *Leucoagaricus*
10a	(7b) On wood (brown, scaly, gills decurrent) *Armillaria* (white, viscid, gills emarginate) *Oudemansiella*	
10b	On soil (grey-brown, gills emarginate) *Tricholoma cingulatum*	
11a	(6b) Mainly medium or large fleshy types	**12**
11b	Mainly small delicate or membraneous types	**23**
12a	(11a) Gills free or emarginate	**13**
12b	Gills not free or emarginate, otherwise attached	**14**
13a	(12a) On soil, several species smelling of meal (spores smooth) *Tricholoma* (most with violet tinges, spores warty) *Lepista* (whitish, vernal) *Calocybe gambosa* (with brighter colours, generally smaller) *Calocybe*	
13b	On wood (with yellow background colour) *Tricholomopsis rutilans*	
14a	(12b) Gills typically somewhat decurrent	**15**
14b	Gills typically adnexed or adnate	**18**
15a	(14a) Cap greasy or viscid, gills thick with waxy feel	**16**
15b	Cap not greasy or viscid, gills not thick with waxy feel	**17**
16a	(15a) Cap more or less convex, mainly in woods	*Hygrophorus*
16b	Cap convex or conical, brightly coloured, in grass	*Hygrocybe*
17a	(15b) Gills strongly forked (wholly orange) *Hygrophoropsis*	
17b	Gills not, or barely, forked (spore smooth) *Clitocybe* (spores warty - see also 37b) *Lepista*	
18a	(14b) Cap brightly coloured, conical, often viscid, gills waxy, adnexed	*Hygrocybe*
18b	Not this combination	**19**
19a	(18b) Gills broadly adnate, thickish and distant	*Laccaria*
19b	Gills adnate but not particularly thick or distant	**20**
20a	(19b) On wood, cap viscid (orange with velvety stem) *Flammulina*	
20b	On soil, cap dry or at most slightly greasy	**21**
21a	(20b) In tufts, gills often blackening	*Lyophyllum*
21b	Solitary or in trooping groups	**22**
22a	(21a) Stem tough and fibrous (some types quite small) *Collybia* (large types) *Megacollybia, Leucopaxillus*	
22b	Stem not tough and fibrous, cap with low umbo	*Melanoleuca*
23a	(11b) Resistant types which can revive after drying out	*Marasmius*
23b	Delicate types which do not revive after drying out	**24**
24a	(23b) Cap campanulate or conical	**25**
24b	Cap not campanulate or conical	**28**
25a	(24a) Cap greasy or viscid, brightly coloured	*Hygrocybe*

25b	Cap not with this combination	**26**
26a	(25b) Smelling distinctly of fish (dark brown) *Macrocystidia*	
26b	Not smelling of fish	**27**
27a	(26b) Cap brown, yellow, grey or pallid, striate *Mycena, Marasmiellus*	
27b	Cap pure white or faintly yellow, striate *Delicatula, Hemimycena*	
28a	(24b) Cap more or less convex or flattened	**29**
28b	Cap somewhat umbilicate or funnel-shaped	**31**
29a	(28a) On remains of other fungi *Nyctalis, Collybia*	
29b	Not on remains of other fungi	**30**
30a	(2b) On wood (sometimes smelling of rotten cabbage) *Micromphale*	
30b	On cones *Baeospora, Strobilurus*	
31a	(28b) Gills decurrent, close *Clitocybe, Tephrocybe* (on fire sites) *Myxomphalia*	
31b	Gills decurrent, distant, in grass or moss	**32**
32a	(31b Gills never dividing *Omphalina, Rickenella*	
32b	Gills sometimes dividing *Gerronema, Phaeotellus, Arrhenia*	
33a	(3b) Spore powder pinkish	**34**
33b	Spore powder some other colour	**38**
34a	(33a) Gills free	**35**
34b	Gills in some way attached to the stem	**37**
35a	(34a) Stem ringed (see also 9a) *Macrolepiota, Lepiota, Leucoagaricus*	
35b	Stem without ring	**36**
36a	(35b) Stem base with membraneous volva	*Volvariella*
36b	Stem base not with membraneous volva, usually on wood	*Pluteus*
37a	(34b) On soil, cap convex, campanulate or umbilicate, gills variously attached, spores angular	*Entoloma*
37b	On soil with more or less emarginate gills, spores distinctly warty (see also 17b)	*Lepista*
38a	(33b) Spores light-, mid- or rust-brown	**39**
38b	Spores chocolate-, purple-brown or almost black	**47**
39a	(38a) Fruiting body with cortina and ring when young *Cortinarius*	
39b	Fruiting body not usually with cortina	**40**
40a	(39b) Spore powder yellowish-brown	**42**
40b	Spore powder darker, not yellowish-brown	**41**
41a	(40b) Spore powder rust-brown	**43**
41b	Spore powder dull- or tobacco-brown	**44**
42a	(40a) Cap greasy, sulcate, on dung and rotten wood etc. *Bolbitius*	
42b	Cap and stem pruinose, cap not sulcate, on soil *Conocybe*	
43a	(41a) Small types, on wood or soil (cap with granules) *Flammulaster* (tough, reviving, cap with erect scales) *Phaeomarasmius* (very delicate, cap smooth) *Galerina, Simocybe* (usually with alders or willows) *Naucoria*	
43b	Larger types, on wood (with or without ring) *Gymnopilus* (cap scaly or viscid, stem ringed or not ringed) *Pholiota*	
44a	(41b) Cap fibrous-scaly or cracked, often umbonate *Inocybe*	
44b	Cap not fibrous-scaly or cracked	**45**
45a	(44b) Cap greasy, stem rarely ringed (some with cortina) *Hebeloma*	
45b	Cap not greasy	**46**
46a	(45b) On wood, cap markedly hygrophanous, stem ringed *Kuehneromyces*	
46b	On soil, not markedly hygrophanous, sometimes with ring	*Agrocybe*
47a	(38b) Gills mottled (spores ripening unevenly) *Panaeolus, Panaeolina*	
47b	Gills not mottled	**48**
48a	(47b) Cap greasy (small and delicate, stem not ringed) *Psilocybe* (smallish, stem ringed) *Stropharia* (fleshy with thick, decurrent gills) *Chroogomphus, Gomphidius*	
48b	Cap dry	**49**
49a	(47b) Stem with ring, fleshy types, gills at first pinkish, then chocolate or black	*Agaricus*
49b	Stem without ring	**50**

50a (49b) Gills typically deliquescent or weeping
Coprinus, Lacrymaria

50b Gills not deliquescent **51**

51a (50b) Stem with a silky appearance *Psathyrella*

51b Stem without silky appearance, often on wood *Hypholoma*

52a (2b) Stem eccentric, lateral or absent, spores white **53**

52b Stem as above, spores coloured differently **55**

53a (52a) Fruiting body soft, decomposing soon after maturity
Pleurotus

53b Fruiting body comparatively tough, not readily
decomposing **54**

54a (53b) Cap scaly *Lentinus, Panellus*

54b Cap more or less smooth (tufted, large, orange, luminous)
Omphalotus

55a (52b) Fruiting body cream or white, spores brown *Crepidotus*

55b Spores pink or beige **56**

56a (55b) On wood (wholly orange-pink) *Rhodotus palmatus*

56b On soil (wholly white) *Clitopilus*
(grey-brown) *Arrhenia acerosa*

57a (1b) Cap and stem dull brown and scaly *Strobilomyces*

57b Cap and stem not dark brown and scaly **58**

58a (57b) Pores whitish, cap dry, stem scaly *Leccinum*

58b Pores in some way coloured, at least at maturity **59**

59a (58b) Pores becoming pinkish-grey, stem netted, bitter *Tylopilus*

59b Pores becoming yellowish-green, not bitter **60**

60a (59b) Stem thickened or ventricose (see also 64a) *Boletus*

60b Stem not thickened or ventricose **61**

61a Cap (damp) generally greasy or slimy **62**

61b Cap (damp) not greasy or slimy **63**

62a Tubes markedly decurrent, yellow bruising blue (alders) *Uloporus*

62b Tubes less markedly decurrent *Suillus*

63a Cap dry, tubes adnate-decurrent (pores yellow) *Pulveroboletus*

63b Cap dry, tubes more or less adnate or free **64**

64a (63b) Flesh turning blue when cut *Boletus, Gyroporus cyanescens*

64b Flesh not turning blue when cut *Xerocomus, Gyroporus castaneus*
(brown with yellow pseudo-gills) *Phylloporus*

COLLECTING IN THE FIELD AND IDENTIFICATION

Specimens should always be collected and taken home for more thorough examination unless their identity is truly obvious at a glance and, as a general first principle, one should never rely on casual observation to determine a species.

Before a specimen is picked have a good look around and, unless equipped with a photographic memory, make some notes. It is useful to have a few cards in your pocket on which to record the more obvious details of the fresh specimens in a methodical progression. In sequence, (1) examine the specimen itself, (2) identify the vegetation with which it is associated and, if possible, (3) establish the soil type.

1. Note if the specimen is growing on its own (solitary); as part of a group (scattered, trooping or clustered); or whether several specimens are fused together in a tuft (caespitose). In some instances species grow in characteristic rings. *Marasmius oreades, Calocybe gambosa* and *Agaricus macrosporus* are good examples of fungi whose fruiting bodies appear at the periphery of regular, radially growing mycelia often revealed by dark, lush circles of grass.

Only at this stage should the specimen be removed for closer inspection. Never be tempted to cut the stem of a fruiting body (unless it is tough and growing on wood) but always lift it with the base and some of the substrate attached because important identification features may rest there. One obvious example is *Megacollybia platyphylla* with its long rooting 'rhizoidal cords' but the basal characteristics of such genera as *Amanita* and *Inocybe* are equally revealing. Try to pick fresh but not immature or 'overblown' specimens for identification and avoid too much handling, particularly of the stem. In the *Inocybe* species, the presence or absence and the position of a powdery but fragile coating (referred to as a *pruina*) on the young stem is highly significant.

Smell the fungus for any kind of distinctive aroma. Very often this will involve crushing and rubbing a small part between your fingers but even then an odour is not always obvious. It will be influenced by weather conditions, age of the material, the sensitivity of your nose and so on. If the smell is faint the fungus can be placed in a closed container for an hour or so. Next, having first made absolutely sure that you are not dealing with one of the deadly species indicated in the text, taste a small piece by nibbling it on the tip of the tongue and spit it out. Hot, bland, bitter, sour, mealy or radishy flavours are of assistance in determining certain species. These flavours are often apparent at once but sometimes may take up to a minute to become obvious.

Check the cap surface texture. Is it slimy, greasy, shiny, smooth, velvety, floury, shaggy, woolly or scaly? Be aware that a cap which is glutinous or viscid in damp weather may take on a dry sheen in arid conditions and that scales can be washed off by rain. Is the cap ridged or creased? Does the margin overhang, or fall short of, the gills? What is its profile shape? See if there is a ring on the stem, if so whether it turns upwards (booted) or downwards (hanging), whether the stem as a whole is club-shaped (clavate) or more or less equal and whether the base of the stem is bulbous. Is the base surrounded by a bag-like sheath (*volva*), or are there other remains of a volva?

Note if there is any colour change on handling or cutting and where the change takes place i.e. if it occurs all over, only in the cap flesh, in the stem, or in the extreme base of the stem. Colour changes in the flesh are particularly significant in the *Boletus* fungi as well as some of the *Inocybe, Russula* and *Agaricus* species. In the case of boletes, always cut the fruiting body in half lengthways and watch for details of colour change.

If you own a small hand lens, make a closer inspection of the cap, stem and gills whilst still in the field. Check the finer texture of the cap and stem surfaces and decide whether the gill edges are smooth or serrated. Check also if there appears to be minute bristling (*cystidia*) on the edges or the faces of the gills and whether the outer edges of gills or pores are coloured differently.

Never, at this stage, unless you are very familiar with the species, rely on a comparison with the photograph as an absolute and final test of identity. There is a natural temptation to turn the pages until a photograph looks familiar. This is all right so long as one accepts that it is only the first stage on a route towards positive determination. A much more constructive exercise is to narrow the field of possibilities by working sequentially through the keys (see pages 18-21).

2. Is the specimen growing on bare soil, amongst moss, grass, on humus, on a plant stem or on wood? In many instances the distinction between coniferous and broad-leaf trees will be an important one but sometimes more precise recognition of the nearest tree species is also significant. Many fungi exist in close biological relationship with specific host plants and whilst the relationship may be obvious – an *Inonotus* growing on the trunk of an ash tree – in other circumstances more careful scrutiny is needed. The deadly *Amanita phalloides* almost invariably grows in close association

with oaks whilst *Lactarius circellatus* is restricted to an association with hornbeams. Fungi growing on old rotten timber can present a particular trap for the unwary. A moss-covered stump in a conifer plantation may be that of a broad-leaf tree which grew there before the conifers were planted! Some genera, such as *Hygrocybe*, prefer life in open grasslands whilst others, like the *Galerina* species, favour mossy ground. Location and habitat details are always important.

3. Try to establish the soil type in the area you are searching i.e. is it acid or alkaline, sand, chalk or clay? This may not be a critical characteristic but, nonetheless, some species do prefer specific soil types.

Take the collected specimens home, with your field notes. The fungi are best laid out, separately, in a flat wicker-type trug or basket but small specimens will be reasonably happy stored in a screw-capped plastic or glass tube. Try to keep the material as cool as possible and, if collecting for eating, only collect specimens which are not over-ripe and have no signs of worm tunnels in the flesh. If further identification is difficult right away, specimens will keep in a fridge, in a sealed container, for a few days.

Once laid out on the table and under a good light, the material can be checked more closely, working through the key and then through the individual descriptions in the text but always proceed methodically, in the order indicated, and never presume to fit the 'cloth' according to the 'coat' you have in mind! Unless you care to haul bottles of chemical around in the field this is also the time to run any recommended chemical tests (see page 26). In the case of species in which it is important to observe changes taking place in the colour of the cut flesh, the specimen should be sliced through from top to bottom using a clean knife and observed over a period of not less than ten minutes.

A spore print is often one of the most important 'first step' tests as it will determine the colour of the spores in a mass. Cut a cap from its stem and lay it, face down, on a piece of glass or a microscope slide, cover it to exclude draughts and to retain moisture (a piece of damp tissue is useful) and leave it undisturbed for several hours, preferably overnight. The spores from a print are also vital at a later stage of identification when measuring spore sizes because the print will include mainly mature spores whilst those retained on the gills will have a high percentage of immature and therefore smaller spores. The spore sizes offered in the text represent the overall range rather than the mean sizes and are extracted from a variety of current published sources, monographs etc. as well as from personal observation and that of experts in the field.

Positive identification of many species, particularly amongst the smaller less conspicuous types does, alas, depend ultimately on the microscopic features and, therefore, on the availability of an instrument equipped with a mechanical stage and a calibrated micrometer, so a substantial capital outlay could be involved. Unfortunately a basic student's or child's microscope will not suffice because there is a need to achieve magnifications of x800 - x1000 with high resolution and then to make very accurate measurements which are impossible without the facility of a mechanical stage. Perhaps the more sensible alternative for a beginner, or anyone who does not intend to delve so deeply, is to join a local foray group (most areas now have BMS Affiliated Groups or a Natural History club) where the finer points of identification may be carried out for you by a local expert.

Microscopic examination is best carried out using fresh material although pieces of dried specimens, which tend to become very tough and brittle, can be softened and reconstituted adequately by placing them in a small drop of 5 per cent KOH (potassium hydroxide).

It is always worth drying a specimen for future reference. It will look nothing like the fresh material but this is relatively unimportant as the microscopic characteristics are retained. Small specimens can be dried whole, larger ones require a vertical median slice taking in a cross-section of the cap and stem. Attempting to dry a whole bolete, for example, will result in a rather nasty mess! The simplest and most effective method is to place the samples in a small sieve, or on a wire rack, over a radiator or domestic boiler. A fan oven on the lowest warming heat sometimes works well. Once the material is wholly dry, wrap it carefully in a sachet of thin paper, or in a small envelope, label it and store in a dry place, in a shoe box, or similar, with one or two mothballs as a preservative to keep predatory insect larvae at bay.

Keeping the records, accompanied by dried specimens, is probably as important as making the initial notes, even for beginners. It helps remind one of species from one season to the next and may be necessary for final validation. It all really depends on how seriously you wish to take your mycology. There are no rules so long as one does not take personal risks by eating specimens that are not 100 per cent properly identified and no one else is affected or misled by one's mistakes.

Never be put off by individuals who announce that mycology is properly for the academic fraternity – it is a discipline occasionally plagued by a degree of unnecessary elitism and academic snobbishness. Ironically, some of those who proclaim elitism most loudly have no academic training in mycology themselves and gained their amateur knowledge through the assistance of more open-minded mycologists.

I freely admit I spent many happy years collecting fungi and putting names to them without the proper equipment – sometimes I was right, sometimes I was wrong. It was all part of the process of learning and I gained enormous pleasure from it.

PHOTOGRAPHY

With a very limited number of exceptions all the illustrations reflect specimens located over a three-year period between 1992 and 1994 and photographed *in vivo* where they were found growing. In a few cases the surrounding vegetation was trimmed to reveal the subjects more clearly or material was transported for convenience and replaced on a similar substrate. Every effort was made to locate material in an optimum condition displaying a species in its mature state and where immature material was available it was also included in the photograph. In the case of rarer species, however, it was sometimes only possible to photograph material in a less than perfect state.

Subject material was largely dried, retained and stored in a herbarium collection which is to be lodged with the Royal Botanic Gardens at Kew, Richmond, Surrey. In instances where the material was examined and determined elsewhere it was not always possible to recover specimens. Unless subject material was recognisable unequivocally from its gross macroscopic features, identity was determined subsequent to photography from microscopic characteristics. The co-operation of many specialist mycologists was also obtained and is gratefully acknowledged elsewhere in this book.

Photography was largely carried out using an SLR 35mm Nikon FM camera body, a standard 50mm Nikon lens and a 2 x macro focusing Vivitar teleconverter. A limited amount of photography was done using a back-up SLR 35mm Exakta Varex IIa with a 50 mm Meyer-Optik lens. Film was Ektachrome 400, Ektachrome 400 Elite and Fujichrome 400 transparency stock. The camera was always tripod-mounted and film was exposed using available natural light at f16 or f11 aperture and, using a remote cable release, at slow shutter speeds. In conditions of very low light intensity, timed exposures were employed within limits beyond which reciprocity failure might have affected colour accuracy. In a few isolated instances where the available natural light was wholly inadequate, electronic flash was incorporated.

CHEMICAL TESTS

Application of a number of chemical reagents may help with identifying species, both in macroscopic and microscopic examination. Various substances, produced within the tissues of individual species of fungi, react by changing colour when they come into contact with certain chemicals known as reagents, and this can distinguish types which are otherwise difficult to tell apart.

The chemicals required are simple compounds which should not be too difficult to obtain through British Mycological Society Affiliated Groups, local pharmacies, schools or laboratories. Preparing the reagents requires little equipment or experience and only small quantities are needed. Normal precautions of labelling and keeping in a safe place should be observed, particularly if children may otherwise gain access to the substances. Special care must be taken when handling the ammonia, and the sulphuric and nitric acids employed in certain tests.

Macroscopic tests

Ammonia. Dilute concentrated NH_3 with deionised water to make a 70 per cent solution. Use with care.

Iron salts. Dissolve 1g ferrous sulphate ($FeSO_4$) crystals in 10ml H_2O and add a few drops of concentrated sulphuric acid (H_2SO_4). Used in the determination of certain species of *Russula* and other fungi; when applied to the cap cuticle a colour change takes place. Note: only use solutions which have not discoloured. Ferrous ammonium sulphate works as well or better. For fieldwork a large crystal of the salt may be more convenient but it is generally less effective than the solution.

Phenol. 2-3 per cent aqueous solution applied to cap or stem gives a colour change in certain species.

Potassium hydroxide solution. Dissolve 2g potassium hydroxide (KOH) in 10ml H_2O (some authors suggest a 30-40 per cent solution is more effective). A distinctive colour change occurs in certain *Cortinarius* species.

Schaeffer's Test. Using a clean glass rod for application, streak aniline across the caps of *Agaricus* species and, using a fresh rod, cross this with concentrated nitric acid. Red coloration at the point of intersection indicates a positive result. Use with care – concentrated nitric acid is highly corrosive.

Microscopic tests

Melzer's Reagent. Add 0.5g iodine and 1.5g potassium iodide to 20ml chloral hydrate. Used in determination of certain groups of Ascomycetes where the tips of the asci turn blue in a positive reaction. Used to distinguish amyloid spores of Basidiomycetes, which turn blue-black in the mass, from non-amyloid spores which remain colourless. Also used to distinguish dextrinoid spores which turn reddish-brown in the mass.

Sulpho-vanillin. Place a few drops of 80 per cent sulphuric acid (H_2SO_4) in a suitable container and add a few grains of vanillin (vanilla). Mix until dissolved and the solution has turned evenly yellow. Used in determination of certain species of *Russula*. When applied to the stem of certain species, a colour change takes place. Note: the solution is unstable and should ideally be used when freshly prepared.

CONSERVATION

Fungi are fascinating organisms, free for everybody's benefit. They are also, like all living things, vunerable: to destruction of habitats, pollution, chemical sprays and, not least, thoughtless over-exploitation, particularly amongst those species prized for their culinary worth. Contrary to some exaggerated press reports there is little firm evidence of decline in numbers or species through picking for eating. (There are, however, indications that some species may be in decline in Continental Europe due to the effects of pollution and loss of habitat.) Nonetheless, stripping a woodland of Boletes or Chanterelles will, sooner or later, take its toll and there are reasons to believe that in certain parts of the country collection is being carried out on a commercial scale. At present there is some debate on how best to curb large-scale harvesting from the wild. Ultimately, licensing may be the only effective recourse unless a voluntary code of practice is accepted by all. For identification purposes no more than three fruiting bodies need be collected, one 'button' specimen, one mature, and a third to preserve for the record. When fungi are picked for the table, an area should never be stripped bare. Take what you need and no more – here, as in all things, moderation is the name of the game.

REFERENCE SECTION

AUTHORS OF FUNGAL NAMES

A. & S.	Albertini von, L.B.	1769-1831
	Schweinertz, L.	1780-1834
Afz.	Afzelius, A.	1750-1837
Arnould	Arnould, L.	fl. 1893
Atk.	Atkinson, G.F.	1854-1918
Ayers	Ayers, T.T.	1900-1967
Balbis	Balbis, G.B.	1765-1831
Banker	Banker, H.J.	1866-1940
Baral	Baral, H.O.	born 1954
Barl.	Barlocher, F.	died 1988
Bat.	Bataille, Fr.	1850-1946
Batsch	Batsch, A.	1861-1902
Baumg.	Baumgartner, J.	1870-1955
Beck	Beck, G.V.M.L.	1856-1913
Bellú	Bellú, F.	fl. 1988
Benedix	Benedix, E.H.	1914-1983
Berk.	Berkeley, M.J.	1803-1889
Boid.	Boidin, J.	born 1893
Boiffard	Boiffard, J.	fl. 1972
Bolt.	Bolton, J.C.	1758-1799
Bon	Bon, M.	born 1925
Bon.	Bonorden, J.	1801-1844
Bond.	Bondartsev, A.S.	1867-1968
Borszcow (Borshchow)	Borchchow, J.G.	fl. 1833-1878
Boud.	Boudier, J.L.E.	1828-1920
Bourd.	Bourdot, H.	1861-1937
Bousset	Bousset, H.	fl. 1939
Br.	Broome, C.E.	1812-1886
Bref.	Brefeld, J.O.	1839-1925
Bres.	Bresadola, G.	1847-1929
Brig.	Briganti, F.	1802-1865
Britz.	Britzelmayr, M.	1839-1909
Buchwald	Buchwald, J.	1869-1927
Bull.	Bulliard, J.B.F.	1752-1793
Burt	Burt, B.D.	1902-1938
Capelli	Capelli, A.	fl. 1982
Carpenter	Carpenter, W.	1787-1874
Ces.	Cesati, V.	1806-1883
Chaill.	de Chaillet, J.F.	1749-1839
Chev.	Chevallier, F.F.	1796-1840
Chod.	Chodat, R.H.	1865-1934
Christ.	Christiansen, M.P.	1889-1975
Cke.	Cooke, M.C.	1825-1914
Clç.	Clémençon, H.	born 1935
Cleland	Cleland, J.F.	1878-1971
Clem.	Clements, F.E.	1874-1945
Coker	Coker, W.C.	1872-1953
Constant.	Constantinesen, O.	born 1933
Corn.	Corner, E.J.H.	born 1906
Courtec.	Courtecuisse, R.	born 1956
Currey	Currey, F.	1819-1888
Curt.	Curtis, M.A.	1808-1872
Dassier	Dassier de la Chassagne, H.G.B.	1748-1816
David	David, A.	1826-1900
DC.	de Candolle, A.P.	1778-1841
Dennis	Dennis, R.W.G.	born 1910
Dermek	Dermek, A.	1925-1989
Desm.	Desmazières, J.B.H.J.	1786-1862
Dharne	Dharne, C.G.	fl. 1965
Dicks.	Dickson, J.J.	1738-1822
Ditm.	Ditmar, L.P. Fr.	fl. 1806-1817
Dom.	Domanśki, S.	born 1916
Donk	Donk, M.A.	1908-1972
Dorfelt	Dorfelt, H.	fl. 1973
Doty	Doty, M.S.	born 1916
Dring	Dring, D.M.	1932-1978
Duby	Duby, J.E.	1798-1885
Dufour	Dufour, J.L.M.	1780-1865
Dur.	Durieu de Maisonneuve, M.C.	1796-1878
Ebb.	Ebben, M.H.	fl. 1961
Ehrenb.	Ehrenberg, C.G.	1795-1876
Ellis	Ellis, J.B.	1829-1905
Erikss.	Eriksson jun. J.	born 1921
Ev.	Everhart, B.M.	1818-1904
Farlow	Farlow, W.G.	1844-1919
Farr.	Farrell, L.	fl. 1930
Fayod	Fayod, V.	1860-1900
Feder.	Federer, Z.	born 1931
Ferd.	Ferdinandsen, C.C.F.	1879-1944
Fischer	Fischer, E.	1861-1939
Fr.	Fries, E.M.	1794-1878
Freeman	Freeman, G.W.	fl. 1979
Fuckel	Fuckel, K.W.G.L.	1821-1876
Galz.	Galzin, A.	1853-1925
Gamundi	Gamundi de A, I.J.	born 1925
Geest.	Maas Geesteranus, R.A.	born 1911
Genevier	Genevier, L.G.	1830-1880
Gibbs	Gibbs, L.S.	1870-1925
Gilb.	Gilbert, E.J.	1888-1954
Gill.	Gillet, C.G.	1806-1896
Ginns	Ginns, J.H.	born 1938
Gmelin	Gmelin, J.F.	1748-1804
Gramberg	Gramberg, E.	fl. 1920
Graddon	Graddon, W.D.	1896-1989
Gray	Gray, S.F.	1766-1828
Grev.	Greville, P.K.	1794-1866
Groves	Groves, J.W.	1906-1970
Haller	Haller, V.A. von	1708-1777
Hahn	Hahn, G.	1889-1968
Harmaja	Harmaja, H.	born 1944

Hedw.	Hedwig, J.	1730-1799	Mass.	Massee, G.E.	1850-1917
Heim	Heim, R.	1900-1979	Maubl.	Maublanc, A.	1880-1958
Hein.	Heinemann, P.	born 1916	Melz.	Melzer, V.	1878-1968
Henn.	Hennings, P.C.	1841-1908	Merat	Merat de Vaumartoise, F.V.	1780-1851
Henry	Henry, R.	1884-1960	Métrod	Métrod, G.	fl. 1938
Hes.	Hesler, L.R.	1888-1977	Mich.	Michael, E.	1849-1920
Hoffm.	Hoffman, G.F.	1760-1826	Miller	Miller, J.D.	fl. 1985
Höhn.	Höhnel von F.X.R.	1852-1923	Möll.	Möller, F.A.G.J.	1887-1964
Hök	Hök, C.T.	fl. 1836	Mont.	Montagne, J.P.F.C.	1784-1866
Holmsk.	Holmskjøld, T.H.	1732-1794	Morg.	Morgan, A.P.	1836-1907
Hook.	Hooker, W.J.	1785-1865	Moser	Moser, M.	born 1924
Horn. (Hornem.)	Horneman, J.W.	1770-1841	Mouton	Mouton, V.	1875-1901
Huds.	Hudson, W.	1730-1793	Müller	Müller, O.	1730-1784
Huijsman	Huijsman, H.S.C.	born 1900	Murr.	Murrill, W.A.	1869-1957
Imazeki	Imazeki, R.	born 1904	Nannf.	Nannfeldt, J.A.F.	1904-1985
Imbach	Imbach, E.J.	1897-1970	Nees	Nees van Esenbeck, C.G.D.	1776-1858
Imler	Imler, L.	fl. 1955	Neuh.	Neuhoff, W.	1891-1971
Itzerott	Itzerott, H.	1912-1983	Niemela	Niemala, T.	born 1940
Jaap	Jaap, O.	1864-1922	Nke.	Nitschke, T.R.J.	1834-1883
Jacq.	Jacquin, N.J. von	1727-1817	Nobles	Nobles, M.K.	born 1903
Joss.	Josserand, M.	born 1900	de Not.	de Notaris, G.	1805-1877
Jül.	Jülich, W.	born 1942	Noullet (Noulet)	Noulet, J.B.	1802-1890
Jungh.	Junghuhn, F.W.	1809-1864	Nyl.	Nylander, W.	1822-1899
Kchb.	Kalchbrenner, K.	1807-1886	Oeder	von Olderburg Oeder, G.C.E.	1728-1791
Kallenb.	Kallenbach, F.	1893-1944	Opat.	Opatowski, W.	1810-1838
Kam.	de Kam, M.	fl. 1969	Orl.	Orlos, H.	fl. 1967
Kanouse	Kanouse, B.B.	1889-1969	Orton P.	Orton, P.D.	born 1916
Karst.	Karsten, P.A.	1834-1917	Paniz.	Panizzi, F.	1817-1893
Kauff.	Kauffmann, C.H.	1869-1931	Parm.	Parmasto, E.	born 1928
Keller	Keller, S.	fl. 1980	Pass.	Passerini, G.	1816-1893
Kickx	Kickx, J.J.	1842-1887	Pat.	Patouillard, N.T.	1854-1926
Klotsch (Klotzsch)	Klotzsch, J.F.	1805-1860	Paulet	Paulet, J.J.	1740-1826
Knapp	Knapp, J.A.	1834-1859	Peck	Peck, C.	1883-1917
Knudsen	Knudsen, H.	born 1948	Pearson	Pearson, A.A.	1874-1954
König	König, J.G.	1728-1785	Perdek	Perdek, A.C.	fl. 1950
Konrad	Konrad, P.	1877-1948	Pers.	Persoon, C.H.	1761-1836
Korf	Korf, R.P.	born 1925	Peters.	Petersen, R.H.	born 1934
Kotl.	Kotlaba, F.	born 1927	Petrak	Petrak, F.	1837-1896
Kreisel	Kreisel, H.	born 1931	Pil.	Pilát, A.	1903-1974
Krombh.	Krombholz, J.V.	1782-1843	Phill.	Phillips, W.	1822-1905
Kühn.	Kühner, R.	born 1904	Pouz.	Pouzar, Z.	born 1932
Kujp.	Kujper, J.	born 1953	Quél.	Quélet, L.	1839-1899
Kummer	Kummer, P.	1834-1912	Raith.	Raithelhuber, J.	fl. 1969
Kuntze	Kuntze, C.E.O.	1843-1907	Ram.	Ramamurthi, B.	fl. 1965
Lambotte	Lambotte, J.B.E.	1832-1905	Rausch.	Rauschert, R.	fl. 1975
Lamoure	Lamoure, X.	born 1928	Rea	Rea, C.	1861-1946
Lange J.	Lange, J.B.E.	1864-1941	Rehm	Rehm, H.	1828-1916
Lange M.	Lange, M.	born 1919	Reid	Reid, D.	born 1927
Lasch	Lasch, W.	1786-1863	Relh.	Relhan, R.	1754-1823
Le Gal	Le Gal, M.L.F.	1895-1979	Retz	de Retz, B.G.G.	born 1910
Lennox	Lennox, J.W.	1867-1948	Richon	Richon, C.E.	1820-1893
Lenz.	Lenz, H.	1798-1870	Ricken	Ricken, A.	1851-1921
Letellier	Letellier, J.B.L.	1817-1898	Rifai	Rifai, M.A.	born 1940
Lév.	Léveillé, J.H.	1796-1870	Rocques (Roques)	Roques, J.	1772-1850
Leyss.	Leyssen, F.W. von	1731-1815	Rogers	Rogers, J.D.	born 1937
Lib.	Libert, M.A.	1782-1865	Rolland	Rolland, L.	1841-1912
Limm.	Limminghe, A.M.A.	1834-1861	Romagn.	Romagnesi, H.C.L.	born 1912
Link	Link, L.H.F.	1767-1851	Romell	Romell, L.G.T.	born 1891
L.	Linnaeus, C.	1707-1778	Rostk.	Rostkovius, F.W.G.T.	1770-1848
Litsch.	Litschauer, V.	1879-1939	Rostrup	Rostrup, O.G.F.	1864-1933
Lloyd	Lloyd, C.G.	1859-1926	Roth	Roth, A.W.	1757-1834
Locq.	Locquin, M.S.	fl. 1943	Roze	Roze, E.	1833-1900
Lund.	Lundell, S.	1892-1966	Ryv.	Ryvarden, L.	born 1935
Macbr.	Macbride, T.H.	1848-1934	S.	(see A. & S.)	
Maire	Maire, R.	1878-1949	Sacc.	Saccardo, P.A.	1854-1921
Martin	Martin, F.N.	fl. 1987	Sadler	Sadler, J.	1781-1849

Santi	Santi, G.	1746-1822
Schäff, J.	Schäffer, Julius	1882-1944
Schaeff.	Schaeffer, J. Ch.	1718-1790
Schleich.	Schleicher, J.C.	1768-1834
Schmidt	Schmidt, D.	fl. 1990
Schrad.	Schrader, H.A.	1767-1836
Schroet.	Schroeter, J.	1837-1894
Schulz.	Schulzer, M. von	1802-1892
Schum.	Schumacher, H.C.F.	1757-1830
Schw.	(see A. & S.)	
Scop.	Scopoli, J.A.	1723-1788
Seav.	Seaver, F.J.	1877-1970
Sécr.	Sécrétan, L.	1758-1839
Shear	Shear, C.L.	1865-1956
Sing.	Singer, R.	born 1906
Smarda	Smarda, F.	1902-1976
Smith A.H.	Smith, A.H.	1904-1986
Smotl.	Smotlacha, F.	fl. 1921
Snell	Snell, W.H.	1889-1980
Somm.	Sommerfelt, S.C.	1794-1838
Sow.	Sowerby, J.	1752-1822
Speg.	Speggazini, C.L.	1858-1926
Spooner	Spooner, B.M.	born 1951
Stalpers	Stalpers, J.A.	born 1947
Stangl	Stangl, J.	fl. 1963
Steyaert	Steyaert, R.L.A.G.J.	1905-1978
Strid	Strid, A.	born 1932
Studer	Studer-Steinhäuslin, S.B.	1847-1910
Sutara	Sutara, J.	fl. 1987
Swartz	Swartz, O.	1760-1818
Thom	Thom, C.	1872-1956
von Thuemen	von Thumen, K.A.E.J.	1839-1892
Tode	Tode, H.J.	1733-1797
de la Torre	Ruiz de la Torre, J.	born 1927
Tournef.	de Tournefort, J.P.	1656-1708
Trav.	Traverso, G.B.	1878-1955
Tul, C.	Tulasne, C.	1816-1884
Tul.	Tulasne, L.R.	1815-1885
Underwood	Underwood, L.M.	1853-1907
Vaill.	Vaillant, S.	1669-1722
Vel.	Velenovsky, J.	1858-1949
Venturi	Venturi, A.	1805-1864
Veselsky	Veselsky, J.	fl. 1975
Vilgalis (Vilgalys)	Vilgalys, R.	born 1958
Vitt.	Vittadini, C.	1800-1865
Wahl.	Wahlenberg, G.	1780-1851
Wallr.	Wallroth, C.F.W.	1792-1857
Wasser	Wasser, S.P.	born 1946
Watl.	Watling, R.	born 1938
Weberb.	Weberbauer, O.	1846-1881
Weinm.	Weinmann, J.A.	1782-1858
Wells	Wells, K.	born 1927
Wettst.	Wettstein, R. von	1863-1931
Wichansky	Wichansky, E.	fl. 1958
Wigg.	Wiggers, F.H.	1746-1811
Willd.	Willdenow, C.L.	1765-1812
Winge	Winge, O.	1886-1964
With.	Withering, W.	1741-1799
Woron.	Woronidain, N.N.	born 1882
Wulf.	Wulfen von, F.X.	1728-1805
Wunsch	Wunsch	fl. 1988
Yao	Yao, Y.-J.	fl. 1989
Zopf	Zopf, F.W.	1846-1909
Zv	Zvara, J.I.	fl. 1922

(fl. = date(s) of publication of relevant work(s) by the authors indicated.)

BMS DATABASE

It may be useful for a reader to refer, at some time, to the British Mycological Society Database. At the time of going to press the relevant ordering of the Database is constructed as follows (the numbers being the codes used in the Database). It should be appreciated that not all orders included in the list are represented in this volume or reflect the nomenclature found here:

Basidiomycotina

01	Agaricales	15	Podaxales	
02	Aphyllophorales	16	Russulales	
03	Auriculariales	17	Sclerodermatales	
04	Boletales	18	Septobasidiales	
05	Brachybasidiales	19	Sporidiales	
06	Cantharellales	20	Tremellales	
07	Dacrymycetales	21	Tulasnellales	
08	Exobasidiales	22	Tulostomatales	
09	Gauteriales	23	Uredinales	
10	Hymenogastrales	24	Ustilaginales	
11	Lycoperdales	25	Sundry Basidiomycetes not included above	
12	Melanogastrales	26	Arthoniales	
13	Nidulariales	27	Caliciales	
14	Phallales	28	Clavicipitales	

29	Diaporthales	53	Rhytismatales
30	Diatrypales	54	Sordariales
31	Dothideales	55	Sphaeriales
32	Elpahomycetales	56	Taphrinales
33	Endomycetales	57	Teloschistales
34	Erysiphales	58	Verrucariales
35	Eurotiales		
36	Graphidales		
37	Gyalectales		
38	Gymnoascales		
39	Helotiales		
40	Hypocreales		
41	Laboulbeniales		
42	Lecanidiales		
43	Lecanorales		
44	Microascales		
45	Opegraphales		
46	Ophiostomatales		
47	Ostropales		
48	Peltigerales		
49	Pertusariales		
50	Pezizales		
51	Polystigmatales		
52	Pyrenulales		

BIBLIOGRAPHY

Agaricales in Modern Taxonomy, Singer, R. (ed. Cramer), Weinheim 1962

Authors of Plant Names, Brummitt, R.K. and Powell, C.E., Kew 1992

Die Blätterpilze Deutschlands etc., Ricken, A., Leipzig 1915

British Ascomycetes, Dennis R.W.G. (repr. Cramer), 2nd ed. 1981

British Basidiomycetes, Rea, C., Cambridge University Press 1922

British Cup Fungi and their Allies, Dennis, R.W.G., London 1960.

British Fungal Flora: Agarics and Boleti, Vols 1-7, authors various (publ. HMSO), Royal Botanic Gardens, Edinburgh 1970-1993

British Fungal Flora: Colour Identification Chart, (publ. HMSO), Royal Botanic Gardens, Edinburgh 1969

British Fungi, Ellis, E.A., London 1976 etc.

British Fungi and Lichens, Massee, G., London 1911

British Mycena Species, Emmett, E.E. (series publ. in *The Mycologist*), London 1992-1993

Champignons d'Europe, (ed. N. Boubée & Cie), Paris 1969

Champignons de nos Pays, Romagnesi, H., Paris 1977

Champignons du Nord et du Midi, Marchand A. Soc. Mycologique des Pyrénées Mediterranéennes et Hachette, Perpignan 1971-1980

Clavaria, Marriott, J.V.R. (paper in *Keys* publ. BMS), London 1990

Clavaria and Allied Genera (monograph), Corner, E.J.H., Oxford 1950

Conspectus of the Mycenas of the Northern Hemisphere, Maas Geesteranus R.A., Netherlands 1982-1989 etc.

Corticiaceae of northern Europe, Eriksson, J. & Ryvarden, L., Oslo 1985.

Dictionary of the Fungi, Ainsworth, G.C. & Bisby, G.R., 7th edition (first publ. CMI London 1961)

Flora Agaricina Danica, Lange, J.E., Copenhagen 1935-1941

Flora Agaricina Neerlandica, Noordeloos M.E. etc., (publ. Balkema) Rotterdam 1987 etc.

Flore Analytique des Champignons Supérieurs, Kühner, R. & Romagnesi, H., Paris 1978

I Funghi dal Vero, Cetto, B., Arti Grafiche Saturnia, 1970-1987

Fungi of Britain and Europe, Buczacki, S., London 1989

Fungi of Switzerland, Vols. 1-3, Breitenbach, J. & Kränzlin, F., Lucerne 1984-1991

De Fungi van Nederland, Maas Geesteranus R.A., 1964 etc.

Fungi without Gills, Ellis, M.B. & J.P., London 1990

Genre (le) Mycena, Kühner, R., *Encyclopedia Mycologique* (10), Paris 1938

Hyménomycetes de France, Bourdot & Galzin, Paris 1927

Icones Selectae Fungorum, Konrad, P. & Maublanc, A., (repr. G. Biella) 1985

Iconographia Mycologica, Bresadola, J., (repr. Mediolani) 1981

Fungi without Gills (Hymenomycetes and Gasteromycetes – an Identification Handbook), Ellis & Ellis, London 1990

A Guide to Temperate Myxomycetes, Nannenga-Bremekamp, N.E. (tr. Feest A. & Burrgraaf Y. Bristol 1991), Netherlands 1989.

Guides for the Amateur Mycologist, Marriott J.V.R., (publ. BMS) London 1993 onwards

Keys to Agarics and Boleti, Moser, M., 4th edition (publ. Phillips, Tonbridge) 1983

Key to British Ramariopsis Species, Marriott, J.V.R., (paper in *Keys* publ. BMS), London 1993

Keys to the British Species of Russula, Rayner, R.W., 3rd edition (publ. BMS) London 1985

Keys to Naucoria Species of Damp Ground, Marriott, J.V.R., (paper in *Keys* publ. BMS) London 1992

Manual of British Discomycetes, Phillips, W., London 1893

Microfungi on miscellaneous substrates, Ellis, M.B. & J.P., 1988

Die Milchlinge (monograph), Neuhoff, W., Band IIb: *Die Pilze Mitteleuropas*, 1956 index under '2'

Mushrooms and other fungi of Great Britain and Europe, Phillips, R., London 1981

Mushrooms and Toadstools (Guide to), Lange, M. & Hora, F.B., London 1963

Mushrooms and Toadstools of Britain and North West Europe, Bon, M., London 1987

Nouvel Atlas des Champignons, Romagnesi, H., Bordas 1970

Polyporaceae of North Europe, Ryvarden, L., Oslo 1976-78

Polypores, Pegler, D.N., 2nd edition (publ. BMS) London 1973

Ramaria, Marriott, J.V.R. (paper in *Keys* publ. BMS) London 1990

A Revision of the Genus Inocybe in Europe, Kuyper, T.W., Persoonia Supplement Vol. 3

Die Russulae (monograph), Schäffer, J., Band III: *Die Pilze Mitteleuropas*, 1952 (rep. Cramer 1979) index under '12'

Zeitschrift für Mykologie: Studien zur Gattung Coprinus etc. Bender H., Enderle M. & Krieglsteiner G.J., 1984

GLOSSARY

adnate
(of a gill profile) in which there is broad attachment to the stem (see diagram page 13)

adnexed
(of a gill profile) in which there is narrow attachment to the stem (see diagram page 13)

adpressed
lying closely against a surface

Agaricales
an order of true gill-bearing fungi

allantoid
(of a spore) curved into a sausage-shape

amorphous
lacking a clear structure

amyloid
make-up of certain cell walls (particularly of spores) which causes them to turn blue in Melzer's reagent

anastomosing
forming cross connections

annulus
ring around the stem, the remnant of a partial veil covering the young emergent sporophore

Aphyllophorales
a group of basidiomycete fungi (the title is applied only as a heading in this book) including certain non-gilled fungi

apical
towards the apex

apiculus
(of a spore) a short projection at the base

ascomycete
a member of the Ascomycotina class of fungi characterised by the development of asci in which sexual ascospores are produced

ascospore
a spore of the ascomycete fungi

ascus
microscopic flask-shaped structure from which the ascospores (most typically 8) are ejected

basal
the area at the base of a stem

basidiomycete
member of the Basidiomycotina class of fungi characterised by the development of basidia on which sexual basidiospores are produced

basidiospore
a spore of the basidiomycete fungi

basidium
microscopic structure on which basidiospores (usually 4 or 2) develop

bulbous
referring to the abruptly swollen basal region of a stem

caespitose
(of fruiting bodies) crowded together in tufts, partly fused at the stem bases

calcareous
(of soils) containing a high proportion of calcium e.g. chalk and limestone

campanulate
bell-shaped

capitate
(of cystidia) with a rounded knob-like tip

cartilaginous
firm, tough but readily bent

chlamydospore
thick-walled asexual spore

ciliate
fringed with hairs

clavate
club-shaped

collar
a ring-like structure around the apex of the stem at the attachment of the gills

concave
downwardly curved, reverse of convex

concolorous
possessing the same colour

convex
domed, reverse of concave

cortina
a partial cobweb-like veil

cuticle
surface layers of hyphae in a cap or stem

cystidium
at the surface of a cap, gill or stem, the terminal cell of a hypha which grows into a large and often distinctive shape

decurrent
(of a gill profile) in which there is extensive attachment running down the stem (see diagram page 13)

dentate
possessing tooth-like projections

denticulate
possessing small teeth

depressed
(of a cap profile) when the surface is slightly concave

dextrinoid
reaction when spore (or hyphal) wall turns red with Melzer's reagent

droplets
(of a spore) fluid-filled microscopic cavities

eccentric
(with reference to a stem) off-centre in relation to the cap

effused
spread over the substrate

ellipsoid
(of spores) shaped like an ellipse or flattened circle with conical projections in one plane

emarginate
(of a gill profile) notched just before joining the stem

ephemeral
very short-lived, transient

endoperidium
innermost covering wall of a fruiting body in e.g. gastromycete fungi

farinose
covered with fine mealy particles

fibrillose
surface covered with fine thread-like fibres

filiform
thread-like

floccose
decorated with cottony tufts

free
(of a gill profile) in which the gill is not joined to any area of the stem

fusiform
spindle-shaped, narrow at both ends

Gastromycetes
a group of basidiomycete fungi (the title is applied only as a heading in this book) in which the spore-producing cells (gleba) develop within a hollow body (peridium)

germ pore
a hole in the spore wall, opposite the apiculus from which a germ tube may extend

gill (lamella)
fertile surface in the form of one of a series of radiating plates sub-tending the cap

guttation drops
drops secreted by the fruiting bodies of certain species

Heterobasidiomycetes
a group of non-gilled basidiomycete fungi in which the basidia have longitudinal or transverse septa or are shaped liked tuning forks; including the 'jelly fungi'

Homobasidiomycetes
the major grouping of basidiomycete fungi in which the basidia are neither forked nor septate

hyaline
colourless and transparent

hygrophanous
becoming paler after loss of water content

hymenium
fertile layer containing spore-forming cells

inferior	(of a ring) positioned below mid-height on a stem
infundibuliform	funnel-shaped
iodoform	a substance smelling of iodine
lacerate	(of pores) as if torn away, revealing the tubes like organ-pipes
lageniform	(of cystidia) with a neck-like extension
lamellate	possessing gills or lamellae
lanceolate	(of cystidia) lance-shaped, tapering to a point
lateral	(of a stem) attached at the side of the cap
ligulate	strap-shaped
mammiform	shaped like a breast with nipple
marginate	(of a swollen stem base) having a well-defined upper edge
median	(of a ring) positioned at mid-height on a stem
mycelium	network of vegetative hyphae from which the fruiting body arises
nodulose	(of a spore) decorated with small knobbly projections
ostiole	the microscopic opening of a perithecium
ovoid	egg-shaped
papilla	nipple-like projection
parasite	an organism which sustains itself upon another living organism whilst providing nothing in return
peridium	layer of tissue covering a fruiting body, typically seen in Gastromycetes
perithecia	flask-shaped structures containg spore-bearing cells and embedded within a mass of infertile tissue (stroma)
pore	the mouth of spore-bearing tube
pruina	fine powdery coating on a cap or stem
pruinose	covered with a fine powder (finer in texture than farinose)
punctate	surface decorated with fine dots
reflexed	turning back or upwards
reniform	kidney-shaped
resupinate	closely applied to the substrate and upside down
reticulate	having a net-like pattern of ornamentation
saprophyte	an organism sustaining itself off dead and dying organic material (typical of most fungi)
sclerotium	a resting or overwintering structure, typically buried within the substrate
scurfy	decorated with bran-like flakes
septa	cross-walls deviding hyphae, cystidia, basidia, spores etc.
septate	possessing septa
serrate	having a sharply defined saw-like edge
sessile	possessing no stem
setae	microscopic hair- or awl-like structures often supplying significant determining characteristics
sinuate	(see emarginate)
spathulate	flattened oblong with a narrow base
species	a narrowly defined type of organism, differentiating members of a genus
stellate	shaped like a star
striate	possessing parallel or radiating lines or fine ridges
stroma	sterile mass of hyphal tissue
stuffed	(of a stem) central core of a looser consistency than the outer tissues
subspherical	somewhat less than a perfect sphere (cf. sub-globose)
substrate	matrix on which a fungus grows and in which the mycelium is commonly embedded
sulcate	possessing radiating wrinkles
superior	(of a ring) positioned above mid-height
tube	cylindrical structure bearing the hymenial layer in certain types of fungi
umbilicate	(of a cap) possessing a small central depression, like a navel
umbo	(of a cap) small central mound or boss which may be bluntly rounded or sharply pointed
umbonate	possessing an umbo
utriform	(of cystidia) shaped like a bag
veil	thin protective layer of tissue covering the emerging fruiting body which ruptures and disperses sometimes leaving a variety of remnants. If partial the veil may leave a ring on the stem or, if cobwebby, the cortina
velar	referring to a veil
ventricose	swollen, usually in the mid-region
vermiform	worm-like
vesicular	shaped like a bladder
viscid	sticky, slimy or glutinous
volva	bag-like stucture at the stem base, remains of the universal veil

THE ENCYCLOPEDIA

ASCOMYCOTINA (ASCOMYCETES)

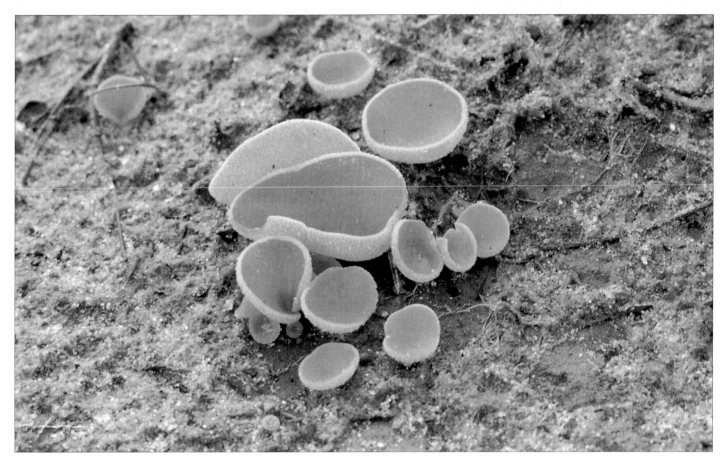

This is a major group with varied morphology, classified according to the structure of their fruiting bodies. Those which possess open, frequently cup-shaped *apothecia* are included in the Discomycetes; those with enclosed spheres, *perithecia*, are included in the Pyrenomycetes and Plectomycetes.

Although the fruiting bodies may grow to several centimetres, many are very small. They are further classified on the structure of the spore-forming organs, the *asci*, within which the spores develop and from which they are expelled at maturity.

The external appearances of ascomycetes may seem bafflingly diverse, but they are probably an easier group to understand and recognise than the basidiomycetes. Approximate numbers of species recognised within Europe are given below in square brackets.

Within the Discomycetes, the Pezizales [350 species] includes all the cup fungi and their allies in which the *hymenium* arises on the inner (upper) side of the cup. In the Helvellas and Morchellas, however, the cup is borne up on a stem and then reflexed in varying degrees of complexity so that the hymenium lines the outer surface. Of twelve families, the six most likely to be encountered and identified are included here. In the Tuberales [30

species] the asci are borne within a specialised tuber-like structure which develops below ground and is designed to be dispersed by animals. The Helotiales (Leotiales) [900 species] includes most of the genera with smaller or microscopic, disc-shaped, fruiting bodies but the group also contains the *Geoglossum* or 'earth tongue' species in which the hymenium is raised in the form of a vertical tongue or strap. The Rhytismatales [50 species] have their fruiting bodies immersed in the tissues of the host plant.

From the vast numbers of Pyrenomycetes only the more significant members of four out of ten orders have been included. The Clavicipitales [20 species] mostly parasitise grasses, insect larvae and pupae and other fungi. The Diatrypales [30 species] includes members whose fruiting bodies are small and cushion-like. The Hypocreales [120 species] includes three families with mostly small, more or less spherical fruiting bodies. The Sphaeriales [750 species] includes a large number of genera and a wide variety of form.

The Plectomycetes includes the powdery mildews and the subterranean 'false truffles', Elaphomycetales, [6 species] of which only the latter are represented here.

Ascobolus carbonarius Karst.　　Ascobolaceae

Small olive brown disc; densely clustered on fire sites.

Dimensions 0.2 - 0.5 cm dia.
Fruiting body apothecial inner (hymenial) surface at first olive yellow or with orange tinge, smooth, becoming dull olive brown and roughened; outer (lower) surface concolorous or more pallid, roughened; cup-shaped becoming shallowly saucer-shaped; sessile. Flesh pallid brown, brittle and thin.
Asci 150-230 x 13-22 µm. **Spores** (8) hyaline but with violaceous brown sheath when mature, coarsely warty, broadly ellipsoid, 22-23 x 13-14 µm. Paraphyses filiform, septate, forked.
Odour not distinctive. **Taste** not distinctive.
Chemical tests asci tips blued with Melzer's reagent.
Occurrence spring to late summer; frequent.
■ Inedible.

Ascobolus denudatus Fr.: Fr.　　Ascobolaceae

Minute pale yellow disc; densely clustered in rotting straw.

Dimensions 0.05 - 0.1 cm dia.
Fruiting body apothecial inner (hymenial) surface yellowish, smooth; outer (lower) surface whitish; shallowly saucer-shaped; sessile. Flesh pallid, brittle and thin.
Asci not recorded. **Spores** (8) hyaline but with purplish sheath when mature, decorated with parallel lines, ellipsoid, 17-20 x 8-9.5 µm. Paraphyses filiform.
Odour not distinctive. **Taste** not distinctive.
Chemical tests asci tips blued with Melzer's reagent.
Occurrence throughout the year but mainly summer to early autumn; infrequent.
■ Inedible.

Ascobolus furfuraceus Pers.: Fr.　　Ascobolaceae

Small yellowish brown disc; densely clustered on dung, favouring cow manure.

Dimensions 0.1 - 0.5 cm dia.
Fruiting body apothecial inner (hymenial) surface at first greenish yellow, smooth, becoming violaceous brown and roughened (darkening spores in projecting asci); outer (lower) surface concolorous, minutely scurfy; cup-shaped becoming shallowly saucer-shaped; sessile. Flesh pallid, yellowish brown, brittle and thin.
Asci 200-230 x 20-22 µm. **Spores** (8) hyaline but with violaceous brown sheath when mature, ornamented with longitudinal ribs, some anastomosing, ellipsoid, 23-26 x 10-14 µm. Paraphyses filiform, septate, non-forked.
Odour not distinctive. **Taste** not distinctive.
Chemical tests asci tips blued with Melzer's reagent.
Occurrence spring to autumn; frequent.
■ Inedible.

Saccobolus glaber (Pers.) Lambotte   Ascobolaceae

Small yellowish olive disc; densely clustered on fire sites.

Dimensions 0.2 - 0.5 cm dia.
Fruiting body apothecial upper (hymenial) surface olive yellow, smooth, becoming dull olive brown and roughened; outer (lower) surface concolorous, minutely roughened; top-shaped; sessile. Flesh pallid olive brown, brittle and thin.
Asci 120-160 x 12-15 μm. **Spores** (8) hyaline, glued together by a viscous matrix when released, smooth, ellipsoid, 20-25 x 13-14 μm. Paraphyses filiform.
Odour not distinctive. **Taste** not distinctive.
Chemical tests asci tips blued with Melzer's reagent.
Occurrence late summer to autumn; infrequent.
■ Inedible.
Note: a rogue species appears in the top left of this illustration.

Pyronema omphalodes (Bull.: Fr.) Fuckel   Pyronemataceae

Very small bright orange disc; densely clustered, on soil on fire sites and burnt plant debris, also on sterilised soil in greenhouses.

Dimensions 0.05 - 0.1 cm dia.
Fruiting body apothecial inner (hymenial) surface orange or salmon pink, smooth; outer (lower) surface concolorous, smooth; at first cushion-like or lens-shaped then flattened saucer-shaped and often wavy; sessile. Flesh orange, brittle and thin.
Asci cylindrical, 150-200 x 13-15 μm. **Spores** (8) hyaline, coarsely reticulated, ellipsoid, non-septate, uniseriate, 11-15 x 6.5-8.5 μm. Paraphyses narrowly cylindrical, slightly swollen at the tips, septate.
Odour not distinctive. **Taste** not distinctive.
Chemical tests asci tips not blued with Melzer's reagent.
Occurrence throughout the year but mainly late summer to late autumn; rare. (Fire site on South Downs near Worthing)
■ Inedible.
Very similar to *P. domesticum* but with smaller spores.

Aleuria aurantia (Pers.: Fr.) Fuckel   Otidiaceae
= *Peziza aurantia* Pers.: Fr.   Orange Peel Fungus

Orange cup or irregular disc; clustered on bare soil in woods or among low grasses, often favouring gravelly ground.

Dimensions 1-10 cm dia x 2-4 cm tall.
Fruiting body apothecial upper (hymenial) surface bright orange; outer (lower) surface whitish covered with very small scales; at first cup-shaped, becoming more saucer-like and irregular with age, often splitting at the wavy margin; sessile. Flesh pallid, brittle and thin.
Asci cylindrical, 185-220 x 10-13 μm. **Spores** (8) hyaline, coarsely reticulated, ellipsoid, non-septate, uniseriate, 2 droplets, 17-24 x 9-11 μm. Paraphyses cylindrical, slightly swollen at the tips, septate, with orange granules.
Odour not distinctive. **Taste** not distinctive.
Chemical test asci tips not blued with Melzer's reagent but paraphyses granules turn green.
Occurrence late summer to late autumn; common.
☐ Edible and moderately good.

Anthracobia macrocystis (Cke.) Boud.   **Otidiaceae**

Small orange-red disc; densely clustered, on burnt ground.

Dimensions 0.1 - 0.3 cm dia.

Fruiting body apothecial upper (hymenial) surface orange-red, smooth; margin and outer (lower) surface more pallid and covered with tufts of minute brown hairs; at first hemispherical becoming saucer-shaped or flattish disc; sessile. Flesh reddish and brittle.

Asci cylindrical, 120-160 x 12-15 µm. **Spores** (8) hyaline, smooth, ellipsoid, non-septate, uniseriate, 2 droplets, 16-18 x 8-9 µm. Paraphyses cylindrical, slender, slightly swollen at the tips, septate.

Odour not distinctive. **Taste** not distinctive.

Chemical tests asci tips not blued with Melzer's reagent.

Occurrence spring to autumn; infrequent.

■ Inedible.

Byssonectria fusispora (Berk.) Rogerson & Korf **Otidiaceae**
= *Inermisia fusispora* (Berk.) Rifai
= *Octospora carbonigena* (Berk.) Dennis

Bright yellowish orange disc; clustered on soil or rotting vegetation, favouring burnt ground.

Dimensions 0.05 - 0.3 cm dia.

Fruiting body apothecial upper (hymenial) surface yellowish or reddish orange, smooth; outer (lower) surface concolorous or more pallid; at first sub-spherical becoming flattened top-shaped, smooth but with minutely flaky margin; sessile. Flesh yellowish, brittle and thin.

Asci cylindrical, 200-250 x 12-16 µm. **Spores** (8) hyaline, smooth, fusiform, non-septate, uniseriate, 2 large and several smaller droplets, 20-25 x 8-10 µm. Paraphyses cylindrical, slightly swollen at the tips, septate.

Odour not distinctive. **Taste** not distinctive.

Chemical tests asci tips not blued with Melzer's reagent.

Occurrence spring to autumn; infrequent.

■ Inedible.

Cheilymenia stercorea (F H Wigg.: Fr.) Boud.   **Otidiaceae**

Small orange-red disc; densely clustered, on cow and occasionally other dung.

Dimensions 0.2 - 0.3 cm dia.

Fruiting body apothecial upper (hymenial) surface orange-red, smooth; margin and outer (lower) surface more pallid and covered with brownish pointed hairs, margin ciliated with longer hairs; conical with saucer-shaped or flattish disc; sessile. Flesh reddish and brittle.

Asci cylindrical, 220-230 x 12-13 µm. **Spores** (8) hyaline, smooth, ellipsoid, non-septate, uniseriate, droplets absent, 17-20 x 8-10 µm. Paraphyses cylindrical, slender, septate, forked. Hairs brown, narrowly awl-shaped, septate.

Odour not distinctive. **Taste** not distinctive.

Chemical tests asci tips not blued with Melzer's reagent.

Occurrence throughout the year; common.

■ Inedible.

Coprobia granulata (Bull.: Fr.) Boud.   **Otidiaceae**

Very small orange-red disc; densely clustered, on cow and other dung.

Dimensions 0.1 - 0.2 cm. dia.
Fruiting body apothecial upper (hymenial) surface orange-red, smooth; margin and outer (lower) surface pallid and granular; saucer-shaped or flattish; sessile. Flesh reddish, brittle and thin.
Asci cylindrical, 170-190 x 10-15 µm. **Spores** (8) hyaline, smooth, ellipsoid, non-septate, uniseriate, droplets absent, 15-18 x 7-8 µm. Paraphyses clavate, markedly swollen at the tips, septate.
Odour not distinctive. **Taste** not distinctive.
Chemical tests asci tips not blued with Melzer's reagent.
Occurrence throughout the year but mainly summer to autumn; common.
■ Inedible.
Note: in the illustration another small Ascomycete appears.

Geopora arenicola (Lév.) Kers.   **Otidiaceae**
= *Sepultaria arenicola* (Lév.) Mass.

Dirty cream, subterranean, rounded fruiting body which breaks the surface in spring, at first like a worm burrow then opening to reveal cream cup; in small trooping groups, in sandy loam.

Dimensions 1 - 3 cm dia x 2 - 3 cm tall.
Fruiting body apothecial inner (hymenial) surface cream, smooth; outer (lower) surface dirty brownish decorated with fine matted hairs; at first almost vesicular, margin splitting open when mature; sessile. Flesh whitish cream, brittle and thick. Hairs 10-15 µm septate, minutely roughened.
Asci cylindrical, 200-250 x 15-18 µm. **Spores** (8) hyaline, ellipsoid, non-septate, uniseriate, 2 droplets, 23-28 x 14-16 µm. Paraphyses cylindrical, barely swollen at the tips, septate.
Odour not distinctive. **Taste** not distinctive.
Chemical test asci tips not blued with Melzer's reagent.
Occurrence mainly summer to autumn but also in spring; rare. (Gower Peninsula, South Wales)
■ Inedible.

Geopora sepulta (Fr.) Korf & Burds.   **Otidicaeae**

Dirty cream, subterranean, rounded fruiting body which breaks the surface in spring, at first like a worm burrow then opening to reveal creamy grey cup; in small trooping groups in sandy loam.

Dimensions 2 - 4.5 cm dia x 2 - 3 cm tall.
Fruiting body apothecial inner (hymenial) surface pallid greyish-lilac, smooth; outer (lower) surface dirty brownish decorated with fine matted hairs; at first almost vesicular, the margin splitting open when mature; sessile. Flesh whitish cream, brittle and thick. Hairs 10-20 µm, and septate, minutely roughened.
Asci cylindrical, 300-350 x 15-20 µm. **Spores** (8) hyaline, broadly ellipsoid, non-septate, uniseriate, mostly I-droplet, 23-28 x 14-19 µm. Paraphyses cylindrical, barely swollen at the tips, septate.
Odour not distinctive. **Taste** not distinctive.
Chemical test asci tips not blued with Melzer's reagent.
Occurrence autumn to early summer but visible above ground in spring; rare. (Swanscombe, Kent - first recorded collection in Britain)
■ Inedible.

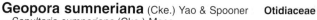

Geopora sumneriana (Cke.) Yao & Spooner   **Otidiaceae**
= *Sepultaria sumneriana* (Cke.) Mass.

Brownish, subterranean, rounded fruiting body which breaks the surface in spring and opens to reveal cream cup; in small trooping groups in soil under cedars.

Dimensions 3 - 7 cm dia x 2 - 5 cm tall.
Fruiting body apothecial upper (hymenial) surface cream with ochraceous tinges, smooth; outer (lower) surface brownish covered with dark hairs; at first almost vesicular, becoming more cup-shaped, the margin splitting in 'petals' and becoming partly reflexed; sessile. Flesh whitish cream, brittle and thick.
Asci cylindrical, 320-380 x 12-15 μm. **Spores** (8) hyaline, fusiform-ellipsoid, non-septate, uniseriate, 2 droplets, 30-40 x 14-16 μm. Paraphyses cylindrical, barely swollen at the tips, septate.
Odour not distinctive. **Taste** not distinctive.
Chemical test: asci tips not blued with Melzer's reagent.
Occurrence submerged autumn to spring but sporulating in early spring; infrequent.
■ Inedible.
Geopora sepulta is indistinguishable apart from the spore size.

Geopyxis carbonaria (A & S.: Fr.) Sacc.   **Otidiaceae**

Small reddish brown cup; clustered on burnt ground.

Dimensions 0.5 - 1.5 cm dia.
Fruiting body apothecial upper (hymenial) surface reddish brown, smooth; outer (lower) surface concolorous, smooth or minutely whitish pruinose; at first sub-spherical becoming cup-shaped; on slender stem embedded in the substrate. Flesh brown, brittle and thin.
Asci cylindrical, 180-250 x 10-11 μm. **Spores** (8) hyaline, smooth, ellipsoid, non-septate, uniseriate, without droplets, 11-17 x 6.5-9 μm. Paraphyses cylindrical, filiform, septate, forked at the tips.
Odour not distinctive. **Taste** not distinctive.
Chemical tests asci tips not blued with Melzer's reagent.
Occurrence summer to autumn; infrequent.
■ Inedible.

Humaria hemisphaerica (F H Wigg.: Fr.) Fuckel **Otidiaceae**
= *Peziza hemisphaerica* F H Wigg.: Fr.

White and brown cup; trooping on soil or very rotten wood.

Dimensions 1 - 3 cm dia x 0.5 - 1 cm tall.
Fruiting body apothecial upper (hymenial) surface greyish white, smooth; margin and outer (lower) surface concolorous, covered densely with dark brown, stiff, septate, sharply pointed hairs; cup-shaped, sessile. Flesh whitish, brittle and thin.
Asci cylindrical, 250-350 x 18-20 μm. **Spores** (8) hyaline, coarsely warty, broadly ellipsoid, non-septate, uniseriate, 2 droplets, 20-27 x 10-13 μm. Paraphyses narrowly cylindrical, markedly swollen at the tips, septate. Hairs brownish, septate, with rounded tips.
Odour not distinctive. **Taste** not distinctive.
Chemical tests asci tips not blued with Melzer's reagent.
Occurrence late summer to autumn; frequent.
■ Inedible.

Melastiza cornubiensis (Berk. & Br.) J Moravec **Otidiaceae**
= *Melastiza chateri* (W G Smith) Boud.
= *Melastiza miniata* (Fuckel) Boud.

Bright orange disc; clustered on sandy and gravelly soils.

Dimensions 0.5 - 1.5 cm dia.
Fruiting body apothecial upper (hymenial) surface bright orange-red or vermilion, smooth; outer (lower) surface concolorous, smooth; at first sub-spherical, becoming irregularly flattened, saucer-shaped and often wavy, covered in minute downy hairs at the margin; sessile. Flesh reddish, brittle and thin.
Asci cylindrical, 275-300 x 12-15 µm. **Spores** (8) hyaline, coarsely reticulated with apical processes, ellipsoid, non-septate, uniseriate, sometimes with small droplets at the ends,17-19 x 9-11 µm. Paraphyses cylindrical, slightly swollen at the tips, septate. Hairs brownish, cylindrical, septate, with blunt tips.
Odour not distinctive. **Taste** not distinctive.
Chemical tests asci tips not blued with Melzer's reagent.
Occurrence spring to autumn; infrequent.
■ Inedible.

Octospora rutilans (Fr.: Fr.) Dennis & Itzerott   **Otidiaceae**
= *Neottiella rutilans* (Fr.: Fr.) Dennis
= *Peziza rutilans* Fr.

Bright yellowish orange disc; clustered amongst *Polytrichum*-type mosses on sandy soils.

Dimensions 0.5 - 1.5 cm dia.
Fruiting body apothecial upper (hymenial) surface bright orange-yellow, smooth; outer (lower) surface more whitish, minutely downy; at first sub-spherical, becoming irregularly flattened, saucer-shaped and often wavy; on a short stem. Flesh yellowish, brittle and thin.
Asci cylindrical, 275-300 x 15-20 µm. **Spores** (8) hyaline, coarsely reticulated, broadly ellipsoid, non-septate, uniseriate, 1 or more droplets, 22-25 x 13-15 µm. Paraphyses cylindrical, slightly swollen at tips, septate, sometimes forked. Hairs hyaline, cylindrical, septate, with blunt tips.
Odour not distinctive. **Taste** not distinctive.
Chemical tests asci tips not blued with Melzer's reagent.
Occurrence autumn; infrequent.
■ Inedible.

Otidea alutacea (Pers.) Mass.   **Otidiaceae**
= *Peziza alutacea* Pers.

Fawn irregular split cup; solitary or in small trooping groups on soil in woods generally.

Dimensions 2 - 4 cm dia x 3 - 6 cm tall.
Fruiting body apothecial inner (hymenial) surface clay or buff, smooth; outer surface pallid fawn, farinose; irregularly cup-shaped, one shorter side split vertically, often wavy; sessile. Flesh pallid yellow, brittle and thick.
Asci cylindrical, 250-300 x 10-15 µm. **Spores** (8) hyaline, smooth, ellipsoid, non-septate, uniseriate, 2 droplets, 12-15 x 6-7 µm. Paraphyses narrowly cylindrical, markedly curved at the tips, branched, septate.
Odour not distinctive. **Taste** not distinctive.
Chemical tests asci tips not blued with Melzer's reagent.
Occurrence late summer to autumn; infrequent.
■ Inedible

Otidea bufonia (Pers.) Boud. **Otidiaceae**

Dark brown irregular split cup; solitary or in small trooping groups on soil in woods generally.

Dimensions 3 - 6 cm dia x 3 - 6 cm tall.
Fruiting body apothecial inner (hymenial) surface dull dark reddish brown, smooth; outer surface rather more pallid, finely downy; irregularly cup-shaped, margin typically remaining incurved with one shorter side split vertically, not ear-shaped; sessile. Flesh brown, brittle and thick.
Asci cylindrical, 160-200 x 10-12 µm. **Spores** (8) hyaline, smooth, ellipsoid, non-septate, uniseriate, 2 droplets, 13-15 x 6.5-7 µm. Paraphyses narrowly cylindrical, markedly curved at the tips, branched, septate.
Odour not distinctive. **Taste** not distinctive.
Chemical tests asci tips not blued with Melzer's reagent.
Occurrence late summer to early autumn; rare. (Culbin Forest, Morayshire)
■ Inedible.

Otidea cantharella (A. & S.: Fr.) Sacc. **Otidiaceae**
= *Flavoscypha cantharella* (Fr.: Fr.) Harmaja

Bright yellow elongated ear; in small trooping groups on soil in mixed woods and parks.

Dimensions 1 - 2 cm dia x 1 - 5 cm tall.
Fruiting body apothecial inner (hymenial) and outer surfaces bright chrome yellow, pallid at the base, smooth; elongated ear-shaped, split vertically along the shorter side; sessile. Flesh yellow, brittle and thin.
Asci cylindrical, 150-175 x 10-12 µm. **Spores** (8) hyaline, smooth, ellipsoid, non-septate, uniseriate, 2 droplets, 10-12 x 5-6 µm. Paraphyses narrowly cylindrical, curved at the tips, septate.
Odour not distinctive. **Taste** not distinctive.
Chemical tests asci tips not blued by Melzer's reagent.
Occurrence late summer to autumn; infrequent or rare. (Thetford Warren, Suffolk)
■ Inedible.

Otidea cochleata (L.: Fr.) Fuckel **Otidiaceae**
= *Peziza cochleata* L.: Fr.

Pale brownish irregular split cup; solitary or in small trooping groups on soil in mixed woods.

Dimensions 3 - 5 cm dia x 3 - 5 cm tall.
Fruiting body apothecial inner (hymenial) surface dull brown, smooth; outer surface pallid buff, farinose; irregularly cup-shaped, margin typically remaining incurved with one shorter side split vertically, not ear-shaped; sessile. Flesh pallid brown, brittle and thick.
Asci cylindrical, 200-225 x 11-14 µm. **Spores** (8) hyaline, smooth, ellipsoid, non-septate, uniseriate, 2 droplets, 16-19 x 9-11 µm. Paraphyses narrowly cylindrical, markedly curved at the tips, branched, septate.
Odour not distinctive. **Taste** not distinctive.
Chemical tests asci tips not blued with Melzer's reagent.
Occurrence spring to late autumn; rare. (Soudley Ponds, Forest of Dean, Gloucestershire)
■ Inedible.

Otidea onotica (Pers.: Fr.) Fuckel   Otidiaceae
= *Peziza onotica* Pers.: Fr.

Pinkish yellow irregular ear; solitary or in small trooping groups on soil in broad-leaf and mixed woods, favouring beech.

Dimensions 4 - 6 cm dia x 4 - 10 cm tall.
Fruiting body apothecial inner (hymenial) surface ochraceous with slight pink tinge, smooth outer surface ochraceous, mealy; ear-shaped with one shorter side split vertically, margin tending to remain incurved; stem pallid, very short and stout. Flesh pallid whitish, brittle and thin.
Asci cylindrical, 225-250 x 8-10 µm. **Spores** (8) hyaline, smooth, broadly ellipsoid, non-septate, uniseriate, 2 droplets, 12-13 x 5-6 µm. Paraphyses narrowly cylindrical, distinctively curved at the tips, branched, septate.
Odour not distinctive. **Taste** not distinctive.
Chemical tests asci not blued at the tips by Melzer's reagent.
Occurrence summer to early autumn; infrequent.
■ Inedible.

Pulvinula constellatio (Berk. & Br.) Boud.   Otidiaceae
= *Pulvinula cinnabarina* (Fuckel) Boud.

Small orange-pink cushion; solitary or clustered in small groups, on sandy soil including fire sites amongst moss.

Dimensions 0.3 - 1 cm dia.
Fruiting body apothecial upper (hymenial) surface orange-pink, smooth; margin and outer (lower) surface more pallid pink, smooth, margin slightly undulating; at first top-shaped, becoming cushion-like, sessile or with rudimentary stem. Flesh pinkish, somewhat gelatinous.
Asci cylindrical, 200-250 x 17-18 µm. **Spores** (8) hyaline, finely smooth, spherical, non-septate, uniseriate, 1 (or more) droplet(s), 13-17 µm. Paraphyses cylindrical, granular contents, forked at the base, bent at the tips, non-septate.
Odour not distinctive. **Taste** not distinctive.
Chemical tests asci tips not blued with Melzer's reagent.
Occurrence summer to early autumn; infrequent.
■ Inedible.

Scutellinia olivascens (Cke.) Kuntze   Otidiaceae
= *Scutellinia ampullacea* (Limm.) Kuntze

Small dull orange disc; densely clustered, on rotten wood or on soil.

Dimensions 0.5 - 1.5 cm dia.
Fruiting body apothecial upper (hymenial) surface dull orange-red, smooth; margin and outer (lower) surface brownish, covered with short, stiff, dark brown hairs; at first sub-spherical becoming saucer-shaped or flattish; sessile. Flesh reddish, brittle and thin.
Asci cylindrical, 250-280 x 20-25 µm. **Spores** (8) hyaline, coarsely roughened or granular, broadly ellipsoid, non-septate, uniseriate, several small droplets, 20-22 x 14-15 µm. Paraphyses cylindrical, markedly swollen (pear-shaped) at the tips, septate. Hairs dark brown, thick-walled, septate, with pointed tips.
Odour not distinctive. **Taste** not distinctive.
Chemical tests asci tips not blued with Melzer's reagent.
Occurrence summer to late autumn; infrequent.
■ Inedible.

Scutellinia scutellata (L.: Fr.) Lambotte   **Otidiaceae**
Eyelash Fungus

Very small scarlet orange disc; densely clustered, on rotten wood or woody debris often obscured by moss.

Dimensions 0.2 - 1 cm dia.
Fruiting body apothecial upper (hymenial) surface scarlet, smooth; margin and outer (lower) surface brownish, covered with long, pointed, dark brown hairs; at first sub-spherical becoming saucer-shaped or flattish; sessile. Flesh reddish, brittle and thin.
Asci cylindrical, 250-300 x 18-25 µm. **Spores** (8) hyaline, finely roughened or granular, broadly ellipsoid, non-septate, uniseriate, several small droplets, 18-20 x 10-12 µm. Paraphyses cylindrical, somewhat swollen at the tips, septate. Hairs dark brown, thick-walled, septate, with pointed tips.
Odour not distinctive. **Taste** not distinctive.
Chemical tests asci tips not blued with Melzer's reagent.
Occurrence summer to late autumn; common.
■ Inedible.

Scutellinia umbrorum (Fr.: Fr.) Lambotte   **Otidiaceae**

Small dull orange disc; densely clustered, on rotten wood or on bare soil.

Dimensions 0.5 - 1.5 cm dia.
Fruiting body apothecial upper (hymenial) surface dull orange red, smooth; margin and outer (lower) surface brownish covered with short, stiff, dark brown hairs; at first sub-spherical becoming saucer-shaped or flattish; sessile. Flesh reddish, brittle and thin.
Asci cylindrical, 250-300 x 20-25 µm. **Spores** (8) hyaline, coarsely roughened or granular, broadly ellipsoid, non-septate, uniseriate, several small droplets, 20-22 x 14-15 µm. Paraphyses cylindrical, slightly swollen at the tips, septate. Hairs dark brown, thick-walled, septate, with pointed tips.
Odour not distinctive. **Taste** not distinctive.
Chemical tests asci tips not blued with Melzer's reagent.
Occurrence summer to late autumn; common.
■ Inedible.

Sowerbyella radiculata (Sow.: Fr.) Nannf.   **Otidiaceae**

Medium-sized yellowish cup; typically in small clusters on soil with coniferous trees.

Dimensions 2-5 cm dia x 1-1.5 cm tall.
Fruiting body apothecial upper (hymenial) surface ochraceous with olivaceous tinge, more pallid when dry, smooth; lower (outer) surface concolorous, finely downy with smooth margin; cup-shaped, irregular, wavy narrowing into a wrinkled stem. Flesh ochraceous, brittle and thin.
Asci cylindrical, 280-350 x 16-20 µm. **Spores** (8) hyaline, coarsely warty, ellipsoid, non-septate, uniseriate, 2 large droplets, 20-24 x 11-13 µm. Paraphyses: narrowly cylindrical, bent, branched near the slightly swollen or lobed tips, non-septate.
Odour not distinctive. **Taste** not distinctive.
Chemical tests asci tips not blued with Melzer's reagent.
Occurrence autumn and winter; very rare. (Westonbirt Arboretum, Gloucestershire).
■ Inedible.

Sowerbyella radiculata (Sow.: Fr.) Nannf.   **Otidiaceae**
var **kewensis** Yao & Spooner (unpublished).

Comparatively large yellowish cup; typically in dense clusters on soil near broad-leaf trees.

Dimensions 4 - 10 cm dia x 2 - 5 cm tall.
Fruiting body apothecial upper (hymenial) surface ochraceous with olivaceous tinges, more pallid when dry, smooth; outer (lower) surface concolorous, finely downy with smooth margin; cup-shaped, irregular, wavy, narrowing into a wrinkled stem. Flesh ochraceous, brittle and thin.
Asci cylindrical, 280-350 x 16-20 µm. **Spores** (8) hyaline, coarsely warty, ellipsoid, non-septate, uniseriate, 2 large droplets, 20-24 x 11-13 µm. Paraphyses narrowly cylindrical, bent, branched near the slightly swollen or lobed tips, non-septate.
Odour not distinctive. **Taste** not distinctive.
Chemical tests asci tips not blued with Melzer's reagent.
Occurrence autumn to winter; very rare. (Carbrooke, Norfolk, under horse chestnut.)
■ Inedible.

Tazzetta catinus (Holmsk. : Fr.) Korf & Rogers   **Otidiaceae**
= *Pustularia catinus* (Holmsk.: Fr.) Fuckel

Smallish cream cup; typically in troops on soil in broad-leaf woods.

Dimensions 2 - 5 cm dia x 0.5 - 1.5 cm tall.
Fruiting body apothecial upper (hymenial) surface pallid ochraceous cream, smooth; outer (lower) surface concolorous, finely downy; cup-shaped, margin denticulate, not remaining incurved, typically partially sunk into the substrate and with rudimentary stem. Flesh pallid cream, brittle and thin.
Asci cylindrical, 280-350 x 16-20 µm. **Spores** (8) hyaline, smooth, ellipsoid, non-septate, uniseriate, 2 large droplets, 20-24 x 11-13 µm. Paraphyses narrowly cylindrical, slightly swollen or lobed at the tips, branched, septate.
Odour not distinctive. **Taste** not distinctive.
Chemical tests asci tips not blued with Melzer's reagent.
Occurrence summer to autumn; infrequent.
■ Inedible.
Distinguished from *T. cupularis* by possessing wider spores and lobed paraphyses.

Tazzetta cupularis (L.: Fr.) Lambotte   **Otidiaceae**
= *Pustularia cupularis* (L.: Fr.) Fuckel

Small, brownish or cream, flask-like cup; in troops on soil in damp woods.

Dimensions 1 - 2 cm dia x 0.5 - 2.5 cm tall.
Fruiting body apothecial upper (hymenial) surface pallid creamy grey, smooth; outer (lower) surface greyish brown, more pallid with age, finely warty; flask-shaped, with persistently incurved and denticulate margin which bears coarse matted hairs when young, partly sunk into the substrate with a rudimentary stem. Flesh pallid cream, brittle and thin.
Asci cylindrical, 250-300 x 15-16 µm. **Spores** (8) hyaline, smooth, ellipsoid, non-septate, uniseriate, 2 large droplets, 20-22 x 13-15 µm. Paraphyses narrowly cylindrical, slightly clavate at the tips but non-lobed, forked at the base, septate.
Odour not distinctive. **Taste** not distinctive.
Chemical tests asci tips not blued by Melzer's reagent.
Occurrence spring to autumn; infrequent.
■ Inedible.
Distinguished from *T. catinus* by possessing narrower spores and non-lobed paraphyses.

Trichophaea hemisphaerioides (Mouton) Graddon
Otidiaceae

Small greyish white cup with brown hairs; solitary or in small groups, on soil on fire sites amongst moss (typically *Funaria*), also on charred wood.

Dimensions 0.5 - 1.5 cm dia.
Fruiting body apothecial upper (hymenial) surface white or greyish white, smooth; margin and outer (lower) surface concolorous covered with brown hairs; at first sub-spherical becoming saucer-shaped or flattish, sessile. Flesh whitish, brittle and thin.
Asci cylindrical, 175-200 x 7-8 μm. **Spores** (8) hyaline, finely roughened or granular, narrowly ellipsoid, non-septate, uniseriate, 2 droplets, 13-18 x 5-7 μm. Paraphyses cylindrical, forked at the base, faintly swollen at the tips, septate. Hairs dark brown, thick-walled, septate, with pointed tips.
Odour not distinctive. **Taste** not distinctive.
Chemical tests asci tips not blued with Melzer's reagent.
Occurrence spring to autumn; infrequent.
■ Inedible.

Trichophaea woolhopeia (Cke. & Phill.) Arnould
Otidiaceae

Very small greyish white cup with brown hairs; solitary or in small groups, on sandy soil in damp places, also on fire sites and amongst moss; often (as with this specimen) in association with *Pulvinula constellatio*.

Dimensions 0.3 - 0.6 cm dia.
Fruiting body apothecial upper (hymenial) surface white or greyish white, smooth; margin and outer (lower) surface concolorous covered with brown hairs; at first sub-spherical, becoming saucer-shaped or flattish; sessile. Flesh whitish, brittle and thin.
Asci cylindrical, 220-300 x 15-20 μm. **Spores** (8) hyaline, finely roughened or granular, broadly ellipsoid, non-septate, uniseriate, 1 large and sometimes 1 small droplet, 21-22 x 13-15 μm. Paraphyses cylindrical, forked at the base, faintly swollen at the tips, septate. Hairs dark brown, thick-walled, septate, with pointed tips.
Odour not distinctive. **Taste** not distinctive.
Chemical tests asci tips not blued with Melzer's reagent.
Occurrence summer to autumn; infrequent.
■ Inedible.

Pseudoplectania nigrella (Pers.: Fr.) Fuckel
Sarcoscyphaceae

Small blackish cup, in small trooping groups, on soil, often amongst moss and short grass near coniferous trees, favouring path embankments on sandy soils.

Dimensions 1 - 2.5 cm dia x 0.6 - 1 cm tall.
Fruiting body apothecial upper (hymenial) surface black with bluish or violaceous tinges, smooth; outer (lower) surface concolorous, finely felty and crusty at the rim; at first deeply cup-shaped becoming more flattened at maturity; sessile. Flesh blackish, brittle, moderate and firm.
Asci cylindrical, 250-300 x 11-13 μm. **Spores** (8) hyaline, smooth, spherical, non-septate, uniseriate, occasionally with 1 droplet, 10-12 μm. Paraphyses cylindrical, branched at the tips, multi-septate.
Odour not distinctive. **Taste** not distinctive.
Chemical test asci tips not blued with Melzer's reagent.
Occurrence winter to spring; rare. (Parkend, Forest of Dean, Gloucestershire. Note there is only one other confirmed collection of the species (in Sutherland) since the early part of the century.)
■ Inedible.

Sarcoscypha austriaca (Beck ex Sacc.) Boud.

Sarcoscyphaceae

Large shallow scarlet cup; solitary or in small groups on damp rotting wood of broad-leaf and coniferous trees (often beneath leaf litter).

Dimensions 1 - 5 cm dia x 0.5 - 1.5 cm tall.
Fruiting body apothecial upper (hymenial) surface scarlet, smooth, outer (lower) surface pallid ochraceous; cup-shaped, margin at first slightly incurved, typically frayed with age, narrowing into short stem. Flesh pallid, brittle and thin.
Asci cylindrical, 400-450 x 14-16 µm. **Spores** (8) hyaline, smooth, elongated-ellipsoid, non-septate, uniseriate, with several small oil droplets at each end, 24-32 x 12-14 µm. Paraphyses very narrowly cylindrical, not swollen at the tips, septate, branched, reddish granular contents.
Odour not distinctive. **Taste** not distinctive.
Chemical tests asci tips not blued with Melzer's reagent.
Occurrence winter to spring; infrequent.
☐ Edible - can be eaten raw.
Formerly identified as *S. coccinea*, a distinct species in which spores and surface hyphae differ microscopically but which has not been recorded in recent times in Britain.

Peziza ammophila Dur. & Mont.

Pezizaceae

Small or medium-sized lobed vase, dull brown, outside paler; in small trooping groups, in sand dunes with marram grass (*Ammophila*).

Dimensions 2 - 4 cm dia x 1 - 5 cm tall.
Fruiting body apothecial upper (hymenial) surface cigar brown, smooth; outer (lower) surface more pallid, mealy, typically dusted with sand grains; at first vase-shaped, the margin becoming divided in lobes or triangular 'petals'; more or less sessile. Flesh brown, brittle and moderately thick.
Asci cylindrical, 200 x 15 µm. **Spores** (8) hyaline, smooth, ellipsoid, non-septate, uniseriate, 2 droplets, 14-16 x 8-10 µm. Paraphyses cylindrical, barely swollen at the tips, barely septate.
Odour not distinctive. **Taste** not distinctive.
Chemical test asci tips blued with Melzer's reagent.
Occurrence late summer to autumn; localised.
■ Inedible.

Peziza ampelina Quél.

Pezizaceae

Small cup, dulll violet, outside pale cream; in small trooping groups, some fused, on soil and on charred wood.

Dimensions 2 - 5 cm dia x 0.5 - 2 cm tall.
Fruiting body apothecial upper (hymenial) surface deep violet with brown tinge, smooth; outer (lower) surface pallid cream, mealy; at first cup-shaped, becoming more flattened, margin smooth but typically split in older specimens; sessile. Flesh pallid violaceous, brittle and thin.
Asci cylindrical, 200-350 x 12-15 µm. **Spores** (8) hyaline, smooth, ellipsoid, non-septate, uniseriate, 2 droplets, 18-22 x 9-11 µm. Paraphyses cylindrical, barely swollen at the tips, septate.
Odour not distinctive. **Taste** not distinctive.
Chemical test asci tips blued with Melzer's reagent.
Occurrence spring to early summer; infrequent or rare. (Woodchester Park, Gloucestershire.)
■ Inedible.
Note: a distinguishing microscopic feature of the species is the large size and smoothness of the spores.

Peziza ampliata Pers.: Fr.   **Pezizaceae**
= *Aleuria ampliata* (Pers.: Fr.) Gill.

Small brown cup, wrinkled inside; in small trooping groups, some fused, on wood chippings and other rotting wood fragments favouring oak.

Dimensions 1 - 3 cm dia x 1 - 2 cm tall.
Fruiting body apothecial upper (hymenial) surface cinnamon brown, smooth; outer (lower) surface more pallid with darker mealy dots; at first almost vesicular, becoming more cup-shaped with a denticulate margin; sessile. Flesh brown, brittle and thin.
Asci cylindrical, 300-330 x 12-15 µm. Spores (8) hyaline, reticulated, ellipsoid, non-septate, uniseriate, 2 droplets, 18-20 x 10-22 µm. Paraphyses cylindrical, barely swollen at the tips, septate.
Odour not distinctive. **Taste** not distinctive.
Chemical test asci tips blued with Melzer's reagent.
Occurrence late autumn to early spring; infrequent or rare. (Williton School, Somerset.)
■ Inedible.

Peziza arvernensis Boud.   **Pezizaceae**
= *Peziza sylvestris* (Boud.) Sacc. & Trav.
= *Aleuria amplissima* Gill.

Medium-sized, sometimes large, brown cup; in small trooping groups, some fused, on soil in broad-leaf woods favouring beech.

Dimensions 3 - 10 cm dia x 1 - 3 cm tall.
Fruiting body apothecial upper (hymenial) surface hazel or chestnut brown, smooth; outer (lower) surface more pallid, particularly at the margin, and finely roughened; at first almost vesicular, becoming more cup-shaped; sessile. Flesh brown, brittle and thin.
Asci cylindrical, 180-250 x 11-13 µm. **Spores** (8) hyaline, finely roughened, broadly ellipsoid, non-septate, uniseriate, droplets absent, 14-19 x 8-11 µm. Paraphyses cylindrical, barely swollen at the tips, septate.
Odour not distinctive. **Taste** not distinctive.
Chemical test asci tips blued with Melzer's reagent.
Occurrence spring to summer; infrequent.
■ Inedible.

Peziza badia Pers.: Fr.   **Pezizaceae**

Small, brown, irregular saucer; in small trooping groups, some fused, on soil in woods generally, particularly favouring bare path sides.

Dimensions 3 - 8 cm dia x 1 - 2 cm tall.
Fruiting body apothecial upper (hymenial) surface dark liver or olive brown, smooth; outer (lower) surface more pallid reddish brown, finely scurfy; shallowly cup-shaped with a wavy irregular margin; sessile. Flesh reddish brown, brittle and thin, yielding a watery juice when broken.
Asci cylindrical, 300-330 x 12-15 µm. Spores (8) hyaline, reticulated, ellipsoid, non-septate, uniseriate, 2 droplets, 17-20 x 9-12 µm. Paraphyses cylindrical, barely swollen at the tips, septate.
Odour not distinctive. **Taste** not distinctive.
Chemical test asci tips blued with Melzer's reagent.
Occurrence late summer to autumn; frequent.
✥ Poisonous when raw; edible when cooked thoroughly.

Peziza cerea Sow. : Fr.   **Pezizaceae**

Medium-sized pale brown cup; solitary or in small trooping groups, on old mortar, also in woods on buried rotting wood and other organic debris.

Dimensions 2 - 5 cm dia x 1 - 3 cm tall.
Fruiting body apothecial upper (hymenial) surface pallid hazel brown, smooth; outer (lower) surface more or less concolorous, finely whitish scurfy; at first cup-shaped, becoming irregularly flattened; sessile or with rudimentary stem. Flesh pallid brown, moderate.
Asci cylindrical, 250-300 x 11-14 µm. **Spores** (8) hyaline, smooth or slightly punctate, ellipsoid, non-septate, uniseriate, droplets absent, 14-16 x 8-11 µm. Paraphyses cylindrical, septate, not in chains.
Odour not distinctive. **Taste** not distinctive.
Chemical tests asci tips blued with Melzer's reagent.
Occurrence late spring to autumn; infrequent.
■ Inedible.

Peziza domiciliana Cke.   **Pezizaceae**
= *Peziza adae* Sadler: Cke.

Smallish, pale brown cup; in small trooping groups, some fused, on damp soil often in association with old masonry.

Dimensions 1.5 - 5 cm dia x 1 - 2 cm tall.
Fruiting body apothecial upper (hymenial) surface pallid brown, smooth; outer (lower) surface ochraceous white; irregularly saucer-shaped, almost smooth; sessile. Flesh pallid, brittle and thin.
Asci cylindrical, 200-250 x 11-12 µm. **Spores** (8) hyaline, more or less smooth, ellipsoid, non-septate, uniseriate, droplets absent, 14-16 x 8-9 µm. Paraphyses cylindrical, barely swollen at the tips, septate.
Odour not distinctive. **Taste** not distinctive.
Chemical test asci tips blued with Melzer's reagent.
Occurrence early summer to autumn; infrequent.
■ Inedible.

Peziza echinospora Karst.   **Pezizaceae**
= *Peziza anthracophila* Dennis

Small or largish cup, dark brown inside, almost white outside; often solitary or in small trooping groups, on damp fire sites, 1-2 years old.

Dimensions 2 - 10 cm dia x 1 - 3 cm tall.
Fruiting body apothecial upper (hymenial) surface dark or hazel brown, smooth; outer (lower) surface pallid whitish, mealy or scurfy; at first cup-shaped with a finely notched, incurved margin, becoming more flattened; sessile. Flesh pallid brown, brittle and thin.
Asci cylindrical, 250-280 x 11-12 µm. **Spores** (8) hyaline, minutely warty (difficult to detect), ellipsoid, non-septate, uniseriate, droplets absent, 14-17 x 7-8 µm. Paraphyses finely cylindrical, barely swollen at the tips, septate.
Odour not distinctive. **Taste** not distinctive.
Chemical test asci tips blued with Melzer's reagent.
Occurrence late spring to early autumn; infrequent.
■ Inedible.

Peziza granulosa Schum.: Fr. ss. Boud.   **Pezizaceae**

Small pale cup; often solitary or in small trooping groups, on soil and rotting vegetation in upland woods and pastures.

Dimensions 1 - 3 cm dia x 1 - 2 cm tall.
Fruiting body apothecial upper (hymenial) surface pallid olivaceous brown, becoming darker on ageing, smooth; outer (lower) surface concolorous with minute darker scales; at first cup-shaped with a finely notched, incurved margin, becoming more flattened and irregular; sessile. Flesh pallid brown, brittle and thin.
Asci cylindrical, 250-300 x 17-20 µm. **Spores** (8) hyaline, smooth, ellipsoid, non-septate, uniseriate, droplets more or less absent, 19-22 x 10-12 µm. Paraphyses finely cylindrical, markedly swollen at the tips, non-septate.
Odour not distinctive. **Taste** not distinctive.
Chemical test asci tips blued with Melzer's reagent.
Occurrence late spring to early autumn; infrequent.
■ Inedible.

Peziza micropus Pers.: Fr.   **Pezizaceae**

Small, brown, irregular saucer; in small trooping groups, some fused, on rotting wood, favouring beech.

Dimensions 2 - 5 cm dia x 1 - 2 cm tall.
Fruiting body apothecial upper (hymenial) surface hazelnut or reddish brown, smooth; outer (lower) surface more pallid brown, finely scurfy; shallowly cup-shaped with a wavy irregular or notched margin; sessile. Flesh reddish brown, brittle and thin.
Asci cylindrical, 230-250 x 12-15 µm. **Spores** (8) hyaline, smooth, ellipsoid, non-septate, uniseriate, droplets absent, 14-15 x 7.5-9.5 µm. Paraphyses cylindrical, barely swollen at tips, septate, sometimes forked near base.
Odour not distinctive. **Taste** not distinctive.
Chemical test asci tips blued with Melzer's reagent.
Occurrence summer to autumn; infrequent.
■ Inedible.

Peziza petersii Berk. & Curt.   **Pezizaceae**
= *Galactinia sarrazinii* Boud.

Small, brownish, irregular saucer; solitary or in trooping groups, often densely fused, typically on bonfire sites or other burnt ground with organic debris.

Dimensions 3 - 5 cm dia x 1 - 2 cm tall.
Fruiting body apothecial upper (hymenial) surface reddish brown, with greyish tinges at the centre, smooth; outer (lower) surface reddish brown greying at the base, finely scurfy; irregularly and shallowly cup-shaped; sessile. Flesh greyish brown, brittle and thin.
Asci cylindrical, 180-200 x 8-10 µm. **Spores** (8) hyaline, finely warty, ellipsoid, non-septate, uniseriate, 2 droplets, 10-12 x 5-6 µm. Paraphyses cylindrical with barely clavate tips, septate.
Odour not distinctive. **Taste** not distinctive.
Chemical test asci tips blued with Melzer's reagent.
Occurrence summer to autumn; infrequent.
■ Inedible.

Peziza repanda Fr.: Fr.   **Pezizaceae**

Regular yellowish brown cup, becoming irregularly saucer-shaped; solitary or in small trooping groups, on soil around stumps and on rotting sawdust and wood.

Dimensions 3 - 12 cm dia x 1 - 4 cm tall.
Fruiting body apothecial upper (hymenial) surface ochraceous brown or darker, smooth; outer (lower) surface cream, finely scurfy; at first cup-shaped, becoming irregularly saucer-shaped with a wavy margin; sessile. Flesh pallid yellowish and brittle.
Asci cylindrical, 280-330 x 12-14 µm. **Spores** (8) hyaline, smooth, ellipsoid, non-septate, uniseriate, droplets absent, 15-17 x 8-10 µm. Paraphyses cylindrical, slightly swollen at the tips, septate.
Odour not distinctive. **Taste** not distinctive.
Chemical tests asci tips blued with Melzer's reagent.
Occurrence late spring to late autumn; infrequent.
■ Inedible.

Peziza succosa Berk.   **Pezizaceae**

Comparatively large, brown, irregular saucer; in small trooping groups on soil in mixed woods, particularly favouring pathsides and banks.

Dimensions 2 - 10 cm dia x 1 - 3 cm tall.
Fruiting body apothecial upper (hymenial) surface dull hazel brown, smooth, outer (lower) surface more pallid brown, finely scurfy; shallowly cup-shaped, often somewhat wrinkled at the centre and with a wavy irregular margin; sessile. Flesh brown, brittle and thin, yielding a watery juice when broken which turns yellow in air.
Asci cylindrical, 300-330 x 12-15 µm. **Spores** (8) hyaline, coarsely warty or sub-reticulated, ellipsoid, non-septate, uniseriate, 2 droplets, 17-22 x 9-12 µm. Paraphyses cylindrical, barely swollen at the tips, septate, occasionally forked.
Odour not distinctive. **Taste** not distinctive.
Chemical test asci tips blued with Melzer's reagent.
Occurrence late summer to autumn; infrequent.
■ Inedible

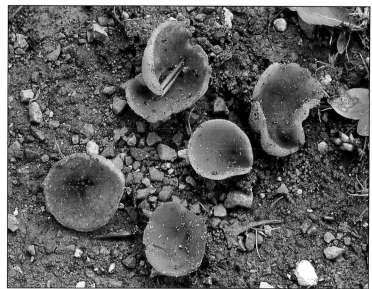

Peziza varia (Hedw.) Fr.   **Pezizaceae**

Medium-sized pale brown cup; solitary or in small trooping groups, in woods on buried rotting wood and other organic debris.

Dimensions 2 - 6 cm dia x 1 - 3 cm tall.
Fruiting body apothecial upper (hymenial) surface pallid hazel brown, smooth; outer (lower) surface more or less concolorous, finely whitish scurfy; at first cup-shaped, becoming irregularly flattened; sessile or with rudimentary stem. Flesh pallid brown, layered and moderate.
Asci cylindrical, 250-300 x 11-14 µm. **Spores** (8) hyaline, smooth, ellipsoid, non-septate, uniseriate, droplets absent, 14-16 x 8-11 µm. Paraphyses cylindrical, septate but often constricted at the septa and looking like chains of cells.
Odour not distinctive. **Taste** not distinctive.
Chemical tests asci tips blued with Melzer's reagent.
Occurrence spring to autumn; common.
■ Inedible.

Peziza vesiculosa Bull.: Fr.   **Pezizaceae**

Large, pale straw, wrinkled cup; generally in dense trooping groups, on manure, rotting straw and rich soil.

Dimensions 4 - 10 cm dia x 3 - 6 cm tall.
Fruiting body apothecial upper (hymenial) surface ochraceous brown, smooth; outer (lower) surface pallid buff, coarsely mealy; bowl-shaped, typically wrinkled, with incurved margin; sessile. Flesh pallid buff, brittle and thick.
Asci cylindrical, 320-380 x 18-25 µm. **Spores** (8) hyaline, smooth, ellipsoid, non-septate, uniseriate, droplets absent, 20-24 x 11-14 µm. Paraphyses cylindrical, slightly swollen at the tips, septate.
Odour not distinctive. **Taste** not distinctive.
Chemical tests asci tips blued with Melzer's reagent.
Occurrence throughout the year but mainly late spring to autumn; common.
✛ Poisonous.
One of the few Pezizas to withstand frost. Toxins thermolabile and may be destroyed by thorough cooking.

Peziza violacea Pers.   **Pezizaceae**

Small, brownish, irregular saucer; solitary or in trooping groups, often densely fused, typically on bonfire sites or other burnt ground with organic debris.

Dimensions 1 - 4 cm dia x 1 - 1.5 cm tall.
Fruiting body apothecial upper (hymenial) surface delicate violaceous or mauve brown, smooth; outer (lower) surface con-colorous, finely scurfy; shallow cup-shaped or irregularly expanded; sessile. Flesh thin, mauve and brittle.
Asci cylindrical, 200-250 x 8-10 µm. **Spores** (8) hyaline, finely warty, ellipsoid, non-septate, uniseriate, 2 droplets, 11-13 x 6-8 µm. Paraphyses clavate, swollen at the tips, branched, septate.
Odour not distinctive. **Taste** not distinctive.
Chemical test asci tips blued with Melzer's reagent.
Occurrence summer to autumn; infrequent.
■ Inedible.
Note: the colour of the specimen illustrated is washed out and therefore a-typical.

Plicaria anthracina (Cke.) Boud.   **Pezizaceae**

Medium-sized blackish brown cup; generally in trooping groups, on burnt ground including bonfire sites.

Dimensions 1 - 3 cm dia x 0.5 - 1 cm tall.
Fruiting body apothecial upper (hymenial) surface blackish brown, smooth; dull, with brownish grey margin outer (lower) surface con-colorous with sooty flecks; bowl-shaped, becoming irregularly expanded at maturity; sessile. Flesh brown, brittle and thick, not yielding yellow juice when cut.
Asci cylindrical, 200-270 x 12-20 µm. **Spores** (8) hyaline, coarsely warty, spherical, non-septate, uniseriate, occasionally with droplet, 11-14 µm. Paraphyses finely cylindrical, slightly swollen at the tips, septate.
Odour not distinctive. **Taste** not distinctive.
Chemical tests asci tips blued with Melzer's reagent.
Occurrence spring to autumn; common.
■ Inedible.

Plicaria leiocarpa (Currey) Boud.   Pezizaceae

Medium-sized dark cup; generally in trooping groups, on burnt ground including bonfire sites.

Dimensions 2 - 6 cm dia x 1 - 3 cm tall.
Fruiting body apothecial upper (hymenial) surface blackish brown, smooth; outer (lower) surface concolorous with sooty flecks; bowl-shaped, becoming more expanded at maturity; sessile. Flesh brown, brittle and thick, sometimes with yellow juice when cut.
Asci cylindrical, 180-200 x 8-12 μm. **Spores** (8) hyaline, smooth, spherical, non-septate, uniseriate, occasionally with droplet, 8-10 μm. Paraphyses finely cylindrical, slightly swollen at the tips, septate.
Odour not distinctive. **Taste** not distinctive.
Chemical tests asci tips blued with Melzer's reagent.
Occurrence spring to autumn; common.
■ Inedible.

Sarcosphaera coronaria (Jacq.: Fr.) Boud.   Pezizaceae
= *Sarcosphaera crassa* (Santi) Pouz.
= *Sarcosphaera eximia* (Dur.: Lév.) Maire

Large white and violet irregular cup; scattered or clustered, on calcareous soils in broad-leaf woods favouring beech, less frequently with conifers.

Dimensions 3 - 20 cm dia x 3 - 15 cm tall.
Fruiting body apothecial upper (hymenial) surface violaceous, either pallid or with deep brownish tinge, smooth; outer (lower) surface whitish with grey tinges, slightly roughened; at first vesicular, embedded in the substrate, then breaking through the surface and opening, more expanded at maturity, splitting into star-shaped rays; sessile. Flesh white, brittle and thick.
Asci cylindrical, 300-360 x 10-13 μm. **Spores** (8) hyaline, smooth, broadly ellipsoid, non-septate, uniseriate, 2 droplets, 13.5-18 x 7-8.5 μm. Paraphyses finely cylindrical, slightly swollen at the tips, forked at the base, septate.
Odour not distinctive. **Taste** not distinctive.
Chemical tests asci tips blued with Melzer's reagent.
Occurrence late spring to summer; rare but locally more abundant. (Buckholt Wood (SSSI), Gloucestershire).
✧ Poisonous when raw.

Gyromitra esculenta (Pers.: Fr.) Fr.   Helvellaceae
False morel

Irregular, brown brain-like structure on a pale stem; solitary or in small trooping groups near conifers (pines), typically on acid, sandy soils.

Dimensions 5 - 15 cm dia x 5 - 12 cm tall, including stem.
Fruiting body apothecium outer (hymenial) surface reddish brown, irregularly lobed and convoluted, brain-like; stem white, short, stout, deeply furrowed and finely granular. Flesh yellowish ochre in cap, white in stem, brittle, medium and hollow.
Asci cylindrical, 325-350 x 16-20 μm. **Spores** (8) hyaline, smooth, ellipsoid, non-septate, uniseriate, usually 2 droplets, 16-22 x 7-12 μm. Paraphyses cylindrical, slightly swollen at the tips, septate, branched.
Odour not distinctive. **Taste** not distinctive.
Chemical tests asci tips not blued with Melzer's reagent.
Occurrence spring; infrequent.
✚ Deadly poisonous.
Toxins allegedly lost after thorough cooking but, if ingested, the effect is probably cumulative with no immediate symptoms.

Helvella acetabulum (L.: Fr.) Quél.   Helvellaceae
= *Paxina acetabulum* (L.: Fr.) Kuntze

Large brown cup on paler wrinkled stem; generally in trooping groups on sandy or calcareous soil in woodlands and woodland margins.

Dimensions 4 - 6 cm dia x 4 - 6 cm tall (stem 1 - 4 cm).
Fruiting body apothecium upper (hymenial) surface smooth tan or chestnut brown; outer (lower) surface concolorous or more pallid, finely downy, at first deeply cup-shaped, becoming more expanded, tapering into a stout stem; stem whitish, sculptured with ribs extending into base of cup. Flesh pallid, brittle and thin.
Asci cylindrical, 270-320 x 15-20 µm. **Spores** (8) hyaline, smooth, broadly ellipsoid, non-septate, uniseriate, 1 droplet, 18-22 x 12-14 µm. Paraphyses cylindrical, swollen at the tips, septate, branched.
Odour not distinctive. **Taste** not distinctive.
Chemical tests asci tips not blued with Melzer's reagent.
Occurrence spring to early summer; infrequent.
✣ Poisonous. The toxins are thermolabile and destroyed by thorough cooking.

Helvella crispa Fr.   Helvellaceae

Contorted whitish cap on a chambered stem; solitary or in small trooping groups, on soil in broad-leaf and mixed woods.

Dimensions cap 3 - 6 cm dia; stem 6 - 12 cm tall.
Fruiting body apothecium whitish, irregularly lobed, more or less saddle-shaped with 2 or 3 lobes, smooth or wrinkled with wavy margin not fused with stem; stem creamy white, stout, tapering upwards, deeply grooved or furrowed with lacunae to hollow interior. Flesh white, brittle and medium.
Asci cylindrical, 250-300 x 14-18 µm. **Spores** (8) hyaline, smooth, broadly ellipsoid, non-septate, uniseriate, 1 droplet, 17-20 x 10-13 µm. Paraphyses cylindrical, slightly swollen towards the tips, barely septate, branched.
Odour not distinctive. **Taste** not distinctive.
Chemical tests asci tips not blued with Melzer's reagent.
Occurrence summer to autumn; frequent.
☐ Edible but uninspiring.

Helvella corium (Weberb.) Mass.   Helvellaceae
= *Cyathipodia corium* (Weberb.) Boud.

Black cup on concolorous slender stem; solitary or in small trooping groups, on sandy soil or in dunes.

Dimensions cap 1 - 3 cm dia; stem 0.6 - 1.8 cm tall.
Fruiting body apothecium black, cup-shaped, often somewhat compressed, becoming expanded; upper (hymenial) surface smooth; outer surface covered with downy hairs and slightly roughened; margin finely notched; stem black, cylindrical, slender, covered in greyish felty hairs. Flesh black, brittle and thin.
Asci cylindrical, 250-300 x 14-18 µm. **Spores** (8) hyaline, smooth, broadly broadly ellipsoid, non-septate, uniseriate, 1 large droplet, 20-22 x 12-14 µm. Paraphyses cylindrical, slightly swollen towards the tips, non-septate.
Odour not distinctive. **Taste** not distinctive.
Chemical tests asci tips not blued with Melzer's reagent.
Occurrence early summer to autumn; infrequent, but locally abundant.
■ Inedible.

Helvella elastica Bull.: Fr.   Helvellaceae
= *Leptopodia elastica* (Bull.: Fr.) Boud.

Irregularly saddle-shaped yellow-brown cap on a whitish stem; solitary or in small trooping groups, on soil in mixed woods.

Dimensions cap 2 - 4 cm dia; stem 7 - 10 cm tall.
Fruiting body apothecium upper (hymenial) surface yellowish brown, smooth; lower surface ochraceous, smooth, irregularly saddle-shaped, convoluted or wavy, margin not fused with the stem; stem whitish, smooth or finely downy at the apex, slender, cylindrical, hollow. Flesh whitish, brittle and thin.
Asci cylindrical 300-330 x 15-20 µm. **Spores** (8) hyaline, smooth, broadly ellipsoid, non-septate, uniseriate, 1 large droplet, 19-22 x 11-13 µm. Paraphyses cylindrical, barely septate, swollen towards the tips.
Odour not distinctive. **Taste** not distinctive.
Chemical tests asci tips not blued with Melzer's reagent.
Occurrence early summer to autumn; infrequent.
☐ Edible but uninspiring.

Helvella ephippium Lév.   Helvellaceae
= *Leptopodia ephippium* (Lév.) Boud.

Saddle-shaped brownish grey cap on a whitish stem; solitary or in small trooping groups, on soil in mixed woods.

Dimensions cap 1.5 - 2.5 cm dia; stem 4 - 5 cm tall.
Fruiting body apothecium upper (hymenial) surface brownish grey, smooth; lower surface concolorous but with ochraceous tinge, finely downy, more or less regularly saddle-shaped, margin not fused with the stem; stem whitish with grey or ochraceous tinges, finely downy along the whole length, cylindrical or slightly compressed, hollow. Flesh whitish, brittle and thin.
Asci cylindrical, 250-300 x 12-15 µm. **Spores** (8) hyaline, smooth, broadly ellipsoid, non-septate, uniseriate, 1 droplet, 19-22 x 11-13 µm. Paraphyses cylindrical, barely septate, swollen towards the tips.
Odour not distinctive. **Taste** not distinctive.
Chemical tests asci tips not blued with Melzer's reagent.
Occurrence early summer to autumn; rare. (Alfriston Forest, East Sussex.)
☐ Edible but uninspiring.

Helvella lacunosa Afz.: Fr.   Helvellaceae
                                       Black Helvella

Contorted blackish-grey cap on a chambered stem; solitary or in small trooping groups, on soil in broad-leaf, coniferous and mixed woods.

Dimensions cap 2 - 5 cm dia; stem 3 - 10 cm tall.
Fruiting body apothecium greyish black, irregularly lobed and convoluted, sometimes 2-lobed and saddle-shaped, smooth or wrinkled, margin fused to the stem; stem pallid grey, stout, tapering upwards, grooved and furrowed with lacunae to hollow interior. Flesh grey, brittle and thickish.
Asci cylindrical, 240-350 x 14-16 µm. **Spores** (8) hyaline, smooth, broadly ellipsoid, non-septate, uniseriate, 1 large central drop, 18-22 x 11-13 µm. Paraphyses clavate or cylindrical, slightly swollen towards the tips, septate.
Odour not distinctive. **Taste** not distinctive.
Chemical tests asci tips not blued with Melzer's reagent.
Occurrence autumn; infrequent.
☐ Edible but uninspiring.

Helvella leucomelaena (Pers.: Fr.) Nannf.   **Helvellaceae**
= *Paxina leucomelaena* (Pers.: Fr.) Kuntze
= *Acetabula vulgaris* Fuckel

Largish greyish brown cup; generally in trooping groups on sandy soil in woodlands and in short grass on heaths, often favouring poor soils.

Dimensions 2 - 4 cm dia x 2 - 4 cm tall (stem 1 - 1.5 cm).
Fruiting body apothecium upper (hymenial) surface dark greyish-brown, smooth; outer (lower) surface pallid greyish buff, finely downy, at first deeply cup-shaped, becoming expanded, usually splitting at the margin, tapering into a stout stem; stem whitish, sculptured, with ribs extending into base of cup. Flesh pallid, brittle and thin.
Asci cylindrical, 250-300 x 13-16 µm. **Spores** (8) hyaline, smooth, broadly ellipsoid, non-septate, uniseriate, 1 droplet, 8-22 x 12-14 µm. Paraphyses cylindrical, swollen at the tips, septate, unbranched.
Odour not distinctive. **Taste** not distinctive.
Chemical tests asci tips not blued with Melzer's reagent.
Occurrence spring to early summer; infrequent.
✛ Poisonous. The toxins are thermolabile and destroyed by thorough cooking.

Helvella macropus (Pers. : Fr.) Karst.   **Helvellaceae**
= *Macroscyphus macropus* (Pers.: Fr.) S F Gray
= *Macropodia macropus* (Pers. : Fr.) S F Gray
= *Cyathipodia macropus* (Pers.: Fr.) Dennis

Small greyish cup on a pale stem; solitary or in small groups, on soil in broad-leaf or, very occasionally, coniferous woods.

Dimensions cap 1.5 - 4 cm dia; stem 2 - 5 cm tall.
Fruiting body apothecium upper (hymenial) surface grey with brownish tinges, smooth; outer (lower) surface more pallid , cup-shaped, sometimes with slightly wavy margin, densely felty-hairy (use lens); stem concolorous, cylindrical, slender, tapering upwards, felty -hairy. Flesh white, brittle and thin.
Asci cylindrical, 220-350 x 15-20 µm. **Spores** (8) hyaline, very finely warty, broadly ellipsoid or fusiform, non-septate, uniseriate, with 1 central and 2 parietal droplets, 18-21 x 10-12 µm. Paraphyses cylindrical, swollen towards the tips, non-septate.
Odour not distinctive. **Taste** not distinctive.
Chemical tests asci tips not blued with Melzer's reagent.
Occurrence early summer to early autumn; infrequent.
■ Inedible.

Helvella stevensii Peck   **Helvellaceae**
= *Leptopodia stevensii* (Peck) Le Gal

Saddle-shaped greyish cap on a pale stem; solitary or in small trooping groups, on soil in broad-leaf woods.

Dimensions cap 2 - 4 cm dia; stem 4 - 5 cm tall.
Fruiting body apothecium upper (hymenial) surface greyish-brown, smooth; lower surface greyish-brown, irregularly saddle-shaped, margin wavy, involute but not fused with the stem; stem whitish, cylindrical, coarsely downy, hollow. Flesh brownish, brittle and thin.
Asci cylindrical, 225-280 x 14-18 µm. **Spores** (8) hyaline, smooth, broadly ellipsoid, non-septate, uniseriate, 1 droplet, 18-19 x 12-13 µm. Paraphyses cylindrical, swollen towards the tips, septate.
Odour not distinctive. **Taste** not distinctive.
Chemical tests asci tips not blued with Melzer's reagent.
Occurrence late summer to autumn; rare. (Alfriston Forest, East Sussex.)
■ Inedible.

Rhizina undulata Fr.: Fr.   Helvellaceae
= *Rhizina inflata* (Schaeff.) Quél.

Large irregular brown or black cushion; in fused clusters, on conifer debris, particularly favouring old fire sites.

Dimensions 2 - 10 cm dia
Fruiting body apothecium dark brown ageing black, irregular cushion-like, the margin more pallid; underside ochraceous with root-like attachments; sessile. Flesh reddish brown, tough and thick.
Asci cylindrical, 300-400 x 15-20 μm. **Spores** (8) hyaline, finely warty, fusiform with small appendage at each end, non-septate, more or less uniseriate, 2 droplets, 30-40 x 8-11 μm. (excluding appendages). Paraphyses cylindrical but swollen at the tips, septate, smooth but encrusted brownish at the tips.
Odour not distinctive. **Taste** not distinctive.
Chemical tests asci tips not blued with Melzer's reagent.
Occurrence early summer to early autumn; infrequent.
■ Inedible.
Causes 'Group Dying' disease in conifer plantations.

Disciotis venosa (Pers.: Fr.) Boud.   Morchellaceae

Large, brown, irregular and distinctively wrinkled saucer with paler outer surface, on a short thick stalk; solitary or in trooping groups in open mixed woodland or edges of woods in damp locations, favouring bare soil or light grasses, mosses and other vegetation.

Dimensions 3 - 20 cm dia x 2 - 6 cm tall.
Fruiting body apothecium upper (hymenial) surface dull brown, smooth but with radial wrinkling or puckering towards point of attachment; lower surface pallid, finely granular; at first irregularly cup-shaped, expanding to a shallow irregular saucer; stem pallid, short, stout, wrinkled and finely granular. Flesh pallid brown, brittle and medium.
Asci cylindrical, 300-350 x 18-23 μm. **Spores** (8) hyaline, smooth, broadly ellipsoid, non-septate, uniseriate, sometimes with 1 droplet at each end outside the spore wall, 19-25 x 12-15 μm. Paraphyses cylindrical, slightly swollen at the tips, septate, branched.
Odour of chlorine. **Taste** not distinctive.
Chemical tests asci tips not blued with Melzer's reagent.
Occurrence spring; infrequent.
✣ Poisonous if uncooked. The toxins are destroyed on thorough cooking. Recommended for eating if correctly prepared.

Morchella conica Fr.   Morchellaceae

Largish, brown, conical, honeycombed cap on a whitish stem; solitary on soil, often amongst short grass, in or close to broad-leaf or coniferous woods favouring calcareous soils.

Dimensions 2 - 4 cm dia x 5 - 15 cm tall.
Fruiting body apothecium yellowish or greyish brown becoming darker brown at maturity, with dark brown ridges forming irregular honeycomb-like pits; conical, tapering downwards into creamy-white stem, finely granular, sometimes furrowed or wrinkled, more or less equal. Flesh white, brittle, thin and hollow.
Asci cylindrical, 300-375 x 25-28 μm. **Spores** (8) hyaline, smooth but granulated at the ends, broadly ellipsoid, non-septate, uniseriate, droplets absent, 20-25 x 11-16 μm. Paraphyses cylindrical, swollen towards the tips, septate, branched.
Odour not distinctive. **Taste** not distinctive.
Chemical tests asci tips not blued with Melzer's reagent..
Occurrence spring; rare. (Soudley, Forest of Dean, Gloucestershire.)
☐ Edible
Considered by some authorities to be a variety of *Morchella esculenta*.

Morchella elata Fr.: Fr.   **Morchellaceae**

Smallish, brown, conical, honeycombed cap on a whitish stem; solitary on soil, often amongst short grass, in or close to coniferous woods, favouring calcareous soils.

Dimensions 2 - 4 cm dia x 5 - 15 cm tall.
Fruiting body apothecium yellowish brown, with darker parallel longitudinal ribs and connecting ridges forming honeycomb-like pits; conical, tapering downwards into creamy-white stem, coarsely granular, sometimes furrowed, more or less equal. Flesh white, brittle, thin and hollow.
Asci cylindrical, 250-300 x 18-20 µm. **Spores** (8) hyaline, smooth but granulated at the ends, broadly ellipsoid, non-septate, uniseriate, droplets absent, 18-25 x 11-16 µm. Paraphyses cylindrical, swollen towards the tips, septate, branched.
Odour not distinctive. **Taste** not distinctive.
Chemical tests asci tips not blued with Melzer's reagent.
Occurrence spring; rare, generally more northern. (Hurn near Ringwood, Hampshire.)
☐ Edible.

Morchella esculenta (L.) Pers. : Fr.   **Morchellaceae**

Large or massive, yellowish brown, more or less rounded honey-combed cap on a stout whitish stem; solitary on soil, in scrub or open woodland.

Dimensions 6 - 9 cm dia x 6 - 25 cm tall.
Fruiting body apothecium yellowish brown with concolorous irregular ridges forming honeycomb-like pits, variable in shape from sub-spherical to ovoid or bluntly conical, tapering abruptly downwards into creamy-white stem, coarsely granular, often distorted and bulbous. Flesh white, brittle, thin and hollow.
Asci cylindrical, 330-380 x 17-22 µm. **Spores** (8) cream, smooth but granulated at the ends, broadly ellipsoid, non-septate, uniseriate, droplets absent, 18-23 x 11-14 µm. Paraphyses cylindrical, slightly swollen towards the tips, septate, branched.
Odour not distinctive. **Taste** not distinctive.
Chemical tests asci tips not blued with Melzer's reagent.
Occurrence spring; infrequent.
☐ Edible and good.

Morchella esculenta (L.) Pers.: Fr. forma **vulgaris** (Pers.: Fr.) Boud.   **Morchellaceae**

Large, brown, egg-shaped, honeycombed cap on a whitish stem; solitary or in scattered troops, on soil, often amongst short grass, favouring gardens and calcareous soil.

Dimensions 3 - 6 cm dia x 5 - 12 cm tall.
Fruiting body apothecium greyish brown becoming pallid brown at maturity, with ridges forming very irregular honeycomb-like pits; ovoid, merging abruptly into creamy-white stem, finely granular, tapering slightly upwards. Flesh white, brittle, thin and hollow.
Asci cylindrical, 300-380 x 17-22 µm. **Spores** (8) hyaline, smooth but granulated at the ends, broadly ellipsoid, non-septate, uniseriate, droplets absent, 18-23 x 11-14 µm. Paraphyses cylindrical, swollen towards the tips, septate, branched.
Odour not distinctive. **Taste** not distinctive.
Chemical tests asci tips not blued with Melzer's reagent..
Occurrence spring; infrequent.
☐ Edible.
Not recognised as a distinct form by Royal Botanic Gardens, Kew.

Morchella semilibera DC.: Fr.   **Morchellaceae**
= *Mitrophora semilibera* (DC.: Fr.) Lév.

Smallish, olive, brown, conical honeycombed cap on a whitish stem, often quite tall; solitary on soil, typically amongst mosses or other vegetation, frequently close to hawthorn.

Dimensions 1.5 - 3.0 cm dia x 3 - 13 cm tall.
Fruiting body apothecium olivaceous brown, becoming darker brown at maturity, with dark brown more or less longitudinal ridges forming irregular honeycomb-like pits; conical, lower part free of the stem; stem creamy white, granular, sometimes finely furrowed, more or less equal. Flesh white, brittle, thin and hollow.
Asci cylindrical, 300-450 x 17-20 µm. **Spores** (8) hyaline, smooth but granulated at the ends, broadly ellipsoid, non-septate, uniseriate, droplets absent, 22-30 x 12-18 µm. Paraphyses cylindrical, swollen towards the tips, septate, branched.
Odour not distinctive. **Taste** not distinctive.
Chemical tests asci tips not blued with Melzer's reagent..
Occurrence spring; infrequent.
☐ Edible but poor.

Verpa conica (Müller: Fr.) Swartz   **Morchellaceae**

Dull brown irregular conical cap on a whitish stem; solitary on soil, close to hawthorn favouring calcareous soils.

Dimensions 2 - 4 cm dia x 4 - 9 cm tall.
Fruiting body apothecium brown with olivaceous tinges, at first ovoid, becoming campanulate and somewhat irregular, the margin incurved and fitting closely to the stem; stem white, with coarse brown granules arranged in horizontal bands, cylindrical, more or less equal. Flesh white, brittle, thin and hollow.
Asci cylindrical, 250-350 x 18-23 µm. **Spores** (8) hyaline, smooth but sometimes with granules at the ends, broadly ellipsoid, non-septate, uniseriate, droplets absent,18-25 x 11-15 µm. Paraphyses cylindrical, swollen towards the tips, septate, branched.
Odour not distinctive. **Taste** not distinctive.
Chemical tests asci tips not blued with Melzer's reagent.
Occurrence spring; rare. (Near Abergavenny, Gwent.)
☐ Edible.

Tuber aestivum Vitt.   **Tuberaceae**
Summer Truffle

Subterranean dark brown warty ball, typically under beech but also other broad-leaf trees on calcareous soils.

Dimensions 3 - 7 cm dia.
Fruiting body apothecium blackish-brown, irregularly sub-spherical, covered in 5-6 sided pyramidal warts. Flesh at first whitish, becoming grey brown with white marbled veins.
Asci sub-spherical, 60-90 x 50-70 µm. **Spores** (2-5) at first hyaline, becoming yellowish brown, reticulated, broadly ellipsoid or sub-spherical, non-septate, irregularly arranged, 25-50 x 17-35 µm.
Paraphyses: absent
Odour pleasantly sweet. **Taste** nutty.
Chemical tests none.
Occurrence summer to autumn; rare. (Highdown near Worthing, West Sussex.)
☐ Edible and good - can be eaten raw.

Geoglossum cookeianum Nannf.   **Geoglossaceae**

Black tongue-like spindle; in small trooping groups, amongst short turf, favouring sandy soils.

Dimensions 0.5 - 1.0 cm dia x 3 - 7 cm tall.
Fruiting body apothecium outer (hymenial) surface black, more or less fusiform or slightly clavate, smooth, often compressed in one plane, tapering into a short concolorous stem. Flesh black, and brittle.
Asci cylindrical or clavate, 150-180 x 18-20 μm. **Spores** (8) brown, smooth, sub-cylindrical, 7-septate at maturity, parallel, 50-90 x 5-7 μm. Paraphyses filiform, extended at the tips into short barrel-like segments.
Odour not distinctive. **Taste** not distinctive.
Chemical tests asci tips blued with Melzer's reagent.
Occurrence late summer to autumn; infrequent.
■ Inedible.

Geoglossum glutinosum Pers.   **Geoglossaceae**

Black, slimy, tongue-like spindle; in small trooping groups on soil amongst grass.

Dimensions 0.3 - 0.5 cm dia x 2.5 - 5 cm tall.
Fruiting body apothecium outer (hymenial) surface black or blackish-brown, more or less fusiform or slightly clavate, often compressed in one plane, tapering into a slender cylindrical stem, smooth, viscid. Flesh black or blackish-brown and brittle.
Asci cylindrical, 130-180 x 18-25 μm. **Spores** (8) brown, smooth, sub-cylindrical, 3- to 7-septate at maturity, parallel, 60-110 x 4.5-5.5 μm.
Note: the ascospores may appear, erroneously, 15-septate on casual inspection. Paraphyses: filiform with swollen pyriform tips, non-septate, tips brown.
Odour not distinctive. **Taste** not distinctive.
Chemical tests asci tips blued with Melzer's reagent.
Occurrence late summer to autumn; rare. (Abergavenny, Gwent).
■ Inedible.

Microglossum viride (Pers.: Fr.) Gill.   **Geoglossaceae**

Smallish green tongue-like spindle; in small trooping groups on soil amongst mosses in broad-leaf woodlands.

Dimensions 0.3 - 0.7 cm dia x 3 - 6 cm tall.
Fruiting body apothecium outer (hymenial) surface olive green, more or less fusiform or slightly clavate, smooth or furrowed, often compressed in one plane, tapering into a comparatively long, more pallid, stem. Flesh greenish and brittle.
Asci cylindrical or clavate, 130-150 x 9-11 μm. **Spores** (8) brown, smooth, sub-cylindrical, 3 or 4-septate at maturity, irregularly biseriate, 16-20 x 5-6 μm. Paraphyses filiform, slightly swollen at the tips, forked.
Odour not distinctive. **Taste** not distinctive.
Chemical tests asci tips blued with Melzer's reagent.
Occurrence late summer to autumn; infrequent.
■ Inedible.

Mitrula paludosa Fr.   Geoglossaceae

Small yellowish-orange swollen head on white stem; in trooping groups in ditches and other wet places attached to rotting plant debris.

Dimensions 0.75 - 1.25 cm dia x 1 - 4 cm tall.
Fruiting body apothecium outer (hymenial) surface bright yellowish orange, irregular, clavate, swollen, smooth, tapering sharply into white, smooth, more or less cylindrical stem. Flesh pallid and fragile.
Asci cylindrical or clavate, 90-150 x 8-9 μm. **Spores** (8) hyaline, smooth, elongated-cylindrical, non-septate, biseriate, 10-15 x 2.5-3.0 μm. Paraphyses filiform, branched, septate.
Odour not distinctive. **Taste** not distinctive.
Chemical tests asci tips faintly blued with Melzer's reagent.
Occurrence spring to early summer; rare. (New Forest near Ringwood, Hampshire.)
■ Inedible.

Spathularia flavida Pers.: Fr.   Geoglossaceae

Small yellow fan-shaped head on pale whitish stem; in small trooping groups on soil amongst moss and other dwarf vegetation with coniferous trees .

Dimensions 1 - 3 cm dia x 2 - 8 cm tall.
Fruiting body apothecium outer (hymenial) surface bright yellow, compressed and encircling the upper part of the whitish, smooth stem which tapers towards the base. Flesh yellowish and elastic.
Asci cylindrical or clavate, 100-105 x 11.5-13 μm. **Spores** (8) hyaline, smooth, elongated-cylindrical, septate, lying parallel, 40-50 x 2.0-2.5 μm. Paraphyses filiform, branched, tips curled in short spirals.
Odour not distinctive. **Taste** not distinctive.
Chemical tests asci tips not blued with Melzer's reagent.
Occurrence late summer to autumn; rare. (Forest of Dean, Gloucestershire.)
■ Inedible.

Trichoglossum hirsutum (Pers.: Fr.) Boud.
Geoglossaceae

Black tongue-like velvety spindle with club-shaped head; in small trooping groups, on soil or amongst short grass and moss, and favouring acid soils.

Dimensions 0.3 - 0.8 cm dia x 3 - 7 cm tall.
Fruiting body apothecium outer (hymenial) surface black, cylindrical, with clavate upper section, downy, often compressed in one plane,
the fertile head tapering into a concolorous, compressed and grooved stem. Flesh black and brittle.
Asci cylindrical or clavate, 150-220 x 20-25 μm. **Spores** (8) brown, smooth, sub-cylindrical, 15-septate at maturity, parallel, 100-150 x 6-7 μm. Paraphyses filiform, brown at the tips, curved, slightly swollen at the tips. Setae black, thick-walled and stiffly pointed, numerous.
Odour not distinctive. **Taste** not distinctive.
Chemical tests asci tips blued with Melzer's reagent.
Occurrence late summer to autumn; infrequent.
■ Inedible.

Callorina fusarioides (Berk.) Korf   **Dermateaceae**
= *Callorina neglecta* (Lib.) Hein.

Minute orange-red discs; clustered on dead stems of stinging nettle.

Dimensions 0.05 - 0.1 cm dia.
Fruiting body apothecium upper (hymenial) surface orange, lower surface concolorous; erumpent, sessile. Flesh concolorous.
Asci cylindrical or clavate, 65-85 x 7-10 μm. **Spores** (8) hyaline, smooth, cylindrical-ellipsoid, 1-septate at maturity, uni- and bi-seriate, droplets absent, 11-15 x 3-4 μm. Paraphyses filiform, barely longer than asci, forked, with swollen tips.
Odour not distinctive. **Taste** not distinctive.
Chemical tests asci tips not blued with Melzer's reagent.
Occurrence late winter to spring; common.
■ Inedible.
The conidial stage on *Urtica* stems is more commonly encountered.

Mollisia cinerea (Batsch : Fr.) Karst.   **Dermateaceae**

Minute pale grey disc; clustered and often densely massed on dead wood of beech and oak.

Dimensions 0.05 - 0.2 cm dia.
Fruiting body apothecium grey, sometimes with ochraceous tinge and with pallid margin; upper (hymenial) surface smooth, lower surface finely downy; saucer-shaped, sessile. Flesh pallid and thin.
Asci cylindrical, 50-70 x 5-6 μm. **Spores** (8) hyaline, smooth, ellipsoid, non-septate, uniseriate, frequently with droplets, 7-9 x 2-2.5 μm. Paraphyses filiform with blunt tips, occasionally septate.
Odour not distinctive. **Taste** not distinctive.
Chemical tests asci tips blued with Melzer's reagent.
Occurrence throughout the year; common.
■ Inedible.

Mollisia ligni (Desm.) Karst.   **Dermateaceae**

Minute pale grey disc; clustered and often densely massed on fallen 'mast' of beech and other barkless woody debris of broad-leaf trees, including oak and chestnut.

Dimensions 0.05 - 0.1 cm dia.
Fruiting body apothecium grey with pallid margin, upper (hymenial) surface smooth, lower surface finely downy; saucer-shaped, sessile. Flesh pallid and thin.
Asci cylindrical, 45-60 x 5-6 μm. **Spores** (8) hyaline, smooth, fusiform-clavate, non-septate, biseriate, droplets absent, 6-10 x 2-2.5 μm. Paraphyses filiform, slightly swollen at the tips, forked.
Odour not distinctive. **Taste** not distinctive.
Chemical tests asci tips blued with Melzer's reagent.
Occurrence throughout the year; common.
■ Inedible.

Mollisia revincta (Karst.) Rehm   Dermateaceae
= *Mollisia minutella* (Sacc.) Rehm

Very small dull grey disc; clustered and often densely massed on a
variety of rotting herbaceous stems.

Dimensions 0.03 - 0.08 cm dia.
Fruiting body apothecium dull grey, upper (hymenial) surface
smooth; lower surface smooth; saucer-shaped, sessile. Flesh grey
and thin.
Asci cylindrical, 40-50 x 4-5 µm. **Spores** (8) hyaline, smooth,
ellipsoid, non-septate, irregularly biseriate, droplets absent, 7-8 x
1-2 µm. Paraphyses filiform, non-septate, not forked.
Odour not distinctive. **Taste** not distinctive.
Chemical tests asci tips blued with Melzer's reagent.
Occurrence summer to early autumn; common.
■ Inedible.

Tapesia fusca (Pers.: Fr.) Fuckel   Dermateaceae

Minute greyish white disc; clustered and often densely massed, on
rotting wood and other debris of broad-leaf and, less frequently,
coniferous trees.

Dimensions 0.05 - 0.2 cm dia.
Fruiting body apothecium upper (hymenial) surface greyish with
blue tinges and with more pallid margin, at first cup-shaped,
becoming more saucer-shaped or flat, smooth arising from dark
brown felty stroma, sessile. Flesh pallid and thin.
Asci clavate, 45-50 x 5-7 µm. **Spores** (8) hyaline, smooth, fusiform-
cylindrical, non-septate, biseriate, droplets absent, 7-12 x 1.8-2.2 µm.
Paraphyses filiform, septate, branched.
Odour not distinctive. **Taste** not distinctive.
Chemical tests asci tips blued with Melzer's reagent.
Occurrence throughout the year; common.
■ Inedible.
Note: the felty stroma may be extremely difficult to see in the field.

Orbilia delicatula (Karst.) Karst.   Orbiliaceae

Minute yellow disc with a waxy or glassy appearance; clustered and
often densely massed, on rotting wood of broad-leaf trees.

Dimensions 0.05 - 0.15 cm dia.
Fruiting body apothecium upper (hymenial) surface yellow, smooth;
lower surface smooth; saucer-shaped or flat with indistinct margin,
small white hyphae anchoring the disc from the margin, sessile.
Flesh pallid and thin.
Asci cylindrical or clavate, 30-40 x 4-4.5 µm. **Spores** (8) hyaline,
smooth, reniform, non-septate, uniseriate, 2 droplets, 5-11 x 1-1.5
µm. Paraphyses filiform, swollen at the tips, not forked.
Odour not distinctive. **Taste** not distinctive.
Chemical tests asci tips not blued with Melzer's reagent.
Occurrence spring to autumn; common.
■ Inedible.

Lachnum brevipilosum Baral   Hyaloscyphaceae
= *Dasyscyphus brevipilus* Le Gal.

Minute pure white fringed cup on concolorous stem; clustered on rotting wood and woody remains.

Dimensions 0.05 - 0.1 cm dia.
Fruiting body apothecium upper (hymenial) surface pure white or tinged cream; lower surface and margin densely white, hairy; cup-shaped on comparatively short stem. Flesh concolorous.
Asci cylindrical or clavate, 45-55 x 5-6 µm. **Spores** (8) hyaline, smooth, fusiform, non-septate, uniseriate, droplets absent, 6-10 x 1.5-2.5 µm. Paraphyses narrowly lanceolate, barely longer than asci. Hairs septate.
Odour not distinctive. **Taste** not distinctive.
Chemical tests asci tips blued with Melzer's reagent.
Occurrence early spring to summer; common.
■ Inedible.

Lachnum fuscescens (Pers.) Rehm var **fagicola**
(Phill.) Dennis   Hyaloscyphaceae
= *Dasyscyphus fuscescens var fagicola* (Phill.) Dennis

Minute pale brownish fringed cup on short concolorous stem; clustered inside fallen 'mast' of beech.

Dimensions 0.02 - 0.1 cm dia
Fruiting body apothecium upper (hymenial) surface pallid brown; lower surface and margin densely covered with darker brown hairs; cup-shaped on short stem. Flesh concolorous.
Asci cylindrical or clavate, 40-45 x 4-5 µm. **Spores** (8) hyaline, smooth, fusiform, non-septate, irregularly biseriate, droplets absent, 7-10 x 2-2.5 µm. Paraphyses lanceolate, markedly longer than asci. Hairs septate, encrusted at the tips.
Odour not distinctive. **Taste** not distinctive.
Chemical tests asci tips blued with Melzer's reagent.
Occurrence summer to early autumn; infrequent.
■ Inedible.

Lachnum virgineum S F Gray   Hyaloscyphaceae
= *Dasyscyphus virgineus* S F Gray

Minute pure white fringed cup on concolorous stem; clustered on fallen 'mast' of beech and, less frequently, other plant debris including stems of blackberry.

Dimensions 0.05 - 0.1 cm dia.
Fruiting body apothecium upper (hymenial) surface pure white or tinged cream; lower surface and margin densely white, hairy; cup-shaped on comparatively long stem. Flesh concolorous.
Asci cylindrical or clavate, 50-55 x 4-5 µm. **Spores** (8) hyaline, smooth, fusiform, non-septate, uniseriate, droplets absent, 6-10 x 1.5-2.5 µm. Paraphyses lanceolate, notably longer than asci. Hairs septate.
Odour not distinctive. **Taste** not distinctive.
Chemical tests asci tips blued with Melzer's reagent.
Occurrence early spring to summer; common.
■ Inedible.

Lachnellula occidentalis (Hahn & Ayers) Dharne
= *Dasyscyphus calycinus* (Schum.) Fuckel
= *Lachnellula hahniana* (Seav.) Dennis   **Hyaloscyphaceae**

Small or minute yellowish-orange cup or disc; clustered on dead twigs of larch and spruce.

Dimensions 0.1 - 0.4 cm dia.
Fruiting body apothecium upper (hymenial) surface yellowish orange, erumpent, at first sub-spherical becoming saucer-shaped or cup-shaped; lower surface and margin densely white, hairy; stem rudimentary. Flesh concolorous.
Asci cylindrical or clavate, 100-120 x 10-11 µm. **Spores** (8) hyaline, smooth, fusiform-ellipsoid, non-septate, uniseriate, with faint droplets, 12-18 x 4-6 µm. Paraphyses fusiform, swollen between septa, forked.
Odour not distinctive. **Taste** not distinctive.
Chemical tests asci tips not blued with Melzer's reagent.
Occurrence early spring to autumn; infrequent.
■ Inedible.

Lachnellula subtilissima (Cke.) Dennis
= *Trichoscyphella calycina* (Schum.: Fr.) Nannf.   **Hyaloscyphaceae**

Small or minute yellowish orange cup or disc; clustered on dead twigs of fir and, less frequently, pine and spruce.

Dimensions 0.1 - 0.3 cm dia
Fruiting body apothecium upper (hymenial) surface yellowish-orange, erumpent, at first sub-spherical becoming saucer-shaped or cup-shaped; lower surface and margin densely white, hairy; stem rudimentary. Flesh concolorous.
Asci cylindrical or clavate, 45-50 x 4-5 µm. **Spores** (8) hyaline, smooth, elongated ellipsoid, non-septate, irregularly biseriate, droplets absent, 5-9 x 1.8-2 µm. Paraphyses filiform, swollen between septa, sometimes forked at the base.
Odour not distinctive. **Taste** not distinctive.
Chemical tests asci tips not blued with Melzer's reagent.
Occurrence late autumn to early spring; infrequent.
■ Inedible.

Ascocoryne cylichnium (Tul. & C Tul.) Korf   **Leotiaceae**
= *Coryne cylichnium* (Tul. & C Tul.) Boud.

Small gelatinous pinkish purple saucer on short stem; in clusters on dead wood (trunks and branches) favouring beech.

Dimensions 0.5 - 2 cm dia.
Fruiting body apothecium pinkish purple, at first spherical becoming top-shaped and finally concave or cup-shaped (more so than in *A. sarcoides*) with a slightly wavy margin; upper (hymenial) surface smooth; lower surface finely granular; sessile or with rudimentary stem. Flesh pinkish purple and gelatinous.
Asci narrowly clavate or cylindrical, 210-220 x 10-12 µm. **Spores** (8) hyaline, smooth, ellipsoid, biseriate, septate and sometimes budding at maturity, 2 droplets when young, 20-24 x 5.5-6 µm. Paraphyses cylindrical, slightly swollen at the tips, non-septate, unbranched.
Odour not distinctive. **Taste** not distinctive.
Chemical tests asci tips blued with Melzer's reagent.
Occurrence late summer to early winter; frequent.
■ Inedible.
Distinguished from *A. sarcoides* by the slightly larger and more cup-shaped apothecia and by larger spores.

Ascocoryne sarcoides (Jacq.: Fr.) Groves & Wilson
= *Coryne sarcoides* (Jacq.: Fr.) Tul. & C Tul.   **Leotiaceae**

Small gelatinous pinkish purple saucer on short stem; in clusters on dead wood (trunks and branches) favouring beech.

Dimensions 0.5 - 1.5 cm dia.
Fruiting body apothecium pinkish purple, darker at the margin, at first spherical, becoming top-shaped, then flattened, and finally shallowly cup-shaped (less so than in *A. cylichnium*), slightly wavy margin, upper (hymenial) surface smooth; lower surface finely granular; sessile or with rudimentary stem. Flesh pinkish purple and gelatinous.
Asci cylindrical, 115-125 x 8-10 µm. **Spores** (8) hyaline, smooth, ellipsoid, 1-septate at maturity, uniseriate or biseriate, 2 droplets, 12-16 x 3-5 µm. Paraphyses cylindrical, slightly swollen at the tips, sparsely septate, sparsely branched.
Odour not distinctive. **Taste** not distinctive.
Chemical tests asci tips blued with Melzer's reagent.
Occurrence late summer to early winter; common.
■ Inedible.
Distinguished from *A. cylichnium* by the slightly smaller and more flattened apothecia and by smaller spores.

Ascotremella faginea (Peck.) Seav.   **Leotiaceae**

Gelatinous pinkish purple lobes on short stems, collectively appearing brain-like; in clusters on dead wood (trunks and branches) favouring beech and alder .

Dimensions 2 - 4 cm dia overall x 1 - 2 cm tall.
Fruiting body apothecium pinkish purple, darker when dry, individual lobes sub-spherical, shiny and gelatinous when damp, dull and tough when dry, smooth; sessile or with rudimentary stem. Flesh pinkish purple and gelatinous.
Asci cylindrical, 70-80 x 7-8 µm. **Spores** (8) hyaline, smooth with faint longitudinal striations, ellipsoid, uni- or bi-seriate, 2 droplets, 7-9 x 4-5 µm. Paraphyses filiform, non-septate, unbranched.
Odour not distinctive. **Taste** not distinctive.
Chemical tests asci tips not blued with Melzer's reagent.
Occurrence summer to early autumn; rare. (Forest of Dean, Gloucestershire.)
■ Inedible.
Note: illustration shows fruiting body in the dry state.

Bisporella citrina (Batsch : Fr.) Korf & Carpenter   **Leotiaceae**
= *Calycella citrina* ([Hedw.] Fr.) Boud.

Very small bright yellow saucer; densely clustered on dead wood of broad-leaf trees.

Dimensions 0.05 - 0.3 cm dia.
Fruiting body apothecium bright yellow, ageing to more orange-yellow, saucer-shaped, smooth, sessile. Flesh pallid.
Asci narrowly clavate or cylindrical, 100-135 x 8-9 µm. **Spores** (8) hyaline, smooth, ellipsoid, 1-septate at maturity, uniseriate or biseriate, 2 droplets, 9-14 x 3-5 µm. Paraphyses filiform, somewhat swollen at the tips, non-septate.
Odour not distinctive. **Taste** not distinctive.
Chemical tests asci tips blued with Melzer's reagent.
Occurrence throughout the year; common.
■ Inedible.

Bisporella subpallida (Rehm) Dennis   **Leotiaceae**

Very small ochre yellow saucer; densely clustered on dead wood of broad leaf trees favouring cut surfaces of beech , ash and hazel.

Dimensions 0.05 - 0.15 cm dia.
Fruiting body apothecium ochraceous, top-shaped; upper (hymenial) surface flat or saucer-shaped, smooth; lower surface slightly roughened; sessile. Flesh pallid.
Asci narrowly clavate or cylindrical, 50-65 x 4-5 µm.
Spores (8) hyaline, smooth, ellipsoid, 1-septate at maturity, uniseriate or biseriate, 6-7 x 2.5-3 µm. Paraphyses cylindrical, barely swollen at the tips, non-septate.
Odour not distinctive. **Taste** not distinctive.
Chemical tests asci tips barely blued with Melzer's reagent.
Occurrence throughout the year; frequent.
■ Inedible.

Bulgaria inquinans (Pers.: Fr.) Fr.   **Leotiaceae**

Black rubbery disc, reminiscent of liquorice; clustered on dead wood of broad leaf trees, favouring oak but also less commonly found on ash and beech.

Dimensions 1 - 4 cm dia.
Fruiting body apothecium upper (hymenial) surface black, smooth, shiny when wet; lower surface dark brown, finely granular; at first top-shaped, then shallowly saucer-shaped; sessile. Flesh blackish brown and rubbery.
Asci narrowly clavate or cylindrical, 95-125 (200) x 8-9 µm.
Spores (8) upper 4 dark brown, smooth, reniform, non-septate, uniseriate, 11-14 x 6-7 µm.; lower 4 hyaline (immature), otherwise as above, 5-7 x 2-4µm. Paraphyses filiform, slightly swollen at the darker tips, non-septate, branched.
Odour not distinctive. **Taste** not distinctive.
Chemical tests asci tips blued with Melzer's reagent.
Occurrence late summer to autumn; common.
■ Inedible.
Possibly can be confused with *Exidia plana* but distinguished by microscopic features.

Chlorociboria aeruginascens (Nyl.) Kanouse : Ram.
= *Chlorosplenium aeruginascens* (Nyl.) Kanouse **Leotiaceae**

Very small green saucer-shaped disc; in small clusters, on rotting wood of broad-leaf trees favouring oak.

Dimensions 0.2 - 0.5 cm dia.
Fruiting body apothecium blue-green, saucer-shaped, becoming flattened and wavy; upper (hymenial) surface smooth; lower surface finely downy, tapering into a very short slender stem. Flesh concolorous, thin and rather papery.
Asci narrowly clavate or cylindrical, 60-70 x 4.5-5 µm.
Spores (8) hyaline, smooth, fusiform, non-septate, uni- or biseriate, 2 droplets, 6-10 x 1.5-2 µm. Paraphyses cylindrical, slightly swollen towards the tips, smooth, septate, branched.
Odour not distinctive. **Taste** not distinctive.
Chemical tests asci tips blued with Melzer's reagent.
Occurrence late summer to winter; fruiting bodies infrequent, dyed wood common.
■ Inedible.

Cudoniella acicularis (Bull.: Fr.) Schroet.   **Leotiaceae**

Very small white top-like structure; densely massed on old rotting wood and stumps of broad-leaf trees favouring oak.

Dimensions 0.1 - 0.4 cm dia x 0.5 - 1 cm tall.
Fruiting body apothecium white, becoming brownish grey with age, top-shaped becoming more cushion-like, smooth, on long slender cylindrical stem. Flesh white.
Asci broadly cylindrical, 110-120 x 10-13 µm. **Spores** (8) hyaline, smooth, fusiform, sometimes 1-septate at maturity, irregularly biseriate, droplets absent, 15-22 x 4-5 µm. Paraphyses filiform, gradually thickening towards the tips, septate.
Odour not distinctive. **Taste** not distinctive.
Chemical tests asci tips not blued with Melzer's reagent
Occurrence late summer to winter; infrequent.
■ Inedible.

Cudoniella clavus (A. & S. : Fr.) Dennis   **Leotiaceae**
= *Cudoniella aquatica* (Lib.) Sacc.

Small cream top-like structure; solitary or in small groups on woody remnants in wet places.

Dimensions 0.4 - 1.2 cm dia x 1 - 3 cm tall.
Fruiting body apothecium cream with violaceous or grey tinges, top-shaped, becoming more cushion-like, smooth, on stem which may become quite long in partly submerged specimens. Flesh white.
Asci broadly cylindrical, 90-115 x 9-10 µm. **Spores** (8) hyaline, smooth, ellipsoid-fusiform, more or less uniseriate, droplets absent, 8-15 x 3.5-5 µm. Paraphyses filiform, septate, occasionally forked.
Odour not distinctive. **Taste** not distinctive.
Chemical tests asci tips not blued with Melzer's reagent
Occurrence spring to summer; infrequent.
■ Inedible.
Inset shows long-stalked form.

Cyathicula cyathoidea (Bull.: Fr.) de Thuemen **Leotiaceae**
= *Phialea cyathoidea* (Bull.: Fr.) Gill.

Small cream cup- or disc-like structure, stalked; singly or clustered on old stems of umbellifers and other herbaceous plants lying on ground beneath undergrowth.

Dimensions 0.05 - 0.15 cm dia x 0.05 - 0.15 cm tall.
Fruiting body apothecium whitish cream, at first cup-shaped, becoming flattened, smooth, on long, cylindrical, concolorous stem. Flesh white.
Asci cylindrical, 40-55 x 4-5 µm. **Spores** (8) hyaline, smooth, narrowly fusiform, uniseriate or biseriate, 1 droplet at each end, 6-12 x 1.5-2.5 µm. Paraphyses cylindrical, non-septate.
Odour not distinctive. **Taste** not distinctive.
Chemical tests asci tips blued with Melzer's reagent
Occurrence spring to summer; common.
■ Inedible.

Encoelia furfuracea (Roth : Fr.) Karst.   Leotiaceae

Small dull or hazel brown saucer; densely clustered, erumpent, on living and, less frequently, dead hazel trunks; very occasionally on dead alder.

Dimensions 0.5 - 2.5 cm dia.
Fruiting body apothecial upper (hymenial) surface dull brown or more ochraceous tan, smooth; lower surface, darker than hymenial surface when damp, more pallid when dry, coarsely mealy; at first more or less vesicular, becoming cup-shaped and finally flattened, splitting into petal-like lobes; sessile or with rudimentary stem. Flesh dull brown and thin.
Asci clavate, 90-120 x 6-7µm. **Spores** (8) hyaline, smooth, cylindrical or allantoid, non-septate, biseriate, small droplet at each end, 6-11 x 2.0-2.5 µm. Paraphyses narrowly clavate or cylindrical, swollen at the tips, non-septate, smooth.
Odour not distinctive. **Taste** not distinctive.
Chemical tests asci tips blued with Melzer's reagent.
Occurrence winter to spring; infrequent.
■ Inedible.

Hymenoscyphus epiphyllus (Pers.: Fr.) Rehm ex Kauffman
Leotiaceae

Small bright yellow disc; typically in large groups, on mossy rotting wood and other organic debris in damp shady places.

Dimensions cap 0.1 - 0.4 cm dia.
Fruiting body apothecium bright chrome or lemon yellow, outer surface more whitish, cushion-shaped becoming more flattened; upper surface smooth; lower surface felty, with short stalk or more or less sessile. Flesh yellow and rubbery.
Asci narrowly clavate or cylindrical, 90-120 x 8-12 µm. **Spores** (8) hyaline, smooth, fusiform, septate only at maturity, uni- or weakly bi-seriate, with 2 large and several smaller droplets, 15-18 x 3.5-5 µm. Paraphyses filiform, slightly swollen at the tips, unbranched, occasionally septate, smooth.
Odour not distinctive. **Taste** not distinctive.
Chemical tests asci tips blued with Melzer's reagent.
Occurrence late summer to autumn; frequent.
■ Inedible.
In microscopic characteristics apparently more appropriate to the genus *Phaeohelotium* and may be re-named in future.

Hymenoscyphus fructigenus (Bull.: Fr.) S F Gray
Leotiaceae

Very small white saucer on distinct stem; densely clustered, on fallen husks 'mast' of beech, occasionally also on fruits of hazel and oak .

Dimensions 0.1 - 0.4 cm dia.
Fruiting body apothecial upper (hymenial) surface white, occasionally tinged ochraceous, smooth; lower (outer) surface concolorous, smooth; at first cup-shaped, becoming more flattened on white, cylindrical, stem. Flesh white and thin.
Asci clavate, 100-180 x 7-8µm. **Spores** (8) hyaline, smooth, irregularly fusiform, occasionally 1-septate, uniseriate or biseriate, 2 or more droplets, 13-19 x 3-5 µm. Paraphyses cylindrical, barely swollen at the tips, septate, occasionally forked, smooth.
Odour not distinctive. **Taste** not distinctive.
Chemical tests asci tips blued with Melzer's reagent.
Occurrence from spring to winter but more frequently in summer and autumn; common.
■ Inedible.

Leotia lubrica (Scop.: Fr.) Pers.   **Leotiaceae**
Jelly Babies

Small gelatinous dull yellow cap on tapering stem; typically in small groups, often tufted, on soil in damp woods and under other vegetation including bracken.

Dimensions cap 1 - 1.5 cm dia; stem 1 - 5 cm tall.
Fruiting body apothecium outer (hymenial) surface olivaceous ochre, convex with irregularly lobed margin, smooth and often viscid; stem ochraceous covered with minute greenish granules, rounded in cross section, tapering towards the base, occasionally grooved. Flesh yellow and rubbery-gelatinous.
Asci narrowly clavate or cylindrical, 130-180 x 8-12 µm. **Spores** (8) hyaline, smooth, fusiform and sometimes slightly curved, 3-5 septate at maturity, 5- to 7- droplets, 20-25 x 5-6 µm. Paraphyses filiform, slightly swollen at the tips, branched, septate, smooth.
Odour not distinctive. **Taste** not distinctive.
Chemical tests asci tips not blued with Melzer's reagent.
Occurrence late summer to autumn; infrequent.
■ Inedible.

Neobulgaria pura (Fr.: Fr.) Petrak   **Leotiaceae**

Pinkish gelatinous top-shaped structure; in groups, on dead wood of broad-leaf trees favouring beech.

Dimensions 1 - 4 cm dia
Fruiting body apothecium buff or pallid pink, sometimes with violaceous tinges, at first sub-spherical then top-shaped or cushion-like, smooth, sessile. Flesh pallid buff, rubbery or gelatinous.
Asci narrowly clavate or cylindrical, 70-95 x 8-9 µm. **Spores** (8) hyaline, smooth, ellipsoid, uniseriate, 2 droplets, sometimes with germination tube, 6-9 x 3-4.5 µm. Paraphyses cylindrical, slightly swollen at the tips, non-septate.
Odour not distinctive. **Taste** not distinctive.
Chemical tests asci tips blued with Melzer's reagent.
Occurrence summer to autumn; infrequent.
■ Inedible.

Neobulgaria pura var foliacea (Bres.) Dennis & Gamundi
    Leotiaceae

Brownish gelatinous lobes, collectively joining into a brain-like structure; in groups on dead wood of beech.

Dimensions 1 - 5 cm dia.
Fruiting body apothecium brown with lilaceous tinges, individually sub-spherical, smooth, sessile. Flesh pallid brown, rubbery or gelatinous.
Asci narrowly clavate or cylindrical, 70-95 x 8-9 µm. **Spores** (8) hyaline, smooth, ellipsoid, uniseriate, 2 droplets, sometimes with germination tube, 6-9 x 3-4.5 µm. Paraphyses cylindrical, slightly swollen at the tips, non-septate.
Odour not distinctive. **Taste** not distinctive.
Chemical tests asci tips blued with Melzer's reagent.
Occurrence late summer to autumn; infrequent.
■ Inedible.
Possibly confused with *Ascotremella faginea* but spore ornamentation differs.

Phaeohelotium subcarneum (Schum.: Fr.) Dennis
Leotiaceae

Small gelatinous creamy buff disc; typically in large groups, on bark-less rotting wood of broad-leaf trees in damp shady places.

Dimensions cap 0.1 - 0.3 cm dia.
Fruiting body apothecium pallid cream or with buff tinges, flat or saucer-shaped, smooth, more or less sessile. Flesh whitish and rubbery-gelatinous.
Asci narrowly clavate or cylindrical, 60-70 x 6-7 μm. **Spores** (8) hyaline, smooth, fusiform, non-septate, uniseriate, with droplets, 9 - 12 x 2.5-4 μm. Paraphyses filiform, slightly swollen at the tips, occasionally branched, occasionally septate, smooth.
Odour not distinctive. **Taste** not distinctive.
Chemical tests asci tips blued with Melzer's reagent.
Occurrence late summer to autumn; infrequent.
■ Inedible.

Ciboria amentacea (Balbis : Fr.) Fuckel   Sclerotiniaceae

Very small light brown cup on slender stem; usually solitary, on old rotting (previous season's) male catkins of alder, more rarely hazel and willow, often submerged at water's edge.

Dimensions cap 0.5 - 1 cm dia; stem 1- 3 cm tall.
Fruiting body apothecium light brown or greyish brown, upper (hymenial) surface smooth; lower surface smooth or slightly pruinose; cup-shaped, becoming more flattened in older specimens, on dark brown, comparatively long, slender, stem. Flesh brown and very thin.
Asci cylindrical, 100-135 x 6-10 μm. **Spores** (8) hyaline, smooth, ellipsoid, non-septate, uniseriate, droplets absent, 7.5-10 x 4.5-6 μm. Paraphyses cylindrical, slightly swollen at the tips, non-septate.
Odour not distinctive. **Taste** not distinctive.
Chemical tests asci tips blued with Melzer's reagent.
Occurrence spring to early summer; infrequent.
■ Inedible.

Rutstroemia echinophila (Bull.: Fr.) von Höhnel
Sclerotiniaceae

Small brown cup, sometimes on short stem; solitary or in small groups, on inner surface of husks of sweet chestnut.

Dimensions cup 0.2 - 0.8 cm dia; stem 0.2 - 0.3 cm tall (when present).
Fruiting body apothecium reddish brown, often with slightly brighter, denticulate, margin, smooth, cup-shaped on concolorous short cylindrical stem, or sessile. Flesh brown and thin.
Asci cylindrical, 110-120 x 10-13 μm. **Spores** (8) hyaline, smooth, allantoid, 3-septate, uni- or bi-seriate, with droplets and sometimes part-spores at the ends, 15-20 x 5-6 μm. Paraphyses filiform, slightly swollen at the tips, non-septate.
Odour not distinctive. **Taste** not distinctive.
Chemical tests asci tips blued with Melzer's reagent.
Occurrence late summer to winter; infrequent.
■ Inedible.

Rutstroemia firma (Pers.: Fr.) Karst.   **Sclerotiniaceae**

Small brown cup on short stem; solitary or in small groups, on old twigs and branches of oak.

Dimensions cup 0.5 - 1.5 cm dia; stem 0.5 x 1 cm tall.
Fruiting body apothecium reddish brown, often with slightly darker margin, cup-shaped on concolorous short cylindrical stem, smooth. Flesh brown and thin.
Asci cylindrical, 125-150 x 7-10 µm. **Spores** (8) hyaline, smooth, cylindrical, non-septate, uniseriate, with droplets and sometimes part-spores at the ends 13-17 x 3-5 µm. Paraphyses cylindrical, slightly swollen at the tips, non-septate.
Odour not distinctive. **Taste** not distinctive.
Chemical tests asci tips blued with Melzer's reagent.
Occurrence late summer to winter; frequent.
■ Inedible.

Sclerotinia tuberosa (Hedw.: Fr.) Fuckel   **Sclerotiniaceae**

Smallish pale brown cup on sometimes tall stem; in scattered clusters, arising from dark irregular structure on old, rotting tubers of *Anemone* or *Ranunculus* species.

Dimensions cap 1 - 3 cm dia; stem 3 - 10 cm tall.
Fruiting body apothecium brown, upper (hymenial) surface tan or chestnut, smooth; lower surface more pallid, smooth; at first cup- or goblet-shaped, becoming more expanded in older specimens, on more or less concolorous, sometimes long, slender stem arising from a sclerotium attached to the tuber of the host. Flesh brown and thin.
Asci cylindrical, 150-170 x 8-10 µm. **Spores** (8) hyaline, smooth, ellipsoid, non-septate, uniseriate, occasionally with 2 droplets, 12-17 x 6-9 µm. Paraphyses cylindrical, slightly swollen at the tips, septate.
Odour not distinctive. **Taste** not distinctive.
Chemical tests asci tips blued with Melzer's reagent.
Occurrence late summer to autumn; infrequent.
■ Inedible.

Rhytisma acerinum (Pers.: Fr.) Fr.   **Rhytismataceae**

Small black encrusted discs; clustered on attached and fallen leaves of sycamore from previous year.

Dimensions 1 - 2 cm dia.
Fruiting body stroma black, closely applied to the substrate, embedded with minute apothecia, microscopically somewhat wrinkled or brain-like, opening through clefts, sessile. Flesh pallid and thin.
Asci clavate, 100-130 x 8-10 µm. **Spores** (8) hyaline, smooth, filiform, non-septate, lying side-by-side, 55-80 x 1.5-2.5 µm. Paraphyses filiform, septate, bent over at the tips, occasionally branched.
Odour not distinctive. **Taste** not distinctive.
Chemical tests asci tips not blued with Melzer's reagent.
Occurrence throughout the year but sporulating during the spring; common.
■ Inedible.

Claviceps purpurea (Fr.: Fr.) Tul. & C Tul. Clavicipitaceae

In two distinct stages, the most obvious being black, spindle-shaped curved structure; arising from the inflorescences of rye and, less commonly, other grass species.

Dimensions sclerotia 0.5 - 1 cm long; perithecial fruiting bodies 0.1-0.4 cm dia.
Fruiting body sclerotium violet-black, sickle-shaped. Flesh white. This falls to the ground in the autumn where it over-winters, low temperature stimulating production of drumstick-shaped perithecia with purplish heads dotted with ostioles (perithecial openings) on pallid purple slender stems. These produce ascospores which in turn infect other grasses and generate new sclerotia.
Asci elongated-cylindrical, 150-160 x 5-7 μm. **Spores** (8) hyaline, smooth, very elongated filiform, parallel, septate after discharge, 100-120 x 1 μm. Paraphyses absent.
Odour not distinctive. **Taste** not distinctive *.
Chemical tests asci tips not blued with Melzer's reagent.
Occurrence sclerotia : summer to autumn; perithecia spring; all stages infrequent.
✚ Dangerously poisonous.

Cordyceps canadensis Ellis & Everhart   Clavicipitaceae

Small brown globular head on tall stout yellowish stem; arising from subterranean *Elaphomyces* species in coniferous woods.

Dimensions fertile head (stroma) 1 - 1.5 cm dia; stem 3 - 8 cm tall.
Fruiting body stroma chestnut brown, darkening with age, sub-spherical; stem olivaceous, cylindrical, tough; perithecia wholly embedded in stromatal tissue. Flesh pallid yellow.
Asci elongated-cylindrical, thick-walled at the apex, 225-250 x 4-8 μm. **Spores** (8) hyaline, smooth, very elongated filiform, parallel, septate after discharge, disintegrating into a large number of part-spores, 20-50 x 3-5 μm. Paraphyses absent.
Odour not distinctive. **Taste** not distinctive.
Chemical tests asci tips not blued with Melzer's reagent.
Occurrence late summer to autumn; infrequent.
■ Inedible.

Cordyceps militaris (L.: Fr.) Link   Clavicipitaceae

Small orange-red club-shaped head with slender stem; typically solitary or in small clusters, appearing to be growing on soil but actually arising from bodies of dead insect (*Lepidoptera*) larvae and pupae.

Dimensions 0.3 - 0.7 cm dia x 2 - 3 cm tall.
Fruiting body stroma bright orange, elongated, cylindrical, finely warty; stem more pallid, short, slender, cylindrical; perithecia wholly embedded in stromatal tissue. Flesh pallid orange-yellow and brittle.
Asci elongated cylindrical, 250-300 x 3-6 μm. **Spores** (8) hyaline, smooth, very elongated filiform, parallel, septate after discharge, disintegrating into large numbers of part-spores, 200-300 x 1-2 μm. Paraphyses absent.
Odour not distinctive. **Taste** not distinctive.
Chemical tests asci tips not blued with Melzer's reagent.
Occurrence summer to autumn; infrequent or rare. (Witley Common near Guildford, Surrey.)
■ Inedible.

Cordyceps ophioglossoides (Ehr.: Fr.) Link
Clavicipitaceae

Small yellow or black club-shaped head on tall slender yellowish stem; arising from subterranean *Elaphomyces* species typically in small groups in coniferous woods.

Dimensions fertile head (stroma) 0.5 - 1 cm dia x 1 - 2 cm long; stem 3 - 8 cm tall.
Fruiting body stroma yellowish, becoming reddish -brown or black with age, clavate and minutely punctate; stem yellow, becoming darkened with age, slender, cylindrical, tough, arising from yellow mycelium; perithecia wholly embedded in stromatal tissue. Flesh pallid yellow.
Asci elongated-cylindrical, thick-walled at the apex, 225-250 x 4-8 µm. **Spores** (8) hyaline, smooth, very elongated filiform, parallel, septate after discharge, disintegrating into a large number of part-spores, 150-200 x 2-3 µm. Paraphyses absent.
Odour not distinctive. **Taste** not distinctive.
Chemical tests asci tips not blued with Melzer's reagent.
Occurrence autumn; infrequent.
■ Inedible.

Diatrype disciformis (Hoffm.: Fr.) Fr.　　**Diatrypaceae**

Small blackish dot-like fruiting body; clustered in dense masses, sometimes fused together, on dead branches and twigs of beech, occasionally on wood of other broad-leaf trees.

Dimensions stroma 0.05 - 0.3 cm dia.; peritheci um 0.02 - 0.04 cm dia.
Fruiting body stroma grey, becoming blackish, flat, circular, cushion-like, covered with minute darker dots (ostioles); perithecia wholly embedded in stromatal tissue. Flesh white.
Asci elongated, clavate with tapering stalk, 30-40 x 2-5 µm.
Spores (8) hyaline or pallid brown, smooth, allantoid, non-septate, irregularly biseriate, 5-8 x 1.5-2 µm. Paraphyses absent.
Odour not distinctive. **Taste** not distinctive.
Chemical tests asci tips not blued with Melzer's reagent.
Occurrence throughout the year; common.
■ Inedible.

Hypocrea rufa (Pers.: Fr.) Fr.　　**Hypocreaceae**

Small reddish cushion, often fused together; on damp rotten wood.

Dimensions 0.5 - 1 cm dia.
Fruiting body stroma reddish-brown, covered with minute dark ostioles of embedded perithecia, cushion-like discs typically fused into larger masses. Flesh whitish and tough.
Asci cylindrical, 65-70 x 3-5 µm. **Spores** (16) hyaline, finely warty, sub-spherical, 3.5-6 x 3.5-4.5 µm. Paraphyses absent.
Odour not distinctive. **Taste** not distinctive.
Chemical tests asci tips not blued with Melzer's reagent.
Occurrence throughout the year but sporulating in autumn; infrequent.
■ Inedible.

Hypomyces aurantius (Pers.:Fr.) Tul.   Hypocreaceae

Very small orange-yellow pinheads; in dense clusters on the upper surface and hymenium of old polypores.

Dimensions stroma variable dia; perithecium 0.03 - 0.04 cm dia.
Fruiting body orange-yellow, spherical with a rounded papilla, embedded in a dense golden-yellow felty stroma. Flesh yellowish.
Asci clavate or cylindrical, 120-130 x 4-5 μm. **Spores** (8) hyaline, smooth, fusiform (one side flattened), punctate, 1-septate, uniseriate, 24-30 x 5-6 μm. Paraphyses absent.
Odour not distinctive. **Taste** not distinctive.
Chemical tests asci tips not blued with Melzer's reagent.
Occurrence spring to early summer; rare. (Hope Wood, Forest of Dean, Gloucestershire.)
■ Inedible.

Nectria cinnabarina (Tode : Fr.) Fr.   Hypocreaceae

Very small brownish-red pinheads, with a more distinctive pink conidial stage; in dense clusters spread along dead amd dying twigs and logs of broad-leaf, and very rarely coniferous, trees.

Dimensions stroma 0.1 - 0.4 cm dia; perithecium 0.03 - 0.04 cm dia.
Fruiting body conidial stage which forms the stroma, coral pink, bearing dense masses of perithecia, dark red, flask-shaped, rough; each with single apical ostiole. Flesh red and hard.
Asci clavate or cylindrical, 70-90 x 9-12 μm. **Spores** (8) hyaline, smooth, ellipsoid or cylindrical, 1-septate, more or less biseriate, 12-15 x 4-9 μm. Paraphyses absent.
Odour not distinctive. **Taste** not distinctive.
Chemical tests asci tips not blued with Melzer's reagent.
Occurrence throughout the year; common.
■ Inedible.

Nectria peziza (Tode : Fr.) Fr.   Hypocreaceae

Very small yellowish-red pinheads; in dense clusters spread on the surface of rotting polypores and wood.

Dimensions stroma 0.02 - 0.04 cm dia; perithecium 0.02 - 0.03 cm dia
Fruiting body perithecium reddish-yellow, flask-shaped then collapsing when dry, smooth; each with single apical ostiole. Flesh red.
Asci clavate or cylindrical, 70-90 x 8-10 μm. **Spores** (8) hyaline, smooth, ellipsoid or cylindrical, 1-septate, more or less biseriate, 12-15 x 4-7 μm. Paraphyses absent.
Odour not distinctive. **Taste** not distinctive.
Chemical tests asci tips not blued with Melzer's reagent.
Occurrence throughout the year; frequent.
■ Inedible.

Peckiella lateritia (Fr.) Maire   **Hypocreaceae**

Very small olive-brown pinheads; in dense clusters on the gills of *Psathyrella* and other agaric fungi.

Dimensions stroma variable dia; perithecium 0.02 - 0.03 cm dia.
Fruiting body olivaceous-brown, flask-shaped, rough, embedded in a dense olivaceous felty stroma. Flesh olivaceous-brown.
Asci clavate or cylindrical, 160-170 x 7-11 μm. **Spores** (8) hyaline, smooth, fusiform, 1-septate, more or less biseriate, 26-35 x 5-6 μm. Paraphyses absent.
Odour not distinctive. **Taste** not distinctive.
Chemical tests asci tips not blued with Melzer's reagent.
Occurrence late summer to autumn; rare. (Lyme Park, Cheshire.)
■ Inedible.
Note: the photograph shows the gross morphology including the host plant.

Podostroma alutaceum (Pers.: Fr.) Atk.   **Hypocreaceae**

Small cream club-shaped fruiting body, often fused together; on damp rotten wood amongst needle litter under conifers, less frequently on wood of broad-leaf trees.

Dimensions 0.5 - 0.6 cm dia x 1.5 - 4.0 cm tall.
Fruiting body cream, darkening to brownish with age, covered with minute dark ostioles of embedded perithecia; cylindrical or bluntly clavate, typically fused into small groups. Flesh whitish and tough.
Asci cylindrical, 80-90 x 4-5 μm. **Spores** (16) hyaline, finely warty, sub-spherical, 3-5 x 3-4 μm. Paraphyses absent.
Odour not distinctive. **Taste** not distinctive.
Chemical tests asci tips not blued with Melzer's reagent.
Occurrence late summer to autumn; infrequent.
■ Inedible.

Daldinia concentrica (Bolt.: Fr.) Ces. & de Not.
Xylariaceae
Cramp Balls

Hard greyish-brown or black ball; encrusting dead wood, favouring beech and ash but also appearing on a wide variety of other wood.

Dimensions stroma 2 - 7 cm dia; perithecium 0.03 - 0.05 cm dia.
Fruiting body stroma reddish-brown, with lead-grey tinge when young, quickly becoming black, sub-spherical, sometimes shiny, punctured by ostioles; perithecia black, sub-spherical, fully embedded in stromatal tissue in a single layer just below the surface; sessile. Flesh blackish-grey with silvery sheen, hard, brittle and fibrous, arranged in concentric zones.
Asci cylindrical, 200-210 x 10-12 μm. **Spores** (8) black, smooth, ellipsoid with cleft to one side, non-septate, uniseriate, 1 droplet, 12-17 x 6-9 μm.
Odour not distinctive. **Taste** not distinctive.
Chemical tests asci tips blued with Melzer's reagent.
Occurrence throughout the year but sporulating in the spring
■ Inedible.

Daldinia vernicosa (Schw.) Ces. & de Not.   **Xylariaceae**

Hard greyish-brown or black ball on short stalk; on burnt wood of gorse but also occasionally on burnt oak in exposed situations.

Dimensions stroma 1 - 3 cm dia; perithecium 0.03 - 0.05 cm dia
Fruiting body stroma reddish-brown, with lead-grey tinge when young, quickly becoming black, sub-spherical, sometimes shiny, punctured by ostioles; perithecia black, sub-spherical, fully embedded in stromatal tissue in a single layer just below the surface; on short stalk. Flesh blackish-grey with silvery sheen, hard, brittle and fibrous, arranged in concentric zones.
Asci cylindrical, 200-210 x 10-12 μm. **Spores** (8) black, smooth, ellipsoid with cleft to one side, non-septate, uniseriate, 12-17 x 6-9 μm. Paraphyses filiform.
Odour not distinctive. **Taste** not distinctive.
Chemical tests asci tips blued with Melzer's reagent.
Occurrence throughout the year but sporulating in the spring; infrequent.
■ Inedible.

Hypoxylon fragiforme (Pers.: Fr.) Kickx   **Xylariaceae**

Smallish globular fruiting body, ranging from pink, through brown, to black; clustered on logs and dead branches of beech.

Dimensions stroma 0.5 - 1 cm dia; perithecium 0.03 - 0.05 cm dia
Fruiting body stroma bright salmon-pink when young, darkening to muddy-brown and then black, warty, hemispherical; perithecia black, flask-shaped, wholly embedded in stromatal tissue. Flesh blackish and hard.
Asci cylindrical, 130-150 x 6-9 μm. **Spores** (8) dark brown, smooth, sub-fusiform, flattened on one side with distinct cleft, non-septate, uniseriate, 11-15 x 5-7 μm. Paraphyses filiform.
Odour not distinctive. **Taste** not distinctive.
Chemical tests asci tips not blued with Melzer's reagent.
Occurrence throughout the year but generally sporulating in the autumn; common.
■ Inedible.
Inset: mature fruiting bodies.

Hypoxylon fuscum (Pers.: Fr.) Fr.   **Xylariaceae**

Brownish cushion-like crust; clustered on dead hazel and alder branches and twigs.

Dimensions stroma of variable dia; perithecium 0.03 - 0.05 cm dia.
Fruiting body stroma brown with purple or grey tinges, effused, encrusting, extending to a variable extent over the surface of the substrate, more or less matt; perithecia greyish black, flask-shaped, wholly embedded in the stromatal tissue. Flesh black and hard.
Asci cylindrical, 130-140 x 6-8 μm. **Spores** (8) dark brown, smooth, ellipsoid, flattened on one side with distinct cleft, non-septate, uniseriate, 12-15 x 5-7 μm. Paraphyses apparently absent.
Odour not distinctive. **Taste** not distinctive.
Chemical tests asci tips blued with Melzer's reagent.
Occurrence throughout the year but generally sporulating in the autumn; common.
■ Inedible.

Hypoxylon mammatum (Wahl.) Miller   Xylariaceae

Small patches of blackish crust, unevenly knobbly, with distinctive papillate perithecia; clustered on dead branches, with bark, of willow and poplar.

Dimensions stroma up to 1 cm dia; perithecium 0.1 - 0.2 cm dia.
Fruiting body stroma whitish, pruinose, soon blackish, effused, encrusting, extending to a variable extent over the surface of the substrate in rounded patches, more or less matt, each with about 5 large protruding perithecia; perithecia greyish-black, flask-shaped, papillate, partly embedded in the stromatal tissue. Flesh black and hard.
Asci cylindrical, 80-90 x 6-7 μm. **Spores** (8) mid-brown or dark brown, smooth, ellipsoid, flattened on one side but without obvious cleft, non-septate, uniseriate, 20-36 x 6-14 μm. Paraphyses filiform, poorly defined.
Odour not distinctive. **Taste** not distinctive.
Chemical tests asci tips blued with Melzer's reagent.
Occurrence throughout the year but generally sporulating in the autumn; rare in Europe, very rare in Britain.
■ Inedible.
Note: the large papillate perithecia and the very large ascospores are characteristic.

Hypoxylon multiforme (Fr.: Fr.) Fr.   Xylariaceae
= *Hypoxylon crustaceum* (Sow.: Fr.) Nke.

Blackish cushion-like crust, often elongate across the long axis of the wood; clustered on dead branches, with bark, of birch and, occasionally, other broad-leaf trees.

Dimensions stroma of variable dia; perithecium 0.04 - 0.08 cm dia.
Fruiting body stroma at first reddish-brown, soon blackish, effused, encrusting, extending to a variable extent over the surface of the substrate, tending to be arranged in elongated masses across the girth of the substrate, more or less matt; perithecia greyish-black, flask-shaped, wholly embedded in the stromatal tissue. Flesh black and hard.
Asci cylindrical, 80-90 x 6-7 μm. **Spores** (8) dark brown, smooth, ellipsoid, flattened on one side with distinct cleft, non-septate, uniseriate, 9-11 x 4.5-5.0 μm. Paraphyses filiform, barely forked, poorly defined.
Odour not distinctive. **Taste** not distinctive.
Chemical tests asci tips blued with Melzer's reagent.
Occurrence throughout the year but generally sporulating in autumn; infrequent.
■ Inedible.

Hypoxylon nummularium Bull.: Fr.   Xylariaceae

Shiny black cushion-like crust; clustered on dead wood of beech.

Dimensions stroma of variable dia; perithecium 0.03 - 0.05 cm dia.
Fruiting body stroma black, effused, encrusting, often shiny, extending to a variable extent over the surface of the substrate; perithecia black, flask-shaped, wholly embedded in the stromatal tissue. Flesh black and hard.
Asci cylindrical, 100-125 x 10-12 μm. **Spores** (8) dark-brown, smooth, sub-spherical with distinct cleft on one side, non-septate, uniseriate, 11-14 x 7-10 μm. Paraphyses filiform, occasionally branched, poorly defined.
Odour not distinctive. **Taste** not distinctive.
Chemical tests asci tips blued with Melzer's reagent.
Occurrence throughout the year but generally sporulating in the autumn; common.
■ Inedible.

Poronia punctata (L.: Fr.) Fr.   **Xylariaceae**

Small whitish flattened disc covered with darker pin-head dots, on long stem; in clusters on horse dung.

Dimensions stroma 0.5 - 1.5 cm dia; perithecium 0.02 - 0.03 cm dia.
Fruiting body stroma pallid greyish-white attached to a black, elongated stem running down into the substrate; perithecia black, sub-spherical with distinct papilla, rough, each with single apical ostiole. Flesh pallid and tough.
Asci elongated-cylindrical, 160-180 x 15-18 µm. **Spores** (8) hyaline, finely smooth, bean-shaped, 1-septate, uniseriate, 18-26 x 7-12 µm.
Odour not distinctive. **Taste** not distinctive.
Chemical tests asci tips not blued with Melzer's reagent.
Occurrence late summer to autumn; very rare. (New Forest near Brockenhurst, Hampshire.)
■ Inedible.

Ustulina deusta (Hoffm.: Fr.) Petrak   **Xylariaceae**

Irregular greyish white cushion becoming black like charred wood when old; on rotting stumps and roots of broad-leaf trees favouring beech.

Dimensions stroma 5 - 10 cm dia; perithecium approx. 0.15 cm dia.
Fruiting body stroma greyish-white, becoming black with age, effused and loosely attached to the substrate, very brittle particularly when old; perithecia black, flask-shaped, almost wholly embedded in the stromatal tissue. Flesh white, becoming discoloured with age.
Asci cylindrical,300-335 x 12-15 µm. **Spores** (8) chocolate, smooth, fusiform, flattened on one side with distinct cleft, non-septate, uniseriate, 3-4 droplets, 28-34 x 7-10 µm. Paraphyses filiform.
Odour not distinctive. **Taste** not distinctive.
Chemical tests asci tips blued with Melzer's reagent.
Occurrence throughout the year but mainly summer and autumn; common.
■ Inedible.

Xylaria carpophila (Pers.: Fr.) Fr.   **Xylariaceae**

Similar in appearance to *X. hypoxylon* but much more slender and delicate; on rotting 'mast' of beech, often buried under leaf litter.

Dimensions stroma 0.2 - 0.3 cm dia x 2 - 5 cm tall; perithecium 0.03 - 0.05 cm dia.
Fruiting body upper stromatal surface whitish becoming black-tipped at maturity, lower sterile parts black and downy, antler-shaped, compressed; perithecia black, sub-spherical, fully embedded in the stromatal tissue and arranged in a single dense layer just beneath the surface towards the apex. Flesh white and hard.
Asci cylindrical, 100-150 x 6-8 µm. **Spores** (8) chocolate, smooth, reniform, with distinct cleft on one side, non-septate, uniseriate, 1-2 droplets, 11-14 x 5-6 µm. Paraphyses filiform.
Odour not distinctive. **Taste** not distinctive.
Chemical tests asci tips blued with Melzer's reagent.
Occurrence throughout the year but mainly summer and autumn; common.
■ Inedible.

Xylaria hypoxylon (L.: Fr.) Grev.   **Xylariaceae**
Candle snuff

Antler-shaped structure with characteristic black and white appearance like an extinguished candle-wick; on dead wood of broad-leaf and, less commonly, coniferous trees.

Dimensions stroma 3 - 5 cm tall; perithecium 0.03 - 0.05 cm dia.
Fruiting body upper stromatal surface white, powdery, but becoming black-tipped at maturity, lower sterile parts black, downy; sub-cylindrical or flattened and branching into antler-shape; perithecia black, sub-spherical and arranged in a single dense layer close to the surface of the stromatal tissue towards the apex. Flesh white and hard.
Asci cylindrical, 100-150 x 6-8 μm. **Spores** (8) black, smooth, reniform, with distinct cleft on one side, non-septate, uniseriate, 1-2 droplets, 11-14 x 5-6 μm. Paraphyses filiform-cylindrical.
Odour not distinctive. **Taste** not distinctive.
Chemical tests asci tips blued with Melzer's reagent.
Occurrence throughout the year but mainly summer and autumn; common.
■ Inedible.

Xylaria longipes Nke.   **Xylariaceae**

Irregularly club-shaped, looking like a small, more delicate, version of *X. polymorpha*; on dead wood of sycamore, less frequently other broad-leaf trees.

Dimensions stroma 0.25 - 0.75 cm dia x 2 - 6 cm tall; perithecium 0.03 - 0.05 cm dia.
Fruiting body upper stromatal surface black, warty, often somewhat bent over, narrowing slightly into a brownish-black sterile stem; cylindrical or clavate, smooth or slightly downy; perithecia black, sub-spherical, fully embedded in stromatal tissue and arranged in a single dense layer just below the surface towards the apex. Flesh white and hard.
Asci cylindrical, 125-140 x 6-8 μm. **Spores** (8) dark brown, smooth, ellipsoid or reniform, flattened on one side with distinct cleft, non-septate, uniseriate, 1-2 droplets, 12-16 x 5-7 μm. Paraphyses filiform.
Odour not distinctive. **Taste** not distinctive.
Chemical tests asci tips blued with Melzer's reagent.
Occurrence throughout the year but mainly summer and autumn; frequent.
■ Inedible.

Xylaria polymorpha (Pers.: Fr.) Grev.   **Xylariaceae**
Dead man's fingers

Black, irregularly club-shaped structure; arising in small tufts from stumps of broad-leaf trees, favouring beech.

Dimensions stroma 1 - 3 cm dia x 3 - 8 cm tall; perithecium 0.06 - 0.08 cm dia.
Fruiting body upper stromatal surface black, irregularly clavate, warty, narrowing into brownish-black, short, stout, more or less cylindrical sterile stem; perithecia black, sub-spherical, fully embedded in stromatal tissue, in a single dense layer just below the surface. Flesh white and hard.
Asci cylindrical, 130-150 x 8-10 μm. **Spores** (8) chocolate, smooth, fusiform, flattened on one side with distinct cleft, non-septate, uniseriate, 20-32 x 5-9 μm. Paraphyses filiform.
Odour not distinctive. **Taste** not distinctive.
Chemical tests asci tips blued with Melzer's reagent.
Occurrence throughout the year but mainly summer and autumn; common.
■ Inedible.

Lasiosphaeria ovina (Fr.) Ces. & de Not.   **Sordariaceae**

Minute white spherical bodies, each with a dark apical point, extending collectively in a thin mat; on rotten wood generally.

Dimensions 0.06 - 0.1 cm dia; groups extending over several centimetres.
Fruiting body whitish-grey, flaky, covered with a felt-like layer of hyphal tissue, the perithecium opening by an ostiole mounted on a blackish apical papilla. Flesh greyish-white.
Asci cylindrical, 180-210 x 18-22 µm. **Spores** (8) yellowish, smooth, irregularly cylindrical, some with pointed hyaline extensions at each end, with droplets, 36-40 x 4-5 µm. Paraphyses absent.
Odour not distinctive. **Taste** not distinctive.
Chemical tests asci tips not blued with Melzer's reagent.
Occurrence throughout the year but mainly late summer to late autumn; infrequent.
■ Inedible.

Elaphomyces granulatus Fr.: Fr.   **Elaphomycetaceae**

Brownish ball-like fruiting body; barely subterranean in coniferous and, very rarely, broad-leaf woods.

Dimensions 2 - 4 cm dia.
Fruiting body cleistothecium brownish, sub-spherical, warty, the spore mass (gleba) enclosed by a very thick duplex wall; subterranean. Flesh pallid, gleba becoming purplish-black when mature.
Asci sub-spherical or pear-shaped, quickly degenerating, 35-60 x 35-45 µm. **Spores** brownish black, warty, spherical, 24-32 µm.
Odour not distinctive. **Taste** not distinctive.
Chemical tests none.
Occurrence throughout the year but sporulating in the summer and autumn; infrequent.
■ Inedible.
Elaphomyces can often be located by finding *Cordyceps canadensis* which is parasitic upon it and appears above ground.

HOMOBASIDIOMYCETES 'APHYLLOPHORALES'

Amongst the Homobasidiomycetes, the Aphyllophorales species, whose reproductive organs, *basidia*, are not divided by partitions or *septa*, are the major group of non-gilled fungi. The term Aphyllophorales is currently out of favour with some taxonomists who prefer to classify according to the smaller divisions listed in this volume.

The section includes the Chanterelles, Fairy Clubs, Coral Fungi, Hedgehog Fungi, Earth Fans and the 'brackets' which range from small encrusting forms to massive projecting structures and most of which grow on wood. Approximate numbers of species recognised within Europe are given below in square brackets.

The Cantharellales [200] include the Chanterelles in which the *hymenium* is spread over the outer surface of a typically funnel-shaped fruiting body, and the Fairy Clubs in which it is raised and spread evenly over the upper regions of club-shaped receptacles. The order also includes the Hedgehog or Tooth Fungi in which there is a mushroom-like cap surmounting a stem and the hymenium is spread over the surface of downwardly projecting spines.

The Gomphales [60] includes the Coral Fungi which may be viewed as providing an elaboration of structure over the Fairy Clubs into dense, antler-like branches, arising from a trunk-like base.

The Hericiales [40] include some of the most beautiful members such as the large white forms which grow on wood as well as the tiny Earpick Fungus.

The simplest forms are found amongst the Stereales [600], including the genus *Corticium* and others, most of which consist of little more than a thin crust of tissue on whose upper surface is borne the hymenium. Slightly more elaborate are the *Stereum* species, typically quite hard and resistant, and the *Phlebia* and *Merulius* species which are soft and rubbery.

As they evolve in complexity, the encrusting forms project from the substrate as brackets (a transition which can be seen in the *Stereum* genus). Some, including many of the Hymenochaetales [70], may be difficult to determine without microscopic examination. The larger forms, including the Fistulinales [1], the Ganodermatales [12] and the Polyporales [200], are species with hoof-like or plate-like fruiting bodies consisting of layers of tissue, sometimes spongy, but more often hard and woody, below which arises a compact layer of tubes bearing the hymenium and opening by pores on the under surface.

Cantharellus cibarius Fr.
Cantharellaceae
Chantarelle

Egg-yellow, shallowly funnel-shaped fruiting body with gills replaced by forked ridges; on soil in woods, generally with broad-leaf trees.

Dimensions cap 3 - 10 cm dia; stem 3 - 8 cm tall x 0.5 - 1.5 cm dia.
Cap egg-yellow, fading with age, at first flattened with incurved margin becoming somewhat infundibuliform and wavy, smooth and more or less shiny. Gills absent - lower (hymenial) surface yellow bearing vein-like ridges, irregularly forked, decurrent. Stem egg-yellow fading with age, smooth or finely downy, tapering downwards. Flesh yellowish, fading and firm.
Spores pallid ochraceous, smooth, ellipsoid, non-amyloid, 8-10 x 4.5-6.5 μm. Basidia 4-spored. Cystidia absent.
Odour slight, fruity, reminiscent of apricots, or not distinctive.
Taste slight, peppery after a delay.
Chemical tests none.
Occurrence summer to autumn; common.
☐ Edible and excellent.
Confusion must be avoided with poisonous species including *Hygrophoropsis aurantiaca* and *Cortinarius speciosissimus*.

Cantharellus infundibuliformis (Scop.) Fr.
Cantharellaceae

Medium-sized, funnel-shaped dingy-brown cap bearing forked veins instead of gills, yellowish stem; on acid soil in broad-leaf and coniferous woods.

Dimensions cap 1.5 - 6 cm dia; stem 3 - 8 cm tall x 0.4 - 0.9 cm dia.
Cap dark greyish-brown, at first convex with depressed centre, becoming moderately infundibuliform with wavy irregular margin, smooth or slightly scaly. Gills absent - lower (hymenial) surface bearing shallow forked ridges. yellowish becoming more grey with age. Stem yellowish, more or less equal but often flattened and grooved, hollow. Flesh yellowish, tough and thin.
Spores pallid ochraceous, smooth, ellipsoid, non-amyloid, 7.5-10 x 6.5-8 μm. Basidia 4-spored. Cystidia absent.
Odour faint, aromatic. **Taste** bitter.
Chemical tests none.
Occurrence late summer to autumn; infrequent.
☐ Edible.

Clavariadelphus pistillaris (Fr.) Donk **Clavariadelphaceae**
= *Clavaria pistillaris* Fr.

Large yellow, simple club with thick rounded tip; solitary or in small groups, on calcareous soil with beech.

Dimensions 1.5 - 5 cm dia x 10 - 20 cm tall.
Fruiting body yellow or orange-yellow, darker towards the base, brownish when older, unbranched, smooth, massively clavate or tapering downwards, not laterally compressed, bruising violaceous. Flesh yellow, soft and fibrous, turning violaceous-brown where cut.
Spores hyaline, smooth, ellipsoid, non-amyloid, with droplets, 11-13 x 6-7 μm. Basidia 4-spored. Cystidia absent.
Odour not distinctive. **Taste** bitter.
Chemical tests fertile surface saffron with KOH.
Occurrence late summer to autumn; infrequent.
■ Inedible.

Craterellus cornucopioides (L.: Fr.) Pers. **Craterellaceae**
= *Cantharellus cornucopioides* L.: Fr.   Horn of Plenty

Medium-sized, greyish-brown, wrinkled and deeply funnel-shaped or trumpet-like fruiting body; in trooping groups on soil amongst leaf litter of broad-leaf woods.

Dimensions 2 - 8 cm dia x 3 - 10 cm tall.
Fruiting body deeply infundibuliform, greyish-brown, drying more pallid, margin irregularly wavy and lined, finely scaly. Gills absent - lower (hymenial) surface ash-grey, smooth when young, wrinkled at maturity. Stem merging with cap, hollow to the base. Flesh greyish-brown, thin and rather crisped.
Spores hyaline, smooth, ellipsoid, non-amyloid, with droplets, 12-17 x 9-11 µm. Basidia 2-spored. Cystidia absent.
Odour not distinctive. **Taste** not distinctive.
Chemical tests none.
Occurrence summer to autumn; frequent.
☐ Edible and good.

Pseudocraterellus sinuosus (Fr.) Corn.: Hein.
= *Craterellus crispus* Fr.   **Craterellaceae**

Small, greyish-brown, trumpet-shaped and wrinkled cap with stem; in trooping groups on soil amongst leaf litter in broad-leaf woods.

Dimensions cap 1 - 5 cm dia; stem 2 - 3 cm tall x 0.5 - 0.75 cm dia.
Cap greyish-brown, irregularly infundibuliform with lobed and wavy margin. Gills absent - lower (hymenial) surface wrinkled, appearing as vein-like gills. Stem cream, tinged grey towards base, tapering downwards, wrinkled in the upper spore-bearing region. Flesh pallid, thin and fibrous.
Spores hyaline, smooth, ellipsoid, with droplets, 9.5-12 x 7-8 µm. Basidia 2-, 3- or 4- spored. Cystidia absent.
Odour faint, fruity. **Taste** not distinctive.
Chemical tests none.
Occurrence late summer to autumn; rare. (Haldon Forest near Exeter, Devon)
☐ Edible and good.

Clavaria acuta Fr.   **Clavariaceae**

Tall whitish unbranched spindle; usually solitary or scattered, on soil in grassy places including clearings and woodland rides.

Dimensions 0.2 - 0.5 cm dia x 1 - 6 cm tall.
Fruiting body whitish when young, becoming greyish-ochre with age, unbranched, fusiform, laterally compressed and grooved; becoming more rounded at the tip with age, no obvious distinction between fertile upper region and stem. Flesh white, soft and brittle.
Spores hyaline, smooth, broadly ovoid, non-amyloid, with droplets, 7-10 x 5.5-8 µm. Basidia 2- or 4- spored. Cystidia absent.
Odour not distinctive. **Taste** not distinctive.
Chemical tests none.
Occurrence late summer to autumn; rare. (Whitley Common, near Guildford, Surrey)
☐ Edible but of no culinary value.

Clavaria argillacea Fr.   Clavariaceae
= *Clavaria ericetorum* Pers.

Pale creamy-yellow, club-shaped, unbranched spindle; in small clusters on soil on sandy heaths.

Dimensions 0.2 - 0.8 cm dia x 3 - 8 cm tall.
Cap pallid greenish or creamy-yellow, erect, clavate, somewhat compressed laterally, smooth, simple with blunt almost notched tips. Stem more deeply egg-yellow than cap, merging distinctly with cap, cylindrical, smooth. Flesh pallid yellow and brittle.
Spores hyaline, smooth, ellipsoid or cylindrical, non-amyloid, 8-11 x 4-6 µm. Basidia 4-spored. Cystidia absent.
Odour not distinctive. **Taste** slight, tallow-like.
Chemical tests none.
Occurrence summer to early autumn; infrequent.
☐ Edible but of limited culinary value.

Clavaria vermicularis Fr.   Clavariaceae

White worm-like unbranched spindle; in clusters or sometimes solitary on soil in or near woodlands.

Dimensions 0.3 - 0.7 cm dia x 3 - 12 cm tall.
Cap whitish, sometimes with yellow tinge near the apex and base, erect, vermiform or fusiform, rounded or somewhat compressed laterally, smooth, simple with bluntly pointed tips. Stem more deeply egg-yellow than the cap, merging distinctly with cap, cylindrical, smooth. Flesh white and brittle. **Spores** hyaline, smooth, ovoid, non-amyloid, with many small droplets, 5-9 x 3.5-5 µm. Basidia 4-spored. Cystidia absent.
Odour not distinctive. **Taste** not distinctive.
Chemical tests none.
Occurrence summer to early autumn; infrequent.
☐ Edible but of limited culinary value.

Clavulinopsis corniculata (Fr.) Corn.   Clavariaceae
= *Clavaria corniculata* Fr.: Schaeff.

Smallish, bright egg-yellow fruiting body, repeatedly branched antler-like; solitary or in tufts, on soil in short grass or turf, less frequently in open grassy woodlands.

Dimensions variable dia x 2 - 8 cm tall.
Fruiting body egg-yellow or ochraceous, repeatedly branched particularly towards the apices, laterally compressed, incurved at the tips, fork angles rounded not acute, branches merging into a pallid, downy, stout stem. Flesh yellow, firm and tough. **Spores** hyaline, smooth, subspherical, non-amyloid, 1 droplet, 4-7 x 4-6 µm. Basidia 4-spored. Cystidia absent.
Odour rancid or mealy. **Taste** bitter.
Chemical tests none.
Occurrence early summer to autumn; infrequent.
☐ Edible.
Not to be confused with *Calocera viscosa*.

Clavulinopsis fusiformis (Fr.) Corn.   **Clavariaceae**
= *Clavaria fusiformis* Fr.

Bright yellow, small, slender, simple club; typically in dense tufts on soil in grasslands, light open mixed woodlands or clearings.

Dimensions 0.2 - 0.6 cm dia x 2 - 8 cm tall.
Fruiting body bright, clear yellow or sometimes egg-yellow, unbranched clavate, typically laterally compressed and grooved, tips acute; no obvious distinction between fruiting body and stem. Flesh pallid yellow, firm and fibrous. **Spores** hyaline, smooth, sub-spherical with distinct apiculus, non-amyloid, with droplets, 5-6.5 x 4.5-6 µm. Basidia 2- or 4-spored. Cystidia absent.
Odour not distinctive. **Taste** slight, bitter.
Chemical tests none.
Occurrence late summer to autumn; locally common.
■ Inedible.

Clavulinopsis helvola Fr.   **Clavariaceae**
= *Clavaria helvola* Fr.

Bright yellow, small, slender, simple club; solitary or in tufts on soil in light open mixed woodlands or clearings.

Dimensions 0.2 - 0.4 cm dia x 3 - 7 cm tall.
Fruiting body bright, clear yellow or sometimes orange-yellow, unbranched clavate, typically laterally compressed and grooved, tips blunt and very occasionally forked; no obvious distinction between fruiting body and stem. Flesh pallid yellow, firm and fibrous.
Spores hyaline, decorated with blunt spines, sub-spherical or faintly angular, non-amyloid, 1 droplet, 4-7 x 3.5-6 µm. Basidia 2- or 4-spored. Cystidia absent.
Odour not distinctive. **Taste** not distinctive.
Chemical tests none.
Occurrence late summer to autumn; locally common.
■ Inedible.
The coarsely warty spores distinguish *C. helvola* from all related species other than *C. asterospora*.

Clavulinopsis laeticolor (Berk. & Curt.) Peters.
= *Clavulinopsis pulchra* (Peck) Corn.   **Clavariaceae**

Bright golden yellow, small, simple club; solitary or in groups on soil in mixed woodlands.

Dimensions 0.2 - 0.6 cm dia x 2 - 8 cm tall.
Fruiting body bright golden-yellow or sometimes orange-yellow, unbranched clavate or spathulate, typically rounded or laterally compressed and grooved, tips blunt; distinction between fruiting body and stem slight. Flesh yellow and soft. **Spores** yellow, smooth, sub-spherical or broadly ellipsoid, non-amyloid, with droplets, 4-6 x 3.5-5 µm. Basidia 4-spored. Cystidia absent.
Odour not distinctive. **Taste** not distinctive.
Chemical tests none.
Occurrence summer to late autumn; infrequent.
■ Inedible.

Clavulinopsis luteo-alba (Rea) Corn.   **Clavariaceae**
= *Clavaria luteo-alba* Rea

Smallish yellow, simple club with pale tip; solitary or in small groups, on soil in short grass or turf.

Dimensions 0.1 - 0.4 cm dia x 3 - 6 cm tall.
Fruiting body yellow or orange-yellow, unbranched, smooth, slender, more or less clavate or tapering downwards, typically laterally compressed, the apices pallid whitish, blunt or slightly forked. Flesh yellow, soft and fragile. **Spores** hyaline, smooth, ellipsoid, non-amyloid, with droplets, 5-8 x 2.5-4.5 µm. Basidia 2- or 4-spored. Cystidia absent.
Odour slight, musty. **Taste** not distinctive.
Chemical tests none.
Occurrence late summer to autumn; infrequent but locally common.
■ Inedible.

Clavariadelphus fistulosa var **contorta**
(Holmsk.: Fr.) Jül.
= *Clavariadelphus contortus* (Fr.) Pil.   **Typhulaceae**

Similar to *M. fistulosa* but smaller and thicker with stunted, bent-over fruiting bodies; often solitary or sparsely clustered, on dead trunks, branches and twigs of alder and some other broad-leaf trees.

Dimensions 0.2 - 0.8 cm dia x 1 - 4 cm tall.
Fruiting body pallid, dirty-yellow or greyish-ochre, stunted, twisted or bent over with blunt or poorly pointed tips, smooth or finely granular. Flesh yellowish, firm and moderately tough. **Spores** hyaline, smooth, fusiform, non-amyloid, droplets absent, 10-15(24) x 5-8 µm. Basidia 4-spored. Cystidia absent.
Odour not distinctive. **Taste** not distinctive.
Chemical tests none.
Occurrence autumn to early winter; infrequent.
☐ Edible but not worthwhile.
Some authors consider this to be a distinct species, others allege that it is con-specific with *C. fistulosus* developing under unfavourable conditions.

Macrotyphula fistulosa (Fr.) Peters.   **Typhulaceae**
= *Clavaria fistulosa* Fr.
= *Clavariadelphus fistulosus* (Fr.) Corn.

Dirty yellow, tall, very slender, simple, elongated spindle; often solitary or sparsely clustered, on twigs of beech and other broad-leaf trees.

Dimensions 0.2 - 0.8 cm dia x 8 - 25 cm tall.
Fruiting body dirty-yellow or tawny-brown, elongated and narrowly fusiform, at first acute at the tips then slightly blunted, smooth or finely granular. Flesh yellowish, firm and moderately tough.
Spores hyaline, smooth, fusiform, non-amyloid, droplets absent, 10-18 x 5-8 µm. Basidia 4-spored. Cystidia absent.
Odour not distinctive. **Taste** not distinctive.
Chemical tests none.
Occurrence autumn; rare. (Westonbirt Arboretum, Gloucestershire)
☐ Edible but not worthwhile.

Clavulina amethystina (Fr.) Donk   Clavulinaceae
= *Clavaria amethystina* Fr.

Smallish, amethyst or violet branched fruiting body; solitary or in small groups on soil amongst leaf litter in woods generally, but favouring broad-leaf trees.

Dimensions variable dia x 2 - 8 cm tall.
Fruiting body amethyst, repeatedly branched especially towards the apices, branches cylindrical, fairly stout, finely wrinkled merging into more pallid, stout stem. Flesh whitish or tinged violaceous, soft and fragile. **Spores** hyaline, smooth, broadly ellipsoid or sub-spherical, non-amyloid, with droplets, 7-12 x 6-8 µm. Basidia usually 2-spored. Cystidia absent.
Odour not distinctive. **Taste** not distinctive.
Chemical tests none.
Occurrence late summer to autumn; rare. (Haldon Forest, near Exeter, Devon)
■ Inedible.
Illustration shows young specimens with limited branching.

Clavulina cinerea (Fr.) Schroet.   Clavulinaceae
= *Clavaria cinerea* Fr.

Grey, coral-like, extensively branching fruiting body, often tallish, solitary or in small groups, on soil in woods generally, often adjacent to paths.

Dimensions variable dia x 3 - 10 cm tall.
Fruiting body ash-grey, repeatedly branched into dense antler-like tufts, branches narrowly cylindrical or laterally compressed, tips not forked and typically rather blunt; no obvious distinction between fruiting body and stem, but the basal area tending to be more pallid. Flesh greyish-white, soft but tough. **Spores** hyaline, smooth, sub-spherical or broadly ellipsoid, non-amyloid, with droplets, 6.5-11.0 x 6-10 µm. Basidia usually 2-spored, but also with 1 or 4. Cystidia absent. **Odour** not distinctive. **Taste** not distinctive.
Chemical tests none.
Occurrence summer to autumn; common.
☐ Edible and pleasant.
Not to be confused with *Clavulina cristata* infected with *Spadicioides clavariarum* which also becomes grey.

Clavulina cristata (Fr.) Schroet.   Clavulinaceae
= *Clavaria cristata* Fr.

White, densely tufted and repeatedly branching fruiting body, fringed at the tips; solitary or in small groups, on soil in coniferous and, less frequently, broad-leaf woods.

Dimensions variable dia x 2 - 8 cm tall.
Fruiting body white, repeatedly branching, particularly towards the fringed apices. Branches cylindrical or laterally compressed, smooth or slightly grooved, arising from stout laterally compressed basal region. Flesh white, soft and moderately tough. **Spores** hyaline, smooth, sub-spherical or broadly ellipsoid, non-amyloid, 1 droplet, 7-9 x 6-7.5 µm. Basidia 1- or 2-spored. Cystidia absent.
Odour not distinctive. **Taste** not distinctive.
Chemical tests none.
Occurrence summer to early autumn; common.
☐ Edible.

Clavulina rugosa (Fr.) Schroet.   **Clavulinaceae**
= *Clavaria rugosa* Bull.: Fr.

Small, sparsely-branched, antler-like white fruiting body; solitary or more frequently trooping, on soil in woods generally, often adjacent to paths.

Dimensions variable dia x 5 - 12 cm tall.
Fruiting body white or cream, occasionally with grey tinge; erect, generally with limited irregular branching but sometimes wholly unbranched, laterally compressed, the surface wrinkled and uneven or twisted, tips of branches blunt; no obvious distinction between fruiting body and stem. Flesh white, soft, elastic and fragile.
Spores hyaline, smooth, sub-spherical or broadly ellipsoid, non-amyloid, with droplets, 9-14 x 8-12 µm. Basidia 2-spored. Cystidia absent.
Odour not distinctive. **Taste** not distinctive.
Chemical tests none.
Occurrence late summer to early winter; common.
☐ Edible but of limited culinary value.

Hydnum repandum L.: Fr.   **Hydnaceae**
Hedgehog Fungus

Medium-sized cream fruiting body, at first glance like an agaric mushroom but with spines; solitary or in small trooping groups, on soil in broad-leaf or coniferous woods.

Dimensions cap 3 - 10 cm dia; stem 2 -6 cm tall x 1.5 - 3 cm dia .
Fruiting body cream, sometimes with yellowish tinge, cap smooth or faintly downy, at first convex with inrolled margin becoming flattened and often slightly infundibuliform, margin remaining incurved; stem bruising yellowish towards the base, stout, more or less equal, sometimes eccentric, finely downy. Flesh white, soft, thick and rather crumbly. Spines off-white, 0.2-0.6 cm long. **Spores** hyaline, smooth, sub-spherical or broadly ellipsoid, non-amyloid, with droplets, 6-9 x 5-6.5 µm. Basidia 4-spored. Cystidia absent.
Odour not distinctive. **Taste** bitter after a delay.
Chemical tests none.
Occurrence late summer to autumn; infrequent.
☐ Edible and excellent.

Hydnum repandum var **rufescens** L.: Fr.   **Hydnaceae**
= *Hydnum rufescens* Fr.

Small or medium-sized fruiting body similar to that of *H. repandum* but salmon pink ; solitary or in small trooping groups, on soil in broad-leaf or coniferous woods.

Dimensions cap 2 - 6 cm dia; stem 1.5 - 4 cm dia x 0.8 - 1.5 cm tall.
Fruiting body salmon-pink, cap smooth or faintly downy, at first convex with inrolled margin, becoming flattened and often slightly infundibuliform, margin remaining incurved; stem stout, more or less equal, sometimes eccentric, finely downy. Flesh pinkish, soft, thick and rather crumbly. Spines salmon-pink, 0.2-0.4 cm long.
Spores hyaline, smooth, sub-spherical or broadly ellipsoid, non-amyloid, with droplets, 6-9 x 5-6.5 µm. Basidia 4-spored. Cystidia absent.
Odour not distinctive. **Taste** bitter after a delay.
Chemical tests none.
Occurrence late summer to autumn; infrequent.
☐ Edible.

Sparassis crispa Wulf.: Fr.   **Sparassidiaceae**
Cauliflower Fungus

Large cauliflower-like cream mass, associated with wood, parasitic at the base of conifers. The fungus causes a brown cubical rot in infected timber.

Dimensions 20 - 60 cm dia.
Fruiting body cream, pallid ochraceous or buff, darkening with age, sub-spherical comprising large number of flattened wavy lobes arising from a thick, short rooting stem. Flesh ochraceous, crisp and elastic. Spore-bearing surface ochraceous. **Spores** hyaline or cream, smooth, sub-spherical or broadly ellipsoid, non-amyloid, with droplets, 5-7 x 4-5 μm. Basidia 4-spored. Cystidia absent.
Odour sweetish, pleasant. **Taste** not distinctive.
Chemical tests none.
Occurrence summer to autumn; infrequent.
☐ Edible. For culinary purposes the specimens must be collected young and fresh.

Sparassis laminosa Fr.   **Sparassidiaceae**

Large cauliflower-like whitish cream mass, associated with wood, parasitic on soil at the base of broad-leaf and coniferous trees.

Dimensions 20 - 60 (80) cm dia.
Fruiting body pallid cream with buff tinge, darkening with age, sub-spherical, comprising considerable number of large, flattened, slightly wavy lobes, with ragged or fringed tips, arising from a thick, short rooting stem. Flesh ochraceous, crisp and elastic. Spore-bearing surface ochraceous. **Spores** hyaline or cream, smooth, sub-spherical or broadly ellipsoid, non-amyloid, with droplets, 5-7 x 4-5 μm. Basidia 4-spored. Cystidia absent.
Odour sweetish, pleasant. **Taste** not distinctive.
Chemical tests none.
Occurrence late summer to autumn sometimes persisting during mild winters; rare. (Stourhead, Wiltshire)
☐ Edible.
Note: the species is distinguished from *S. crispa* by the larger fronds and more pallid appearance.

Ramaria botrytis (Fr.) Ricken   **Ramariaceae**

Medium-sized, coral-pink or clay-brown, densely branched fruiting body, not unlike a miniature cauliflower; solitary or in small groups, on soil in broad-leaf woods.

Dimensions 6 - 20 cm dia x 7 - 15 cm tall.
Fruiting body at first ochraceous or pallid tan becoming coral-pink, repeatedly branching, with darker pink or purple, forked, pointed tips; branches cylindrical or laterally compressed, arising from a very stout, whitish, finely downy stem. Flesh white, soft and firm.
Spores pallid yellow, longitudinally striate, oblong-ellipsoid, cyano-philic, non-amyloid, 14-17 x 4.5-8 μm. Basidia 4-spored. Cystidia absent.
Odour not distinctive. **Taste** slight, fruity.
Chemical tests none
Occurrence summer to early autumn; very infrequent or rare. (Near Blakeney, Gloucestershire)
☐ Edible.

Ramaria flaccida (Fr.) Ricken   Ramariaceae

Medium-sized, pale brown, repeatedly branching or tufted fruiting body; solitary or in groups, on soil in coniferous woods.

Dimensions 1 - 4 cm dia x 2 - 5 cm tall.
Fruiting body at first ochraceous becoming pallid brown, repeatedly branching, with forked, pointed tips; branches 2- or 3-pointed, cylindrical or laterally compressed, arising from a very stout, whitish, finely downy stem, 0.2-0.5 cm dia. Flesh whitish, fibrous, soft and firm. **Spores** hyaline (yellow in the mass), warty, oblong-ellipsoid, cyanophilic, non-amyloid, 7-8 x 3.5-4 µm. Basidia 4-spored. Cystidia absent.
Odour slight, fruity. **Taste** not distinctive or slight, bitter.
Chemical tests no reaction with KOH.
Occurrence summer to early autumn; infrequent.
■ Inedible.

Ramaria ochraceo-virens (Jungh.) Donk   Ramariaceae
= *Ramaria abietina* (Pers.: Fr.) Quél.

Medium-sized, pale brown, repeatedly branching or tufted fruiting body; solitary or in groups, on needle litter in coniferous woods favouring spruce.

Dimensions 1 - 4 cm dia x 3 - 6 cm tall.
Fruiting body at first yellowish with olive tinge, becoming wholly greenish with age or on handling, repeatedly branching, with forked, pointed tips; branches 2- or 3-pointed, cylindrical or laterally compressed, arising from a stout, whitish, finely downy stem, 0.3-1.4 cm dia. Flesh whitish, fibrous and fairly tough. **Spores** yellowish, warty, pip-shaped, non-amyloid, 9-11 x 3.5-5 µm. Basidia 4-spored. Cystidia absent.
Odour not distinctive. **Taste** not distinctive or slight, bitter.
Chemical tests surface olivaceous with KOH.
Occurrence summer to early autumn; infrequent.
■ Inedible.

Ramaria stricta (Fr.) Quél.   Ramariaceae

Medium-sized, ochre-yellow or flesh-coloured fruiting body, repeatedly branching, coral-like; solitary or more typically in extensive troops, on soil and plant debris in broad-leaf woods favouring beech, less frequently on rotting conifer stumps.

Dimensions 3 - 8 cm dia x 4 - 11 cm tall.
Fruiting body ochraceous tinged buff, becoming darker cinnamon-brown with age, repeatedly branching coral-like with pointed forked tips, at first yellow then concolorous; branches cylindrical or laterally compressed arising from pallid stout stem, 0.5-1.5 cm dia. Flesh white, elastic and tough. **Spores** hyaline (yellow in the mass), finely warty, broadly ellipsoid, non-amyloid, with droplets 7.5-10 x 4-5 µm. Basidia 4-spored. Cystidia absent.
Odour sweet or earthy. **Taste** slight, peppery.
Chemical tests none
Occurrence summer to autumn; infrequent.
☐ Edible.

Ramariopsis kunzei (Fr.) Donk.   **Ramariaceae**
= *Ramaria kunzei* (Fr.) Quél.
= *Clavaria kunzei* Fr.

Small, fragile, cream fruiting body, repeatedly branching, coral-like; solitary or in small groups, on soil amongst grass associated with coniferous trees.

Dimensions 1 - 6 cm dia x 2 - 12 cm tall.
Fruiting body ivory-white or cream, fragile, repeatedly but loosely branching, coral-like, with pointed or bluntly forked tips; branches cylindrical arising from slender stem, minutely downy at the base. Flesh white and elastic. **Spores** hyaline, minutely warty, sub-spherical, non-amyloid, with droplets, 4-5 µm. Basidia 2- or 4-spored. Cystidia absent.
Odour not distinctive. **Taste** not distinctive.
Chemical tests flesh turns green with $FeSO_4$
Occurrence late summer to autumn; infrequent.
■ Inedible.

Ramariopsis pulchella (Boud.) Corn.   **Ramariaceae**
= *Clavaria pulchella* Boud.
= *Clavaria tenuissima* Sacc.

Small, fragile, lilac fruiting body, repeatedly branching, coral-like; solitary or in small groups, on soil amongst grass and other dwarf vegetation in mixed woods.

Dimensions 1 - 2 cm dia x 1 - 2 cm tall.
Fruiting body lilaceous, fragile, with loose, limited branching, coral-like, with slender tips; branches filiform arising from slender stem, white or reddish tinge, minutely downy at the base. Flesh white and elastic. **Spores** hyaline, minutely warty, sub-spherical, non-amyloid, with droplets, 2.5-4.5 µm. Basidia 2- or 4-spored. Cystidia absent.
Odour not distinctive. **Taste** not distinctive.
Chemical tests flesh turns green with $FeSO_4$
Occurrence late summer to autumn; very rare. (Forest of Dean, Gloucestershire)
■ Inedible.

Clavicorona taxophila (Thom) Doty   **Clavicoronaceae**
= *Clavaria taxophila* (Thom) Lloyd

Small, white, simple tongue; sparsely clustered on exposed roots, twigs and other debris, in mixed woodlands typically on or adjacent to damp paths.

Dimensions 0.5 - 0.75 cm dia x 1 - 2 cm tall.
Fruiting body white, but becoming yellowed with age, unbranched, flattened and tapering towards the stem, tips spathulate; stem white or off-white, cylindrical, distinct. Flesh white, soft and elastic.
Spores hyaline, smooth, sub-spherical or broadly ellipsoid, non-amyloid, 3-4 x 2-3 µm. Basidia 4-spored. Cystidia, elongated, thin-walled with oily contents.
Odour not distinctive. **Taste** not distinctive.
Chemical tests none.
Occurrence late summer to autumn; infrequent or rare. (Tor Woods, Wells, Somerset)
■ Inedible.

Auriscalpium vulgare S F Gray   **Auriscalpiaceae**
= *Hydnum auriscalpium* L.

Small fruiting body with brown kidney-shaped cap, spines and slender stem; arising from buried pine cones.

Dimensions cap 1 - 2 cm dia; stem 2 - 6 cm tall x 0.1 - 0.2 cm dia.
Cap brown, pallid at the margin, reniform, with more or less wavy margin, covered with coarsely downy chestnut hairs. Stem dark brown, slender, eccentric, wavy, attenuated upwards, densely bristly. Flesh whitish and tough. Spines dark pinkish-brown, becoming more grey-brown with age. **Spores** hyaline, minutely spiny, broadly ellipsoid, amyloid, cyanophilic, 4.5-5.5 x 3.5-4.5 µm. Basidia 2- or 4-spored. Cystidia elongated, with granular contents.
Odour not distinctive. **Taste** not distinctive.
Chemical tests none.
Occurrence late summer to autumn; frequent. The species is probably common but sometimes overlooked.
■ Inedible.

Creolophus cirrhatus (Pers.: Fr.) Karst.   **Hericiaceae**

Large cream bracket-like caps, with warty, short-spined upper surface and more pendulous spines below; solitary; on dead wood of a variety of broad-leaf trees favouring beech and birch.

Dimensions 4 - 9 cm dia. x 1 - 3 cm deep.
Fruiting body cream, becoming more ochraceous with age, irregular, bracket-like, upper surface warty and with short sterile spines, the fertile hymenium on the under-surface spread over pendulous conical spines; sessile, attached laterally. Flesh pallid cream, elastic and soft. **Spores** hyaline, smooth, sub-spherical, amyloid, with droplets, 3.5-4.5 x 3-3.5 µm. Basidia 2- or 4-spored. Cystidia cylindrical and sinuous.
Odour not distinctive. **Taste** not distinctive.
Chemical tests none.
Occurrence late summer to early autumn; very rare. (Stourhead, Wiltshire)
☐ Edible.
Note: because of the considerable rarity of this species it should not be picked for eating.

Hericium erinaceum Pers.   **Hericiaceae**

Large, pure white fruiting body, of dramatic appearance, cushion-like with long pendulous spines; solitary; growing from wounds on living broad-leaf trees favouring beech, also from ends of felled trunks.

Dimensions 10 - 25 cm dia. x 10 - 20 cm deep.
Fruiting body white, cushion-like, upper surface smooth, the fertile hymenium on the under-surface covering pendulous elongated conical spines; sessile. Flesh white and firm. **Spores** hyaline, finely warty or spiny, sub-spherical, amyloid, some with droplets, 6-6.5 x 4-5.5 µm. Basidia 4-spored. Cystidia cylindrical.
Odour vague, unpleasant. **Taste** not distinctive.
Chemical tests none.
Occurrence late summer to early autumn; rare. (Fynne Court, West Somerset)
☐ Edible.
Note: because of the considerable rarity of this species it should not be picked for eating.

Lentinellus cochleatus (Pers.: Fr.) Karst.   **Lentinellaceae**
= *Lentinus cochleatus* (Pers.: Fr.) Fr.

Fairly small, reddish-brown, funnel-shaped fruiting body; typically caespitose, on stumps of broad-leaf trees.

Dimensions cap 2 - 6 cm dia; stem 2 - 5 cm tall x 0.8 - 1.5 cm dia.
Cap chestnut-brown, deeply infundibuliform or more ear-shaped, smooth and shiny. Gills pallid buff, deeply decurrent, distant. Stem chestnut-brown, darker towards base, eccentric or lateral, attenuated downwards, often rooting. Flesh pinkish-brown, tough.
Spores hyaline, minutely spiny, sub-spherical, 4.5-5 x 3.5-4 µm. Basidia 4-spored. Cystidia-like hyphal ends present.
Stem chestnut-brown, darker towards the base, eccentric or lateral, attenuated downwards, often rooting. Flesh pinkish-brown, tough.
Odour strong, of aniseed. **Taste** not distinctive.
Chemical tests none.
Occurrence summer to autumn; infrequent.
☐ Edible.

Polyporus badius (Pers.) Schw.   **Polyporaceae**
= *Polyporus picipes* Fr.

Medium-sized, bay-brown or chestnut, funnel-shaped cap with white pores on the under-surface, annual; arising from living wood or stumps of broad-leaf trees favouring beech, sometimes appearing to be growing on soil.

Dimensions cap 5 - 20 cm dia; stem 2 - 3.5 cm tall x 0.5 - 1.5 cm dia.
Fruiting body upper surface at first pallid, brownish-grey becoming a rich dark chestnut, at first convex then flattened, becoming shallowly infundibuliform, smooth, with thin wavy margin; stem pallid but black at the base, slender, more or less equal, sometimes rudimentary. Flesh whitish, soft, corky and very thin.
Pores whitish becoming pallid buff, circular, 4-7 per mm. Tubes whitish becoming cream, decurrent, 0.5-2.5 mm deep.
Spores hyaline, smooth, elongated-ellipsoid, or cylindrical, non-amyloid, 6-10 x 3-4 µm. Basidia 4-spored. Cystidia absent.
Odour not distinctive. **Taste** bitter.
Chemical tests none.
Occurrence spring to early winter, but sporulating from summer to autumn; rare. (Bridgham near Thetford, Norfolk)
■ Inedible.

Polyporus brumalis Pers.: Fr.   **Polyporaceae**
Winter Polypore

Small to medium-sized, brownish funnel-shaped cap with whitish or pale buff pore-bearing under-surface, annual; arising from dead wood broad-leaf trees favouring logs and fallen branches lying on soil.

Dimensions cap 2 - 8 cm dia; stem 2 - 6 cm tall x 0.3 - 1 cm dia.
Fruiting body upper surface greyish-brown or cigar-brown, sometimes with faint concentric zonation, convex with shallow central depression, smooth, with wavy, often inrolled margin; stem pallid, tawny, central, slender, more or less equal, typically curved and sometimes thickened at the base. Flesh whitish and leathery.
Pores whitish-cream or pallid buff, circular, elongating with age, 2-3 per mm. Tubes whitish becoming cream, decurrent, 0.5-2 mm deep.
Spores hyaline, smooth, cylindrical or allantoid, non-amyloid, with droplets, 6-8 x 2-2.5 µm. Basidia 4-spored. Cystidia absent.
Odour not distinctive. **Taste** not distinctive.
Chemical tests none.
Occurrence late autumn to spring, but sporulating in winter; infrequent.
■ Inedible.
Readily confused with *Polyporus ciliatus* but distinguished by size of pores.

Polyporus ciliatus Fr.: Fr.   Polyporaceae
= *Polyporus lepideus* Fr.

Small to medium-sized, brownish funnel-shaped cap with white or pale buff pore-bearing under-surface, annual; arising from dead wood of broad-leaf trees favouring logs and fallen branches lying on soil.

Dimensions cap 2 - 12 cm dia; stem 2 - 4 cm tall x 0.2 - 0.7 cm dia.
Fruiting body upper surface greyish-brown or cigar-brown, convex with shallow central depression, smooth, with wavy, often incurved margin; stem pallid, tawny, central, slender, more or less equal, typically curved and sometimes thickened at the base. Flesh whitish and leathery.
Pores whitish-cream to buff, circular, elongating with age, 4-6 per mm. Tubes whitish becoming cream, decurrent, 0.5-2 mm deep.
Spores hyaline, smooth, cylindrical-ellipsoid, non-amyloid, 5-6 x 1.5-2.5 µm. Basidia 4-spored. Cystidia absent.
Odour not distinctive. **Taste** not distinctive.
Chemical tests none.
Occurrence spring to late summer, but sporulating in summer; infrequent.
■ Inedible.
Readily confused with *Polyporus brumalis* but distinguished by size of pores.

Polyporus squamosus Huds.: Fr.   Polyporaceae
Dryad's Saddle

Large, creamy-brown scaly cap with cream pore-bearing under-surface, annual; parasitic on broad-leaf trees, also on stumps, favouring beech, elm and sycamore.

Dimensions cap 10 - 60 cm dia x 0.5 - 5 cm thick.
Fruiting body upper surface cream with more or less concentric bands of cinnamon, fibrillose scales, fan-shaped or semi circular, more or less flattened, regular margin; stem darkening towards the base, lateral or, less commonly eccentric, short. Flesh whitish, thick and leathery.
Pores whitish-cream, irregularly angular, 1-3 x 0.5-1.5 mm dia. Tubes whitish becoming cream, decurrent, 5-10 mm deep.
Spores hyaline, smooth, cylindrical or allantoid, non-amyloid, with droplets, 12-16 x 4-6 µm. Basidia 4-spored. Cystidia absent.
Odour strong, mealy. **Taste** not distinctive.
Chemical tests none.
Occurrence spring to early autumn, but sporulating in summer; common.
☐ Edible.

Polyporus tuberaster Pers.: Fr.   Polyporaceae
= *Polyporus floccipes* Rostk.
= *Polyporus lentus* Berk.
= *Polyporus squamosus var. coronatus* (Rostk.) Pil.

Small to medium-sized, creamy-brown scaly cap with cream pore-bearing under-surface, annual; arising from dead wood of broad-leaf trees.

Dimensions cap 2 - 10 cm dia; stem 5 - 6 cm tall x 0.5 - 1.5 cm dia.
Fruiting body upper surface cream with cinnamon scales, darker at the tips, convex with shallow depression close to the point of attachment, wavy margin; stem pallid, central or eccentric, more or less equal, usually covered at the base with white hairs. Flesh whitish, thick, elastic and leathery.
Pores whitish-cream, irregularly angular and toothed, 2-2.5 x 1-1.5 mm dia. Tubes whitish becoming cream, very short, deeply decurrent, 1-3 mm deep. **Spores** hyaline, smooth, cylindrical-ellipsoid, non-amyloid, with droplets, 12-16 x 4-6 µm. Basidia 4-spored. Cystidia absent.
Odour not distinctive. **Taste** not distinctive.
Chemical tests none.
Occurrence spring to late summer, but sporulating in summer; infrequent.
■ Inedible.

Polyporus varius Pers.: Fr.   **Polyporaceae**

Smallish, ochre-brown funnel-shaped cap with white or pale buff pore-bearing under-surface, annual; arising from dead or dying wood of broad-leaf trees, favouring logs and fallen branches lying on soil.

Dimensions cap 1 - 10 cm dia; stem 0.5 - 3 cm tall x 0.5 - 1.5 cm dia.
Fruiting body upper surface cream, with brown radial streaks or wholly more ochraceous-brown, darkening with age, convex with shallow central depression, smooth, with wavy or wrinkled margin margin; stem pallid above, dark brown in the lower half, central, slender, more or less equal. Flesh whitish, thin and leathery.
Pores whitish-cream ageing ochre-brown, circular, 4-7 per mm. Tubes whitish becoming cream, decurrent, 0.5-2.5 mm deep.
Spores hyaline, smooth, cylindrical, non-amyloid, with droplets, 9-11 x 3.5-4 µm. Basidia 4-spored. Cystidia absent.
Odour not distinctive. **Taste** slight, bitter.
Occurrence spring to late summer, but sporulating in summer; infrequent but locally more common.
Chemical tests none.
■ Inedible.

Abortiporus biennis (Bull.: Fr.) Sing.   **Coriolaceae**
= *Daedalea biennis* Bull.: Fr.
= *Heteroporus biennis* (Bull.: Fr.) Laz.

Rosette-like lobes, whitish then brown, with pore-bearing under-surface, annual; solitary on buried rotting wood.

Dimensions up to 20 cm dia overall, brackets 3 - 9 cm dia x 0.5 - 1.5 cm thick.
Fruiting body at first white, then ochraceous-brown, often appearing as an amorphous cushion, then top-shaped and finally lobed fan-wise; upper surface smooth, finely downy, margin undulating. Stem embedded in the substrate. Flesh white, sometimes producing red guttation droplets when young, red where cut, outer layers soft with a harder inner core.
Pores concolorous, angular-labyrinthine, 1-3 per mm. Tubes concolorous, 2-5 mm deep. **Spores** hyaline (yellow in the mass), smooth, ellipsoid, non-amyloid, with droplets, 4.5-6 x 3.5-4.5 µm. Basidia 4-spored. Cystidia, clavate, sinuous.
Odour not distinctive. **Taste** not distinctive.
Chemical tests none.
Occurrence winter to autumn, but sporulating from spring to late summer; infrequent.
■ Inedible.

Antrodia serialis (Fr.) Donk   **Coriolaceae**

Spreading, whitish, partly encrusting, partly hoof-like, the pore-bearing surface exposed as vertical 'lacerations' between individual brackets, annual; in overlapping tiers on vertical surfaces of dead conifer stumps and other timber.

Dimensions variable dia overall, individual brackets 1 - 2 cm dia x 0.5 - 1.5 cm thick.
Fruiting body upper surface at first white becoming pallid cream, sometimes with pinkish spotting, irregular, lumpy hoof-shaped, margin undulating. Flesh white, soft when damp, chalky when dry.
Pores white, small, sub-circular or angular, lacerate, exposed vertically, 2-4 per mm. Tubes concolorous, 3-5 mm deep.
Spores hyaline, smooth, ellipsoid, non-amyloid, 6.5-9-6.5 x 3-4 µm. Basidia 4-spored. Cystidia slender, elongated.
Odour not distinctive. **Taste** bitter.
Chemical tests none.
Occurrence winter to autumn, but sporulating from spring to late summer; infrequent.
■ Inedible.

Antrodia xantha (Fr.: Fr.) Ryv.   Coriolaceae

Spreading, creamy-yellow, partly encrusting, partly hoof-like, the pore-bearing surface exposed as vertical 'lacerations' between individual brackets, annual; in overlapping tiers on vertical surfaces of dead conifer stumps and other timber.

Dimensions variable dia overall, individual brackets 1 - 3 cm dia x 0.5 - 1 cm thick.
Fruiting body upper surface pallid creamy-yellow or brighter, with slightly iridescent appearance, irregular, lumpy hoof-shaped, margin undulating. Flesh whitish, soft when damp, chalky when dry.
Pores concolorous, small, sub-circular or angular, lacerate, exposed vertically, 4-6 per mm. Tubes white, 3-5 mm deep. **Spores** hyaline, smooth, allantoid, non-amyloid, 4-5 x 1-1.5 µm. Basidia 4-spored. Cystidia absent.
Odour not distinctive. **Taste** bitter.
Chemical tests none.
Occurrence winter to autumn, but sporulating from spring to late summer; infrequent.
■ Inedible.

Bjerkandera adusta (Willd.:Fr.) Karst.   Coriolaceae

= *Polyporus adustus* Willd.: Fr.
= *Gloeoporus adustus* (Willd.: Fr.) Pil.

Small, greyish-white bracket with characteristic grey pore-bearing under-surface, annual; in dense overlapping tiers on dead wood of broad-leaf trees.

Dimensions 2 - 4 cm dia x 0.3 - 0.6 cm thick.
Fruiting body upper surface when young grey-brown with a white margin, becoming darker with age, finely downy then smooth, irregular or lumpy, margin undulating. Flesh whitish, rubbery or leathery.
Pores smoke-grey darkening with age, very small (hand lens), sub-circular or angular, 4-6 per mm. Tubes grey, 0.5-1.5 mm deep. **Spores** hyaline, smooth, elongated-ellipsoid or cylindrical, non-amyloid, 4.5-5.5 x 2-3 µm. Basidia 4-spored. Cystidia absent.
Odour strong, fungoid. **Taste** acidic.
Chemical tests none.
Occurrence throughout the year, but sporulating from late summer to autumn; common.
■ Inedible.

Coriolus hirsutus (Wulf.: Fr.) Quél.   Coriolaceae

= *Trametes hirsuta* (Wulf.: Fr.) Pil.

Smallish thin bracket, pale cream with brownish concentric zoning and dense silvery hairs, under-surface pale, bearing pores, annual; arranged singly or in overlapping tiers, on dead wood of broad-leaf trees favouring trunks and stumps of beech in exposed positions.

Dimensions 3 - 10 cm dia x 0.5 - 0.8 cm thick.
Fruiting body upper surface pallid cream with brownish or greyish tinges, concentrically zoned, covered densely with pallid hairs which appear silvery when young; sessile. Flesh white, tough and leathery.
Pores at first white becoming cream and finally greyish, sub-circular, 2-4 per mm. Tubes white, yellowish when dry, 1-5 mm deep. **Spores** hyaline, smooth, sub-cylindrical or ellipsoid, non-amyloid, 5.5-7.5 x 1.5-2.5 µm. Basidia 4-spored. Cystidia absent.
Odour not distinctive. **Taste** not distinctive.
Chemical tests none.
Occurrence throughout the year, but sporulating late summer to autumn; infrequent or rare. (Mendip Hills, Wookey, Somerset)
■ Inedible.

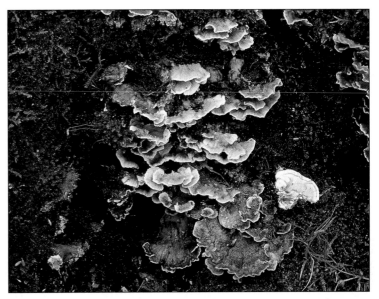

Coriolus versicolor (L.: Fr.) Quél.   Coriolaceae
= *Trametes versicolor* (L.: Fr.) Pil.

Smallish thin bracket, very variable in colour but always concentrically zoned, under-surface pale, bearing pores, annual; arranged in densely overlapping tiers, on dead wood of broad-leaf trees.

Dimensions 3 - 8 cm dia x 0.1 - 0.3 cm thick.
Fruiting body upper surface very variable in colour, ranging through ochres, greens, blues, greys, rusts and black, always concentrically zoned, at first finely downy then smooth; sessile. Flesh white, tough and leathery.
Pores white, yellowish or tawny, circular or irregularly angular, 3-5 per mm. Tubes white, yellowish when dry, 0.5-1 mm deep.
Spores hyaline, smooth, cylindrical or allantoid, non-amyloid, 5-7 x 1.5-2 µm. Basidia 2- or 4-spored. Cystidia absent.
Odour not distinctive. **Taste** not distinctive.
Chemical tests none.
Occurrence throughout the year, but sporulating late summer to autumn; very common.
■ Inedible.

Coriolus zonata (Nees: Fr.) Quél.   Coriolaceae
= *Trametes multicolor* (Schaeff.) Jül.
= *Trametes zonata* (Nees : Fr.) Pil.

Smallish bracket, concentrically zoned with grey and brown or orange, under-surface cream bearing pores, more or less triangular in cross-section, annual; arranged in overlapping tiers, on dead wood of broad-leaf trees favouring oak, hornbeam and ash.

Dimensions 2 - 5 cm dia x 0.5 - 1 cm thick.
Fruiting body greyish-white or cream, zoned concentrically with brown, tinged orange or ochraceous, at first finely downy then smooth; sessile. Flesh white, tough and leathery.
Pores cream or pallid ochraceous, circular or irregularly angular, 3-4 per mm. Tubes concolorous, 0.2-0.4 mm deep.
Spores hyaline, smooth, cylindrical or allantoid, non-amyloid, 5.5-7.5 x 2.5-3 µm. Basidia 4-spored. Cystidia absent.
Odour slight, acidic. **Taste** not distinctive.
Chemical tests none.
Occurrence throughout the year, but sporulating late summer; infrequent.
■ Inedible.

Daedalea quercina L.: Fr.   Coriolaceae
= *Trametes quercina* (L.: Fr.) Pil.   Oak Maze Gill

Medium-sized or large, thick, corky bracket, pale creamy grey, annual; generally solitary but occasionally in limited tiers, on dead oak trunks and, less frequently, other broad-leaf trees.

Dimensions 5 - 20 cm dia x 2 - 5 cm thick.
Fruiting body upper surface creamy-ochraceous with greyish tinge, semi circular and broadly attached, generally warty or lumpy, more rarely smooth; sessile. Flesh pallid (like wood), tough and corky.
Pores pallid ochraceous, very large, irregular, maze- or gill-like in a radial arrangement, 1-3 mm dia. Tubes more or less ochraceous, 10-40 mm deep. **Spores** hyaline, smooth, ellipsoid, non-amyloid, 5.5-7 x 2.5-3.5 µm. Basidia 2- or 4-spored. Cystidia present as thick-walled, fusiform hyphal ends.
Odour faint, acrid and fungoid. **Taste** not distinctive.
Chemical tests none.
Occurrence spring to winter, but sporulating late spring to late autumn;
■ Inedible.

Daedaleopsis confragosa (Bolt.: Fr.) Schroet.**Coriolaceae**
= *Trametes rubescens* (A. & S.: Fr.) Fr.

Small to medium-sized kidney-shaped bracket, reddish-brown with dirty pinkish-white pores on the under-surface, annual; arranged singly or in tiers, on dead wood of broad-leaf trees favouring willow, birch and beech.

Dimensions 5 - 20 cm dia x 1 - 4 cm thick.
Fruiting body upper surface at first ochraceous, becoming reddish-brown towards the point of attachment, sometimes with pinkish flush, more or less reniform, broadly attached, finely downy, zoned and ridged concentrically, wrinkled radially, margin thin and acute. Flesh pallid ochraceous becoming darker with age, tough, corky; sessile.
Pores creamy-ochraceous bruising reddish or pinkish, irregular or maze-like, elongated radially, 1-2 per mm. Tubes pallid ochraceous, 5-10 mm deep. **Spores** hyaline, smooth, cylindrical or slightly allantoid, non-amyloid, some with droplets, 7-11 x 2-3 μm. Basidia 4-spored. Cystidia absent.
Odour not distinctive. **Taste** slight, bitter.
Chemical tests none.
Occurrence throughout the year, but sporulating late summer to autumn; frequent.
■ Inedible.

Daedaleopsis confragosa var **tricolor**
(Bull.) Bond. & Sing. **Coriolaceae**

Small to medium-sized kidney-shaped bracket, narrowly zoned with shades of dark red from first appearance, dirty grey-brown pores on the under-surface, thin-fleshed, annual; arranged singly or in tiers, on dead wood of broad-leaf trees favouring beech, wild cherry and hazel.

Dimensions 5 - 20 cm dia x 1 - 4 cm thick.
Fruiting body upper surface reddish, tinged vinaceous to dark or blackish-red, more or less reniform, broadly attached, finely downy, zoned and ridged concentrically, wrinkled radially, margin thin and acute. Flesh ochraceous becoming reddish-brown with age, tough and corky; sessile.
Pores ochraceous-brown bruising reddish, irregular or maze-like, elongated in a radial arrangement, 1-2 per mm. Tubes ochraceous, 5-10 mm deep. **Spores** hyaline, smooth, cylindrical or slightly allantoid, non-amyloid, 7-10 x 2-3 μm. Basidia 4-spored. Cystidia absent.
Odour not distinctive. **Taste** slight, bitter.
Chemical tests none.
Occurrence throughout the year, but sporulating late summer to autumn; infrequent.
■ Inedible.

Datronia mollis (Somm.) Donk **Coriolaceae**
= *Trametes mollis* (Somm.) Fr.
= *Antrodia mollis* (Somm.) Karst.

Small leathery bracket, dark above and with whitish angular or slot-like tubes below, perennial; in tiers on dead wood of broad-leaf trees favouring beech.

Dimensions 1 - 7 cm dia x 0.2 - 0.6 cm thick.
Fruiting body upper surface at first umber-brown, becoming darker brown and finally more or less black, narrow, partly resupinate, broadly attached, at first downy then smooth, usually undulating; sessile. Flesh pallid buff, tough and leathery, brittle when dry.
Pores pallid greyish, becoming ochraceous-brown when handled, angular or elongated slot-like, 1-2 per mm. Tubes more or less ochraceous, 0.5-5 mm deep. **Spores** hyaline, smooth, cylindrical, non-amyloid, some with droplets, 8-10.5 x 2.5-4 μm. Basidia 4-spored. Cystidia fusiform, short.
Odour not distinctive. **Taste** not distinctive.
Chemical tests none.
Occurrence throughout the year, but sporulating spring to late autumn; common.
■ Inedible.

Dichomitus campestris (Quél.) Dom. & Orl. **Coriolaceae**
= *Coriolellus campestris* (Quél.) Bond.
= *Trametes campestris* Quél.

Smallish cream, woody cushion with irregular tubes, annual or perennial; solitary or in small groups on dead branches of broad-leaf trees favouring hazel and oak.

Dimensions 3 - 7 cm dia x 0.5 - 1.5 cm thick.
Fruiting body upper surface cream or pallid hazel, typically with dark brown or blackish margin, resupinate, broadly attached; sessile. Flesh pallid buff, tough and woody.
Pores covering the entire fruiting body, cream or whitish, becoming ochraceous-brown when handled, angular or irregular, 1-2 per mm. Tubes more or less concolorous, 1-3 mm. deep, layered in perennial specimens. **Spores** hyaline, smooth, narrowly ellipsoid, non-amyloid, some with droplets, 9-12.5 x 3.5-4.5 µm. Basidia 2- or 4-spored. Cystidia absent.
Odour not distinctive. **Taste** not distinctive.
Chemical tests none.
Occurrence throughout the year, but sporulating spring to late autumn; rare. (Cannop Ponds, Forest of Dean, Gloucestershire)
■ Inedible.

Fomes fomentarius (L.:Fr.) Kickx   **Coriolaceae**

Thick, greyish-brown hoof-shaped bracket with pale brownish pores on the under-surface, perennial; generally solitary but sometimes several on the same host, on wood of broad-leaf trees, predominantly in northern regions where it favours birch but also beech and sycamore.

Dimensions 10 - 25 cm dia x 2 - 25 cm thick.
Fruiting body upper surface greyish, concentrically zoned, very thick, almost conical or hoof-like, horny, crusted; sessile. Flesh pallid cinnamon brown, darkening with age, fibrous, leathery or woody.
Pores at first ochraceous, becoming light brownish-grey, darker where bruised or handled, rounded, 2-4 per mm. Tubes pallid ochraceous, in tiers separated by darker brownish bands, each tier 2-7 mm deep. **Spores** hyaline with lemon tinge, smooth, elongated-ellipsoid or cylindrical, non-amyloid, 15-20 x 4.5-7 µm. Basidia 4-spored. Cystidia absent.
Odour slight, fruity. **Taste** acrid.
Chemical tests none.
Occurrence throughout the year, but sporulating spring and early summer; infrequent.
■ Inedible.

Gloeophyllum sepiarium (Wulf.: Fr.) Karst. **Coriolaceae**
= *Lenzites sepiaria* (Wulf.: Fr.) Fr.

Thinnish, dark brown fan-shaped bracket, often clustered, with yellowish brown slotted pores on the under-surface, annual but persistent; in limited tiers, sometimes several arising from a common base, on dead wood of coniferous and, very occasionally, broad-leaf trees. Causes a virulent brown rot.

Dimensions 5 - 12 cm dia x 0.5 - 1 cm thick.
Fruiting body upper surface dark rust-brown, with concentric maroon-tinged zones and more pallid margin, darkening with age almost to black, coarsely concentrically ridged and radially wrinkled, at first softly hairy, becoming more bristly; sessile. Flesh dark brown, fibrous, leathery or woody.
Pores at first ochraceous-rust becoming more tobacco-brown at maturity, maze- or gill-like in a radial arrangement. **Spores** hyaline, smooth, cylindrical, non-amyloid, 9-13 x 3-5 µm. Basidia 4-spored. Cystidia awl-shaped and slender.
Odour not distinctive. **Taste** not distinctive.
Chemical tests none.
Occurrence throughout the year, but sporulating late summer to late autumn; infrequent.
■ Inedible.

Grifola frondosa (Dicks.: Fr.) S F Gray   Coriolaceae

Tongue-like, brownish fruiting bodies, arising in clumps from a common repeatedly branching stem, creating the appearance of an irregular rosette, annual; parasitic on wood of broad-leaf trees, usually arising from the base of the trunk, favouring beech but also on oak, hornbeam and Spanish chestnut. The fungus generates a destructive white rot.

Dimensions individual caps 4-10 cm dia x 0.5 - 1 cm thick; collectively 20 - 50 cm dia.
Fruiting body upper surface tan, with olivaceous tinge, several or many ligulate caps arising from a central branched stem; each cap thick and leathery with undulating, sometimes split, margin, zoned concentrically and wrinkled radially; stem pallid cream or greyish, laterally compressed, 2-5 cm long. Flesh whitish, soft and fibrous.
Pores whitish-cream, angular or rounded, 2-3 per mm. Tubes white, decurrent, 2-3 mm deep. **Spores** hyaline, smooth, broadly ellipsoid, non-amyloid, with droplets, 5-7 x 3.5-5 µm. Basidia 4-spored. Cystidia absent.
Odour reminiscent of mice. **Taste** when young - pleasant; when older - more acrid.
Chemical tests none.
Occurrence late summer to late autumn; infrequent and mainly in the south.
☐ Edible.

Heterobasidion annosum (Fr.) Bref.   Coriolaceae
= Fomes annosus (Fr.) Cke.

Large brown fruiting body, darkening with age, typically with a white margin and with white pores on the under-surface, either encrusting or bracket, perennial; singly or in limited more or less fused tiers, at the extreme base, and on roots, of coniferous trees. Causes acute and prolific decay in coniferous trees.

Dimensions 5 - 30 cm dia x 1 - 2 cm thick.
Fruiting body upper surface light brown, becoming dark brown with age, flattened or slightly convex, resupinate or irregular, broadly attached, at first downy then smooth, unevenly wrinkled or knobbly. Margin rounded, sometimes undulating; sessile. Flesh whitish-cream, fibrous and corky.
Pores whitish-cream, bruising and ageing brown, circular or somewhat angular, 2-3 per mm. Tubes whitish-cream, arranged in tiers, each tier 2-5 mm deep. **Spores** hyaline, minutely warty, sub-spherical or broadly ellipsoid, non-amyloid, cyanophilic, with droplets, 4.5-6 x 4-4.5 µm. Basidia 4-spored. Cystidia absent.
Odour strong, fungoid. **Taste** not distinctive.
Chemical tests none.
Occurrence throughout the year, but sporulating late summer to autumn; very common.
■ Inedible.

Ischnoderma benzoinum (Wahl.: Fr.) Karst.   Coriolaceae

Medium to large dark brown bracket with white pores on the under-surface, annual; single or in small tiers, on dead wood of coniferous trees favouring spruce.

Dimensions 4 - 20 cm dia x 1 - 2 cm. thick.
Fruiting body upper surface dark reddish-brown with concentric zoning, becoming darker with age, downy, wrinkled, undulating, exuding resinous guttation drops when young, margin more or less acute and whitish when young; sessile though not broadly attached. Flesh whitish or pallid brown, soft and succulent when damp, hard when dry.
Pores whitish-cream becoming ochraceous with age, angular, 4-6 per mm. Tubes buff, 5-8 mm deep. **Spores** hyaline, smooth, sub-cylindrical, non-amyloid, 5-6 x 2-2.5 µm. Basidia 4-spored. Cystidia absent.
Odour not distinctive. **Taste** not distinctive.
Chemical tests none.
Occurrence late summer to winter, but sporulating in autumn; rare. (Priddy Forest, Somerset)
■ Inedible.

Ischnoderma resinosum (Fr.) Karst. **Coriolaceae**
= *Polyporus resinosus* Fr.

Medium to large rust-brown bracket with white pores on the under-surface, annual; single or in small tiers, on dead wood of broad-leaf trees favouring beech.

Dimensions 5 - 30 cm dia x 1 - 3 cm thick.
Fruiting body upper surface rust-brown with concentric zoning, becoming darker with age, downy, wrinkled, undulating, exuding resinous guttation drops when young, margin more or less acute. Sessile though not broadly attached. Flesh whitish or pallid brown, soft and succulent when damp, hard when dry.
Pores whitish-cream becoming ochraceous with age, angular, 2-3 per mm. Tubes buff, 6-10 mm deep. **Spores** hyaline, smooth, sub-cylindrical, non-amyloid, 5-6 x 2-2.5 μm. Basidia 4-spored. Cystidia absent.
Odour not distinctive. **Taste** not distinctive.
Chemical tests none.
Occurrence late summer to winter, but sporulating in autumn; rare. (Thursley near Guildford, Surrey)
■ Inedible.

Laetiporus sulphureus (Bull.: Fr.) Murr. **Coriolaceae**
= *Polyporus sulphureus* Bull.: Fr. Chicken of the Woods

Large fleshy cream or egg-yellow, fan-shaped brackets resembling chickens' feet, annual; in tiered clusters on the trunks and stumps of broad-leaf trees favouring oak, less commonly on willow, Spanish chestnut, yew and apple.

Dimensions 10 - 40 cm dia x 3 - 12 cm thick.
Fruiting body upper surface egg-yellow with pink and orange tinges when young and fresh, drying or ageing cream, flattened or slightly convex, irregular wavy or crinkly, smooth without zonation, the margin thick; sessile but often narrowly attached. Flesh yellowish-orange, thick, soft and juicy when damp; cream and more crumbly when dry.
Pores sulphur-yellow, circular or ovoid, 1-3 per mm. Tubes concolorous, 1.5-3 mm deep. **Spores** hyaline, smooth, sub-spherical, non-amyloid, with droplets, 5-7 x 3.5-5 μm. Basidia 4-spored. Cystidia absent.
Odour strong when fresh, fungoid. **Taste** slight, sour.
Chemical tests none.
Occurrence late spring to late summer, mainly summer; infrequent, perhaps locally common.
□ Edible and good.

Lenzites betulina (L.: Fr.) Fr. **Coriolaceae**
= *Trametes betulina* (L.: Fr.) Pil.

Medium-sized distinctively brownish-zoned, semi circular or fan-shaped bracket with whitish slot-like pores on the under-surface, annual; in limited tiers on dead wood of broad-leaf trees favouring birch and willow, often on felled or uprooted trunks.

Dimensions 3 - 8 cm dia x 0.3 - 1.5 cm thick.
Fruiting body upper surface with concentric zones of alternating whitish-grey, tan and brown, downy or finely hairy and concentrically grooved, sometimes with algal growth, typically narrowly attached; sessile. Flesh white, cottony fibrous and tough.
Pores whitish-buff, becoming straw, gill- or maze-like in a radial arrangement and branching. **Spores** hyaline, smooth, sub-cylindrical, non-amyloid, 5-6 x 1.5-3 μm. Basidia 4-spored. Cystidia absent.
Odour not distinctive. **Taste** not distinctive.
Chemical tests none.
Occurrence throughout the year, but sporulating summer to autumn; infrequent.
■ Inedible.

Meripilus giganteus (Pers.: Fr.) Karst.   **Coriolaceae**
= *Polyporus giganteus* Pers.: Fr.
= *Grifola gigantea* (Pers.: Fr.) Pil.

Massive compound rosette of soft, brown, fan-shaped caps with
pores on the under-surface, arising from a common base, annual;
arising at the extreme base of broad-leaf trees and stumps, and often
from shallowly submerged roots running some distance from the
trunk, favouring beech but also with oak.

Dimensions individual caps 10 - 30 cm dia x 1 - 2 cm thick;
collectively 50 - 80 cm dia.
Fruiting body upper surface concentrically zoned, light and dark
brown, radially grooved, more or less flat, covered in soft, fine brown
scales; caps arising on short stems from a common more pallid
brown base. Flesh white, fibrous and soft.
Pores whitish, bruising black, more or less rounded, 3-5 per mm.
Tubes whitish or cream, 4-6 mm deep. **Spores** hyaline, smooth, sub-
spherical, non-amyloid, with droplets, 5.5-6.5 x 4.5-5.5 µm. Basidia
4-spored. Cystidia absent.
Odour not distinctive. **Taste** slight, acidic.
Chemical tests none.
Occurrence summer to late autumn; frequent.
■ Inedible

Oxyporus populinus (Schum.: Fr.) Donk   **Coriolaceae**
= *Fomes populinus* (Schum.: Fr.) Karst.
= *Fomes connatus* (Weinm.) Gill.

Small to medium-sized pale brown bracket with pores on the under-
surface, perennial; arranged in tiers, parasitic on trunks and stumps
of broad-leaf trees, favouring wound sites and knot holes.

Dimensions 3 - 6 cm dia x 1 - 4 cm thick.
Fruiting body upper surface cream or ochraceous-brown and with
algal growth in older specimens, the margin whitish during growth,
slightly downy or smooth, fan-shaped, irregular, broadly attached;
sessile. Flesh when damp whitish and soft; when dry white and
tough.
Pores whitish, circular or slightly angular, 4-7 per mm. Tubes at first
whitish, becoming pallid yellow, arranged in tiers, each tier 2-4 mm
deep. **Spores** hyaline, smooth, sub-spherical, non-amyloid, 3.5-4.5 x
3-4.5 µm. Basidia 4-spored. Cystidia clavate and encrusted at the
apices.
Odour slight, fungoid. **Taste** not distinctive.
Chemical tests none.
Occurrence throughout the year, but sporulating late summer to
autumn; infrequent.
■ Inedible.

Physisporinus sanguinolentus (A. & S.: Fr.) Pil.
= *Rigidoporus sanguinolentus* (Fr.) Donk   **Coriolaceae**

Encrusting creamy-white fruiting body, porous and uneven, bruising
red then brown, annual; spreading over damp dead wood of broad-
leaf and coniferous trees.

Dimensions variable dia x 0.3 - 1 cm thick.
Fruiting body resupinate, creamy-white, bruising red then brown,
somewhat uneven or knobbly, with distinct margin. Flesh creamy-
white, waxy, cartilaginous when damp, hard when dry.
Pores concolorous, small, rounded or angular, 3-5 per mm. Tubes
concolorous, 1-2 mm deep. **Spores** hyaline, smooth, sub-spherical,
non-amyloid, with droplets, 5-6 µm. Basidia 2- or 4-spored. Cystidia
absent but with inconspicuous cystidia-like fusiform hyphal ends.
Odour not distinctive. **Taste** not distinctive.
Chemical tests none.
Occurrence throughout the year, but sporulating summer to autumn;
frequent.
■ Inedible.

Piptoporus betulinus (Bull.: Fr.) Karst.   Coriolaceae
= *Polyporus betulinus* Bull.: Fr.   Razor Strop Fungus

Largish pale grey-brown, rather rounded, annual bracket with white pores on the under-surface, annual but fruiting bodies persistent; arising singly but often with several on the same host, restricted to dead birch trunks.

Dimensions 10 - 20 cm dia x 2 - 6 cm thick.
Fruiting body upper surface pallid brownish-grey, smooth, often crazed, at first sub-spherical then convex fan- or hoof-shaped with a thick margin; sessile. Flesh white, rubbery or corky.
Pores white, becoming pallid grey-brown with age, circular or angular, 3-4 per mm. Tubes white, 1.5-5 mm deep. **Spores** hyaline, smooth, allantoid, non-amyloid, some with 2 droplets, 5-7 x 1.5-2 µm. Basidia 2- or 4-spored. Cystidia absent.
Odour strong, fungoid but pleasant. **Taste** bitter.
Chemical tests none.
Occurrence throughout the year, but sporulating late summer to autumn; very common.
■ Inedible.

Postia caesia (Schrad.: Fr.) Karst.   Coriolaceae
= *Tyromyces caesius* (Schrad.: Fr.) Murr.

Small whitish bracket, becoming progressively more blue-tinged with age, pore-bearing under-surface, annual; in limited tiers on dead and rotten wood of pines and, very occasionally, broad-leaf trees.

Dimensions 1 - 6 cm dia x 0.2 - 1 cm thick.
Fruiting body at first whitish, becoming increasingly blue with age, occasionally with brownish tinge, upper surface finely hairy, with slight concentric zonation and radial wrinkling, margin somewhat wavy; sessile, narrowly attached. Flesh whitish, elastic and tough.
Pores whitish, small, at first rounded or slightly angular, lacerate on vertical surfaces, 4-6 per mm. Tubes concolorous, 4-6 mm deep.
Spores hyaline, smooth, cylindrical-ellipsoid or allantoid, amyloid (only in the mass), with droplets, 4-5 x 1.5-2 µm. Basidia 4-spored. Cystidia absent.
Odour not distinctive. **Taste** not distinctive.
Chemical tests none.
Occurrence throughout the year, but sporulating in late autumn; frequent.
■ Inedible.

Postia stiptica (Pers.: Fr.) Jül.   Coriolaceae
= *Tyromyces stipticus* (Pers. : Fr.) Kotl. & Pouz.

Triangular white bracket with soft juicy flesh, finely bristled upper surface and pore bearing under-surface, annual; in limited tiers on dead and rotten wood of pines and, very occasionally, broad-leaf trees.

Dimensions 3 - 12 cm dia x 1.5 - 4 cm thick.
Fruiting body white, not blueing with age, upper surface sometimes bristly tomentose, with slight concentric zonation and radial wrinkling, margin somewhat wavy; sessile, broadly attached. Flesh white, fibrous and soft.
Pores white, small, at first rounded or slightly angular, lacerate on vertical surfaces, 3-4 per mm. Tubes concolorous, 6-10 mm deep.
Spores hyaline, smooth, cylindrical-ellipsoid or allantoid, non-amyloid, droplets absent, 3.5-5 x 2-2.5 µm. Basidia 4-spored. Cystidia absent.
Odour not distinctive. **Taste** very bitter and astringent.
Chemical tests none.
Occurrence summer to autumn; infrequent.
■ Inedible.

Postia subcaesia (David) Jül. Coriolaceae
= *Tyromyces subcaesius* David

Small whitish bracket, becoming faintly blue-tinged with age, pore-bearing under-surface, annual; in limited tiers on dead and rotten wood of ash, beech and, less commonly, other broad-leaf trees.

Dimensions 1 - 6 cm dia x 0.2 - 0.6 cm thick.
Fruiting body at first whitish, with ochraceous and faint blue tinges with age, upper surface finely hairy, with slight concentric zonation and radial wrinkling, margin somewhat wavy; sessile, narrowly attached. Flesh whitish, elastic and tough.
Pores whitish, small, at first rounded or slightly angular, lacerate on vertical surfaces, 4-6 per mm. Tubes concolorous, 2-5 mm deep.
Spores hyaline, smooth, cylindrical-ellipsoid or allantoid, non-amyloid, with droplets, 4.5-5.5 x 1-1.3 μm. Basidia 4-spored. Cystidia absent.
Odour not distinctive. **Taste** not distinctive.
Chemical tests none.
Occurrence throughout the year, but sporulating in the autumn; infrequent.
■ Inedible.

Pseudotrametes gibbosa (Pers.) Bond. & Sing. Coriolaceae
= *Trametes gibbosa* (Pers.) Fr.

Medium-sized whitish bracket, often greenish due to algal growth on the upper surface, with white slot-like pores on the under-surface, annual; in limited tiers on dead broad-leaf trees favouring beech. The fruiting bodies are frequently attacked by boring beetle larvae which leave considerable granular deposits in the vicinity.

Dimensions 5 - 20 cm dia x 1 - 8 cm thick.
Fruiting body greyish-white or with yellow tinge, semicircular, slightly convex, upper surface at first downy then more or less smooth, with distinct, thick, even margin; sessile. Flesh white, thick, at first soft, becoming tough and very hard when dry.
Pores creamy-white, elongated slot-like in a radial arrangement, 1-2 per mm. Tubes creamy-white, 5-15 mm deep. **Spores** hyaline, smooth, cylindrical or allantoid, non-amyloid, 4-5.5 x 2-2.5 μm. Basidia 4-spored. Cystidia absent.
Odour not distinctive. **Taste** not distinctive.
Chemical tests none.
Occurrence throughout the year, but sporulating spring to early summer; common.
■ Inedible.

Rigidoporus ulmarius (Sow.: Fr.) Imazeki Coriolaceae
= *Fomes ulmarius* (Sow.: Fr.) Gill.

Very large pale bracket, with pinkish-orange or brownish pore-bearing under-surface, perennial; often overlapping in limited tiers, at the base of living trunks and stumps of broad-leaf trees favouring elm.

Dimensions 12 - 50 cm dia x 4 - 8 cm thick.
Fruiting body pallid dirty-white to buff, becoming dirty-brown in older specimens, upper surface typically supporting heavy algal or moss growth, at first finely downy, concentrically ridged, very knobbly, with thick obtuse margin; sessile. Flesh cream or buff, at first fibrous, then very tough and hard, difficult to break.
Pores orange, fading to clay or buff with age, rounded, 5-8 per mm. Tubes pinkish or orange becoming brown with age, in tiers separated by a thin band of white tissue, 1-5 mm deep in each tier.
Spores hyaline or pallid yellow, smooth, spherical, non-amyloid, 6-7.5 μm. Basidia 2- or 4-spored. Cystidia absent.
Odour not distinctive. **Taste** not distinctive.
Chemical tests none.
Occurrence throughout the year, but sporulating late summer and autumn; once common, now rarer.
■ Inedible.

Skeletocutis amorpha (Fr.) Kotl. & Pouz.   **Coriolaceae**
= *Gloeoporus amorphus* (Fr.) Clem. & Shear

Encrusting or slightly bracket-like, whitish becoming more cream with age, with pores on the under-surface (exposed when encrusting), annual; on dead and rotten wood of coniferous trees.

Dimensions 2 - 4 cm dia x 0.1 - 0.2 cm thick.
Fruiting body upper surface whitish becoming cream, at first minutely downy then smooth, delicate; sessile. Flesh whitish-cream, elastic or rubbery, 2-layered.
Pores whitish, becoming cream or with orange tinge, small, rounded or slightly angular, 3-4 per mm. Tubes whitish, very short, 0.5-1 mm long. **Spores** hyaline, smooth, cylindrical or allantoid, non-amyloid, 3-4 x 1-2 μm. Basidia 4-spored. Cystidia fusiform.
Odour not distinctive. **Taste** not distinctive.
Chemical tests none.
Occurrence throughout the year, but sporulating late summer and autumn; rare. (Risbeth Wood, near Thetford, Norfolk)
■ Inedible.

Skeletocutis nivea (Jungh.) Keller   **Coriolaceae**
= *Incrustoporia semipileata* (Peck.) Murr.
= *Tyromyces semipileatus* (Peck.) Murr.
= *Leptotrimitus semipileatus* (Peck.) Pouz.

Small compact bracket, whitish becoming tinged brown on the upper surface, pore-bearing under-surface, annual; on branches of broad-leaf trees.

Dimensions 2 - 4 cm dia x 0.2 - 2 cm thick.
Fruiting body upper surface whitish, becoming infused brown from the base towards the margin which tends to remain white and distinct, distinctively narrow, at first minutely downy, then smooth or pitted and warty; sessile. Flesh white or pallid brown, corky and tough.
Pores whitish becoming cream, very small (lens required), rounded or slightly angular, 6-8 per mm. Tubes whitish, very short, 1-4 mm deep. **Spores** hyaline, smooth, allantoid, non-amyloid, 3-4 x 0.5-1 μm. Basidia 4-spored. Cystidia absent.
Odour not distinctive. **Taste** not distinctive.
Chemical tests none.
Occurrence throughout the year, but sporulating late summer and autumn; frequent.
■ Inedible.

Trametes suaveolens (L.: Fr.) Fr.   **Coriolaceae**

Medium-sized fan-shaped bracket, whitish-cream throughout, with angular pore-bearing under-surface, annual; in limited tiers on living or dead wood of broad-leaf trees, including willow and poplar.

Dimensions 6 - 12 cm dia x 1.5 - 3.5 cm thick.
Fruiting body upper surface whitish, zoned greyish with age, sometimes with green tinge (algal), finely downy, margin sharp, slightly undulating; sessile. Flesh white and tough.
Pores white, becoming cream or buff with age, angular or elongated slot-like, 1-2 per mm. Tubes concolorous, 10-15 mm deep.
Spores hyaline, smooth, cylindrical or slightly allantoid, non-amyloid, 7-11 x 3-4 μm. Basidia 4-spored. Cystidia absent.
Odour strong, of aniseed (when fresh). **Taste** not distinctive.
Chemical tests none.
Occurrence throughout the year, but sporulating in autumn; rare. (Bignor, West Sussex)
■ Inedible.

Trichaptum abietinum (Fr.) Ryv.   **Coriolaceae**
= *Hirschioporus abietinus* (Dicks.: Fr.) Donk
=*Polystictus abietinus* (Dicks.: Fr.) Fr.

Small encrusting bracket, greyish-white with deep violet, pore-bearing under-surface, annual; in tiers on dead and rotten wood of coniferous trees favouring spruce.

Dimensions 1 - 3 cm dia x 0.1 - 0.2 cm thick.
Fruiting body upper surface whitish, zoned, with pinkish and slightly undulating margin, sometimes with green tinge (algal), woolly, hairy; sessile, broadly attached. Flesh pallid purple or brown, elastic and tough.
Pores at first deep violet or purple, becoming more pallid and brownish with age, small, at first rounded or slightly angular then more maze-like, 3-5 per mm. Tubes concolorous, 0.5-0.8 mm deep.
Spores hyaline, smooth, cylindrical or allantoid, non-amyloid, 6-8 x 2-3.5 µm. Basidia 4-spored. Cystidia clavate or fusiform, sometimes encrusted at the apices.
Odour not distinctive. **Taste** not distinctive.
Chemical tests none.
Occurrence throughout the year but sporulating in late autumn; common.
■ Inedible.

Trichaptum fuscoviolaceum (Ehrenb.: Fr.) Ryv.
= *Hirschioporus fuscoviolaceus* (Ehrenb.: Fr.) Donk   **Coriolaceae**

Small encrusting bracket, greyish-white with deep violet, spiny and porous under-surface, annual; in dense overlapping tiers on dead and rotten wood of coniferous trees favouring pine.

Dimensions 1 - 3 cm dia x 0.1 - 0.2 cm thick.
Fruiting body upper surface whitish, zoned with pinkish, slightly undulating margin, sometimes with green tinge (algal), woolly, hairy; sessile, broadly attached. Flesh pallid purple or brown, elastic and tough.
Spines at first deep violet or purple becoming more pallid and brownish with age, tooth-like with angular porous cavities between, spines 0.01-0.05 mm long. Tubes concolorous, 0.5-0.8 mm deep.
Spores hyaline, smooth, cylindrical or allantoid, non-amyloid, 6-8.5 x 2.5-3 µm. Basidia 4-spored. Cystidia clavate or fusiform, sometimes encrusted at the apices.
Odour not distinctive. **Taste** not distinctive.
Chemical tests none.
Occurrence throughout the year, but sporulating in autumn; infrequent or rare . (Thetford Forest at Mundford, Norfolk)
■ Inedible.
This species differs from *T. abietinum* in the presence of spines on the hymenial surface (lens required) and in the choice of substrate.

Tyromyces chioneus (Fr.: Fr.) Karst.   **Coriolaceae**

Smallish white bracket, with brittle juicy flesh, undulating upper surface and pore-bearing under-surface, annual; in tiers on dead and rotten wood of birch, oak and hazel.

Dimensions 3 - 7 cm dia x 1 - 2 cm thick.
Fruiting body upper surface white, yellowish with age, smooth with slight concentric zonation and radial wrinkling, margin somewhat wavy; sessile, narrowly attached. Flesh white, brittle and juicy.
Pores white, small, at first rounded or slightly angular, 3-4 per mm. Tubes concolorous, 4-8 mm deep. **Spores** hyaline, smooth, allantoid, non-amyloid, 4-5 x 1.5-2 µm. Basidia 4-spored. Cystidia absent.
Odour not distinctive. **Taste** very bitter and astringent.
Chemical tests none.
Occurrence summer to autumn; infrequent.
■ Inedible.

Tyromyces kymatodes (Rostk.) Donk   **Coriolaceae**

Small white bracket, with finely velvety upper surface and pore-bearing under-surface, annual; in tiers on dead and rotten wood of coniferous trees.

Dimensions 2 - 5 cm dia x 1.5 - 2.5 cm thick.
Fruiting body upper surface white, not blueing with age, sometimes downy, with slight concentric zonation and radial wrinkling, margin somewhat wavy, narrow, broadly attached, sessile. Flesh white, then brownish, fibrous and tough.
Pores white, small, angular, lacerate on vertical surfaces, 3-5 per mm. Tubes concolorous, 3-4 mm deep. **Spores** hyaline, smooth, cylindrical-ellipsoid or allantoid, non-amyloid, 4-5.5 x 2-3.5 µm. Basidia 4-spored. Cystidia awl-shaped, sometimes encrusted.
Odour not distinctive. **Taste** not distinctive.
Chemical tests none.
Occurrence summer to autumn; infrequent.
■ Inedible.

Ganoderma adspersum (Schulz.) Donk **Ganodermataceae**
= *Ganoderma europaeum* Steyaert

Very large grey-brown bracket with knobbly appearance, underside pore-bearing whitish, tough, perennial; in limited overlapping tiers, parasitic on trunks of broad-leaf trees, favouring beech, often on the lower parts of the trunks.

Dimensions 10 - 60 cm dia x 3 - 30 cm thick.
Fruiting body greyish-brown, often discoloured reddish-brown or cocoa-brown from deposited spores, more or less flattened, radially wavy or wrinkled and concentrically grooved and zoned, broadly attached; sessile. Flesh cinnamon-brown, much thicker than the tube region, very tough and fibrous.
Pores pallid ochraceous, bruising brown, circular, 3-4 per mm. Tubes brown, in annual layers, 7-20 mm deep. **Spores** brown, warty, broadly ellipsoid, flattened at one end with hyaline germ pore, non-amyloid, 8.5-12 x 6-8 µm. Basidia 4-spored. Cystidia absent.
Odour fungoid. **Taste** bitter.
Chemical tests none.
Occurrence throughout the year; very common.
■ Inedible.
Not to be confused with the rarer *G. applanatum* in which the flesh is thinner.

Ganoderma applanatum (Pers.: Wallr.) Pat. **Ganodermataceae**

Very large grey-brown bracket with knobbly appearance, underside pore-bearing whitish, tough, perennial; in limited overlapping tiers, parasitic on trunks of broad-leaf trees, favouring beech.

Dimensions 10 - 60 cm dia x 2 - 8 cm thick.
Fruiting body greyish-brown, often discoloured reddish-brown or cocoa-brown from deposited spores, more or less flattened, radially wavy or wrinkled and concentrically grooved and zoned, broadly attached; sessile. Flesh cinnamon-brown, thinner than the tube region, very tough and fibrous.
Pores white, bruising brown, circular, 4-5 per mm. Tubes brown, in annual layers, 7-20 mm deep. **Spores** brown, warty, broadly ellipsoid, flattened at one end with hyaline germ pore, non-amyloid, 7-9 x 4.5-6 µm. Basidia 4-spored. Cystidia absent.
Odour fungoid. **Taste** bitter.
Chemical tests none.
Occurrence throughout the year; infrequent.
■ Inedible.
Often confused with *G. adspersum* which is more common but in which the flesh is much thicker.

Ganoderma lucidum (Curt.: Fr.) Karst.   **Ganodermataceae**

Large mahogany-brown kidney-shaped bracket with lacquered appearance, underside pore-bearing, whitish, tough, arising from lateral stem, annual; solitary or in small groups, on stumps of broad-leaf trees, favouring oak.

Dimensions 4 - 30 cm dia x 2 - 4 cm thick.
Fruiting body reddish brown, shiny, margin pallid when young, more or less flattened, radially wavy or wrinkled and concentrically grooved and zoned, attached by stout lateral stem. Flesh cinnamon-brown, tough and fibrous.
Pores white, bruising brown, becoming brown with age, circular, 3-4 per mm. Tubes concolorous with pores, 5-20 mm deep.
Spores pallid brown, warty, broadly ellipsoid, truncated at one end with hyaline germ pore, non-amyloid, 10-13 x 6-8 μm. Basidia 4-spored. Cystidia absent.
Odour fungoid. **Taste** bitter.
Chemical tests none.
Occurrence summer to late autumn; rare.
■ Inedible.
Note: when dry the fruiting body is very light in weight.

Ganoderma resinaceum Boud.: Pat.   **Ganodermataceae**

Large orange-brown bracket with yellowish margin and undulating appearance, underside pore-bearing, whitish, soft when young, blackening, perennial; solitary or in limited overlapping tiers, parasitic at the base of trunks of broad-leaf trees, favouring oak.

Dimensions 15 - 35 cm dia x 4 - 8 cm thick.
Fruiting body orange-brown, becoming more copper, dark brown and finally black, margin yellowish when young, more or less flattened, undulating, smooth, exuding a resin which dries eventually to a black crust; sessile, broadly attached. Flesh alternating light and mid-brown concentric zones, relatively thin, tough and corky.
Pores white when young, turning brown where bruised and with age, circular, 2-3 per mm. Tubes concolorous, in indistinct annual layers, each 8-10 mm deep. **Spores** brown, warty, broadly ellipsoid, truncated at one end with hyaline germ pore, non-amyloid, 10-13 x 7-9 μm. Basidia 4-spored. Cystidia absent.
Odour spicy. **Taste** bitter.
Chemical tests none.
Occurrence throughout the year but sporulating late summer to autumn; rare.
■ Inedible.
When the flesh is cut the resin discharging hardens rapidly; also melts in a flame.

Fistulina hepatica Schaeff.: Fr.   **Fistulinaceae**
Beefsteak Fungus

Large blood-red bracket, reminiscent of raw meat, soft, shaped like a broad tongue, under-surface reddish-brown, pore-bearing, annual; in limited tiers, parasitic, typically on the lower parts of trunks of oak and sweet chestnut. Causes brown rot and generates a valued 'Brown Oak' used in cabinet-making.

Dimensions 10 - 25 cm dia x 2 - 6 cm thick.
Fruiting body at first pinkish, becoming blood-red and finally brownish-purple, more or less ligulate, flattened, slightly convex or concave, radially furrowed, very finely warty, tacky; more or less sessile. Flesh pinkish-buff or darker, with more pallid veining, oozing blood-red juice, soft.
Pores at first whitish, becoming rust, circular, 2-3 per mm. Tubes yellowish white, separable, 10-15 mm deep. **Spores** pallid yellow, smooth, sub-spherical or ovoid, non amyloid, 1 droplet, 5-6 x 3.5-4.5 μm. Basidia 4-spored. Cystidia absent.
Odour not distinctive. **Taste** acidic.
Chemical tests none.
Occurrence summer to autumn; common.
□ Edible.

Coltricia perennis (L.: Fr.) Murr.   Hymenochaetaceae
= *Polystictus perennis* (L.: Fr.) Quél.

More or less circular brownish cap, concentrically zoned, looking like the end of a small cut stump, under-surface brownish pore-bearing, on short stem, annual; in small trooping groups, favouring sandy soil, in woods and on heaths, typically by pathsides.

Dimensions cap 2 - 8 cm dia; stem 0.2 - 1 cm dia x 1.3 - 3.5 cm tall.
Fruiting body upper surface zoned concentrically in shades of ochre, grey and rust with maroon tinge, disc-like, at first finely downy, becoming smooth with age; stem rusty-brown, downy, more or less central. Flesh brown, thin, corky, harder when dry.
Pores brown, covered at first with whitish bloom. Tubes umber-brown, decurrent, 0.5-3 mm deep. **Spores** ochre-brown, smooth, ellipsoid, cyanophilic, non-amyloid, 6-7.5 x 4-4.5 μm. Basidia 2- or 4-spored. Cystidia absent. Setae absent.
Odour not distinctive. **Taste** not distinctive.
Chemical tests none.
Occurrence throughout the year but sporulating summer and autumn; frequent.
■ Inedible.

Hymenochaete corrugata Fr.: Fr.   Hymenochaetaceae

Pale encrusting fruiting body, typically cracked when mature; on dead wood of broad-leaf trees, favouring hazel.

Dimensions 3 - 8 cm dia x 0.1 - 0.2 cm thick.
Fruiting body greyish-white, drying more pallid, resupinate, tightly attached to the substrate, smooth, irregularly warty, becoming crazed; sessile. Flesh greyish, hard and brittle. **Spores** hyaline, smooth, elongated ellipsoid or allantoid, non-amyloid, 4-6.5 x 1.5-2.5 μm. Basidia 4-spored. Cystidia absent. Setae elongated-cylindrical, 40-70 x 6-10 μm.
Odour not distinctive. **Taste** not distinctive.
Chemical tests none.
Occurrence throughout the year but sporulating summer to autumn; infrequent.
■ Inedible.
Note: *H. corrugata* is distinctive in that it can fuse small twigs together.

Hymenochaete rubiginosa (Dicks.: Fr.) Lév.
Hymenochaetaceae

Small chestnut bracket, under-surface brownish, annual; in over-lapping tiers, on dead wood favouring barkless oak.

Dimensions 1 - 4 cm dia x 0.3 - 0.7 cm thick.
Fruiting body upper surface chestnut-brown, more pallid at the margin, reflexed or more or less resupinate, attached to the substrate at the centre, downy, wavy, zoned and ridged concentrically; lower (hymenial) surface at first orange-brown then dark brown, with more pallid margin, finely warty; sessile. Flesh dull brown and leathery. **Spores** hyaline, smooth, ellipsoid, non-amyloid, 4.5-6.5 x 2.5-3 μm. Basidia 4-spored. Cystidia absent. Setae dark brown awl-shaped, 40-60 x 5-7 μm.
Odour not distinctive. **Taste** not distinctive.
Chemical tests none.
Occurrence throughout the year but sporulating in autumn; common.
■ Inedible.

Inonotus cuticularis (Fr.) Karst.   **Hymenochaetaceae**

Large, velvety, reddish-ochre or brown bracket, with yellowish pore-bearing or brown under-surface, annual; often solitary or in limited tiers, parasitic on trunks and felled timber, mainly beech and oak. Causes white soft rot.

Dimensions 6 - 22 cm dia x 2 - 5 cm thick.
Fruiting body upper surface at first ochre-brown; then reddish or umber-brown, and finally black in old specimens, felty or downy, becoming more bristly with age, undulating; sessile. Flesh pallid brown, thick, soft and spongy, tougher when dry.
Pores at first pallid ochraceous, later more yellowish-brown, with greenish tinge, angular, 2-4 per mm. Tubes concolorous, 5-15 mm deep. **Spores** rust, smooth, broadly ellipsoid, 6.5-7 x 4.5-5.5 µm. Basidia 4-spored. Cystidia absent. Setae brown, broadly awl-shaped, angled, 15-25 x 6-9 µm.
Odour not distinctive. **Taste** not distinctive.
Chemical tests none.
Occurrence throughout the year, but sporulating in autumn; rare. (Stourhead, Wiltshire)
■ Inedible.

Inonotus dryadeus (Pers.: Fr.) Murr.   **Hymenochaetaceae**
= *Polyporus dryadeus* Pers.

Large corky bracket, ageing through grey, brown and black, with pale grey pore-bearing under-surface, annual; often solitary, parasitic on oak, at the base of the trunk.

Dimensions 5 - 25 cm dia x 2 - 12 cm thick.
Fruiting body upper surface at first pallid grey, becoming brown, more pallid at the margin, and finally black in old specimens; thick, lumpy margin, broadly rounded and exuding reddish-yellow guttation droplets whilst actively growing; sessile. Flesh yellowish or rust, fibrous and soft.
Pores whitish-grey, becoming rust, at first circular then angular, 3-5 per mm. Tubes dull brown, 1-3 tiers, 15-20 mm deep.
Spores pallid yellow, smooth, sub-spherical, non-amyloid, cyanophilic, 7-8.5 x 5.5-6.5 µm. Basidia 4-spored. Cystidia absent. Setae brown, awl-shaped with hooked tip, 20-35 x 10-14 µm.
Odour not distinctive. **Taste** not distinctive.
Chemical tests none.
Occurrence throughout the year, but sporulating late summer and autumn; Infrequent.
■ Inedible.

Inonotus hispidus (Bull.: Fr.) Karst.   **Hymenochaetaceae**

Large, velvety, reddish-ochre or brown bracket, with yellowish or brown pore-bearing under-surface, annual; often solitary or in limited tiers, parasitic on trunks of ash but also appearing on other broad-leaf trees. Causes serious but limited decay.

Dimensions 6 - 25 cm dia x 2 - 10 cm thick.
Fruiting body upper surface at first ochre-brown, then cigar-brown, and finally black in old specimens, felty or downy, becoming more bristly with age and finally more or less smooth; sessile. Flesh yellowish or rust, thick, soft, spongy, tougher when dry.
Pores at first pallid ochraceous, later dull brown, angular, 2-3 per mm. Tubes ochraceous or rust, 1-4 tiers, 10-30 mm deep.
Spores rust, smooth, sub-spherical, thick-walled, non-amyloid, with droplets, 7-11 x 6-9 µm. Basidia 4-spored. Cystidia absent. Setae brown, broadly awl-shaped, 20-30 x 9-10 µm.
Odour not distinctive. **Taste** not distinctive.
Chemical tests none.
Occurrence throughout the year, but sporulating late summer and autumn; common.
■ Inedible.

Inonotus radiatus (Fr.) Karst.   Hymenochaetaceae

Smallish, velvety, reddish-brown bracket, with yellowish border and pore-bearing under-surface, annual; in limited tiers, parasitic on trunks of alder, occasionally on other trees.

Dimensions 3 - 10 cm dia x 1 - 2 cm thick.
Fruiting body upper surface at first rust-brown with yellow margin, then cigar-brown and finally blackish-brown in old specimens, felty or downy, becoming smooth with age, radially wrinkled and concentrically zoned, often producing yellowish guttation drops when actively growing; sessile. Flesh rust, thick, soft, spongy, tougher when dry.
Pores at first whitish, later dull grey-brown, finally dark brown, more or less rounded, 2-4 per mm. Tubes concolorous with pores, 6-10 mm deep. **Spores** yellowish, smooth, broadly ellipsoid, non-amyloid, 4.5-5.5 x 3.5-4.5 μm. Basidia 4-spored. Cystidia absent. Setae brown, irregularly shaped, some hook-like, 20-30 x 3-8 μm.
Odour not distinctive. **Taste** not distinctive.
Chemical tests none.
Occurrence throughout the year, but sporulating late summer and autumn; rare. (Culbin Forest, Morayshire)
■ Inedible.

Phaeolus schweinitzii (Fr.) Pat.   Hymenochaetaceae
= *Polyporus schweinitzii* Fr.

Brownish yellow bracket, fan, or fused cushions, with yellowish pore-bearing under-surface, annual; often in limited overlapping tiers, parasitic on coniferous trees typically arising from the roots. Causes brown cubical decay in older trees.

Dimensions cap 10 - 30 cm dia x 1 - 2 cm thick; stem 2 - 6 cm long.
Fruiting body upper surface dark sulphur-yellow, becoming rust or darker brown, blackish with age, the margin tending to remain yellowish; more or less flat becoming slightly concave, undulating, concentrically zoned, at first downy, then irregularly warty-hairy, finally smooth; stem dark brown, very short, stout, attenuated downwards, cylindrical or laterally compressed, sometimes fusing with others. Flesh yellowish-brown or rust, fibrous and soft.
Pores olivaceous-yellow, then brown, glistening, 0.3-2.5 mm dia. Tubes concolorous, decurrent, 3-6 mm deep. **Spores** hyaline, tinged yellow, smooth, ellipsoid, non-amyloid, 5-8.5 x 3.5-4.5 μm. Basidia 4-spored. Cystidia clavate or conical, with brown droplets. Setae absent.
Odour not distinctive. **Taste** not distinctive.
Chemical tests none.
Occurrence summer to autumn; infrequent.
■ Inedible.

Phellinus contiguus (Pers.: Fr.) Bourd. & Galz.
Hymenochaetaceae

Light brown encrustation with obvious pore-bearing upper surface, perennial; on dead wood of broad-leaf and coniferous trees. Causes destructive white rot.

Dimensions variable dia x 0.5 - 1.5 cm thick.
Fruiting body light brown with more pallid fibrous margin, ageing rust, resupinate, effused, the distinctive pores incomplete at the margin. Flesh concolorous and corky.
Pores concolorous, large, circular or angular, 2-3 per mm. Tubes rust, single tier, 5-10 mm deep. **Spores** hyaline, smooth, ellipsoid, non-amyloid, 6-7 x 3-3.5 μm. Basidia 4-spored. Cystidia absent. Setae brown, awl-shaped, 50-60 x 10-15 μm.
Odour not distinctive. **Taste** not distinctive.
Chemical tests none.
Occurrence throughout the year but sporulating in the autumn; infrequent.
■ Inedible.

Phellinus ferreus (Pers.) Bourd. & Galz. **Hymenochaetaceae**
= *Polyporus ferreus* Pers.

Small cinnamon-brown cushion, perennial; on dead branches of broad-leaf trees, favouring hazel. Causes serious white rot which destroys the wood.

Dimensions 5 - 10 cm dia x 0.5 - 1 cm thick.
Fruiting body cinnamon-yellow, ageing rust, resupinate, effused, smooth or irregularly wavy, porous. Flesh rust, soft when damp, harder when dry.
Pores rust or cinnamon, circular or angular, 4-6 per mm. Tubes rust, 4-5 tiers, 2-5 mm deep. **Spores** hyaline, smooth, narrowly cylindrical, non-amyloid, 6-7.5 x 2-2.5 μm. Basidia 4-spored. Cystidia absent. Setae dark brown, awl-shaped, 25-30 x 6-7 μm.
Odour not distinctive. **Taste** not distinctive.
Chemical tests none.
Occurrence throughout the year but sporulating in the autumn; common.
■ Inedible.

Phellinus ferruginosus (Schrad.: Fr.) Pat.
= *Polyporus ferruginosus* Pers.   **Hymenochaetaceae**

Thin cinnamon-brown encrustation with pore-bearing upper surface, annual but overwintering; on logs and dead branches of broad-leaf trees, favouring beech.

Dimensions variable dia x 0.1 - 0.5 cm thick.
Fruiting body dull cinnamon-brown, resupinate, effused, smooth or irregularly wavy, porous. Flesh concolorous, thin, soft when damp, harder when dry.
Pores rust or cinnamon, circular, 5-6 per mm. Tubes rust, 2-5 mm deep. **Spores** hyaline, smooth, ellipsoid, non-amyloid, 4.5-5.5 x 3-3.5 μm. Basidia 4-spored. Cystidia absent. Setae dark brown, awl-shaped, 25-30 x 6-7 μm.
Odour not distinctive. **Taste** not distinctive.
Chemical tests none.
Occurrence throughout the year but sporulating in the autumn; frequent.
■ Inedible.

Phellinus igniarius (L.: Fr.) Quél.   **Hymenochaetaceae**
= *Fomes igniarius* (L.: Fr.) Gill.

Large, hoof-shaped, greyish-brown bracket, under-surface cinnamon-brown, pore-bearing, perennial: sometimes in limited tiers, parasitic on broad-leaf trees, favouring willow. Causes extensive white rot.

Dimensions 10 - 40 cm dia x 5 - 20 cm thick.
Fruiting body upper surface at first rust, quickly becoming greyish and ageing black, flattened or irregularly convex above, at first finely downy, then smooth, later cracked, margin more pallid, clearly demarcated and obtuse (in older specimens the upper surface frequently develops algal or moss growth); sessile. Flesh rust, tough and brittle.
Pores rust or cinnamon, circular, 4-6 per mm. Tubes rust, indistinctly layered but forming annual tiers, each 4-7 mm deep. **Spores** hyaline, smooth, sub-spherical, non-amyloid, 5.5-7 x 4.5-6 μm. Basidia 4-spored. Cystidia absent. Setae dark brown, awl-shaped, 12-20 x 5-9 μm.
Odour strong, fungoid. **Taste** acidic or bitter.
Chemical tests none.
Occurrence January - December; infrequent.
■ Inedible.

Phellinus punctatus (Fr.) Pil.   **Hymenochaetaceae**

Cinnamon-brown cushion, with pore-bearing upper surface, typically growing along the longitudinal axis of the branch or trunk, perennial; on dead wood of broad-leaf trees, favouring hazel and willow. Causes white rot which destroys the wood.

Dimensions 5 - 10 cm dia x 1 - 2 cm thick.
Fruiting body chestnut or dull brown, resupinate, smooth, finely porous, margin sterile. Flesh rust and woody.
Pores rust or cinnamon, circular or angular, 5-7 per mm. Tubes rust, 2-4 tiers, 1-3 mm deep. **Spores** hyaline, smooth, sub-spherical or ellipsoid, non-amyloid, cyanophilic, with droplets, 7-9 x 6-7.5 μm. Basidia 4-spored. Cystidioles with long hair-like extensions. Setae absent.
Odour not distinctive. **Taste** not distinctive.
Chemical tests none.
Occurrence throughout the year, but sporulating late summer and autumn; frequent.
■ Inedible.

Phellinus tuberculosus (Baumg.) Niemelä
= *Phellinus pomaceus* (Pers.: S F Gray) Maire   **Hymenochaetaceae**

Small, greyish-brown stubby bracket or cushion with pore-bearing under-surface, perennial; on trunks and undersides of attached branches of both living and dead *Prunus* species, favouring wild cherry and blackthorn. Causes serious white rot which destroys the wood.

Dimensions 3 - 10 cm dia x 1 - 2 cm thick.
Fruiting body upper surface dull grey with brown tinge (often developing algal growth), uneven, concentrically ridged, otherwise smooth porous; generally semi-pileate, less commonly resupinate. Flesh rust, soft when damp, harder when dry.
Pores cinnamon-brown, becoming grey-brown with age, circular, 4-5 per mm. Tubes rust, 2-5 tiers, 2-3 mm deep. **Spores** hyaline, smooth, sub-spherical or broadly ellipsoid, non-amyloid, 5-7 x 4-5.5 μm. Basidia 4-spored. Cystidia absent. Setae brown, fusiform, 12-25 x 6-10 μm.
Odour not distinctive. **Taste** not distinctive.
Chemical tests none.
Occurrence throughout the year but sporulating in the autumn; infrequent.
■ Inedible.

Cerocorticium confluens (Fr.: Fr.) Jül. & Stalpers
= *Radulomyces confluens* (Fr.) Christ.   **Hyphodermataceae**

Small or extensive, pale opalescent yellow, encrusting fruiting body, unevenly lumpy; solitary or in coalescent groups, on dead wood of broad-leaf trees, with or without bark, favouring beech.

Dimensions variable dia up to 40cm; 0.3 - 0.8 cm thick.
Fruiting body pallid ochraceous-grey, or cream with bluish tinge, almost opalescent, hygrophanous, drying pallid ochraceous and cracking, resupinate, lumpy-warty with finely fringed margin, lightly attached and, when dry, often partly detached from substrate. Flesh concolorous; when damp, thinnish and waxy; when dry, more brittle.
Spores hyaline, smooth, sub-spherical or broadly ellipsoid, sometimes with granular content, non-amyloid, 7.5-9 x 5-7.5 μm. Basidia 4-spored. Cystidia absent but with hyphal ends that may appear clavate or cylindrical cystidia-like.
Odour not distinctive. **Taste** not distinctive.
Chemical tests none.
Occurrence throughout the year, but sporulating in the autumn; frequent.
■ Inedible.

Cerocorticium molare (Chaill.: Fr.) Jül. **Hyphodermataceae**
= *Radulomyces molaris* (Chaill.: Fr.) Christ.

Yellowish, encrusting toothed cushion, annual; closely applied to attached dead branches of broaf-leaf trees favouring oak, typically high in the crown of the tree and only observed on the ground after felling or storm damage.

Dimensions 2 - 40 cm dia x 0.3 - 0.6 cm thick.
Fruiting body ochraceous, resupinate, at first as distinct rounded patches which later coalesce, decorated with spines which become shorter towards the downy margin. Flesh ochraceous; when damp, soft and waxy; when dry, hard and brittle. Spines concolorous, 2-4 mm long. **Spores** hyaline, smooth, broadly ellipsoid with lateral apiculus, non-amyloid, 8-11 x 5-7 μm. Basidia 4-spored. Cystidia absent.
Odour not distinctive. **Taste** not distinctive.
Chemical tests none.
Occurrence throughout the year, but sporulating spring and summer; rare. (Gower Peninsula, South Wales)
■ Inedible.

Cylindrobasidium evolvens (Fr.) Jül. **Hyphodermataceae**
= *Corticium evolvens* (Fr.) Fr.
= *Corticium laeve* Pers.

Whitish, encrusting fruiting body, tightly attached to substrate; on dead wood of broad-leaf and coniferous trees, favouring cut surfaces in log piles.

Dimensions 2 - 20 cm dia x 0.05 - 0.1 cm thick.
Fruiting body whitish-cream, later pinkish or yellowish tinges, with irregular but sharply defined white fibrous margin, resupinate, effused, tightly attached to the substrate but typically reflexed at the edges in older specimens, or forming delicate brackets when on vertical surfaces, hymenial (upper) surface becoming somewhat warty with age. Flesh damp, when whitish buff, soft and thin; when dry, crisp and cracking. **Spores** hyaline, smooth, pip-shaped or ovoid, contents granular, non-amyloid, 8-12 x 5-7.5 μm. Basidia 2- or 4-spored. Cystidia smooth, fusiform.
Odour not distinctive. **Taste** not distinctive.
Chemical tests none.
Occurrence throughout the year, but sporulating late summer and autumn; common.
■ Inedible.

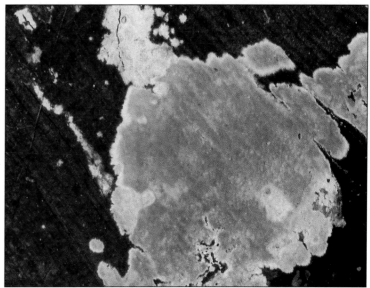

Hyphoderma praetermissum (Karst.) Erikss. & Strid
Hyphodermataceae

White, waxy fruiting body loosely attached to substrate, warty; on rotten wood of broaf-leaf and coniferous trees.

Dimensions variable dia x 0.05 - 0.1 cm thick.
Fruiting body white, with greyish or buff tinges, resupinate, with irregular margin, the hymenial (upper) surface having a waxy consistency, densely warty, decorated with minute teeth. Flesh white, thin, waxy and soft. **Spores** hyaline, smooth, cylindrical-ellipsoid, non-amyloid, 8-10 x 3.5-4.5 μm. Basidia 4-spored. Cystidia cylindrical or slightly capitate, the heads sometimes encrusted, intermingled with distinctive large vesicular cells.
Odour not distinctive. **Taste** not distinctive.
Chemical tests none.
Occurrence throughout the year, but sporulating in the autumn; infrequent.
■ Inedible.

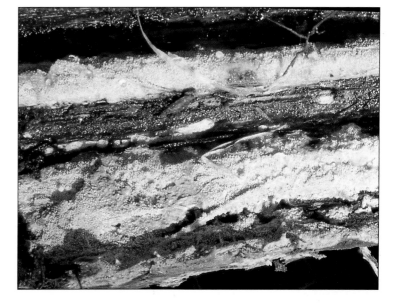

Hyphoderma radula (Fr.) Donk   Hyphodermataceae
= *Basidioradulum radula* (Fr.: Fr.) Nobles
= *Radulum orbiculare* Fr.

White and yellow, encrusting patches, tightly attached, warty or toothed; on dead wood of broaf-leaf and, less frequently, coniferous trees.

Dimensions variable dia x 0.05 - 0.3 cm thick.
Fruiting body white or cream, becoming more ochraceous at maturity, resupinate, with vague concentric zones and regular margin, the hymenial (upper) surface waxy, at first densely warty, then decorated with awl-shaped teeth up to 0.5 cm long. Flesh when damp, whitish, thin, waxy; when dry, crusty. **Spores** hyaline, smooth, cylindrical or allantoid, non-amyloid, 8-10 x 3-4 µm. Basidia 4-spored. Cystidia cylindrical, thin-walled.
Odour not distinctive. **Taste** not distinctive.
Chemical tests none.
Occurrence throughout the year, but sporulating in the autumn; infrequent.
■ Inedible.

Hyphoderma setigerum (Fr.) Donk   Hyphodermataceae

White, encrusting fruiting body tightly attached to substrate, minutely warty; on dead wood of broad-leaf and coniferous trees.

Dimensions variable dia x 0.05 - 0.3 cm thick.
Fruiting body white, with greyish or buff tinges, resupinate, with irregular but clearly defined margin, the hymenial (upper) surface having a waxy consistency, densely warty, decorated with minute bristles. Flesh when damp, white, thin and waxy; when dry, crusty.
Spores hyaline, smooth, elongated-ellipsoid, non-amyloid, 9-12 x 4-6 µm. Basidia 2- or 4-spored. Cystidia elongated-cylindrical, sometimes encrusted.
Odour not distinctive. **Taste** not distinctive.
Chemical tests none.
Occurrence throughout the year, but sporulating in the autumn; common.
■ Inedible.

Hyphodontia sambuci (Pers.) Erikss. Hyphodermataceae
= *Hyphoderma sambuci* (Pers.) Jül.
= *Lyomyces sambuci* (Pers.: Fr.) Karst.
= *Thelephora sambuci* Pers.

White, chalky fruiting body tightly attached to substrate, looking like matt emulsion paint or distemper; on wood of broaf-leaf trees favouring elder.

Dimensions variable dia x 0.05 - 0.1 cm thick.
Fruiting body white, resupinate, with irregular margin, the hymenial (upper) surface having a chalky consistency. Flesh white and extremely thin. **Spores** hyaline, smooth, ellipsoid, non-amyloid, 5-7 x 4-5 µm. Basidia 2- or 4-spored. Cystidia fusiform or narrowly cylindrical, capitate, the heads sometimes encrusted.
Odour not distinctive. **Taste** not distinctive.
Chemical tests none.
Occurrence throughout the year, but sporulating late summer and autumn; common.
■ Inedible.

Schizopora paradoxa (Schrad.: Fr.) Donk   **Hyphodermataceae**

= *Irpex paradoxus* (Schrad.: Fr.) Fr.
= *Irpex obliquus* (Schrad.: Fr.) Fr.
= *Poria versipora* (Pers.) Lloyd
= *Xylodon versipora* (Pers.) Bond.

Encrusting patches, whitish becoming more cream with age, with irregular pores, annual; on dead and rotten wood of broaf-leaf trees, very rarely coniferous.

Dimensions variable dia x 0.2 - 0.6 cm thick.
Fruiting body whitish, becoming cream, resupinate, at first in rounded patches, becoming extensive and irregular; hymenial (upper) surface finely porous. Flesh when damp, whitish, soft; when dry, hard. Pores whitish, circular, angular, elongated or maze-like, 1-4 per mm. Tubes whitish, 1-5 mm deep. **Spores** hyaline, smooth, broadly ellipsoid, non-amyloid, 4-6.5 x 3-4 µm. Basidia 4-spored. Cystidia clavate or fusiform.
Odour not distinctive. **Taste** not distinctive.
Chemical tests none.
Occurrence throughout the year, but sporulating late summer and autumn; common.
■ Inedible.

Peniophora cinerea (Fr.) Cke.   **Peniophoraceae**

Encrusting whitish grey patches; on fallen logs of broad-leaf trees, favouring the undersides.

Dimensions variable dia x 0.1 - 0.5 cm thick.
Fruiting body whitish-grey, margin pallid when young, more distinct in older specimens, resupinate, tightly attached, warty. Flesh when damp, thin and waxy; when dry, crusty and brittle.
Spores hyaline (pink in the mass), smooth, allantoid or cylindrical, non-amyloid, 8-10 x 2-3.5 µm. Basidia 4-spored. Cystidia fusiform or elongated cylindrical, encrusted towards the apex.
Odour not distinctive. **Taste** not distinctive.
Chemical tests none.
Occurrence throughout the year, but sporulating late summer and autumn; common.
■ Inedible.

Peniophora incarnata (Fr.) Karst.   **Peniophoraceae**

= *Thelephora incarnata* Pers.: Fr.

Encrusting orange-pink patches; on dead and dying wood, favouring gorse and tending towards the undersides of attached branches.

Dimensions variable dia x 0.1 - 0.5 cm thick.
Fruiting body orange or salmon-pink, margin fairly sharply defined, resupinate, tightly attached, warty. Flesh reddish-orange; when damp, thin and waxy; when dry, crusty.
Spores hyaline (pink in the mass), smooth, narrowly ellipsoid or cylindrical, non-amyloid, 7.5-10 x 3.5-5 µm. Basidia 4-spored. Cystidia either fusiform, encrusted from half-way, or vermiform.
Odour not distinctive. **Taste** not distinctive.
Chemical tests none.
Occurrence throughout the year, but sporulating late summer and autumn; common.
■ Inedible.

Peniophora lycii (Pers.) Höhn. & Litsch.   Peniophoraceae

Encrusting greyish-blue patches; on dead wood of broad-leaf trees, typically on twigs and branches.

Dimensions variable dia x 0.1 - 0.3 cm thick.
Fruiting body greyish, with pink, violaceous or blue tinges, margin fairly sharply defined, resupinate, very thin, tightly attached, finely warty. Flesh greyish; when damp, thin and waxy; when dry, crusty.
Spores hyaline (pink in the mass), smooth, allantoid or cylindrical, non-amyloid, 8-12 x 3-4.5 µm. Basidia 4-spored. Cystidia broadly clavate, branching, heavily encrusted.
Odour not distinctive. **Taste** not distinctive.
Chemical tests none.
Occurrence throughout the year, but sporulating late summer and autumn; common.
■ Inedible.

Peniophora quercina (Fr.) Cke.   Peniophoraceae

Encrusting pinkish patches; on dead, still fixed, branches of broad-leaf trees, usually the underside, favouring oak.

Dimensions variable dia x 0.1 - 0.5 cm thick.
Fruiting body pinkish-grey, sometimes with ochraceous or violaceous tinges, resupinate, but reflexing at the margins when dry and brittle, revealing dark brown or blackish under-surface, thin, smooth or slightly warty. Flesh pallid; when damp, fairly thick and gelatinous; when dry, crusty, brittle and hyaline other than adjacent to the substrate.
Spores hyaline (pink in the mass), smooth, allantoid or cylindrical, non-amyloid, 9-12 x 3-4 µm. Basidia 4-spored. Cystidia conical, encrusted to halfway.
Odour not distinctive. **Taste** not distinctive.
Chemical tests none.
Occurrence throughout the year, but sporulating late summer and autumn; common.
■ Inedible.

Pulcherricium caeruleum (Fr.) Parm.   Corticiaceae

Encrusting blue patches; on the undersides of fallen logs and branches of broad-leaf trees, favouring ash and hazel.

Dimensions variable dia x 0.1 - 0.5 cm thick.
Fruiting body bright, often intense blue, blackish towards the centre, margin more pallid when young, resupinate, loosely attached, warty. Flesh when damp, thin and waxy; when dry, crusty and brittle.
Spores hyaline (bluish in the mass), smooth, ellipsoid, non-amyloid, 6.5-9 x 4.5-5.5 µm. Basidia 4-spored. Cystidia narrowly clavate or cylindrical and branched at the tips, resembling basidia, also with lateral outgrowths.
Odour not distinctive. **Taste** not distinctive.
Chemical tests none.
Occurrence throughout the year, but sporulating late autumn and early winter; infrequent.
■ Inedible.

Vuilleminia comedens (Nees) Maire   Corticiaceae
= *Corticium comedens* (Nees) *Fr.*

Buff or dull lilac, thin encrustation; beneath rolled-back bark of dead branches of broad-leaf trees. The characteristic rolled-back bark on the substrate is often a give-away to the presence of the fungus.

Dimensions 1 - 13 cm dia x 0.1 - 0.2 cm thick.
Fruiting body resupinate patches, dull lilaceous or buff, slightly viscid when damp, otherwise firm. Flesh pallid; when damp, thin, soft and waxy; when dry, crusty and hard (almost invisible).
Spores hyaline, smooth, allantoid or cylindrical, weakly amyloid, 15-20 x 5-5-6.5 µm. Basidia 4-spored. Cystidia absent.
Odour not distinctive. **Taste** not distinctive.
Chemical tests none.
Occurrence throughout the year, but sporulating late summer and autumn; common but very easily overlooked.
■ Inedible.

Amylostereum laevigatum (Fr.) Boid.   Stereaceae
= *Xerocarpus juniperi* Karst.

Whitish, encrusting, pores absent; in patches on living wood of yew, typically low down on the trunk, very rarely on dead wood and on other coniferous trees.

Dimensions variable dia x 0.05 - 0.1 cm thick.
Fruiting body whitish-buff, irregular, resupinate, smooth, tightly attached, finely crazed when dry. Flesh whitish-buff; when damp, soft and waxy; when dry, brittle. **Spores** hyaline, smooth, cylindrical or narrowly ellipsoid, amyloid, 7-9 x 3-4 µm. Basidia 4-spored. Cystidia brownish, cylindrical or broadly awl-shaped, thick-walled, encrusted from halfway.
Odour not distinctive. **Taste** not distinctive.
Chemical tests none.
Occurrence throughout the year, but sporulating late summer and autumn; infrequent.
■ Inedible.

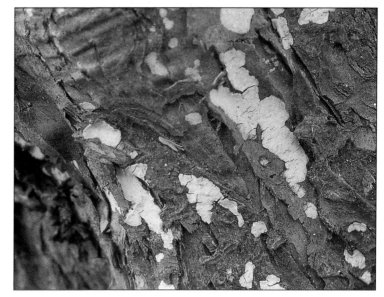

Stereum gausapatum (Fr.) Fr.   Stereaceae

Pale buff encrustation or small bracket; in overlapping tiers on stumps and other dead wood of broad-leaf trees, favouring oak. Note colour change to blood red when cut or scored.

Dimensions 1 - 4 cm dia x 0.2 - 0.4 cm thick.
Fruiting body upper surface (of bracket) ochraceous-brown to grey with more pallid margin, finely downy, concentrically zoned, wavy; lower hymenial surface (outer surface when resupinate) reddish-brown, smooth; resupinate or reflexed. Flesh reddish-ochre, thin; when damp, elastic, tough; when dry, hard and brittle; reddening where cut.
Spores hyaline, smooth, sub-cylindrical, very weakly amyloid, 6.5-9 x 3-4 µm. Basidia 2- or 4-spored. Cystidia-like structures present, cylindrical, smooth, thin-walled with brownish content.
Odour not distinctive. **Taste** not distinctive.
Chemical tests none.
Occurrence throughout the year, but sporulating late summer and autumn; common.
■ Inedible.

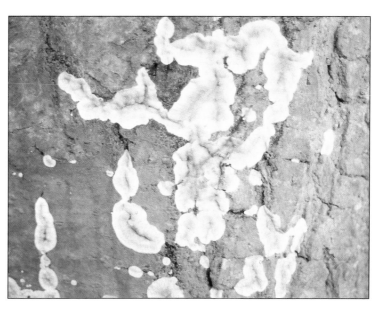

Stereum hirsutum (Willd.: Fr.) S F Gray   **Stereaceae**

Buff or ochre encrustation or small bracket, hairy; in dense over-lapping tiers on stumps and other dead wood of broad-leaf trees. Note: no colour change when cut or scored.

Dimensions 3 - 10 cm dia x 0.3 - 0.5 cm thick.
Fruiting body upper surface (of bracket) ochraceous-brown to grey with more pallid margin, hairy, concentrically zoned, wavy; lower hymenial surface (outer surface when resupinate) yellowish-brown or orange, smooth; resupinate or reflexed. Flesh ochraceous, thin; when damp, elastic, tough; when dry, hard and brittle; not reddening when cut.
Spores hyaline, smooth, ellipsoid, weakly amyloid, 6-7.5 x 2-3.5 μm. Basidia 4-spored. Cystidia-like structures present, cylindrical, smooth, thick-walled.
Odour not distinctive. **Taste** not distinctive.
Chemical tests none.
Occurrence throughout the year, but sporulating autumn and early winter; common.
■ Inedible.
Inset: resupinate fruiting bodies.

Stereum rameale (Pers.: Fr.) Burt   **Stereaceae**
= *Stereum ochraceo-flavum* (Schw.) Ellis

Buff encrustation or small bracket; typically in linear tiers on dead wood of broad-leaf trees, favouring oak. Note: no colour change when cut or scored.

Dimensions 0.5 - 2 cm dia x 0.1 - 0.3 cm thick.
Fruiting body upper surface (of bracket) ochraceous-brown to grey, with pallid margin, finely downy, barely zoned, wavy, margin fringed; lower hymenial surface (outer surface when resupinate) ochraceous or buff, smooth; resupinate or reflexed. Flesh pallid ochraceous, thin; when damp, elastic and tough; when dry, hard and brittle; not reddening when cut.
Spores hyaline, smooth, sub-cylindrical or ellipsoid, amyloid, 7-9 x 2-3 μm. Basidia 4-spored. Cystidia-like structures present, cylindrical, smooth, thick-walled.
Odour not distinctive. **Taste** not distinctive.
Chemical tests none.
Occurrence throughout the year, but sporulating summer and autumn; infrequent or overlooked.
■ Inedible.
Distinguished from *S. hirsutum* by smoother upper surface and the presence of eyelash-like fringing hairs.

Stereum rugosum (Pers.: Fr.) Fr.   **Stereaceae**

Buff encrustation or small bracket; in overlapping tiers on stumps and other dead wood of broad-leaf trees, favouring oak. Note: colour change to blood red when cut or scored.

Dimensions 1 - 4 cm dia x 0.2 - 0.4 cm thick.
Fruiting body upper surface (of bracket) ochraceous, sometimes with pinkish tinge and with distinct margin, slightly warty; lower surface (outer surface when resupinate) ochraceous, smooth; resupinate or, less commonly, reflexed. Flesh ochraceous, thin; when damp, elastic and tough; when dry, hard and brittle; reddening where cut.
Spores hyaline, smooth, ellipsoid, amyloid, 7-10 x 3.5-5 μm. Basidia 4-spored. Cystidia-like structures present, cylindrical, smooth, thick-walled.
Odour not distinctive. **Taste** not distinctive.
Chemical tests none.
Occurrence throughout the year, but sporulating late summer and autumn; common.
■ Inedible.

Stereum sanguinolentum ([A. & S.] Fr.) Fr.   **Stereaceae**

Violet-buff encrustation or small bracket; in overlapping tiers on stumps and other dead wood of coniferous trees. Note colour change to crimson or dark red when cut or scored.

Dimensions 1 - 4 cm dia x 0.2 - 0.4 cm thick.
Fruiting body upper surface (of bracket) brownish-buff, with violaceous tinge and with distinct pallid margin, slightly wavy; lower surface (outer surface when resupinate) pallid, brownish with violaceous tinge, smooth; resupinate or reflexed. Flesh pallid ochre, thin; when damp, elastic and tough; when dry, hard and brittle; dark red or bleeding where cut or bruised.
Spores hyaline, smooth, cylindrical or ellipsoid, amyloid, 6.5-8 x 2.5-3 µm. Basidia 2- or 4-spored. Cystidia-like structures present, cylindrical, smooth, thick-walled.
Odour not distinctive. **Taste** not distinctive.
Chemical tests none.
Occurrence throughout the year, but sporulating late summer and autumn; frequent.
■ Inedible.

Stereum subtomentosum Pouz.   **Stereaceae**

Greyish orange, thin, somewhat curled bracket with yellow fertile surface and narrow attachment, much larger than other *Stereum* species; typically in tiers on dead wood of broad-leaf trees favouring beech. Note: colour change when cut or scored.

Dimensions 3 - 7 cm dia x 0.1 - 0.2 cm thick.
Fruiting body upper surface greyish-brown, with orange-rust concentric zones, margin more pallid, sometimes with greenish tinge due to algal growth, spotting yellow where damaged, finely downy, wavy, margin typically lobed; lower (hymenial) surface ochraceous bruising more rust, smooth; typically reflexed and arising from narrow, darker brown point of attachment. Flesh pallid ochraceous, thin; when damp, elastic and tough; when dry, hard and brittle; not reddening where cut.
Spores hyaline, smooth, sub-cylindrical or ellipsoid, amyloid, 5.5-6.5 x 2-3 µm. Basidia 4-spored. Cystidia-like structures present, cylindrical, smooth, thick-walled with brownish content.
Odour not distinctive. **Taste** not distinctive.
Chemical tests none.
Occurrence throughout the year, but sporulating late summer and autumn; frequent in the southernmost counties of England, rare elsewhere. (High Salvington, West Sussex)
■ Inedible.

Podoscypha multizonata (Berk. & Br.) Pat.
= *Thelephora multizonata* Berk. & Br.   **Podoscyphaceae**

Pale buff rosette of thin lobes; on soil in broad-leaf woods.

Dimensions 5 - 25 cm dia overall; lobes 0.2 - 0.4 cm thick.
Fruiting body upper surface (of individual lobes) buff or brown, with darker brown concentric zones, smooth, wavy; lower (hymenial) surface pallid greyish-brown, smooth; lobes erect, rosette-like. Flesh concolorous, thin; when damp, elastic, tough; when dry, brittle.
Spores hyaline, smooth, sub-cylindrical, amyloid, 4.5-6.5 x 4-5 µm. Basidia 4-spored. Cystidia vermiform with swollen base.
Odour not distinctive. **Taste** not distinctive.
Chemical tests none.
Occurrence late summer to autumn; rare. (New Forest, Hampshire)
■ Inedible.

Chrondrostereum purpureum (Fr.) Pouz. **Meruliaceae**
= *Stereum purpureum* (Fr.) Fr.

Brownish, with smooth violet under-surface, sometimes encrusting but more typically as small bracket, pores absent; in dense overlapping tiers on a variety of broad-leaf trees but favouring plum, causing 'silver leaf' disease'.

Dimensions 1.5 - 3 cm dia x 0.2 - 0.5 cm thick.
Fruiting body upper surface (of bracket) brownish, more pallid at the margin, covered with fine downy hairs, concentrically zoned and radially wrinkled; lower surface (outer surface when resupinate) dark violaceous, becoming browner in older specimens; resupinate or reflexed. Flesh brownish, thin; when damp, tough; when dry, brittle.
Spores hyaline, smooth, sub-cylindrical, non-amyloid, 5-8 x 2.5-3.5 µm. Basidia 4-spored. Cystidia of 2 types; fusiform with narrower apex, sometimes encrusted; clavate with markedly swollen tips, smooth.
Odour not distinctive. **Taste** not distinctive.
Chemical tests none.
Occurrence throughout the year, but sporulating summer to autumn; common.
■ Inedible.

Meruliopsis corium (Fr.) Ginns   **Meruliaceae**
= *Byssomerulius corium* (Fr.) Parm.
= *Merulius papyrinus* (Bull.: Fr.) Quél.

Whitish, rubbery, encrusting and forming small bracket-like projections; on dead branches of broad-leaf trees.

Dimensions variable dia x 0.1 - 0.3 cm thick.
Fruiting body whitish with ochraceous tinge, more brown when old, closely applied to the substrate except where reflexed; upper surface whitish, tinged ochraceous, fibrous downy; hymenial (lower) surface whitish, becoming tinged ochraceous, at first more or less smooth, becoming warty, finely pseudo-porous (netted). Flesh when damp, whitish, thin, rubbery-leathery; when dry, hard and corky.
Spores hyaline, smooth, ellipsoid-cylindrical, non-amyloid, 5-6 x 2.5-3.5 µm. Basidia 4-spored. Cystidia absent.
Odour not distinctive. **Taste** not distinctive.
Chemical tests none.
Occurrence late summer to spring; infrequent.
■ Inedible.

Merulius tremellosus Fr.   **Meruliaceae**

Pale, rubbery, spreading radially when encrusted on substrate, otherwise forming small bracket; on stumps and dead branches of broad-leaf, and less frequently, coniferous trees.

Dimensions 1 - 5 cm dia but often merging into larger areas when resupinate x 0.2 - 0.5 cm thick.
Fruiting body resupinate form - whitish, rubbery-gelatinous, hairy, closely applied to the substrate, spreading radially, margin covered in silky radiating fibres; reflexed form - thin, flexible, typically in tiers, hymenial (lower) surface ranging from orange through pinkish to buff, puckered and appearing almost porous. Flesh when damp, whitish, thin, and rubbery-gelatinous; when dry, hard and corky.
Spores hyaline, smooth, allantoid, non-amyloid, 3.5-4.5 x 1-1.5 µm. Basidia 4-spored. Cystidia narrowly cylindrical, sparsely encrusted.
Odour not distinctive. **Taste** not distinctive.
Chemical tests none.
Occurrence late summer to spring; infrequent.
■ Inedible.
Inset: resupinate fruiting bodies.

Mycoacia uda (Fr.) Donk   Meruliaceae
= *Acia uda* (Fr.) Karst.

Encrusting yellow patches; on dead logs and branches of broad-leaf trees, favouring the undersides.

Dimensions variable dia x 0.1 - 0.3 cm thick.
Fruiting body ochraceous, resupinate, tightly attached, at first granular, then decorated with densely crowded spines, margin smooth. Flesh when damp, thin and waxy; when dry, crusty and brittle.
Spores hyaline, smooth, ovoid, non-amyloid, 1 - 2 droplets, 5-6 x 2-3 µm. Basidia 4-spored. Cystidia fusiform, inconspicuous.
Odour not distinctive. **Taste** not distinctive.
Chemical tests none.
Occurrence throughout the year, but sporulating late summer and autumn; infrequent.
■ Inedible.

Phlebia cornea (Bourd. & Galz.) Parm.   Meruliaceae

Encrusting cream warty patches on bark of dead broad-leaf and coniferous trees.

Dimensions variable dia x 0.05 - 0.2 cm thick.
Fruiting body cream with grey and ochraceous tinges, resupinate, irregular patches with distinct margins, the hymenial surface irregularly warty. Flesh cream; when damp, fairly thin, waxy and soft; when dry, membraneous and tough.
Spores hyaline, smooth, ellipsoid-cylindrical, non-amyloid, with numerous droplets, 7.5-12 x 3.5-5.5 µm. Basidia 4-spored. Cystidia smooth, vermiform with blunt tips.
Odour not distinctive. **Taste** not distinctive.
Chemical tests none.
Occurrence throughout the year, mainly autumn; infrequent.
■ Inedible.

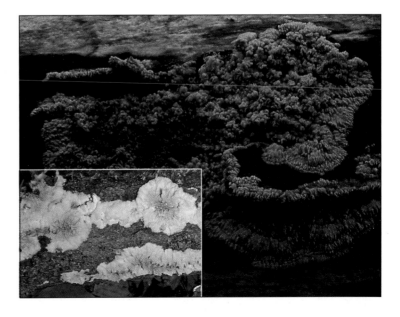

Phlebia radiata Fr.   Meruliaceae
= *Phlebia aurantiaca* (Sow.) Karst.
= *Phlebia merismoides* Fr.

Encrusting, rubbery, gelatinous, bright orange or dull buff patches on bark of dead broad-leaf trees, favouring beech.

Dimensions 8 - 10 cm dia x 0.2 - 0.3 cm thick.
Fruiting body fluorescent orange, sometimes with purple tinge, or dull buff, resupinate, irregular or more or less rounded patches with fringed margins, the hymenial surface radially wrinkled and puckered, appearing more warty in older specimens. Flesh pallid pink; when damp, when fairly thin, gelatinous and soft; when dry, tough.
Spores hyaline, smooth, allantoid or elongated cylindrical, non-amyloid, 2 droplets, 3.5-6 x 1.5-2.5 µm. Basidia 4-spored. Cystidia smooth, fusiform or slightly clavate.
Odour not distinctive. **Taste** not distinctive.
Chemical tests none.
Occurrence throughout the year, mainly autumn; frequent.
■ Inedible.
Inset: lighter form.

Phlebia rufa (Fr.) Christ.   Meruliaceae
= *Merulius rufus* Pers.: Fr.

Encrusting, rubbery, gelatinous, orange or buff patches on bark of dead broad-leaf trees, favouring oak.

Dimensions 8 - 10 cm dia x 0.2 - 0.3 cm thick.
Fruiting body fluorescent orange, or dull buff, resupinate, irregular patches with fringed margins, the hymenial surface irregularly wrinkled and puckered, appearing more warty in older specimens. Flesh pallid reddish; when damp, fairly thin, gelatinous and soft; when dry, tough.
Spores hyaline, smooth, allantoid or elongated cylindrical, non-amyloid, 2 droplets, 4.5-6.5 x 2-2.5 µm. Basidia 4-spored. Cystidia smooth, clavate.
Odour not distinctive. **Taste** not distinctive.
Chemical tests none.
Occurrence late summer to spring; infrequent or rare. (Lyme Park, Cheshire)
■ Inedible.

Phlebiopsis gigantea (Fr.) Jül   Meruliaceae
= *Phlebia gigantea* (Fr.) Donk
= *Peniophora gigantea* (Fr.) Mass.

White encrusting patches on dead wood of coniferous trees, typically extending over stumps.

Dimensions 10 - 100 cm dia x 0.2 - 0.3 cm thick.
Fruiting body white, resupinate, effused, irregular, often covering a large area, tightly attached, margin sharp and distinct but reflexing away from the substrate when dry, hymenial surface smooth or finely warty. Flesh white; when damp, thin, soft and waxy; when dry, tough and brittle.
Spores hyaline, smooth, ellipsoid, non-amyloid, 5-7 x 2.5-3.5 µm. Basidia 4-spored. Cystidia broadly fusiform, encrusted to halfway.
Odour not distinctive. **Taste** not distinctive.
Chemical tests none.
Occurrence summer and autumn; infrequent.
■ Inedible.
Introduced in some forests to counter *Heterobasidium annosum* through antagonistic control.

Sistotrema brinkmanii (Bres.) Erikss.   Sistotremataceae
= *Corticium coronilla* Höhn.

Whitish, thin, warty encrustation on dead wood of broad-leaf and coniferous trees, occasionally on old fungal fruiting bodies.

Dimensions up to 10 cm or more dia x 0.1 - 0.2 cm thick.
Fruiting body resupinate patches, whitish with ochraceous or grey tinges, warty or coarsely granular. Flesh white; when damp, thin, soft and waxy; when dry, more powdery.
Spores hyaline, smooth, ellipsoid or allantoid, non-amyloid, 4-5.5 x 2-2.5 µm. Basidia 4-, 6-, or 8-spored. Cystidia absent.
Odour not distinctive. **Taste** not distinctive.
Chemical tests none.
Occurrence throughout the year, mainly late summer and autumn; common.
■ Inedible.

Botryohypochnus isabellinus (Fr.: Schleich.) Erikss.
Botryobasidiaceae

Yellowish, very thin fruiting body, loosely attached to substrate; on rotten wood of broad-leaf and coniferous trees.

Dimensions variable dia x 0.03 - 0.05 cm thick.
Fruiting body ochraceous, resupinate, with irregular margin, the hymenial (upper) surface having a downy consistency with a narrow, finely fringed margin. Flesh yellowish, very thin and cottony.
Spores yellowish, spiny, sub-spherical, non-amyloid, weakly cyano-philic, 6-10 x 5.5-9 µm. Basidia 4-spored. Cystidia absent.
Odour not distinctive. **Taste** not distinctive.
Chemical tests none.
Occurrence throughout the year, mainly summer to autumn; infrequent.
■ Inedible.

Bankera fuligineo-alba (Schmidt: Fr.) Pouz.   Bankeraceae
= *Hydnum fuligineo-album* Schmidt

Medium-sized yellowish brown cap with spiny under-surface; solitary or in groups, often fused, on soil in pine woods.

Dimensions cap 5 - 10 cm dia; stem 2 - 5 cm tall x 0.5 - 2 cm thick.
Fruiting body cap at first pallid, then brown with yellow or red tinges, remaining more pallid at the margin, wavy, flattish, rosette-shaped, downy; stem white at the apex becoming dull brownish below, equal or tapered towards the base, downy, sometimes eccentric. Flesh white with pinkish tinge at maturity, soft and spongy.
Spines at first white or pallid then grey or with russet-brown tinge, slightly decurrent, coarse. **Spores** hyaline, minutely spiny, sub-spherical or ovoid, non-amyloid, 4.5-5.5 x 2.5-3.5 µm. Basidia 4-spored. Cystidia absent.
Odour fresh, not distinctive; dry, spicy. **Taste** not distinctive.
Chemical tests none.
Occurrence late summer to autumn; rare. (Scottish Highlands)
■ Inedible.

Phellodon confluens (Pers.) Pouz.   Bankeraceae
= *Phellodon amicus* (Quél.) Banker

Medium-sized, blackish-brown, irregular plate-like cap, with white margin and spiny under-surface; solitary or in groups, typically fused, on soil in both broad-leaf and coniferous woods.

Dimensions cap 3 - 10 cm dia; stem 2 - 4 cm tall x 1 - 2 cm thick.
Fruiting body cap at first whitish, soon becoming greyish or blackish-brown with whitish margin, indistinctly zoned, shallowly infundibuliform, rosette-shaped, warty, radially wrinkled and downy, margin smooth, sometimes with secondary caps; stem concolorous with cap or darker, more or less equal, tomentose, typically several arising fused together. Flesh greyish-brown, duplex in the stem, soft and corky.
Spines at first whitish or pallid then grey, more or less decurrent.
Spores hyaline, finely spiny, sub-spherical, non-amyloid, 3.5-4.5 x 3-4 µm. Basidia 4-spored. Cystidia absent.
Odour fresh, not distinctive; dry, spicy. **Taste** slight, bitter.
Chemical tests no flesh reaction with KOH.
Occurrence late summer to autumn; rare. (Scottish Highlands)
■ Inedible.

Phellodon melaleucus (Swarz apud Fr.: Fr.) Karst.
= *Hydnum melaleucum* Swarz apud Fr.　　**Bankeraceae**

Medium-sized, greyish-brown, irregular plate-like cap, with white margin and spiny under-surface; solitary or in groups, often fused, on soil in coniferous and mixed woods, often associated with bilberry *(Vaccinium)*.

Dimensions cap 2 - 8 cm dia; stem 1 - 3 cm tall x 0.3 - 1 cm thick.
Fruiting body cap grey, with brown or yellow tinges, darkening towards the centre and with whitish margin, shallowly infundibuliform, rosette-shaped, radially wrinkled and downy, margin smooth; stem concolorous with cap or darker brown, more or less equal, fibrillose, typically several arising fused together. Flesh reddish brown, darker in the stem, tough and corky.
Spines at first whitish or pallid, then brown, more or less decurrent.
Spores hyaline, spiny, sub-spherical, non-amyloid, 3.5-4.5 x 3-4 µm. Basidia 4-spored. Cystidia absent.
Odour fresh, not distinctive; dry, spicy. **Taste** slight, bitter.
Chemical tests flesh turns green with KOH.
Occurrence late summer to autumn; rare. (Scottish Highlands)
■ Inedible.

Phellodon tomentosus (L.: Fr.) Banker　　**Bankeraceae**
= *Calodon tomentosus* (L.: Fr.) Maire

Smallish, brown, irregular plate-like cap with white margin and spiny under-surface; solitary or in groups, often fused, on soil in coniferous and mixed woods, often associated with bilberry *(Vaccinium)*.

Dimensions cap 2 - 6 cm dia; stem 1 - 3 cm tall x 0.3 - 0.8 cm thick.
Fruiting body cap reddish or hazel-brown, darkening towards the centre and with whitish margin, shallowly infundibuliform, rosette-shaped, radially wrinkled and downy, margin smooth; stem concolorous with cap, more or less equal, matted fibrillose, typically several arising fused together. Flesh reddish-brown, darker in the stem, tough and corky.
Spines at first whitish or pallid then grey, more or less decurrent.
Spores hyaline, spiny, sub-spherical, non-amyloid, 3.5-4 x 2.5-3.5 µm. Basidia 4-spored. Cystidia absent.
Odour fresh, not distinctive; dry, of fenugreek. **Taste** slight, bitter.
Chemical tests no flesh reaction with KOH.
Occurrence late summer to autumn; rare. (Scottish Highlands)
■ Inedible.

Sarcodon imbricatus (L.: Fr.) Karst.　　**Bankeraceae**
= *Hydnum imbricatum* L.: Fr.

Large coarsely scaly brown cap with greyish spiny under-surface; solitary or in scattered groups, on soil in coniferous woods, favouring pine and spruce.

Dimensions cap 10 - 25 cm dia; stem 4 - 7 cm tall x 2 - 4 cm thick.
Fruiting body cap yellowish-brown, decorated with coarse darker brown scales, erect at the centre, more flattened towards the incurved margin, arranged in concentric rows; at first shallowly convex, then flattened or slightly depressed at the centre; at first whitish, becoming brown, remaining pallid at the base, equal or somewhat clavate towards the base, downy. Flesh white, thick in the cap centre, firm, full in the stem.
Spines at first white or pallid, then grey with purplish-brown tinge, decurrent. **Spores** brown, coarsely tuberculate, sub-spherical, non-amyloid, 6.5-8 x 5-6 µm. Basidia 4-spored. Cystidia absent.
Odour spicy. **Taste** mild or slightly bitter.
Chemical tests none.
Occurrence late summer to autumn; rare. (Culbin Forest, Morayshire)
■ Inedible.

Hydnellum caeruleum (Horn.: Pers.) Karst. **Thelephoraceae**

Small, bluish-grey, funnel-shaped cap, with whitish margin and spiny under-surface; solitary or in groups, often fused, on soil in broad-leaf and coniferous woods.

Dimensions cap 3 - 7 cm dia; stem 2 - 5 cm tall x 1 - 2 cm thick.
Fruiting body cap bluish-grey, darkening towards the centre, margin whitish at first, shallowly infundibuliform, rosette-shaped, radially wrinkled, at first downy then smooth; stem brownish, equal or tapered towards the base, downy, wrinkled. Flesh tinged bluish in the cap, otherwise orange-brown, tough and corky.
Spines at first bluish, then more pallid, finally brown, decurrent, coarse. **Spores** light brown, flattened, tuberculate, sub-spherical, non-amyloid, 5-6 x 4-4.5 µm. Basidia 4-spored. Cystidia absent.
Odour faint, of meal. **Taste** not distinctive.
Chemical tests none.
Occurrence late summer to autumn; rare. (Abernethy Forest, Scottish Highlands)
■ Inedible.

Hydnellum concrescens (Pers.: Schw.) Banker
= *Hydnellum zonatum* (Fr.) Karst.   **Thelephoraceae**

Medium-sized, reddish-brown, funnel-shaped cap with pale margin and spiny under-surface; solitary or in groups, often fused, on soil in broad-leaf woods.

Dimensions cap 2 - 7 cm dia; stem 1 - 3 cm tall x 0.3 - 0.8 cm thick.
Fruiting body cap reddish-brown, darkening towards the centre, margin pallid, shallowly infundibuliform, rosette-shaped, radially wrinkled and with small erect scales at the centre; stem concolorous with cap, equal or tapered towards the base (sometimes bulbous), downy, wrinkled. Flesh dark vinaceous red, blackish towards the stem base, tough and corky.
Spines at first pallid then pinkish or reddish-brown, decurrent, coarse. **Spores** light brown, tuberculate (some double), sub-spherical, non-amyloid, 4.5-6 x 3.5-4.5 µm. Basidia 4-spored. Cystidia absent.
Odour faint, of meal. **Taste** slight, bitter.
Chemical tests no reaction with KOH.
Occurrence late summer to autumn; infrequent.
■ Inedible.

Hydnellum peckii Banker apud Peck   **Thelephoraceae**
= *Hydnellum diabolum* Peck

Smallish, red or brown funnel-shaped cap, with white margin and spiny under-surface; solitary or in groups, often fused, on soil in broad-leaf and coniferous woods.

Dimensions cap 3 - 10 cm dia; stem 1 - 5 cm tall x 1 - 3 cm thick.
Fruiting body cap at first pallid, then wine-red with brighter red guttation drops which dry dull brown, margin white when young, shallowly infundibuliform, rosette-shaped, radially wrinkled and with small spiky projections, coarsely roughened at the centre; stem reddish-brown, equal or tapered towards the base, downy or smooth, wrinkled. Flesh brown with pink tinge, often darker spotted, tough and corky.
Spines at first pallid, then russet-brown, decurrent, coarse.
Spores light brown, bluntly tuberculate, sub-spherical, non-amyloid, 5-6 x 4-4.5 µm. Basidia 4-spored. Cystidia absent.
Odour faint, sour. **Taste** hot, acrid.
Chemical tests none.
Occurrence late summer to autumn; infrequent. (Abernethy Forest, Scottish Highlands)
■ Inedible.

Hydnellum scrobiculatum (Fr.: Sécr.) Karst. **Thelephoraceae**
= *Hydnum scrobiculatum* Fr.

Small, pinkish-brown funnel-shaped cap with spiny under-surface; solitary or in groups, often fused, on soil in broad-leaf and coniferous woods.

Dimensions cap 2 - 4.5 cm dia; stem 1 - 2 cm tall x 0.2 - 1 cm thick.
Fruiting body cap at first pallid, then pinkish-brown, darkening towards the centre, shallowly infundibuliform, rosette-shaped, radially wrinkled and with small spiky projections, coarsely roughened at the centre; stem concolorous with cap, equal or tapered towards the base (sometimes bulbous), downy or matted, wrinkled. Flesh dark russet-brown, often darker spotted, tough and corky.
Spines at first pallid then russet-brown, decurrent, coarse.
Spores light brown, tuberculate, sub-spherical, non-amyloid, 5.5-6.5 x 4.5-5.5 µm. Basidia 4-spored. Cystidia absent.
Odour of meal or spice. **Taste** not distinctive.
Chemical tests none.
Occurrence late summer to autumn; rare. (Haldon Forest near Exeter, Devon)
■ Inedible.

Hydnellum spongiosipes (Peck) Pouz. **Thelephoraceae**
= *Hydnellum velutinum var spongiosipes* (Peck) Geest.

Small, brown, irregular cap with pale margin, spiny under-surface and dark spongy stem; solitary or in groups, often fused, on soil in broad-leaf woods, favouring oak and beech.

Dimensions cap 2 - 19 cm dia; stem 2 - 6 cm tall x 1 - 3 cm thick.
Fruiting body cap pallid or cinnamon-brown, at first top-shaped, becoming shallowly infundibuliform, rosette-shaped, downy, unevenly wrinkled and tuberculate; stem dark brown with whitish dots, stout, equal or tapered towards the base (sometimes bulbous), densely downy. Flesh dark russet-brown, thick and spongy.
Spines at first pallid then russet-brown, decurrent, coarse.
Spores light brown, coarsely tuberculate, sub-spherical, non-amyloid, 5.5-6.5 x 4.5-5.5 µm. Basidia 4-spored. Cystidia absent.
Odour not distinctive. **Taste** not distinctive.
Chemical tests none.
Occurrence late summer to autumn; rare. (New Forest, Hampshire)
■ Inedible.

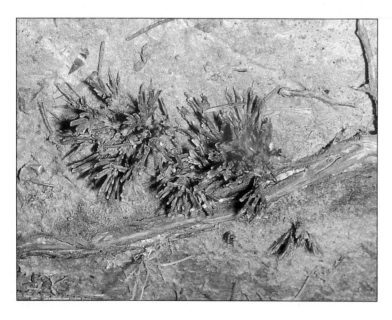

Thelephora pencillata (Pers.) Fr. **Thelephoraceae**

Vertical or horizontal dull brown spiky fan; in clusters on soil, under or near broad-leaf or coniferous trees.

Dimensions overall 3 - 6 cm dia; stem 1 - 1.5 cm tall x 0.1 - 0.3 cm thick.
Fruiting body olivaceous-brown spiky branches, with persistently whitish tips, smooth, cylindrical or laterally compressed, arising from a common stem; stem concolorous, stout or rudimentary. Flesh brown and leathery.
Spores brown, spiny, sub-spherical, angular, 7-10 x 5-8 µm. Basidia 4-spored. Cystidia absent.
Odour not distinctive. **Taste** not distinctive.
Chemical tests none.
Occurrence summer to autumn; rare. (Laughton Woods, East Sussex)
■ Inedible.

Thelephora spiculosa (Fr.) Burt   Thelephoraceae
= *Phylacteria spiculosa* (Fr.) Bourd. & Maire

Whitish spiky rosette, often forming extensive patches; in clusters amongst leaf litter and other debris in broad-leaf and coniferous woods.

Dimensions overall 3 - 15 cm dia x 1 - 4 cm high.
Fruiting body dull purplish-brown in the lower part of the branches, white at the tips, offering an overall whitish appearance, fan-shaped rosettes of branching cottony spicules; sessile or with very rudimentary, concolorous stem. Flesh brown, cottony and thin.
Spores umber-brown, with tiny spines, broadly ellipsoid, 7-12 x 6-7 µm. Basidia 4-spored. Cystidia absent.
Odour not distinctive. **Taste** not distinctive.
Chemical tests none.
Occurrence late summer to autumn; rare. (Stourhead, Wiltshire)
■ Inedible.

Thelephora terrestris (Scop.) Fr.   Thelephoraceae

Brown fan-shaped lobes or rosette, often forming extensive patches; in clusters on soil, amongst leaf litter and other debris in broad-leaf and coniferous woods, and in short turf or heath adjacent to conifers, favouring sandy soils.

Dimensions overall 3 - 6 cm dia x 1 - 4 cm high.
Fruiting body chocolate-brown or reddish-brown, more pallid at the margin, fan-shaped lobes or rosettes, concentrically zoned, downy; sessile or with very rudimentary, concolorous, stout stem. Flesh brown, fibrous and tough.
Spores purple-brown, warty, broadly ellipsoid, angular or tuberculate, 8-9 x 6-7.5 µm. Basidia 4-spored. Cystidia absent.
Odour faint, earthy. **Taste** not distinctive.
Chemical tests none.
Occurrence late summer to autumn; frequent.
■ Inedible.

Thelephora terrestris var resupinata Christ.
Thelephoraceae

Brown fan- or rosette-shaped encrusting patch; attached tightly to dead wood and other debris of coniferous trees.

Dimensions 2 - 5 cm dia x 1 - 3 cm high.
Fruiting body chocolate-brown or reddish-brown, more pallid at the margin, fan-shaped lobes or rosette, tightly attached to the substrate, concentrically zoned, warty, downy; sessile. Flesh brown, fibrous and tough.
Spores purple-brown, warty, broadly ellipsoid, angular or tuberculate, 8-9 x 6-7.5 µm. Basidia 4-spored. Cystidia absent.
Odour faint, earthy. **Taste** not distinctive.
Chemical tests none.
Occurrence late summer to autumn; rare. (Bridgham near Thetford, Norfolk)
■ Inedible.

Schizophyllum commune Fr.   **Schizophyllaceae**

Small, greyish white, fan-shaped bracket; attached to stumps, sawn timber and other dead wood or broad-leaf trees, also on silage, emerging through perforations in polythene wrapping.

Dimensions 1 - 4 cm dia x 0.2 - 0.3 cm thick.
Fruiting body upper surface ochraceous-cream, densely white, downy, sometimes with lilaceous tinge, irregularly lobed and wavy; lower (hymenial) surface bearing ochraceous pseudo-gills radiating from the point of basal attachment, split lengthwise and reflexed back on themselves. Flesh at first brownish, then more pallid, thin and fairly tough.
Spores hyaline, smooth, narrowly ellipsoid or sub-cylindrical, non-amyloid, 6-7 x 2.5-3 µm. Basidia 4-spored. Cystidia absent.
Odour not distinctive. **Taste** not distinctive.
Chemical tests none.
Occurrence throughout the year, mainly summer to autumn; infrequent or rare. (Brereton, Cheshire)
■ Inedible.

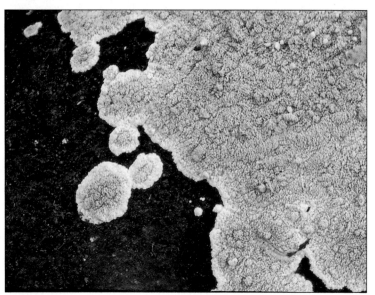

Coniophora puteana (Schum.) Karst.   **Coniophoraceae**
= *Coniophora cerebella* Pers.
= *Corticium puteanum* (Schum.) Fr.

Pale brownish-green, tightly attached encrusting patch, usually with distinct white margin; on decaying trunks and other dead wood, including structural timbers. One of the main causes of wet rot in buildings.

Dimensions 4 - 20 cm dia x 0.05 - 0.1 cm thick.
Fruiting body at first creamy-white, becoming yellowish and finally pallid olivaceous, with distinct broad, whitish, radiating marginal zone, irregular resupinate, the surface rough and warty. Flesh whitish, very thin and soft. **Spores** light brown, smooth, broadly ellipsoid, non-amyloid, strongly cyanophilic, 10-13 x 7-8 µm. Basidia 4-spored. Cystidia absent.
Odour not distinctive. **Taste** not distinctive.
Chemical tests none.
Occurrence throughout the year, mainly late summer and autumn; common.
■ Inedible.

Serpula himantioides (Fr.: Fr.) Karst.   **Coniophoraceae**
= *Merulius himantioides* Fr.: Fr.
= *Gyrophana himantioides* (Fr.: Fr.) Pat.

Brown encrusting wrinkled patch with white margin; on dead wood of coniferous trees, favouring the undersides of logs.

Dimensions 2 - 5 cm dia x 0.2 - 1.0 cm thick.
Fruiting body at first lilaceous becoming umber-brown, resupinate, loosely attached, at first smooth becoming irregularly wrinkled, pseudo-porous, greasy, margin white and silky or downy. Flesh whitish, thin, silky fibrous and soft.
Spores yellowish-brown, smooth, ellipsoid, 9-10 x 5.5-6 µm. Basidia 4-spored. Cystidia smooth, more or less fusiform.
Odour not distinctive. **Taste** not distinctive.
Chemical tests none.
Occurrence throughout the year, mainly autumn; rare. (Lyme Park, Cheshire)
■ Inedible.

Serpula lacrimans (Fr.) Schroet.
= *Merulius lacrimans* (Jacq.: Fr.) Schum.

Coniophoraceae
Dry Rot

Rust-brown encrusting wrinkled patch, with cream margin, sometimes forming small bracket; on dead wood of coniferous, and occasionally, broad-leaf trees used in building construction, in cellars and other poorly ventilated areas, sometimes spreading over walls, fabrics etc.

Dimensions variable dia x 0.2 - 1.2 cm thick.
Fruiting body upper surface (reflexed) cream, with pinkish tinge, discolouring brownish where bruised; under (hymenial) surface reddish-brown, margin cream; generally resupinate, loosely attached, hymenial surface at first smooth, becoming irregularly wrinkled, pseudo-porous, greasy, margin white and silky or downy. Flesh whitish, thick, spongy fibrous and soft.
Spores yellowish, smooth, ellipsoid, 11-14 x 5.5-8 µm. Basidia 4-spored. Cystidia smooth, more or less fusiform.
Odour strong, fungoid. **Taste** not distinctive.
Chemical tests none.
Occurrence throughout the year; once common, becoming rarer.
■ Inedible.

HOMOBASIDIOMYCETES
'AGARICALES AND CORTINARIALES'

The Agaricales, a major order of gill-bearing fungi consisting of nine families in which spore and veil characteristics are of considerable significance, are typified by a cap or *pileus* raised up from the substrate on a stem or *stipe* and bearing the gills or *lamellae* on its under surface. Approximate numbers of species recognised within Europe are given below in square brackets.

The Hygrophoraceae [150], characterised by brightly coloured, rather conical caps, are often slimy or sticky with thick, waxy gills and produce white spores. Most grow in grasslands though some are associated with trees.

The Tricholomataceae [750] constitutes a large, diverse group including at least forty genera which typically possess normal mushroom shape though some have lateral or eccentric stems. The flesh is generally whitish and the spore print can be white but also pale pink or lilac. Mostly saprophytic and growing on soil, some rely on dead or dying trees as substrates and a few are parasitic.

The Amanitaceae [60] constitutes a compact group with distinctive features of velar patches on the cap, a ringed stem and white spores. Many members are dangerously poisonous.

The Entolomataceae [300] are characterised by the production of pink or brownish-pink spores that are either angular or smooth and include the *Entoloma* and *Pluteus* genera, the former having assimilated several smaller groups including *Rhodocybe*, *Leptonia* and *Nolanea*. The order also includes small genera such as *Clitopilus* and *Volvariella*. The *Entoloma* species grow on soil whilst those of *Pluteus* are restricted to wood and woody remains.

The Agaricaceae [200] includes the *Agaricus* genus which contains the common-or-garden edible mushrooms and whose spores, when mature, are chocolate brown, and the *Lepiota* genus. With a wide range of fruiting body sizes, the latter includes the Parasol fungi with scaly caps, ringed stems and white spores.

The Coprinaceae [270] with typically blackish spores includes the Ink Caps whose spores are liberated through autodigestion, the small grassland *Panaeolus* species with characteristic mottled gills and the small, fragile, mainly sylvan *Psathyrella* species.

The Bolbitiaceae [120] with brown spores includes the *Agrocybe*, *Bolbitius* and *Conocybe* genera. The Strophariaceae (160) with brown or chocolate-brown spores includes the *Hypholoma*, *Pholiota*, *Psilocybe* and *Stropharia* genera.

The Kew system varies from the European in establishing the Cortinariales as a separate order. It includes the Crepidotaceae [50], a small group with eccentric or rudimentary stems, pastel shell-shaped caps and gills that at first appear pale but then become brown with pinkish or clay-brown spores. The species grow mostly on wood and other plant debris, rarely on soil. The Cortinariaceae [750] is a diverse group, inadequately investigated, including the *Cortinarius*, *Hebeloma*, *Naucoria*, *Gymnopilus* and *Inocybe* groups as well as several smaller collections such as *Phaeogalera*. Apart from *Gymnopilus* species on wood, most grow on soil, all possess brown spores and many are characterised by a cobwebby veil or *cortina* which may only be apparent in young fruiting bodies.

Camarophyllus niveus (Scop.: Fr.) Wünsche **Hygrophoraceae**
= *Hygrophorus nivea* (Fr.) Fr.

Small white agaric, with slightly greasy feel; in small trooping groups on soil, often acidic, amongst short mown or cropped grass, in pastures, and in open woodlands.

Dimensions cap 1 - 3 cm dia; stem 2 - 5 cm tall x 0.2 - 0.4 cm dia.
Cap white but tinged ivory with age; at first convex becoming more flattened or depressed, greasy and slightly striate at the margin when damp. Flesh whitish, generally thin but thick at centre.
Gills whitish, decurrent, broad, distant. **Spores** hyaline, smooth, ellipsoid, non-amyloid, 7-9 x 4-5.5 µm (basidia 4-spored); 10-12 x 5-6 µm (basidia 2-spored). Cystidia absent.
Stem concolorous with cap, slender, tapering slightly downwards, smooth. Ring absent. Flesh whitish, stuffed.
Odour not distinctive. **Taste** not distinctive.
Chemical tests none.
Occurrence late summer to autumn; common.
☐ Edible.

Camarophyllus pratensis (Pers.: Fr.) Kummer
= *Hygrophorus pratensis* (Pers.: Fr.) Fr.   **Hygrophoraceae**
Meadow Wax Cap

Medium-sized tawny orange agaric, with waxy feel; in widespread troops and clusters on soil, amongst short mown or cropped grass, in pastures and meadows.

Dimensions cap 3 - 8 cm dia; stem 2 - 5 cm tall x 1 - 1.5 cm dia.
Cap tawny buff with orange or ochraceous tinges; at first convex, becoming more flattened and broadly umbonate, and finally unevenly wavy and finely crazed when dry, waxy. Flesh ochraceous, thick, particularly at the centre.
Gills pallid buff or straw, deeply decurrent, broad, distant.
Spores hyaline, smooth, broadly ellipsoid or sub-spherical, non-amyloid, droplets, 5-7 x 4.5-5.5 µm. Basidia 4-spored. Cystidia absent.
Stem concolorous with cap but more pallid, stout, tapering slightly downwards, smooth. Ring absent. Flesh buff, stuffed.
Odour not distinctive. **Taste** not distinctive.
Chemical tests none.
Occurrence late summer to autumn; frequent.
☐ Edible.

Camarophyllus russocoriaceus (Berk. & Miller)
J Lange   **Hygrophoraceae**
= *Hygrophorus russocoriacea* Berk. & Miller

Small ivory-white agaric with slightly sticky cap; in trooping groups, on soil, often acidic, amongst short mown or cropped grass.

Dimensions cap 0.8 - 2 cm dia; stem 2 - 3.5 cm tall x 0.2 - 0.3 cm dia.
Cap ivory-white, tinged with yellow; at first convex, becoming more flattened, slightly greasy, at margin faintly striate. Flesh whitish.
Gills white, decurrent, broad, distant. **Spores** hyaline, smooth, ellipsoid, non-amyloid, 7-9 x 4-5 µm. Basidia 4-spored. Cystidia absent.
Stem concolorous with cap, slender, more or less equal or tapering slightly downwards, typically wavy. Ring absent. Flesh whitish, stuffed, becoming hollow.
Odour of sandalwood. **Taste** not distinctive.
Chemical tests none.
Occurrence late summer to autumn; infrequent.
■ Inedible.

Camarophyllus virgineus (Wulf.: Fr.) Kummer
Hygrophoraceae

Small ivory-white agaric with slightly greasy cap; in small trooping groups on soil, often acidic, amongst short grass in meadows, pastures and open wooded areas.

Dimensions cap 2 - 6 cm dia; stem 3 - 7 cm tall x 0.2 - 0.6 cm dia.
Cap ivory-white, tinged with yellow; at first convex, becoming more expanded with a broad umbo, slightly greasy, faintly striate at the margin. Flesh whitish.
Gills white, decurrent, broad, distant. **Spores** hyaline, smooth, ellipsoid, non-amyloid, 9-12 x 5-6 µm. Basidia 4-spored. Cystidia absent.
Stem concolorous with cap, slender, more or less equal or tapering slightly downwards, typically wavy. Ring absent. Flesh whitish, stuffed, becoming hollow.
Odour not distinctive. **Taste** not distinctive.
Chemical tests none.
Occurrence late summer to autumn; infrequent.
■ Inedible.

Hygrocybe acutopunicea Haller   **Hygrophoraceae**

Small reddish-yellow agaric with broadly conical greasy cap; typically in small trooping groups on soil amongst short mown or cropped grass.

Dimensions cap 1 - 4 cm dia; stem 3 - 5 cm tall x 0.5 - 0.8 cm dia.
Cap at first blood-red but becoming yellowish, particularly at the margin; at first conical, becoming expanded with pointed umbo, slightly greasy, finely radially fibrillose, striate one third to halfway to centre. Flesh yellowish and thin.
Gills pallid yellow, whitish at the margin, more cap-coloured at the base, adnexed or more or less free, broad, close. **Spores** hyaline, smooth, cylindrical-ellipsoid, non-amyloid, occasionally with granules, 8.5-10 x 4.5-5 µm. Basidia 4-spored. Some cylindrical marginal cells.
Stem orange-red, yellowish towards the base, fibrillose, more or less equal or tapering towards the base. Ring absent. Flesh yellowish, fibrous, stuffed, becoming hollow, splitting.
Odour not distinctive. **Taste** not distinctive.
Chemical tests none.
Occurrence late summer to autumn; rare. (Westonbirt Arboretum, Gloucestershire)
✠ Poisonous.

Hygrocybe calyptriformis (Berk. & Br.) Fayod
= *Hygrophorus amoenus* (Lasch) Quél.   **Hygrophoraceae**

Smallish pinkish-lilac agaric with narrowly conical greasy cap; typically solitary on soil amongst short mown cropped grass.

Dimensions cap 2.5 - 6 cm dia; stem 5 - 8 cm tall x 0.5 - 1.0 cm dia.
Cap rose with lilaceous tinge; at first narrowly conical, becoming more campanulate, slightly greasy, finely radially fibrillose, the margin splitting in older specimens. Flesh white, pink tinged below the cuticle, thin.
Gills rose when young, becoming more pallid with age, adnexed, broad, with crenulate edges, close. **Spores** hyaline, smooth, broadly ellipsoid, non-amyloid, occasionally with droplets, 5.5-7.5 x 4.5-5.5 µm. Basidia 4-spored. Cystidia fusiform or cylindrical, sparse.
Stem concolorous with cap or more pallid, finely fibrillose, stoutish, more or less equal. Ring absent. Flesh white, stuffed, becoming hollow
Odour not distinctive. **Taste** not distinctive.
Chemical tests none.
Occurrence late summer to autumn; infrequent.
✠ Poisonous.

Hygrocybe cantharellus (Schw.) Murr. **Hygrophoraceae**
= *Hygrophorus cantharellus* (Schw.) Fr.

Smallish scarlet agaric with greasy cap and pale yellow gills; solitary or in scattered groups on soil amongst short mown or cropped grass.

Dimensions cap 0.5 - 4 cm dia; stem 3 - 6 cm tall x 0.2 - 0.3 cm dia.
Cap scarlet or vermilion with more yellow-tinged margin; at first shallowly convex, becoming wavy-expanded or slightly depressed, finely scurfy. Flesh orange and thin.
Gills pallid yellow, becoming more egg-yellow at maturity, adnexed or decurrent, edges smooth, close. **Spores** hyaline, smooth, ellipsoid, non-amyloid, 8-11 x 5-6 µm. Basidia 4-spored. Cystidia absent.
Stem concolorous with cap, finely fibrillose, slender, more or less equal. Ring absent. Flesh orange, stuffed, becoming hollow.
Odour not distinctive. **Taste** not distinctive.
Chemical tests none.
Occurrence late summer to autumn; rare. (Stourhead, Wiltshire)
■ Inedible.

Hygrocybe chlorophana (Fr.) Wünsche **Hygrophoraceae**

Small yellowish-orange agaric with sticky cap; in small trooping groups on soil amongst short mown or cropped grass.

Dimensions cap 2 - 4 cm dia; stem 4 - 7 cm tall x 0.3 - 0.5 cm dia.
Cap yellow or lemon-yellow; at first convex or campanulate, becoming more flattened and broadly umbonate, very viscid particularly when young, margin striate. Flesh yellow and thin.
Gills at first pallid yellowish, becoming lemon-yellow, adnate, broad, fairly distant. **Spores** hyaline, smooth, ellipsoid, non-amyloid, 6.5-9 x 4-6 µm. Basidia 2- or 4-spored. Cystidia absent.
Stem concolorous with cap, slender, more or less equal, viscid. Ring absent. Flesh yellow and stuffed.
Odour not distinctive. **Taste** not distinctive.
Chemical tests none.
Occurrence late summer to autumn; frequent.
□ Edible.
Very similar to *H. flavescens*.

Hygrocybe citrina (Rea) J Lange **Hygrophoraceae**

Smallish lemon-yellow agaric with narrowly greasy cap; solitary or in scattered groups on soil amongst short mown or cropped grass.

Dimensions cap 1 - 2.5 cm dia; stem 3 - 5 cm tall x 0.2 - 0.3 cm dia.
Cap lemon-yellow; at first shallowly convex, becoming expanded, thinly viscid when damp, otherwise dull, striate more than halfway to centre. Flesh concolorous, watery and thin.
Gills pallid yellow, adnexed or emarginate, broad, with smooth edges, close. **Spores** hyaline, smooth, narrowly ellipsoid, non-amyloid, droplets, 5-8 x 2.5-3.5 µm. Basidia 4-spored. Cystidia absent.
Stem concolorous with cap or more pallid, finely fibrillose, slender, more or less equal. Ring absent. Flesh white, stuffed, becoming hollow.
Odour not distinctive. **Taste** not distinctive.
Chemical tests none.
Occurrence late summer to autumn; rare. (Stourhead, Wiltshire)
✚ Poisonous.

Hygrocybe coccinea (Schaeff.: Fr.) Kummer
= *Hygrophorus coccineus* (Schaeff.: Fr.) Fr.　　**Hygrophoraceae**

Small agaric with blood-red domed cap and stem, pale yellow or red gills, in small trooping groups on soil amongst mown or cropped grass.

Dimensions cap 2 - 4 cm dia; stem 2 - 5 cm tall x 0.3 - 0.8 cm dia.
Cap scarlet or blood-red; hemispherical becoming more convex, faintly fibrillose, greasy or matt. Flesh reddish-orange, unchanging and thin.
Gills pallid yellow then blood-red with age, broadly adnate with decurrent tooth, somewhat distant. **Spores** hyaline, smooth, elongated-ellipsoid, non-amyloid, droplets, 7.5-11 x 4-5.5 μm. Basidia 4-spored. Cystidia absent.
Stem concolorous with cap or more orange, more pallid at base, longitudinally fibrillose, more or less equal or compressed towards base. Ring absent. Flesh orange and fibrous.
Odour not distinctive. **Taste** not distinctive.
Chemical tests none.
Occurence late summer to autumn; infrequent.
■ Inedible

Hygrocybe conica (Scop.: Fr.) Kummer　**Hygrophoraceae**
= *Hygrophorus conicus* (Scop.: Fr.) Fr.

Smallish agaric with orange-red acutely conical cap and yellowish stem; in small trooping groups on soil amongst short mown or cropped grass.

Dimensions cap 2 - 5 cm dia; stem 2 - 6 cm tall x 0.8 - 1 cm dia.
Cap yellowish-orange with scarlet tinge, slowly blackening with age or on bruising; acutely conical and irregularly lobed, not greasy. Flesh yellowish, blackening when cut, thin.
Gills pallid yellow with greyish tinge, adnexed or free, broad, fairly crowded. **Spores** hyaline, smooth, ellipsoid, non-amyloid, droplets, 7-9 x 4-5 μm. (basidia 4-spored); 8.5-9.5 x 5-6 μm. (basidia 2-spored). Cystidia absent.
Stem yellowish but blackening, particularly at the base, slender, more or less equal, finely striate. Ring absent. Flesh yellowish, fibrous, and hollow.
Odour not distinctive. **Taste** not distinctive.
Chemical tests none.
Occurrence late summer to autumn; common.
■ Inedible.

Hygrocybe conicoides (P Orton) Orton & Watling
= *Hygrophorus conicoides* P Orton　　**Hygrophoraceae**

Smallish agaric with orange-red greasy conical cap and egg-yellow stem; in small trooping groups on sandy soil and coastal dunes amongst short grass.

Dimensions cap 1 - 4 cm dia; stem 2 - 7 cm tall x 0.2 - 0.5 cm dia.
Cap at first reddish-orange, slowly blackening with age; convex or campanulate becoming more flattened and broadly umbonate, greasy when moist, finely silky when dry. Flesh reddish, blackening when cut, thin.
Gills egg-yellow with pink tinge, more so next to the cap, blackening at the edges, adnate or adnexed, broad, fairly crowded.
Spores hyaline, smooth, elongated-ellipsoid, non-amyloid, droplets, 10-13 x 4-5 μm. Basidia 4-spored. Cystidia absent.
Stem yellowish but blackening more rapidly than the cap, slender, more or less equal, finely striate. Ring absent. Flesh yellowish-orange, fibrous, and hollow.
Odour not distinctive. **Taste** not distinctive.
Chemical tests none.
Occurrence late summer to autumn; infrequent.
■ Inedible.

Hygrocybe flavescens (Kauff.) A H Smith & Hes.
Hygrophoraceae

Small yellow agaric with slimy cap; in small trooping groups on soil often calcareous, amongst short mown or cropped grass favouring dry meadows.

Dimensions cap 3 - 5 cm dia; stem 5 - 7 cm tall x 0.5 - 1 cm dia.
Cap pallid lemon-yellow or orange-yellow; convex, becoming more flattened, viscid particularly when young, somewhat striate from the margin. Flesh yellow or orange-yellow and thin.
Gills pallid yellow with orange tinge, adnate or emarginate, broad, close. **Spores** hyaline, smooth, broadly ellipsoid, non-amyloid, droplets, 6-8.5 x 5-7 μm. Basidia 4-spored. Cystidia absent.
Stem yellow or orange-yellow, more pallid at the apex, slender, more or less equal, dry. Ring absent. Flesh yellow and hollow.
Odour unpleasant. **Taste** not distinctive.
Chemical tests none.
Occurrence late summer to autumn; frequent.
■ Inedible.

Hygrocybe glutinipes (J Lange) P Orton   **Hygrophoraceae**
= *Hygrocybe vitellina* ss. Bres.

Small bright yellow agaric with flattish cap and decurrent gills, slimy in damp weather; in trooping groups, sometimes clustered, on soil, amongst moss and damp grass.

Dimensions cap 0.5 - 2.5 cm dia; stem 2 - 3 cm tall x 0.1 - 0.3 cm dia.
Cap bright yellow, somewhat translucent, with striate margin; convex then flattened or slightly depressed at the centre, viscid. Flesh yellowish and thin.
Gills concolorous with cap, adnate with steeply decurrent tooth, fairly narrow, distant. **Spores** hyaline, smooth, ellipsoid, non-amyloid, droplets, 5-8 x 3.5-4 μm. Basidia 2- or 4-spored. Cystidia absent.
Stem concolorous with cap, more or less equal, viscid. Ring absent. Flesh whitish, soft, stuffed, becoming hollow.
Odour not distinctive. **Taste** not distinctive.
Chemical tests none.
Occurrence late summer to autumn; infrequent.
■ Inedible.

Hygrocybe intermedia (Pass.) Fayod   **Hygrophoraceae**
= *Hygrophorus intermedius* Pass.

Medium-sized agaric with golden-yellow conical cap and stout stem; solitary or scattered on soil amongst short grass in pastures and meadows.

Dimensions cap 2 - 7 cm dia; stem 6 - 9 cm tall x 0.9 - 1.8 cm dia.
Cap deep ochraceous or golden-yellow; conical, becoming more flattened with an irregular wavy margin, finely fibrillose, not greasy. Flesh yellowish.
Gills pallid, becoming flushed with cap colour, adnexed or free, broad, fairly crowded. **Spores** hyaline, smooth, ellipsoid, non-amyloid, 8-11 x 5-6 μm. Basidia 4-spored. Cystidia absent.
Stem concolorous with cap but pallid at the base, surface coarsely fibrillose, stout, tapering downwards. Ring absent. Flesh yellowish, stuffed, becoming hollow.
Odour not distinctive. **Taste** not distinctive.
Chemical tests none.
Occurrence late summer to autumn; infrequent or rare. (Lyme Park, Cheshire)
■ Inedible.

Hygrocybe konradii Haller   Hygrophoraceae

Small reddish-orange agaric with greasy cap; in small trooping groups on soil amongst short mown or cropped grass.

Dimensions cap 3 - 7 cm dia; stem 3 - 7 cm tall x 0.4 - 0.8 cm dia.
Cap yellowish towards the margin, more orange or scarlet at the centre; bluntly conical or convex, flattened, greasy particularly when young. Flesh sulphur yellow and thin.
Gills at first pallid yellow, becoming more orange-yellow, adnexed almost free, broad, close. **Spores** hyaline, smooth, ellipsoid, non-amyloid, droplets, 9-12 x 6-9 μm. Basidia 4-spored. Cystidia absent.
Stem concolorous with cap, whitish at the base, slender, more or less equal, dry. Ring absent. Flesh yellow, fibrillose, hollow, splitting.
Odour not distinctive. **Taste** not distinctive.
Chemical tests none.
Occurrence late summer to autumn; infrequent.
■ Inedible.

Hygrocybe laeta (Pers.: Fr.) Kummer   Hygrophoraceae

Small orange-brown agaric with very slimy cap and stem; in small trooping groups on soil amongst short mown or cropped grass, often with mosses in moorland locations.

Dimensions cap 1.5 - 3.5 cm dia; stem 3 - 7 cm tall x 0.3 - 0.5 cm dia.
Cap orange-brown sometimes with olivaceous tinge; at first hemispherical, becoming convex and then flattened, somewhat hygrophanous, striate two-thirds to centre, margin slightly toothed, very viscid. Flesh orange-brown, thin and watery.
Gills at first pallid greyish-white with olivaceous tinge, becoming more salmon, slightly decurrent, broad, close. **Spores** hyaline, smooth, ellipsoid, non-amyloid, droplets, 5.5-7.5 x 4-5 μm. Basidia 2- or 4-spored. Marginal cells filiform.
Stem concolorous with cap with olivaceous tinge at the apex when young, whitish at the base, slender, more or less equal, very viscid. Ring absent. Flesh concolorous, fibrous, stuffed or full.
Odour not distinctive. **Taste** not distinctive.
Chemical tests none.
Occurrence late summer to autumn; infrequent.
■ Inedible.

Hygrocybe langei Kühn.   Hygrophoraceae
= *Hygrocybe acutoconica* (Clem.) Sing.
= *Hygrophorus langei* (Kühn.) Pearson

Smallish golden-yellow conical agaric with slimy cap; in small trooping groups on soil amongst short mown or cropped grass.

Dimensions cap 3 - 6 cm dia; stem 3 - 5 cm tall x 0.6 - 1 cm dia.
Cap golden-yellow, tinged orange; sharply conical, viscid particularly when young, faintly striate at the margin. Flesh yellow and thin.
Gills pallid yellow, adnexed, broad, close. **Spores** hyaline, smooth, ellipsoid, non-amyloid, droplets, 10-14 x 5.5-8 μm. Basidia 4-spored. Cystidia absent.
Stem concolorous with cap, whitish at the baser, more or less equal, not viscid. Ring absent. Flesh yellow and stuffed.
Odour not distinctive. **Taste** not distinctive.
Chemical tests none.
Occurrence late summer to autumn; infrequent.
■ Inedible.

Hygrocybe marchii (Bres.) Sing.   **Hygrophoraceae**

Small scarlet-orange agaric; in small trooping groups on soil amongst short mown or cropped grass.

Dimensions cap 2 - 4.5 cm dia; stem 2 - 4 cm tall x 0.2 - 0.5 cm dia.
Cap scarlet or orange, sometimes more ochraceous; convex becoming more flattened, more or less dry, radially fibrillose. Flesh yellow, more cap colour beneath cuticle, thin.
Gills pallid when young, becoming more orange, yellowish at the edges, adnate, broad, distant. **Spores** hyaline, smooth, ellipsoid, non-amyloid, droplets, 6-10 x 4-6 μm. Basidia 4-spored. Cystidia filiform, sparse.
Stem orange-yellow, more pallid towards the base, fibrillose, slender, more or less equal. Ring absent. Flesh yellow and stuffed.
Odour not distinctive. **Taste** not distinctive.
Chemical tests none.
Occurrence late summer to autumn; infrequent.
■ Inedible.

Hygrocybe miniata (Fr.) Kummer   **Hygrophoraceae**
= *Hygrophorus miniatus* (Fr.) Fr.

Small scarlet; agaric; in small trooping groups on soil amongst short mown or cropped grass in pastures, heaths and open woodlands.

Dimensions cap 0.5 - 1.5 cm dia; stem 2 - 5 cm tall x 0.2 - 0.5 cm dia.
Cap bright scarlet, at first convex, becoming more flattened, finely scurfy, not greasy. Flesh reddish-orange and thin.
Gills reddish-orange with pallid yellow edge, adnate, broad, fairly distant. **Spores** hyaline, smooth, ellipsoid, non-amyloid, occasionally with droplets, 7.5-10 x 5-6 μm. Basidia 4-spored. Cystidia absent.
Stem concolorous with cap, slender, more or less equal, smooth, shiny. Ring absent. Flesh reddish-orange and stuffed.
Odour not distinctive. **Taste** not distinctive.
Chemical tests none.
Occurrence late summer to autumn; infrequent.
□ Edible.
Distinguished from other small red *Hygrocybe* species by the dry scurfy cap.

Hygrocybe nigrescens (Quél.) Kühn.   **Hygrophoraceae**
= *Hygrocybe pseudoconica* J Lange
= *Hygrophorus nigrescens* (Quél.) Quél.

Medium-sized orange-red broadly conical agaric, blackening with age; in trooping groups on soil amongst short grass in pastures and open woodlands.

Dimensions cap 3.5 - 5.5 cm dia; stem 3 - 7 cm tall x 0.6 - 1 cm dia.
Cap at first orange or scarlet, becoming markedly blackened with age; convex or campanulate, typically irregularly lobed, becoming more flattened, waxy. Flesh pallid, yellowish and thin.
Gills at first pallid yellow, but blackening with age, adnexed or free, fairly broad, distant. **Spores** hyaline, smooth, ellipsoid, non-amyloid, droplets, 8-12 x 5-6 μm. Basidia 4-spored. Cystidia absent.
Stem yellowish with scarlet tinge, pallid at the base, becoming streaked black with age, stout, equal, greasy. Ring absent. Flesh white, but blackening when cut, stuffed.
Odour not distinctive. **Taste** not distinctive.
Chemical tests none.
Occurrence late summer to autumn; infrequent.
□ Edible.

Hygrocybe persistens (Britz.) Britz.   **Hygrophoraceae**

Smallish orange-yellow agaric with greasy cap; in small trooping groups on soil amongst short mown or cropped grass, often adjacent to coniferous trees.

Dimensions cap 3 - 8 cm dia; stem 6 - 10 cm tall x 0.5 - 0.8 cm dia.
Cap orange with ochraceous tinge; conical or campanulate, becoming more flattened and broadly umbonate, greasy, particularly when young. Flesh pallid, tinged yellow and thin.
Gills yellowish, adnate, broad, fairly distant. **Spores** hyaline, smooth, cylindrical-ellipsoid, non-amyloid, 9-14 x 5-6 μm. Basidia 1-, 2- or 3-spored. Cystidia absent.
Stem concolorous with cap, slender, more or less equal, greasy, fibrillose. Ring absent. Flesh pallid, yellowish and stuffed.
Odour not distinctive. **Taste** not distinctive.
Chemical tests none.
Occurrence late summer to autumn; infrequent.
■ Inedible.

Hygrocybe psittacina (Schaeff.: Fr.) Wünsche
= *Hygrophorus psittacinus* (Schaeff.: Fr.) Fr.   **Hygrophoraceae**
Parrot Wax Cap

Small greenish-yellow agaric with slimy cap; in small trooping groups on soil amongst short mown or cropped grass.

Dimensions cap 1 - 3 cm dia; stem 2 - 4 cm tall x 0.2 - 0.5 cm dia.
Cap at first greenish, becoming markedly more yellow with age and on drying; convex or campanulate, becoming more flattened and broadly umbonate, viscid with green gluten particularly when young, margin striate. Flesh pallid, tinged greenish-yellow and thin.
Gills yellowish towards the edge, more greenish towards the cap, adnate, broad, fairly distant. **Spores** hyaline, smooth, ellipsoid, non-amyloid, 8-9.5 x 4-5.5 μm. Basidia 4-spored. Cystidia absent.
Stem greenish towards the apex, otherwise more yellow, slender, more or less equal, very viscid. Ring absent. Flesh pallid, tinged greenish-yellow and stuffed.
Odour not distinctive. **Taste** not distinctive.
Chemical tests none.
Occurrence late summer to autumn; infrequent.
■ Inedible.
Confusion can arise because of the marked colour variations. Both illustrations are typical. Inset: older fruiting bodies.

Hygrocybe punicea (Fr.) Kummer   **Hygrophoraceae**
= *Hygrophorus puniceus* (Fr.) Fr.

Medium-sized blood-red agaric with greasy cap and stout stem; in trooping groups, sometimes clustered, on soil amongst short mown or cropped grass in pastures and on heaths.

Dimensions cap 3 - 7 cm dia; stem 5 - 13 cm tall x 0.6 - 2 cm dia.
Cap at first deep blood-red, with a whitish bloom when young, fading to orange-red with age; convex or campanulate, becoming more flattened and broadly umbonate with an irregular wavy margin, greasy. Flesh yellowish and thin.
Gills at first yellowish, becoming tinged with cap colour, adnexed or free, broad, fairly distant. **Spores** hyaline, smooth, ellipsoid, non-amyloid, droplets, 9-12 x 5-6 μm. Basidia 2- or 4-spored.
Stem yellowish or tinged cap colour, but pallid at the base, fibrillose, stout, more or less equal but tapering at the base, greasy. Ring absent. Flesh whitish throughout, very fibrous, stuffed, becoming hollow.
Odour not distinctive. **Taste** not distinctive.
Chemical tests none.
Occurrence late summer to autumn; infrequent.
■ Inedible.
Distinguished from *H. splendidissima* which possesses yellowish flesh colour in the stem

Hygrocybe splendidissima (P Orton) Moser **Hygrophoraceae**
= *Hygrocybe punicea forma splendidissima* (P Orton) Reid

Medium-sized blood-red agaric with greasy cap and stout stem; in trooping groups, sometimes clustered, on soil amongst short mown or cropped grass in pastures and on moorland.

Dimensions cap 2 - 10 cm dia; stem 4 - 10 cm tall x 0.6 - 2 cm dia.
Cap at first deep blood-red with a whitish bloom when young, fading to orange-red with age; convex or campanulate, becoming more flattened and broadly umbonate, with an irregular wavy margin, greasy. Flesh yellowish and thin.
Gills at first yellowish, becoming tinged with cap colour, adnexed or free, broad, fairly distant. **Spores** hyaline, smooth, ellipsoid, non-amyloid, droplets, 7-10 x 4-5.5 µm. Basidia 2- or 4- spored.
Stem yellowish or tinged cap colour, but pallid at the base, smooth, silky, stout, more or less equal but tapering at the base, greasy. Ring absent. Flesh yellowish throughout except at the base, fibrous, stuffed, becoming hollow.
Odour not distinctive. **Taste** not distinctive.
Chemical tests none.
Occurrence late summer to autumn; infrequent, generally in northern Britain.
■ Inedible.
Distinguished from *H. punicea* which possesses wholly whitish flesh colour in the stem

Hygrocybe strangulata P Orton   **Hygrophoraceae**
= *Hygrophorus strangulatus* P Orton

Smallish scarlet-orange agaric with slimy cap, yellow gills and orange stem; in trooping groups, sometimes clustered, on soil, often sandy, amongst short mown or cropped grass in pastures and in open woodlands.

Dimensions cap 0.5 - 3.5 cm dia; stem 2 - 5 cm tall x 0.3 - 0.6 cm dia.
Cap scarlet-orange with yellowish margin; convex, becoming more flattened, irregular and somewhat depressed, viscid. Flesh orange and thin.
Gills at first yellow, becoming more orange, adnate or slightly decurrent, fairly broad, distant. **Spores** hyaline, smooth, ellipsoid but waisted in one plane, non-amyloid, droplets, 7-9 x 4-5 µm. Basidia 4-spored. Cystidia absent.
Stem reddish-orange or concolorous with cap, more pallid towards the base, more or less equal but tapering at the base, slightly wavy. Ring absent. Flesh concolorous, stuffed, becoming hollow.
Odour not distinctive. **Taste** not distinctive.
Chemical tests none.
Occurrence late summer to autumn; infrequent.
■ Inedible.

Hygrocybe unguinosa (Fr.) Karst.   **Hygrophoraceae**
= *Hygrophorus unguinosus* (Fr.) Fr.

Smallish grey agaric with slimy cap and wavy stem; in trooping groups, sometimes clustered, on soil amongst short mown or cropped grass in pastures and in open woodlands.

Dimensions cap 2 - 5 cm dia; stem 3 - 6 cm tall x 0.3 - 0.6 cm dia.
Cap pallid greyish-brown; convex or campanulate, becoming more flattened and broadly umbonate, with an irregular, wavy, striate margin, viscid. Flesh concolorous, watery and thin.
Gills whitish, adnate, broad, strongly interveined, fairly distant.
Spores hyaline, smooth, broadly ellipsoid, non-amyloid, droplets, 7-8.5 x 4-5 µm. Basidia 4-spored. Cystidia absent.
Stem concolorous with cap, whitish towards the base, grooved, compressed, wavy, more or less equal, but tapering at the base, viscid. Ring absent. Flesh concolorous, stuffed, becoming hollow.
Odour not distinctive. **Taste** not distinctive.
Chemical tests none.
Occurrence autumn to early winter; rare. (Westonbirt Arboretum, Gloucestershire)
■ Inedible.

Hygrophorus cossus (Sow.: Berk.) Fr.   **Hygrophoraceae**
= *Limacium eburneum var cossus* (Sow.: Berk.) J Lange

Smallish white agaric with slimy cap, smelling strong and unpleasant; in trooping groups on soil, often calcareous, amongst grass in mixed woodland, favouring beech.

Dimensions cap 3 - 7 cm dia; stem 4 - 7 cm tall x 0.5 - 1 cm dia.
Cap whitish, sometimes with buff tinge towards the centre; at first convex, becoming more flattened, viscid. Flesh white and thin.
Gills white, decurrent, broad, fairly distant. **Spores** hyaline, smooth, ellipsoid, non-amyloid, droplets, 8-10 x 4-5 μm. Basidia 4-spored. Cystidia absent.
Stem whitish, slender, more or less equal, finely pruinose towards the apex, viscid. Ring absent. Flesh white and stuffed.
Odour unpleasant, reminiscent of the Goat Moth larva. **Taste** not distinctive.
Chemical tests all parts chrome yellow with KOH.
Occurrence autumn; infrequent.
■ Inedible.

Hygrophorus eburneus (Bull.: Fr.) Fr.   **Hygrophoraceae**
= *Limacium eburneum* (Bull.: Fr.) Kummer

Smallish ivory-white and very slimy agaric; in small trooping groups on soil amongst grass in mixed woodland, favouring oak and beech.

Dimensions cap 3 - 7 cm dia; stem 4 - 7 cm tall x 0.5 - 1 cm dia.
Cap pure ivory-white; at first convex, becoming more flattened, viscid. Flesh white and thin.
Gills white, decurrent, broad, fairly distant. **Spores** hyaline, smooth, ellipsoid, non-amyloid, 6-8 x 4-5 μm. Basidia 4-spored. Cystidia absent.
Stem concolorous with cap, tapering slightly downwards, coarsely pruinose towards the apex, viscid. Ring absent. Flesh white and stuffed.
Odour not distinctive. **Taste** not distinctive.
Chemical tests cap yellowish with KOH.
Occurrence autumn; infrequent or rare. (Leigh Woods, Bristol)
☐ Edible.

Hygrophorus hypothejus (Fr.: Fr.) Fr.   **Hygrophoraceae**
= *Limacium hypothejum* (Fr.: Fr.) Kummer   Herald of Winter

Smallish slimy agaric with dull brown cap and distinctive yellowish gills and stem; in trooping groups on soil in pinewoods, often favouring the edges of grassy rides.

Dimensions cap 3 - 7 cm dia; stem 4 - 7 cm tall x 0.7 - 1.4 cm dia.
Cap dull olive-brown, slightly more pallid at the margin; at first convex, becoming more flattened and sometimes slightly depressed, viscid. Flesh pallid yellow, bruising more orange, thin.
Gills pallid yellow but darkening with age, decurrent, broad, fairly distant. **Spores** hyaline, smooth, ellipsoid, non-amyloid, 7-10 x 4-5 μm. Basidia 4-spored. Cystidia absent.
Stem pallid yellow, sometimes tinged with orange, more or less equal or tapering slightly downwards, viscid below ring zone. Ring absent but superior ring-like swelling. Flesh pallid yellow, bruising more orange, stuffed.
Odour not distinctive. **Taste** not distinctive.
Chemical tests no reaction with KOH.
Occurrence autumn to winter; frequent.
☐ Edible.
More or less the only species of the Hygrophoraceae to appear and thrive after the first frosts, hence the common name.

Hygrophorus leucophaeus ([Scop.] Fr.) Karst.
= *Limacium leucophaeum* ([Scop.] Fr.) Hein.   **Hygrophoraceae**

Small slimy agaric with whitish buff cap, gills and stem; in trooping groups on soil in broad-leaf woods, favouring beech.

Dimensions cap 2 - 5 cm dia; stem 3 - 7 cm tall x 0.4 - 0.7 cm dia.
Cap cream with yellowish-brown or buff centre; at first convex, becoming more flattened and umbonate, viscid. Flesh whitish, thin except at centre.
Gills buff, decurrent, narrow, fairly distant. **Spores** hyaline, smooth, ellipsoid, non-amyloid, 6-8 x 4-5 µm. Basidia 4-spored. Cystidia absent.
Stem concolorous with cap, slender, more or less equal or tapering slightly downwards, viscid. Ring absent. Flesh whitish and stuffed.
Odour not distinctive. **Taste** not distinctive.
Chemical tests no reaction with KOH.
Occurrence late summer to autumn; infrequent.
☐ Edible.

Armillaria mellea (Wahl.: Fr.) Kummer   **Tricholomataceae**
= *Clitocybe mellea* (Wahl.: Fr.) Ricken   Honey Fungus

Distinctive, large, fleshy, tawny, finely scaly agaric with ring on stem; parasitic, in dense caespitose clusters on and around the stumps of broad-leaf and coniferous trees.

Dimensions cap 3 - 15 cm dia; stem 6 - 15 cm tall x 0.5 - 1.5 cm dia.
Cap variable through ochraceous, tawny and dark brown, covered in darker fibrillose scales, more so towards the centre; at first convex, becoming flattened, wavy and shallowly infundibuliform. Flesh white, thin and firm.
Gills at first white, becoming yellowish and then brown, with darker spots in old specimens, decurrent, broad, crowded. **Spores** hyaline, smooth, ellipsoid, non-amyloid, droplets, 8-10 x 5-7 µm. Basidia 4-spored. Marginal cells cylindrical or clavate.
Stem at first whitish, becoming yellowish or reddish-brown, more or less equal or tapering towards the base, finely woolly; with long black bootlace-like rhizomorphs. Ring yellowish, cottony or woolly, superior, fairly persistent. Flesh white, fairly firm, stuffed or full.
Odour faint, acidic. **Taste** strong.
Chemical tests none.
Occurrence summer to autumn; very common.
☐ Edible if cooked.
Inset: alternative form.

Armillaria tabescens (Scop.: Fr.) Emel   **Tricholomataceae**
= *Clitocybe tabescens* (Scop.: Fr.) Bres.

Fleshy, tawny, finely scaly agaric; in dense caespitose clusters on and around the stumps of broad-leaf trees, favouring oak.

Dimensions cap 4 - 10 cm dia; stem 5 - 8 cm tall x 0.8 - 1.2 cm dia.
Cap variable through ochraceous, tawny and dark brown, covered in darker fibrillose scales, more so towards the centre; at first convex, becoming flattened, wavy and shallowly infundibuliform. Flesh white, thin and firm.
Gills at first white, becoming pinkish-brown, adnate-decurrent, broad, crowded. **Spores** pallid cream, smooth, ellipsoid, non-amyloid, droplets, 8-10 x 5-7 µm. Basidia 4-spored. Marginal cells cylindrical or clavate.
Stem pallid ochraceous-brown, more or less equal or tapering towards the base, finely woolly. Ring absent. Flesh white, fairly firm, stuffed or full.
Odour faint, acidic. **Taste** strong.
Chemical tests none.
Occurrence summer to autumn; rare. (Reydon Wood, Suffolk)
☐ Edible if cooked.

Baeospora myosura (Fr.) Sing.   Tricholomataceae
= *Collybia myosura* (Fr.) Quél.
= *Collybia conigena* (Pers.: Fr.) Kummer

Smallish tan or brown agaric; solitary, rooting on buried pine cones and other coniferous debris.

Dimensions cap 1 - 3 cm dia; stem 3 - 5 cm tall x 0.1 - 0.2 cm dia.
Cap pallid date-brown or tan; at first convex, becoming flattened. Flesh concolorous and thin.
Gills whitish, adnate-adnexed, narrow, very crowded.
Spores hyaline, smooth, ellipsoid, amyloid, 3-4.5 x 1.5-2 µm. Basidia 4-spored. Cystidia fusiform or clavate.
Stem concolorous with cap, but more pallid, slender, more or less equal, extending into a rooting base covered with hairs. Ring absent. Flesh pallid brown and fairly tough.
Odour not distinctive. **Taste** not distinctive.
Chemical tests none.
Occurrence late summer to early winter; infrequent.
■ Inedible.

Calocybe carnea (Bull.: Fr.) Donk   Tricholomataceae
= *Tricholoma carneum* (Bull.: Fr.) Kummer

Smallish fleshy agaric with flesh-pink cap and white gills; solitary or grouped on soil amongst grass in parks and mixed woods, often in open grasslands.

Dimensions cap 3 - 5 cm dia; stem 3 - 5 cm tall x 0.3 - 0.8 cm dia.
Cap flesh-pink, sometimes with slight brown tinge at the centre; at first convex becoming expanded with somewhat irregular margin, smooth but pruinose towards the margin, dry. Flesh white with pink tinge beneath cuticle, thin.
Gills white, emarginate, narrow, crowded. **Spores** hyaline, smooth, cylindrical-ellipsoid, non-amyloid, droplets, 4-6 x 2-4 µm. Basidia 4-spored. Cystidia absent.
Stem pallid, concolorous with cap, finely fibrillose, tapering slightly upwards. Ring absent. Flesh whitish, firm, full.
Odour faint, of fruit. **Taste** not distinctive.
Chemical tests none.
Occurrence early summer to early autumn; frequent. (Stourhead, Wiltshire)
□ Edible.

Calocybe chrysenteron (Bull.: Fr.) Sing.   Tricholomataceae
= *Tricholoma crysenteron* (Bull.: Fr.) Sing.

Medium-sized fleshy agaric, reddish-orange cap with yellow margin, yellowish gills; in trooping groups or clusters on soil in coniferous woods.

Dimensions cap 3 - 6 cm dia; stem 3 - 6 cm tall x 0.3 - 0.5 cm dia.
Cap reddish-orange, merging to golden-yellow at the margin, smooth; at first convex, becoming expanded, smooth, dry. Flesh white and thin.
Gills whitish, emarginate, narrow, crowded. **Spores** hyaline, smooth, ellipsoid or sub-spherical, non-amyloid, droplets, 3-4.5 x 2-3 µm. Basidia 4-spored. Cystidia absent.
Stem whitish, smooth. Ring absent. Flesh white, firm, full.
Odour faint, of meal. **Taste** of meal.
Chemical tests none.
Occurrence late summer to autumn; infrequent or rare.
■ Inedible.

Calocybe gambosa (Fr.) Sing.

Tricholomataceae

= *Tricholoma gambosum* (Fr.) Kummer   St George's Mushroom

Medium to large fleshy cream agaric; typically in rings or troops on soil, most commonly found in pastures but also less frequently appearing in mixed woods.

Dimensions cap 5 - 15 cm dia; stem 3 - 7 cm tall x 2 - 3 cm dia.
Cap whitish-cream, often with brownish tinge, smooth; at first sub-spherical becoming expanded and irregularly convex with an incurved margin. Surface smooth. Flesh white and firm.
Gills whitish, emarginate, narrow, crowded. **Spores** hyaline, smooth, ellipsoid, non-amyloid, droplets absent, 5.5-6 x 3-4 μm. Basidia 4- and less frequently 2-spored. Cystidia absent.
Stem whitish, smooth. Ring absent. Flesh white, firm, full.
Odour faint, of meal. **Taste** of meal.
Chemical tests none.
Occurrence spring, traditionally fruiting on St George's Day; occasional.
☐ Edible and excellent.
Inset: growing in woodland.

Calocybe ionides (Bull.: Fr.) Donk   **Tricholomataceae**

= *Tricholoma ionides* (Bull.: Fr.) Kummer

Small to medium-sized fleshy agaric, with bluish-violet cap and yellowish gills; in small trooping groups on soil in broad-leaf and mixed woods.

Dimensions cap 3 - 6 cm dia; stem 3 - 8 cm tall x 0.8 - 1.4 cm dia.
Cap bluish-violet, smooth; at first convex, becoming expanded with slightly incurved margin, smooth, dry. Flesh white with violaceous tinges beneath cuticle, thin.
Gills at first white, soon becoming pallid yellow, emarginate, narrow, crowded. **Spores** hyaline, smooth, ellipsoid, non-amyloid, droplets, 5-6 x 3-4 μm. Basidia 4-spored. Cystidia absent.
Stem whitish, with violaceous tinge at base, finely fibrillose, tapering slightly upwards. Ring absent. Flesh white, with violaceous tinge in stem base, firm, full.
Odour faint, of meal. **Taste** not distinctive.
Chemical tests none.
Occurrence late summer to autumn; rare. (Noxon Park, Gloucestershire)
☐ Edible.

Calocybe persicolor (Fr.) Sing.   **Tricholomataceae**

Smallish fleshy agaric with flesh-pink cap and white gills; often caespitose on soil amongst grass in parks and mixed woods, less frequently in open grasslands.

Dimensions cap 3 - 5 cm dia; stem 3 - 5 cm tall x 0.3 - 0.8 cm dia.
Cap dirty-pink, sometimes with slight ochraceous tinge at the centre; at first convex, becoming expanded with somewhat irregular margin, smooth but pruinose towards the margin, dry. Flesh white with pink tinge beneath cuticle, thin.
Gills white, emarginate, narrow, crowded. **Spores** hyaline, smooth, ellipsoid, non-amyloid, droplets, 4-6 x 2-4 μm. Basidia 4-spored. Cystidia absent.
Stem pallid, concolorous with cap, finely fibrillose, tapering slightly upwards, hairy at the base. Ring absent. Flesh whitish, firm, full.
Odour faint, of fruit. **Taste** not distinctive.
Chemical tests none.
Occurrence autumn; rare. (Stourhead, Wiltshire)
☐ Edible.

Pseudoclitocybe cyathiformis (Bull.: Fr.) Sing.
= *Clitocybe cyathiformis* (Bull.: Fr.) Kummer   **Tricholomataceae**

Smallish fleshy agaric, dark grey-brown, with funnel-shaped cap and decurrent gills; solitary or in small troops on soil in mixed woods amongst grass and debris.

Dimensions cap 2 - 7 cm dia; stem 4 - 8 cm tall x 0.5 - 1 cm dia.
Cap greyish-brown or umber, smooth; infundibuliform, with strongly inrolled margin. Flesh pallid and thin.
Gills pallid, greyish, becoming tinged brown, decurrent, fairly broad, crowded. **Spores** hyaline, smooth, ellipsoid, amyloid, droplets, 8-11 x 5-6 µm. Basidia 4-spored. Cystidia absent.
Stem concolorous with cap, but more pallid, silky fibrillose, swollen and whitish downy at the base. Ring absent. Flesh pallid and stuffed.
Odour not distinctive. **Taste** not distinctive.
Chemical tests none.
Occurrence late autumn to spring; infrequent.
☐ Edible.

Clitocybe barbularum (Romagn.) P Orton **Tricholomataceae**

Small umbilicate brown agaric with decurrent gills; solitary or in small troops, sometimes tufted, amongst mosses in sand dunes.

Dimensions cap 1.5 - 3.5 cm dia; stem 2 - 3 cm tall x 0.2 - 0.3 cm dia.
Cap dark grey-brown, smooth; umbilicate, margin tending to be inrolled, surface slightly gelatinous when damp. Flesh brown, thin, and fairly tough.
Gills grey-brown, decurrent, fairly broad, crowded. **Spores** hyaline, smooth, ellipsoid, non-amyloid, droplets, 5-7 x 3.5-4 µm. Basidia 4-spored. Cystidia absent.
Stem concolorous with cap, smooth, slender. Ring absent. Flesh brown, cartilaginous, full.
Odour slight, of meal. **Taste** not distinctive.
Chemical tests none.
Occurrence autumn; localised.
■ Inedible.

Clitocybe candicans (Pers.: Fr.) Kummer **Tricholomataceae**

Small fleshy agaric, whitish with shallowly funnel-shaped cap and decurrent gills; solitary or in small troops, on soil in broad-leaf woods, favouring beech.

Dimensions cap 2 - 5 cm dia; stem 2 - 5 cm tall x 0.7 - 1 cm dia.
Cap whitish, smooth; at first convex or bun-shaped, later shallowly infundibuliform or umbilicate, with slightly inrolled margin. Flesh white and thick.
Gills white, adnate-decurrent, fairly broad, crowded.
Spores hyaline, smooth, ellipsoid, non-amyloid, droplets absent, 4-5 x 3-3.5 µm. Basidia 4-spored. Cystidia absent.
Stem concolorous, shiny, with white hairs at the base. Ring absent. Flesh white, cartilaginous, full, becoming stuffed.
Odour not distinctive. **Taste** not distinctive.
Chemical tests none.
Occurrence summer to autumn; rare. (Bridgham near Thetford, Norfolk)
■ Inedible.

Clitocybe clavipes (Pers.: Fr.) Kummer   Tricholomataceae

Medium-sized agaric with brownish-grey funnel-shaped cap and white decurrent gills; solitary or in small troops, on soil in broad-leaf woods, favouring beech.

Dimensions cap 4 - 8 cm dia; stem 3 - 7 cm tall x 1 - 1.5 cm dia.
Cap brownish, sometimes with grey or olivaceous tinges, smooth; at first flattened convex with slight umbo, later shallowly infundibuliform. Flesh white, thick and watery.
Gills pallid, creamy-yellow, deeply decurrent, fairly broad, close.
Spores hyaline, smooth, sub-spherical or ellipsoid, non-amyloid, droplets, 4.5-5 x 3.5-4 µm. Basidia 2- or 4-spored. Marginal cells clavate or branched, sparse.
Stem more pallid but otherwise concolorous with cap, silky fibrous, tapering upwards strongly from clavate base. Ring absent. Flesh white, thick, watery and stuffed.
Odour strong, sweet. **Taste** not distinctive.
Chemical tests none.
Occurrence summer to autumn; frequent.
■ Inedible.

Clitocybe dealbata (Sow.: Fr.) Kummer   Tricholomataceae
= *Clitocybe rivulosa subsp. dealbata* (Sow.: Fr.) Konrad & Maubl.

Small fleshy agaric with whitish-grey, shallowly depressed cap and white decurrent gills; trooping in rings on soil in lawns and pastures.

Dimensions cap 2 - 4 cm dia; stem 2 - 3.5 cm tall x 0.5 - 1 cm dia.
Cap buff with white pruinose dusting, otherwise smooth; at first convex with slightly inrolled margin, later shallowly infundibuliform. Flesh white and thin.
Gills white, adnate-decurrent, fairly broad, crowded. **Spores** hyaline, smooth, ellipsoid, non-amyloid, 4.5-5 x 2-3 µm. Basidia 4-spored. Cystidia absent.
Stem concolorous with cap, slightly pruinose at the apex, otherwise silky, more or less equal, may be fused. Ring absent. Flesh white, tough, full or stuffed.
Odour faint, of meal. **Taste DO NOT ATTEMPT TO TASTE ANY PART OF THE SPECIMEN.**
Chemical tests none.
Occurrence summer to autumn; frequent.
✚ Lethally poisonous.

Clitocybe dicolor (Pers.) J Lange   Tricholomataceae
= *Clitocybe metachroa* (Fr.: Fr.) Kummer

Smallish agaric with buff or tan funnel-shaped cap and greyish-white decurrent gills; solitary or in small troops on soil in broad-leaf woods.

Dimensions cap 3 - 5 cm dia; stem 3 - 6 cm tall x 0.4 - 0.8 cm dia.
Cap buff or tan, hygrophanous, drying more pallid; at first convex, later infundibuliform, smooth with finely striate margin. Flesh white or greyish and thin.
Gills whitish-grey, decurrent, fairly broad, crowded. **Spores** hyaline, smooth, ellipsoid, non-amyloid, droplets, 5-7 x 3-4 µm. Basidia 4-spored. Cystidia absent.
Stem concolorous with cap, smooth, silky, more or less equal, but with swollen woolly base. Ring absent. Flesh white or greyish, thin, hollow or stuffed.
Odour not distinctive. **Taste** not distinctive.
Chemical tests none.
Occurrence late summer to autumn; infrequent.
■ Inedible.

Clitocybe ditopus (Fr.: Fr.) Gill.   Tricholomataceae

Smallish agaric with dark grey-brown funnel-shaped cap and greyish decurrent gills; solitary or in small troops occasionally tufted, on soil in coniferous woods.

Dimensions cap 2 - 4 cm dia; stem 2 - 4 cm tall x 0.4 - 0.8 cm dia.
Cap dark grey-brown, hygrophanous, drying more pallid and with a persistently pallid margin; at first convex, later infundibuliform, smooth. Flesh brown and thin.
Gills dark grey, adnate-decurrent, fairly broad, crowded.
Spores hyaline, smooth, sub-spherical, non-amyloid, droplets, 3-3.5 x 2.5-3 µm. Basidia 2- or 4-spored. Cystidia absent.
Stem concolorous with cap, smooth, more or less equal, but white woolly or cottony towards the base. Ring absent. Flesh dark grey, stuffed, becoming hollow.
Odour of meal. **Taste** not distinctive.
Chemical tests none.
Occurrence late summer to autumn; rare. (Forest of Dean, Gloucestershire)
■ Inedible.

Clitocybe ericetorum (Bull.) Quél.   Tricholomataceae

Small agaric, whitish-cream, depressed cap with decurrent gills; in trooping groups, sometimes more or less tufted, on heathland.

Dimensions cap 2 - 5 cm dia; stem 3 - 4 cm tall x 0.3 - 0.5 cm dia.
Cap whitish, becoming more cream; at first flattened convex, becoming centrally infundibuliform. Flesh white and thin.
Gills white becoming cream, decurrent, fairly broad, close.
Spores hyaline, smooth, sub-spherical or broadly ellipsoid, non-amyloid, droplets, 4-5 x 2-3 µm. Basidia 4-spored. Cystidia absent.
Stem concolorous, downy, more or less equal or tapering slightly upwards. Ring absent. Flesh white, fragile and stuffed.
Odour not distinctive. **Taste** mild or slightly bitter.
Chemical tests none.
Occurrence summer to early autumn; rare. (Brereton Heath, Cheshire)
■ Inedible.

Clitocybe geotropa (Bull.) Quél.   Tricholomataceae
= *Clitocybe maxima* ([Fl. Wett.] Fr.) Kummer

Very large distinctive agaric with creamy-buff funnel-shaped cap, gills and stem; typically in troops or rings, on soil in broad-leaf and mixed woods, often amongst grass in woodland clearings.

Dimensions cap 4 - 20 cm dia; stem 5 - 15 cm tall x 2 - 3 cm dia.
Cap at first pallid buff, with yellow tinge, becoming more buff at maturity, smooth; at first convex with a distinct broad umbo, later infundibuliform. Flesh white, thick and firm.
Gills concolorous with cap, deeply decurrent, broad, crowded.
Spores hyaline, smooth, sub-spherical, non-amyloid, droplets, 5.5-8 x 5-6 µm. Basidia 4-spored. Cystidia absent.
Stem slightly more pallid, but otherwise concolorous with cap, smooth, tapering upwards strongly from swollen base. Ring absent. Flesh white, thick, full or densely stuffed.
Odour faint, sweet. **Taste** not distinctive.
Chemical tests none.
Occurrence late summer to autumn; infrequent.
☐ Edible and good.

Clitocybe harmajae Lambotte   Tricholomataceae

Small, cream, agaric, smelling strongly aromatic; solitary or in small troops, on soil in broad-leaf and coniferous woods and in short grass near bushes.

Dimensions cap 2 - 5 cm dia; stem 2 - 5 cm tall x 0.3 - 0.5 cm dia.
Cap cream, sometimes with tan tinge at the centre, slightly hygrophanous; at first convex, becoming shallowly infundibuliform at the centre. Flesh concolorous, thin.
Gills concolorous with cap, barely decurrent, narrow, fairly broad.
Spores hyaline, minutely roughened, ellipsoid, non-amyloid, 6-10 x 3-4.5 µm. Basidia 4-spored. Cystidia absent.
Stem concolorous with cap, smooth but finely woolly at the base, more or less equal but sometimes curved at the base. Ring absent. Flesh concolorous and hollow.
Odour strong, of coumarin. (Coumarin smell is similar to aniseed but more fruity) **Taste** not distinctive.
Chemical tests none.
Occurrence spring to autumn; infrequent.
■ Inedible.

Clitocybe infundibuliformis (Schaeff.) Quél.
Tricholomataceae

Medium-sized agaric with pale buff funnel-shaped cap and whitish decurrent gills; solitary or in small troops on soil in broad-leaf woods and on heaths.

Dimensions cap 3 - 8 cm dia; stem 3 - 8 cm tall x 0.5 - 1 cm dia.
Cap pallid pinkish-buff, sometimes with ochraceous tinge, smooth, silky; infundibuliform, typically with wavy margin. Flesh whitish-buff, thick and soft.
Gills white tinged buff, deeply decurrent, narrow, crowded.
Spores hyaline, smooth, ellipsoid or pip-shaped, non-amyloid, with droplets, 6-7 x 3.5-4 µm. Basidia 4-spored. Cystidia absent.
Stem concolorous with cap, smooth, more or less equal apart from slightly swollen base. Ring absent. Flesh white, tinged buff, tougher than in cap, partly stuffed or hollow.
Odour faint, of almonds. **Taste** not distinctive.
Chemical tests none.
Occurrence summer to autumn; common.
□ Edible.

Clitocybe langei Sing. ex Hora   Tricholomataceae

Smallish agaric with brownish-grey funnel-shaped cap and grey decurrent gills; typically tufted or in small troops on soil in coniferous and mixed woods.

Dimensions cap 2 - 5 cm dia; stem 3 - 4 cm tall x 0.2 - 0.5 cm dia.
Cap brownish-grey, hygrophanous, drying more pallid, finely striate at the margin, otherwise smooth; at first convex becoming infundibuliform. Flesh pallid brownish and thin.
Gills greyish, decurrent, fairly broad, crowded. **Spores** hyaline, smooth, ellipsoid, non-amyloid, 5-6.5 x 3-3.5 µm. Basidia 4-spored. Cystidia absent.
Stem concolorous with cap, more or less smooth, often curved at the base. Ring absent. Flesh pallid brownish, thin, stuffed, becoming hollow.
Odour faint, of meal or cucumber. **Taste** not distinctive.
Chemical tests none.
Occurrence late summer to autumn; common.
■ Inedible.

Clitocybe odora (Bull.: Fr.) Kummer   **Tricholomataceae**
= *Clitocybe viridis* (With.: Fr.) Gill.

Smallish agaric with characteristic blue-green coloration, wavy cap and decurrent gills, smelling strongly aromatic; solitary or in small troops on soil in broad-leaf woods, favouring beech.

Dimensions cap 3 - 8 cm dia; stem 3 - 6 cm tall x 0.5 - 1 cm dia.
Cap bluish or greyish-green, smooth; at first convex with slight broad umbo, later expanded-wavy. Flesh whitish, thin and tough.
Gills more pallid but tinged with cap colour, adnate-decurrent, fairly broad and distant. **Spores** hyaline, smooth, ellipsoid, non-amyloid, droplets, 6-8 x 3-4 µm. Basidia 4-spored. Cystidia absent.
Stem concolorous with cap, silky fibrous, white woolly towards the base, tapering upwards from slightly swollen base. Ring absent. Flesh whitish, tough and full.
Odour strong, of aniseed. **Taste** strong, of aniseed.
Chemical tests none.
Occurrence summer to autumn; infrequent.
☐ Edible. Best used as a flavouring agent.

Clitocybe phyllophila (Fr.) Kummer   **Tricholomataceae**

Medium-sized agaric with whitish or pale flesh-coloured funnel-shaped cap and decurrent gills; solitary or in small troops, on soil in broad-leaf woods, favouring beech.

Dimensions cap 4 - 8 cm dia; stem 4 - 6.5 cm tall x 0.8 - 1.5 cm dia.
Cap pallid whitish, becoming more buff, with dull pruinose dusting; at first convex, becoming depressed with wavy margin. Flesh white or buff and thin. Cap surface made up from hyphae with distinctive short outgrowths, sometimes knot-like.
Gills whitish or pallid buff, decurrent, fairly broad, crowded.
Spores pallid ochraceous-clay, smooth, ellipsoid, non-amyloid, 4-6 x 2.5-3.5 µm. Basidia 4-spored. Cystidia absent.
Stem concolorous with cap, smooth but white downy at the base, tapering upwards from swollen base. Ring absent. Flesh whitish or pallid buff, thin, full or stuffed.
Odour sweet. **Taste** not distinctive.
Chemical tests none.
Occurrence late summer to autumn; infrequent.
■ Inedible.

Clitocybe quercina Pearson   **Tricholomataceae**

Smallish pale brown agaric with characteristic stiffly elastic texture, depressed cap and decurrent gills; solitary or in small troops on soil with oak but also reported on conifer needles.

Dimensions cap 2 - 6 cm dia; stem 4 - 6 cm tall x 0.4 - 0.5 cm dia.
Cap pallid olivaceous-brown, sometimes with pink tinge, hygrophanous, drying dirty-white with darker brownish centre, smooth, striate at the margin; at first convex becoming depressed or umbilicate. Flesh whitish, thin, and rigidly elastic.
Gills more pallid but tinged cap colour, decurrent, fairly broad, crowded. **Spores** hyaline, smooth, ellipsoid, non-amyloid, 6.5-8 x 3.5-4.5 µm. Basidia 4-spored. Cystidia absent.
Stem concolorous with cap, grey-brown below, smooth, more or less equal. Ring absent. Flesh whitish, tough and full.
Odour faint, of meal. **Taste** of meal or slightly bitter.
Chemical tests none.
Occurrence late summer to autumn; infrequent and localised.
■ Inedible.

Clitocybe rivulosa (Pers.: Fr.) Kummer   **Tricholomataceae**

Small fleshy agaric with whitish grey, shallowly depressed cap and white decurrent gills; trooping in rings, on soil, often sandy, in lawns and other grasslands.

Dimensions cap 2 - 5 cm dia; stem 2 - 4 cm tall x 0.4 - 1 cm dia
Cap buff with white pruinose dusting, typically in concentric rings, otherwise smooth; at first convex with slightly incurved margin, later shallowly infundibuliform. Flesh white or buff, thin.
Gills white, adnate-decurrent, fairly broad, crowded. **Spores** hyaline, smooth, ellipsoid, non-amyloid, 4-5.5 x 2-5 µm. Basidia 4-spored. Cystidia absent.
Stem concolorous with cap, slightly pruinose at apex and woolly at base, otherwise smooth, more or less equal. Ring absent. Flesh white, tough, full or stuffed.
Odour faint, sweet. **Taste DO NOT ATTEMPT TO TASTE ANY PART OF THE SPECIMEN.**
Chemical tests none.
Occurrence late summer to late autumn; frequent.
✚ Lethally poisonous.

Clitocybe sinopica (Fr.: Fr.) Kummer   **Tricholomataceae**

Small agaric with bright reddish-brown cap and stem, cream and barely decurrent gills; tufted or in small troops, on acid soil at open margins of coniferous woods and on heaths.

Dimensions cap 2 - 6 cm dia; stem 2 - 4.5 cm tall x 0.4 - 0.7 cm dia.
Cap light chestnut or bright reddish-brown, smooth; flattened or somewhat infundibuliform, often irregular. Flesh white or cap colour beneath cuticle, moderate or thin.
Gills cream, adnate-decurrent, fairly broad, crowded.
Spores hyaline, smooth, ellipsoid, non-amyloid, droplets, 8-10 x 4.5-6.5 µm. Basidia 4-spored.Cystidia absent.
Stem concolorous with cap, more or less equal, covered with fine, slightly darker fibrils. Ring absent. Flesh white, stuffed or hollow.
Odour of meal. **Taste** not distinctive.
Chemical tests none.
Occurrence spring; rare. (Stourhead, Wiltshire)
☐ Edible.

Clitocybe suaveolens (Schum.: Fr.) Kummer **Tricholomataceae**

Small agaric with pale yellowish or cream cap, and whitish, barely decurrent gills, strongly aromatic; tufted, sometimes caespitose, or in small troops on soil in broad-leaf woods favouring patches of grass or moss.

Dimensions cap 2 - 5 cm dia; stem 3 - 6 cm tall x 0.3 - 0.6 cm dia.
Cap pallid yellowish-brown, hygrophanous, drying cream with darker centre, smooth; at first flattened convex, later markedly infundi-buliform at the centre. Flesh white or buff, thin.
Gills whitish buff, adnate-decurrent, fairly broad, crowded.
Spores hyaline, smooth, ellipsoid, non-amyloid, 6-8 x 3.5-4 µm. Basidia 4-spored. Cystidia absent.
Stem concolorous with cap, more or less equal, sometimes curved at the base, silky or pruinose towards apex, slightly woolly at base. Ring absent. Flesh white or buff, thin, stuffed or hollow.
Odour strong, of aniseed. **Taste** not distinctive.
Chemical tests none.
Occurrence late summer to autumn; infrequent.
☐ Edible. **Caution** *C. suaveolens* may be confused with dangerous species including *C. rivulosa* and *C. dealbata*.

Clitocybe subalutacea (Batsch: Fr.) Kummer
Tricholomataceae

Small fleshy agaric with leather-coloured, slightly depressed cap and decurrent gills; in troops on soil in or near broad-leaf woods favouring grassy areas.

Dimensions cap 3 - 5 cm dia; stem 3 - 5 cm tall x 0.3 - 0.6 cm dia.
Cap yellowish leather, more ochraceous-brown with age, non-striate, smooth; at first convex, becoming shallowly infundibuliform at the centre. Flesh pallid and medium.
Gills pallid, decurrent, fairly broad, close. **Spores** hyaline, smooth, sub-spherical, non-amyloid, 3-5 x 3-4 µm. Basidia 4-spored. Cystidia absent.
Stem concolorous with cap or more pallid, more or less equal, smooth. Ring absent. Flesh pallid, stuffed or hollow.
Odour faint, of aniseed. **Taste** not distinctive.
Chemical tests none.
Occurrence autumn; infrequent.
■ Inedible.

Clitocybe truncicola (Peck) Sacc.
Tricholomataceae

Smallish white agaric with wavy, shallowly depressed cap, barely decurrent gills, and eccentric stem; in small troops, or more or less clustered, on dead wood of broad-leaf trees.

Dimensions cap 3 - 6 cm dia; stem 3 - 5 cm tall x 0.3 - 0.6 cm dia.
Cap white; at first convex, becoming expanded or shallowly infundibuliform and with wavy margin, finely downy. Flesh white and thin.
Gills concolorous, tinged ivory with age, adnate or slightly decurrent, fairly broad, crowded. **Spores** hyaline, smooth, sub-spherical, non-amyloid, 4.5-6.5 x 3.5-5.5 µm. Basidia 4-spored. Cystidia absent.
Stem concolorous with cap, more or less equal, longitudinally fibrillose, curved. Ring absent. Flesh white, thin, full or stuffed, fairly tough.
Odour not distinctive. **Taste** not distinctive.
Chemical tests none.
Occurrence autumn; very rare. (Edford Wood, Somerset)
■ Inedible.
Introduced from North America and first described for Europe in 1988.

Clitocybe vibecina (Fr.) Quél.
Tricholomataceae

Small agaric with pale, greyish-brown, slightly depressed cap and decurrent gills; tufted or in small troops on soil in coniferous or mixed woods favouring patches of bracken with grass or moss.

Dimensions cap 1.5 - 5 cm dia; stem 3 - 5 cm tall x 0.3 - 0.6 cm dia.
Cap greyish-brown, hygrophanous, drying more pallid, striate at the margin otherwise smooth; at first convex, becoming infundibuliform at the centre. Flesh pallid, brownish and thin.
Gills concolorous, decurrent, fairly broad, close. **Spores** hyaline, smooth, ellipsoid, non-amyloid, droplets, 6.5-8 x 3.5-4.5 µm. Basidia 4-spored. Cystidia absent.
Stem concolorous with cap, more or less equal, sometimes silky and curved at the base. Ring absent. Flesh pallid, brownish, thin, stuffed or hollow.
Odour of meal. **Taste** not distinctive.
Chemical tests none.
Occurrence autumn; infrequent.
■ Inedible.

Collybia butyracea (Bull.: Fr.) Kummer **Tricholomataceae**

Medium-sized agaric, with distinctive greasy brownish cap, drying paler from the centre, whitish gills and tough stem; solitary or in small troops or tufted on soil in broad-leaf and, occasionally, coniferous woods.

Dimensions cap 3 - 7 cm dia; stem 4 - 8 cm tall x 0.5 - 1.5 cm dia.
Cap dark reddish-brown with ochraceous tinge, hygrophanous, drying ivory; at first convex, becoming expanded and bluntly umbonate, smooth, greasy. Flesh pallid buff and firm.
Gills whitish, adnexed or free, broad, crowded. **Spores** hyaline, smooth, ellipsoid, non-amyloid, droplets, 6.5-8 x 3-3.5 µm. Basidia 4-spored. Cystidia absent.
Stem more or less concolorous with cap, more pallid towards apex, tapering upwards from bulbous base, smooth, base covered with white woolly hairs. Ring absent. Flesh concolorous, tough, stuffed, becoming hollow.
Odour not distinctive. **Taste** not distinctive.
Chemical tests none.
Occurrence autumn and early winter; common.
☐ Edible but poor.

Collybia butyracea var **asema** Fr. **Tricholomataceae**

Medium-sized agaric, with greasy horn-grey cap and stem, whitish gills; solitary or in small troops or tufted, on soil amongst needle litter in coniferous and, less frequently, broad-leaf woods.

Dimensions cap 3 - 7 cm dia; stem 4 - 8 cm tall x 0.5 - 1.5 cm dia.
Cap horn-grey, darker at the centre, hygrophanous, drying more ochraceous; at first convex, becoming expanded and bluntly umbonate, smooth, greasy. Flesh pallid buff and firm.
Gills whitish, adnexed or free, broad, crowded. **Spores** hyaline, smooth, ellipsoid, non-amyloid, droplets, 5-7 x 3-3.5 µm. Basidia 4-spored. Cystidia absent.
Stem more or less concolorous with cap, more pallid towards the apex, tapering upwards from bulbous base, smooth, base covered with white woolly hairs. Ring absent. Flesh concolorous, tough, stuffed, becoming hollow.
Odour not distinctive. **Taste** not distinctive.
Chemical tests none.
Occurrence summer to autumn; common.
☐ Edible but poor.

Collybia confluens (Pers.: Fr.) Kummer **Tricholomataceae**
= *Marasmius confluens* (Pers.: Fr.) Karst.

Smallish agaric, with whitish cap and gills, stem darker; in clusters and rings amongst leaf litter in broad-leaf or mixed woods.

Dimensions cap 3 - 5 cm dia; stem 3 - 6 cm tall x 0.2 - 0.5 cm dia.
Cap whitish-buff, hygrophanous, drying almost wholly white; convex, becoming partly expanded, particularly thin at the margin, wrinkled when aged, otherwise smooth. Flesh white and thin.
Gills pallid, concolorous with cap, adnexed, narrow, crowded.
Spores hyaline, smooth, ellipsoid, non-amyloid, 7-9 x 3-4 µm. Basidia 4-spored. Gill edge cystidia irregularly cylindrical; gill face cystidia cylindrical.
Stem darker buff than cap, more or less equal, compressed, covered with delicate white down. Ring absent. Flesh concolorous, tough, stuffed, becoming hollow.
Odour not distinctive. **Taste** not distinctive.
Chemical tests none.
Occurrence late summer to early winter; common.
☐ Edible but poor.

Collybia dryophila (Bull.: Fr.) Kummer   Tricholomataceae
= *Collybia aquosa* (Bull.: Fr.) Kummer
= *Marasmius dryophilus var aquosus* (Bull.: Fr.) Rea

Smallish agaric, with pale tan cap, whitish gills, and stem flushed tan; trooping or more or less tufted on soil and leaf litter under broad-leaf and coniferous trees.

Dimensions cap 2 - 6 cm dia; stem 2 - 6 cm tall x 0.2 - 0.5 cm dia.
Cap buff or tan, hygrophanous, drying pallid and sometimes more or less white; convex, becoming expanded, flattened or slightly depressed with wavy margin, smooth. Flesh whitish and thin.
Gills at first white becoming buff, adnexed, narrow, crowded.
Spores hyaline, smooth, ellipsoid, non-amyloid, 4-7 x 3-4 µm. Basidia 4-spored. Cystidia absent.
Stem more or less concolorous with cap, more or less equal but slightly bulbous at base, sometimes partly rooting. Ring absent. Flesh whitish, tough, stuffed, becoming hollow.
Odour not distinctive. **Taste** not distinctive.
Chemical tests none.
Occurrence spring to winter, mainly summer and autumn; common.
☐ Edible but poor.

Collybia erythropus (Pers.: Fr.) Kummer Tricholomataceae
= *Marasmius bresadolae* Kühn. & Romagn.
= *Marasmius erythropus* (Pers.: Fr.) Fr.

Smallish agaric, with pale tan cap, buff gills, and dark reddish stem; more or less tufted on leaf litter and rotting wood in broad-leaf and mixed woods.

Dimensions cap 1 - 3 cm dia; stem 4 - 7 cm tall x 0.2 - 0.4 cm dia.
Cap buff or tan, darker towards the centre, hygrophanous, drying cream; convex, becoming expanded, flattened or slightly depressed with wavy margin, smooth, wrinkled when dry. Flesh whitish and thin.
Gills at first white, becoming buff, adnexed or free, narrow, close.
Spores hyaline, smooth, ellipsoid, non-amyloid, 6-8 x 3.5-4 µm. Basidia 4-spored. Marginal cells clavate, irregular.
Stem dark red towards base, more pallid towards apex, more or less equal, compressed, covered in pinkish woolly hairs at base. Ring absent. Flesh reddish-brown, tough, stuffed, becoming hollow.
Odour not distinctive. **Taste** not distinctive.
Chemical tests none.
Occurrence summer to autumn; infrequent or rare. (Bridgham, near Thetford, Norfolk)
☐ Edible but poor.

Collybia fusipes (Bull.: Fr.) Quél.   Tricholomataceae
= *Collybia lancipes* (Fr.) Gill.

Medium-sized agaric, with reddish-brown cap, paler gills, and distinctively rooting stem flushed-cap colour; caespitose at the base of trunks of broad-leaf trees favouring oak and beech.

Dimensions cap 3 - 7 cm dia; stem 4 - 9 cm tall x 0.7 - 1.5 cm dia.
Cap reddish-brown, hygrophanous, drying pallid tan; convex, becoming expanded, bluntly umbonate with wavy margin, slightly viscid or smooth. Flesh whitish tinged cap colour and thin.
Gills at first whitish, becoming tinged reddish-brown, free or emarginate, broad, crowded. **Spores** hyaline, smooth, ellipsoid or pip-shaped, non-amyloid, occasionally with droplets, 4-6 x 2-4 µm. Basidia 2- or 4-spored. Marginal cells cylindrical.
Stem pallid cap colour above, darkening reddish-brown towards base, typically fusiform with a rooting base which often fuses with others, twisted and grooved. Ring absent. Flesh whitish, tough, and more or less full.
Odour not distinctive. **Taste** not distinctive.
Chemical tests none.
Occurrence early summer to late autumn; common.
■ Inedible.

Rhodocollybia maculata (A. & S.: Fr.) Sing. **Tricholomataceae**
= *Collybia maculata* (A. & S.: Fr.) Kummer

Medium to large agaric, with whitish cap, gills and stem, soon spotting rusty; trooping or more or less tufted and often in rings on soil and leaf litter under broad-leaf and coniferous trees, also on heaths.

Dimensions cap 4 - 10 cm dia; stem 5 - 10 cm tall x 0.8 - 1.2 cm dia.
Cap white, becoming spotted rust-brown; convex, becoming expanded, flattened or slightly depressed with wavy margin, smooth. Flesh white and fairly thick.
Gills at first white, becoming spotted rust-brown, free, fairly narrow, crowded. **Spores** hyaline, smooth, sub-spherical, non-amyloid, droplets, 4-6 x 3-5 μm. Basidia 4-spored. Marginal cells cylindrical with outgrowths.
Stem white, discolouring as in cap, more or less equal or tapering slightly upwards, stoutish, sometimes partly rooting. Ring absent. Flesh whitish, firm or tough, stuffed, becoming hollow.
Odour not distinctive. **Taste** bitter.
Chemical tests none.
Occurrence early summer to late autumn; common.
■ Inedible.

Collybia ocior (Pers.) Vilgalis & Miller   **Tricholomataceae**
= *Collybia extuberans* (Fr.) Quél.

Smallish agaric, with tan or darker brown cap, whitish gills, and stem flushed tan; more or less tufted on bark, rotting stumps, and litter.

Dimensions cap 1.5 - 5 cm dia; stem 2 - 7 cm tall x 0.2 - 0.6 cm dia.
Cap reddish-ochre, more chestnut towards the centre, hygrophanous, drying more pallid; convex , becoming expanded, flattened or slightly depressed with wavy margin, smooth, faintly striate at the margin when damp. Flesh whitish-cream, thin except at the centre.
Gills at first white, becoming cream, adnexed, margin smooth or slightly serrated, fairly broad, crowded. **Spores** hyaline, smooth, ellipsoid, non-amyloid, 4-7 x 3-4 μm. Basidia 4-spored. Marginal cells clavate or knobbly.
Stem concolorous with cap, but more pallid, darker towards apex, more or less equal but slightly bulbous and finely hairy at base, sometimes partly rooting. Ring absent. Flesh whitish, elastic, stuffed, becoming hollow.
Odour not distinctive. **Taste** not distinctive.
Chemical tests none.
Occurrence spring to summer; rare. (Parkend, Forest of Dean, Gloucestershire)
■ Inedible.

Collybia peronata (Bolt.: Fr.) Kummer   **Tricholomataceae**
= *Marasmius peronatus* (Bolt.: Fr.) Fr.
= *Marasmius urens* (Bull.: Fr.) Fr.

Smallish agaric, with tan cap, characteristically wood-coloured gills, and dense, pale, woolly hairs at the base of the stem, in small trooping groups, sometimes tufted, on leaf litter under broad-leaf and, less frequently, coniferous trees.

Dimensions cap 3 - 6 cm dia; stem 3 - 6 cm tall x 0.3 - 0.6 cm dia.
Cap dull tan or brownish; convex, becoming expanded, flattened, often with a blunt umbo, smooth but then characteristically wrinkled and leathery with age. Flesh whitish with yellow tinge and thin.
Gills at first cream becoming wood-coloured, adnexed or free, narrow, fairly close. **Spores** hyaline, smooth, ellipsoid or pip-shaped, non-amyloid, 7-10 x 3-4 μm. Basidia 4-spored. Cystidia fusiform.
Stem pallid, dirty-yellow, more or less equal, covered thickly with long woolly hairs towards base. Ring absent. Flesh whitish, with yellow tinges, tough, and stuffed.
Odour not distinctive. **Taste** acrid.
Chemical tests none.
Occurrence summer to winter, mainly autumn; common.
■ Inedible.
The only commonly occurring *Collybia* with wood-coloured gills.

Collybia tuberosa (Bull.: Fr.) Kummer   **Tricholomataceae**
= *Microcollybia tuberosa* (Bull.: Fr.) Lennox

Small whitish agaric, arising from tuberous structure; densely clustered on hardened, atrophied remains of previous year's fruiting bodies of *Russula* and *Lactarius* species.

Dimensions cap 0.5 - 1.5 cm dia; stem 3 - 5 cm tall x 05 - 0.2 cm dia.
Cap whitish, becoming tan at the centre; convex, becoming expanded, flattened, often slightly umbilicate, smooth. Flesh white and thin.
Gills whitish, adnexed or emarginate, broad, close. **Spores** hyaline, smooth, ellipsoid or pip-shaped, non-amyloid, 3.5-5 x 2-3 µm. Basidia 4-spored. Cystidia absent.
Stem white or pallid brown, more or less equal, with fine woolly hairs towards base; mycelial threads link to dark brown shiny sclerotium. Ring absent. Flesh white and firm.
Odour not distinctive. **Taste** not distinctive.
Chemical tests none.
Occurrence late summer to autumn; infrequent, more common in Scottish Highlands.
■ Inedible.

Delicatula integrella (Pers.: Fr.) Fayod   **Tricholomataceae**

Very small, delicate, white agaric with bell-shaped cap; trooping on rotting wood and humus.

Dimensions cap 0.3 - 0.6 cm dia; stem 1.5-2 cm tall x 0.5 - 0.1 cm dia.
Cap wholly white; at first conical-campanulate, irregular and wavy, depressed at the centre. Flesh white and very thin.
Gills concolorous, adnate, veined, narrow, often branching, crowded, disappearing at the margin. **Spores** hyaline, smooth, broadly ellipsoid, amyloid, droplets, 7-9 x 4-5 µm. Basidia 2-, 3- or 4-spored. Cystidia absent.
Stem white, smooth, slender, more or less equal. Ring absent. Flesh white.
Odour not distinctive. **Taste** not distinctive.
Chemical tests none.
Occurrence summer to autumn; infrequent.
☐ Inedible.

Flammulina velutipes (Curt.: Fr.) Karst. **Tricholomataceae**
= *Collybia velutipes* (Curt.: Fr.) Kummer   Velvet Shank

Medium-sized agaric, with yellowish tan slimy cap, and distinctive dark velvety stem; in tufts on trunks, stumps and branches of dead and diseased broad-leaf trees.

Dimensions cap 2 - 10 cm dia; stem 3 - 10 cm tall x 0.4 - 0.8 cm dia.
Cap yellowish tan; at first convex, becoming flattened, smooth, viscid. Flesh concolorous and thin.
Gills pallid yellow, adnexed, broad, crowded. **Spores** hyaline, smooth, cylindrical-ellipsoid, non-amyloid, 7.5-10 x 3.5-4 µm. Basidia 4-spored. Gill-edge cystidia clavate.
Stem concolorous with cap at the apex, otherwise chocolate-brown and finely downy, more or less equal, typically curved. Ring absent. Flesh pallid above, dirty-brown below, tough, cartilaginous, hollow or stuffed.
Odour not distinctive. **Taste** not distinctive.
Chemical tests none.
Occurrence autumn to winter; common.
☐ Edible.
One of the few agaric fungi to survive heavy frosts.

Hemimycena cucullata (Pers.: Fr.) Sing. Tricholomataceae

Small, delicate, chalk-white agaric with bell-shaped cap; solitary or in trooping groups on litter with broad-leaf and coniferous trees.

Dimensions cap 0.75 - 2.5 cm dia; stem 2 - 7 cm tall x 0.1 - 0.2 cm dia.
Cap chalk-white, tinged cream with age, smooth or faintly striate; at first campanulate, becoming expanded and sometimes umbonate. Flesh white and very thin.
Gills concolorous, sub-decurrent, broad, crowded. **Spores** hyaline, smooth, fusiform or narrowly ellipsoid, non-amyloid, droplets, 8.5-12.5 x 3.5-4.5 µm. Basidia 4-spored. Gill edge cystidia fusiform.
Stem white, minutely pruinose or shiny, slender, more or less equal, hairy at base. Ring absent. Flesh white and hollow.
Odour faint, spicy. **Taste** not distinctive.
Chemical tests none.
Occurrence autumn; infrequent.
■ Inedible.

Hemimycena lactea (Pers.: Fr.) Sing.   Tricholomataceae
= *Hemimycena delicatella* (Peck) Sing.
= *Mycena delicatella* (Peck) Sing.

Small, delicate, chalk-white agaric with bell-shaped cap; in trooping groups on coniferous debris, less frequently with broad-leaf trees.

Dimensions cap 0.75 - 2.5 cm dia; stem 2 - 5 cm tall x 0.1 - 0.25 cm dia.
Cap chalk-white, minutely pruinose, sometimes with cream tinge at the centre; at first hemispherical or campanulate, becoming expanded and sometimes umbonate. Flesh white and very thin.
Gills concolorous, adnate or with decurrent tooth, broad, crowded.
Spores hyaline, smooth, ellipsoid, non-amyloid, droplets, 7.5-10.5 x 2.5-4 µm. Basidia 2- or 4-spored. Gill edge cystidia fusiform or cylindrical.
Stem white, minutely pruinose, slender, more or less equal, hairy at base. Ring absent. Flesh white.
Odour not distinctive. **Taste** not distinctive.
Chemical tests none.
Occurrence autumn; infrequent.
■ Inedible.

Hemimycena pithya (Fr.) Dorfelt   Tricholomataceae
= *Hemimycena gracilis* (Quél.) Sing.

Small, delicate, chalk-white agaric with bell-shaped irregularly grooved cap; in large troops or somewhat tufted on coniferous debris, favouring spruce.

Dimensions cap 0.5 - 1.5 cm dia; stem 2 - 5 cm tall x 05 - 0.1 cm dia.
Cap chalk-white, sometimes with cream tinge at the centre; at first hemispherical or campanulate, becoming expanded and sometimes umbonate. Flesh white and very thin.
Gills concolorous, adnate, narrow, crowded. **Spores** hyaline, smooth, cylindrical, non-amyloid, 6-11 x 1.5-2.5 µm. Basidia 4-spored. Gill cystidia absent.
Stem white, smooth, slender, more or less equal, hairy at base. Ring absent. Flesh white.
Odour not distinctive. **Taste** not distinctive.
Chemical tests none.
Occurrence summer to autumn; frequent.
■ Inedible.

Hemimycena pseudogracilis (Kühn.) Sing.
Tricholomataceae

Small, delicate, white agaric with bell-shaped cap; in large troops on coniferous debris favouring pine.

Dimensions cap 0.5 - 1 cm dia; stem 2 - 4 cm tall x 0.05 - 0.1 cm dia

Cap white, sometimes with cream tinge at the centre; campanulate, becoming expanded and sometimes umbonate, smooth or slightly pruinose, sulcate. Flesh white, very thin and soft.

Gills concolorous, adnate, narrow, somewhat distant. **Spores** hyaline, smooth, cylindrical, non-amyloid, droplets, 7-9 x 2.5-4 µm. Basidia 4-spored. Gill cystidia absent.

Stem white, smooth, slender, more or less equal, hairy at base. Ring absent. Flesh white.

Odour not distinctive. **Taste** not distinctive.

Chemical tests none.

Occurrence summer to autumn; frequent.

■ Inedible.

Laccaria amethystea (Bull.) Murr.   **Tricholomataceae**
= *Laccaria laccata var amethystina* ([Huds.] Cke.) Rea

Smallish agaric, deep lilac (damp) or buff (dry) throughout, with distant thickish gills and fibrous stem; in scattered trooping groups on soil in coniferous and broad-leaf woods, favouring beech.

Dimensions cap 1.5 - 6 cm dia; stem 4 - 10 cm tall x 0.5 - 1 cm dia.

Cap deep purplish-lilac, drying buff; at first convex, becoming flattened and finally depressed with wavy margin, smooth or slightly scurfy at the centre in older specimens. Flesh concolorous and thin.

Gills concolorous with cap, often remaining lilac when the cap has become buff, but becoming powdered white with spores, adnate with decurrent tooth, broad, distant. **Spores** hyaline or very pale lilac, spiny, sub-spherical, non-amyloid, 9-11 x 7.5-9.5 µm. Basidia 4-spored. Marginal cells filiform.

Stem concolorous with cap, farinose near the apex, otherwise covered in whitish fibres and woolly white at base, often twisted. Ring absent. Flesh concolorous, fibrous, tough, becoming hollow.

Odour not distinctive. **Taste** not distinctive.

Chemical tests none.

Occurrence early summer to winter; very common.

☐ Edible.

Laccaria bicolor (Maire) P Orton   **Tricholomataceae**

Smallish agaric, cinnamon (damp) or pale yellowish (dry) throughout, with distant thickish gills and fibrous stem with lilac tinge at the base; in scattered trooping groups on poor soil in mixed woods and on heaths.

Dimensions cap 1.5 - 6 cm dia; stem 5 - 10 cm tall x 0.6 - 1 cm dia.

Cap cinnamon or pinkish-brown, drying pallid ochraceous; at first convex, becoming flattened and finally depressed with wavy margin, smooth or slightly scurfy at the centre in older specimens. Flesh concolorous and thin.

Gills pinkish or with lilaceous tinge, but becoming powdered white with spores, adnate, thick, broad, distant. **Spores** hyaline, spiny, broadly ellipsoid or sub-spherical, non-amyloid, occasionally with droplets, 7-10 x 5.5-7 µm. Basidia 4-spored. Marginal cells cylindrical.

Stem concolorous with cap, covered in whitish fibres, pallid lilaceous woolly at base. Ring absent. Flesh concolorous, fibrous, tough, full becoming hollow.

Odour not distinctive. **Taste** not distinctive.

Chemical tests none.

Occurrence early summer to early winter; infrequent.

☐ Edible.

Laccaria laccata (Scop.: Fr.) Cke.   **Tricholomataceae**
= *Clitocybe laccata* (Scop.: Fr.) Kummer   The Deceiver

Smallish agaric, tawny (damp) or pale yellowish (dry) throughout,
with distant thickish gills and fibrous stem; in scattered trooping
groups on soil in mixed woods and on heaths.

Dimensions cap 1.5 - 6 cm dia; stem 5 - 10 cm tall x 0.6 - 1 cm dia.
Cap tawny or brick red, drying pallid ochraceous; at first convex,
becoming flattened and finally depressed with wavy margin, smooth
or slightly scurfy at the centre in older specimens. Flesh concolorous
and thin.
Gills concolorous with cap, or with pinkish tinge, but becoming
powdered white with spores, adnate, thick, broad, distant.
Spores hyaline, spiny, sub-spherical, non-amyloid, occasionally with
droplets, 7-10 x 5.5-7.5 μm. Basidia 4-spored. Marginal cells cylindrical.
Stem concolorous with cap, farinose near the apex, otherwise
covered in whitish fibres and white and woolly at base, often twisted
and laterally compressed. Ring absent. Flesh concolorous, fibrous,
tough, at first full, then hollow.
Odour not distinctive. **Taste** not distinctive.
Chemical tests none.
Occurrence early summer to early winter; very common.
☐ Edible. Very variable in appearance.

Laccaria proxima (Boud.) Pat.   **Tricholomataceae**
= *Laccaria laccata var proxima* (Boud.) Maire

Smallish agaric with reddish-brown cap, distant, thick, pale pink gills
and fibrous stem; scattered or in small trooping groups on poor acid
soil, on heaths, moors and adjacent to boggy areas.

Dimensions cap 2 - 7 cm dia; stem 3 - 12 cm tall x 0.2 - 0.5 cm dia.
Cap reddish-brown when damp, drying ochraceous-buff; at first convex,
becoming flattened and finally depressed, smooth or more or
less scurfy. Flesh pallid, tinged cap colour and thin.
Gills pallid pink, adnate, thick, broad, distant. **Spores** hyaline, spiny,
broadly ellipsoid or sub-spherical, non-amyloid, 7-10 x 6-8 μm.
Basidia 4-spored. Marginal cells cylindrical.
Stem concolorous with cap, fibrillose, tapering slightly towards the
apex, and finely woolly and white at base. Ring absent. Flesh con-
colorous, fibrous, tough, stuffed, becoming hollow.
Odour not distinctive. **Taste** not distinctive.
Chemical tests none.
Occurrence autumn; infrequent.
☐ Edible.

Laccaria tortilis (Bolt.) Cke.   **Tricholomataceae**
= *Laccaria echinospora* (Speg.) Sing.
= *Omphalina tortilis* (Bolt.) S F Gray
= *Clitocybe tortilis* (Bolt.) Gill.

Small, pinkish-brown, slightly funnel-shaped agaric with distant gills;
in close trooping groups on bare soil, typically in damp heavy soil
beneath shrubs and thicket.

Dimensions cap 0.5 - 1.5 cm dia; stem 0.2 - 1 cm tall x 0.1 - 0.2 cm
dia.
Cap brown with pink tinge; flattened, becoming depressed with wavy
margin, drying more pallid and slightly scurfy. Flesh concolorous and
thin.
Gills pink, adnate with decurrent tooth, broad, distant.
Spores hyaline, notably spiny, spherical, non-amyloid, 8-16 μm.
Basidia 2-spored. Marginal cells cylindrical.
Stem concolorous with cap, white and woolly at base. Ring absent.
Flesh concolorous, thin, fibrous, becoming hollow.
Odour not distinctive. **Taste** not distinctive.
Chemical tests none.
Occurrence early summer to early winter; infrequent.
☐ Edible.

Lepista fasciculata Harmaja   Tricholomataceae

Smallish, pale, creamy-brown agaric, with flattish cap depressed at the centre, and decurrent gills; solitary or in small often caespitose tufts on soil in coniferous or broad-leaf woods.

Dimensions cap 3 - 8 cm dia; stem 3 - 7 cm tall x 0.5 - 1.5 cm dia.
Cap pallid brown, not hygrophanous, smooth or slightly scurfy; at first convex, becoming flattened-umbilicate with wavy margin. Flesh concolorous and thin.
Gills pallid brown, shallowly decurrent, narrow, crowded.
Spores hyaline, minutely warty, broadly ellipsoid, non-amyloid, 5-6.5 x 3-4 µm. Basidia 4-spored. Cystidia absent.
Stem pallid, concolorous with cap, smooth, finely woolly towards base, more or less equal. Ring absent. Flesh concolorous, elastic and hollow.
Odour not distinctive. **Taste** not distinctive.
Chemical tests none.
Occurrence late summer to autumn; infrequent.
■ Inedible.

Lepista fragrans (Sow.: Fr.) Harmaja   Trichomolataceae
= *Clitocybe fragrans* (Sow.: Fr.) Kummer

Small agaric, with pale yellowish-brown or cream slightly depressed cap, and whitish, barely decurrent gills, strongly aromatic; tufted or in small troops on soil in broad-leaf woods, favouring patches of grass or moss.

Dimensions cap 1.5 - 4 cm dia; stem 3 - 6 cm tall x 0.3 - 0.6 cm dia.
Cap pallid, yellowish-brown, hygrophanous, drying cream with darker centre, smooth; at first flattened, convex, later shallowly infundibuliform. Flesh white or buff, thin.
Gills whitish buff, adnate-decurrent, broad, crowded.
Spores hyaline, smooth, ellipsoid, non-amyloid, 6-8 x 3.5-5 µm. Basidia 4-spored. Cystidia absent.
Stem concolorous with cap, more or less equal, sometimes curved at base, silky or pruinose towards apex, slightly woolly at base. Ring absent. Flesh white or buff, thin, stuffed or hollow.
Odour strong, of aniseed. **Taste** not distinctive.
Chemical tests none.
Occurrence summer to autumn; infrequent.
□ Edible. **Caution:** the species may be confused with dangerous species including *C. rivulosa* and *C. dealbeata*.

Lepista gilva (Pers.: Fr.) Pat.   Tricholomataceae
= *Clitocybe splendens* (Pers.: Fr.) Gill.

Medium to large fleshy agaric, pale, tawny, funnel-shaped cap, with flesh-coloured decurrent gills; solitary or in small troops, on soil in coniferous or broad-leaf woods.

Dimensions cap 4 - 10 cm dia; stem 3 - 6 cm tall x 1 - 1.5 cm dia.
Cap at first pallid ochraceous-buff, becoming more tawny at maturity, sometimes spotted, smooth; at first convex, becoming shallowly depressed with incurved margin. Flesh whitish-buff and thin.
Gills pallid ochraceous buff, deeply decurrent, broad, crowded.
Spores hyaline, minutely roughened, sub-spherical, non-amyloid, 3.5-5 x 3.5-4 µm. Basidia 2- or 4-spored. Cystidia absent.
Stem pallid, concolorous with cap, smooth, finely woolly at base, more or less equal. Ring absent. Flesh whitish-buff, thick and hollow.
Odour faint, acrid. **Taste** not distinctive.
Chemical tests none.
Occurrence autumn; infrequent.
■ Inedible.
Considered by some authors to be a pale variety of *L. inversa*.

Lepista inversa (Scop.: Fr.) Pat.   Tricholomataceae
= *Clitocybe flaccida* (Sow.: Fr.) Kummer
= *Clitocybe inversa* (Scop.: Fr.) Quél.
= *Lepista flaccida* (Sow.: Fr.) Pat.

Medium-sized fleshy agaric, yellowish, tawny, funnel-shaped cap with creamy decurrent gills; solitary or in small troops, on soil in coniferous woods or, less frequently, broad-leaf woods.

Dimensions cap 5 - 9 cm dia; stem 2 - 5 cm tall x 0.5 - 1 cm dia.
Cap at first ochraceous-buff, becoming tawny at maturity, smooth; at first flattened, convex, later infundibuliform. Flesh pallid cream and thin.
Gills pallid creamy-yellow, deeply decurrent, broad, crowded.
Spores hyaline, minutely roughened, sub-spherical, non-amyloid, 4-5 x 3-4 µm. Basidia 2- or 4-spored. Cystidia absent.
Stem concolorous with cap, smooth at the apex, finely woolly at the base, equal. Ring absent. Flesh pallid cream and hollow.
Odour not distinctive. **Taste** not distinctive.
Chemical tests none.
Occurrence summer to autumn; frequent.
☐ Edible.

Lepista luscina (Fr.: Fr.) Sing.   Tricholomataceae
= *Lepista panaeola* (Fr.) Maire
= *Rhodopaxillus nimbatus* (Batsch) Konrad & Maubl.

Medium-sized fleshy agaric, with beige, funnel-shaped cap and whitish or dirty-pink gills; typically in rings on soil in pastures and other grassy places.

Dimensions cap 5 - 12 cm dia; stem 3 - 6 cm tall x 1 - 2.5 cm dia.
Cap at first pallid beige, with darker brown small spots or blotches, concentrically arranged, more obvious when damp, becoming grey-brown at maturity, smooth; at first hemispherical then flattened, convex, later slightly depressed. Flesh whitish and thick.
Gills pallid cream, becoming greyish-pink with age, adnate, narrow, crowded. **Spores** hyaline, minutely roughened, broadly ellipsoid, non-amyloid, occasionally with droplets, 4.5-6 x 3-4.5 µm. Basidia 4-spored. Cystidia absent.
Stem concolorous with cap, smooth or finely fibrillose, equal. Ring absent. Flesh whitish, stuffed, or full.
Odour faint, of meal. **Taste** not distinctive.
Chemical tests none.
Occurrence autumn; rare.
☐ Edible.

Lepista nebularis (Batsch : Fr.) Harmaja   Tricholomataceae
= *Clitocybe nebularis* (Batsch : Fr.) Kummer   Clouded Agaric

Medium or large agaric, with cloud-grey, funnel-shaped cap and white decurrent gills; typically in troops or rings on soil in broad-leaf or coniferous woods.

Dimensions cap 5 - 20 cm dia; stem 5 - 10 cm tall x 1.5 - 2.5 cm dia.
Cap soft cloud-grey, darker at the centre, sometimes with brownish tinge, smooth but whitish pruinose; at first flattened, convex, later flattened or shallowly depressed. Flesh whitish and thick.
Gills white or buff, decurrent, broad, crowded. **Spores** hyaline, smooth, broad, ellipsoid, non-amyloid, 5.5-8 x 3.5-5 µm. Basidia 4-spored. Cystidia absent.
Stem concolorous with cap, fibrillose, stout, tapering upwards. Ring absent. Flesh white, thick, brittle, stuffed, becoming hollow.
Odour strong, sweet or fruity. **Taste** not distinctive.
Chemical tests none.
Occurrence late summer to early winter; common.
■ Inedible. Note the species has been listed as edible but it is now generally considered better to be avoided.

Lepista nuda (Bull.: Fr.) Cke.   **Tricholomataceae**
= *Tricholoma nudum* (Bull.: Fr.) Kummer   Wood Blewit
= *Rhodopaxillus nudus* (Bull.: Fr.) Maire

Medium-sized fleshy agaric, with brownish cap and lilac gills; in trooping groups, often in rings, on soil in mixed woods, hedgerows, parks and gardens.

Dimensions cap 6 - 12 cm dia; stem 5 - 9 cm tall x 1.5 - 2.5 cm dia.
Cap at first lilaceous, becoming brownish and drying more pallid; at first convex and slightly umbonate, becoming flattened and finally shallowly depressed and wavy, smooth. Flesh bluish-lilac and thick.
Gills lilaceous, fading to buff or brownish with age, emarginate, fairly narrow, crowded. **Spores** pink, minutely roughened, ellipsoid, non-amyloid, 6-8 x 4-5 μm. Basidia 4-spored. Cystidia absent.
Stem concolorous with cap, more or less equal, fibrillose and often slightly thickened at the base. Ring absent. Flesh bluish-lilac, thick, firm, full.
Odour strong, aromatic. **Taste** strong.
Chemical tests none.
Occurrence autumn to winter; common.
☐ Edible and good.
One of the few large agarics which continues to emerge after the first frosts.

Lepista saeva (Fr.) P Orton   **Tricholomataceae**
= *Tricholoma saevum* (Fr.) Gill.   Field Blewit
= *Rhodopaxillus saevus* (Fr.) Maire

Medium-sized fleshy agaric with pale brownish cap, flesh-coloured gills and bluish-lilac stem; in trooping groups, often in rings, on soil in pastures and other grassy places.

Dimensions cap 6 - 10 cm dia; stem 3 - 6 cm tall x 1.5 - 2.5 cm dia.
Cap brownish or pallid; at first convex and slightly umbonate, becoming flattened and finally shallowly depressed and wavy, smooth. Flesh whitish or pinkish-buff, thick.
Gills pinkish-buff, emarginate, fairly narrow, crowded. **Spores** pink, minutely roughened, ellipsoid, non-amyloid, 7-9 x 4-5 μm. Basidia 4-spored. Cystidia absent.
Stem bluish-lilac, more or less equal, fibrillose and often slightly thickened at base. Ring absent. Flesh whitish or pinkish-buff, thick, firm and full.
Odour strong, aromatic. **Taste** strong.
Chemical tests none.
Occurrence autumn to winter; infrequent.
☐ Edible and good.

Lepista sordida (Fr.) Sing.   **Tricholomataceae**
= *Rhodopaxillus sordidus* (Fr.) Maire
= *Tricholoma sordidum* (Fr.) Kummer

Small to medium agaric, more or less lilac, with brownish tinges throughout; in trooping groups on soil, typically amongst plant debris in hedgerows, scrub, parks and gardens and on compost heaps.

Dimensions cap 3 - 8 cm dia; stem 4 - 6 cm tall x 0.5 - 0.8 cm dia.
Cap lilaceous or brownish-lilac; at first convex and slightly umbonate, becoming flattened, and finally shallowly depressed and wavy, smooth. Flesh greyish-lilac, and moderately thick.
Gills lilaceous, becoming more brownish with age, adnate or emarginate, broad, crowded. **Spores** pink, minutely roughened, ellipsoid, non-amyloid, 6-7 x 3.5-4 μm. Basidia 4-spored. Cystidia absent.
Stem concolorous with cap, more or less equal, fibrillose and often slightly thickened at base. Ring absent. Flesh greyish-lilac, thick and full.
Odour aromatic. **Taste** not distinctive.
Chemical tests none.
Occurrence summer to autumn; infrequent.
☐ Edible.

Leucopaxillus giganteus (Sow.: Fr.) Sing. Tricholomataceae
= *Clitocybe gigantea* (Sow.: Fr.) Quél.

Very large ivory-white agaric with decurrent gills; solitary or in trooping groups, typically in rings, amongst grass in pastures, by roadside hedges and in woodland clearings.

Dimensions cap 8 - 30 cm dia; stem 4 - 7 cm tall x 2.5 - 3.5 cm dia.
Cap ivory white, with tan tinge at the centre; at first more or less flattened, with incurved margin, becoming deeply infundibuliform, smooth or cracked and scaly, dry. Flesh white, thick, firm or tough.
Gills concolorous with cap or buff, decurrent, narrow, very crowded.
Spores hyaline, smooth, ellipsoid, amyloid, droplets, 6-8 x 3-4 μm. Basidia 4-spored. Cystidia absent.
Stem concolorous with cap, stout, equal or tapering slightly downwards into more or less bulbous base. Ring absent. Flesh white, firm and full.
Odour not distinctive. **Taste** not distinctive.
Chemical tests none.
Occurrence late summer to autumn; infrequent.
☐ Edible.

Lyophyllum connatum (Schum.: Fr.) Sing. Tricholomataceae
= *Clitocybe connata* (Schum.: Fr.) Gill.

Medium-sized, fleshy agaric, pure white; densely tufted, on soil, typically in grass, in clearings or close by broad-leaf trees and in mixed woods.

Dimensions cap 3 - 7 cm dia; stem 3 - 6 cm tall x 0.8 - 1.5 cm dia.
Cap white; at first convex, becoming expanded and often with a wavy margin, smooth, dry. Flesh white, moderately thick and firm.
Gills white, adnate-decurrent, broad, crowded. **Spores** hyaline, smooth, ellipsoid, non-amyloid, droplets, 5-6 x 2-4 μm. Basidia 4-spored. Cystidia absent.
Stem concolorous with cap, more or less equal but often swollen towards base, then tapered. Ring absent. Flesh white, firm, stuffed, or full.
Odour not distinctive. **Taste** not distinctive.
Chemical tests flesh and gills purple with $FeSO_4$.
Occurrence autumn; common.
■ Inedible.

Lyophyllum decastes (Fr.: Fr.) Sing. Tricholomataceae
= *Clitocybe decastes* (Fr.: Fr.) Kummer
= *Tricholoma aggregatum* (Schaeff.) Constant. & Dufour

Medium-sized, fleshy agaric, with grey-brown cap and greyish gills; densely tufted, on soil in open woodland.

Dimensions cap 4 - 10 cm dia; stem 3 - 6 cm tall x 1 - 2 cm dia.
Cap greyish-brown, with radiating silvery streaks; at first convex or bun-shaped, becoming expanded and often with a wavy margin, smooth, dry. Flesh white, moderately thick and firm.
Gills whitish-grey or grey, sometimes with ochraceous tinges, adnate, broad, crowded. **Spores** hyaline, smooth, sub-spherical, non-amyloid, 5-7 x 5-6 μm. Basidia 4-spored. Cystidia absent.
Stem pallid at apex, becoming more brownish-grey towards base, often eccentric, more or less equal but typically swollen towards base, then tapered. Ring absent. Flesh white, tough, fibrous, stuffed, or full.
Odour not distinctive. **Taste** not distinctive.
Chemical tests none.
Occurrence late summer to autumn; infrequent.
☐ Edible.

Lyophyllum leucophaeatum (Karst.) Karst.
= *Lyophyllum fumatofoetens* (Sécr.) J Schaff.   **Tricholomataceae**

Medium-sized, fleshy agaric, with brownish, finely downy cap and greyish gills; clustered on soil in broad-leaf and coniferous woodland.

Dimensions cap 4 - 8 cm dia; stem 4 - 7 cm tall x 0.6 - 1 cm dia.
Cap greyish-brown, with beige tinge, darker radial streaks and spotting; at first convex or bun-shaped, becoming expanded, umbonate, downy, dry. Flesh whitish, moderately thick, firm, turning rapidly blue then black where cut.
Gills whitish-grey or grey, sometimes with ochraceous tinge, adnate, fairly narrow, crowded. **Spores** hyaline, finely roughened, cylindrical, non-amyloid, 5-7 x 3-4 µm. Basidia 4-spored. Marginal cells filiform.
Stem pallid or slightly ochraceous, decorated with darker fibrils, slightly pruinose at apex, more or less equal or slightly swollen at base. Ring absent. Flesh whitish, reacting as in cap, tough, stuffed or full.
Odour not distinctive. **Taste** not distinctive.
Chemical tests none.
Occurrence late summer to autumn; infrequent.
✚ Poisonous.

Lyophyllum loricatum (Fr.) Kühn.   **Tricholomataceae**
= *Tricholoma cartilagineum* Bull. non Fr.

Medium-sized, hard, fleshy agaric, with dark brown cap and greyish white gills; densely clustered on soil, in open woodland.

Dimensions cap 3 - 12 cm dia; stem 4 - 9 cm tall x 0.7 - 1.5 cm dia.
Cap dark chestnut or olivaceous-brown, hygrophanous, more pallid when dry; at first convex or bun-shaped, becoming expanded and often with a more or less even margin, satiny. Flesh whitish, moderately thick, elastic and tough.
Gills whitish grey or grey, adnate or sub-decurrent, broad, crowded.
Spores hyaline, smooth, sub-spherical, non-amyloid, 5-6 x 4.5-5.5 µm. Basidia 4-spored. Cystidia absent.
Stem pallid brownish, fibrillose, more or less equal. Ring absent. Flesh whitish, tough, cartilaginous and full.
Odour not distinctive. **Taste** not distinctive.
Chemical tests none.
Occurrence late summer to autumn; frequent.
☐ Edible.

Lyophyllum ulmarium (Bull.: Fr.) Kühn. **Tricholomataceae**
= *Pleurotus ulmarius* (Bull.: Fr.) Quél.

Medium-sized, fleshy agaric with creamy-yellow cap and gills; densely tufted, caespitose, on living or dead trunks of broad-leaf trees, favouring elm but also poplar.

Dimensions cap 7 - 10 cm dia; stem 8 - 12 cm tall x 1 - 2.5 cm dia.
Cap ivory, with ochraceous or greyish tinges, more yellow with age, radially fibrillose; at first bun-shaped, becoming expanded, smooth, dry. Flesh white, moderately thick and firm.
Gills at first whitish, becoming pallid ochraceous at maturity, adnate or with decurrent tooth, broad with wavy edges, crowded.
Spores hyaline, smooth, sub-spherical, non-amyloid, droplets, 5.5-7.5 x 4.5-6 µm. Basidia 4-spored. Cystidia absent.
Stem concolorous with cap, more or less equal but typically swollen towards base, then tapered. Ring absent. Flesh white, tough, fibrous and full.
Odour sour. **Taste** not distinctive.
Chemical tests none.
Occurrence late summer to autumn; infrequent.
■ Inedible.

Tephrocybe anthracophila (Lasch) P Orton
= *Lyophyllum anthracophilum* (Lasch) M Lange & Sivertsen
Trichomolataceae

Small delicate, agaric with dark-brown cap and stem, white or grey gills; in small trooping groups on soil on fire sites.

Dimensions cap 2 - 3 cm dia; stem 1 - 3 cm tall x 0.1 - 0.2 cm dia.
Cap dark brown or blackish-brown, more pallid when dry; convex and often slightly umbilicate, margin faintly wrinkled. Flesh greyish-white and thin.
Gills at first white, becoming pallid grey-brown, adnate, broad, close or somewhat distant. **Spores** hyaline, smooth, spherical, non-amyloid, droplets, 4.5-6 µm. Basidia 4-spored. Cystidia absent.
Stem concolorous with cap, slightly pruinose at the apex, otherwise smooth. Ring absent. Flesh greyish, cartilaginous and hollow.
Odour faint, of meal. **Taste** not distinctive.
Chemical tests none.
Occurrence autumn to early winter; infrequent or rare. (Stourhead, Wiltshire)
■ Inedible.
Species differs microscopically from *T. atrata* which has ellipsoid spores.

Tephrocybe atrata (Fr.: Fr.) Donk   **Tricholomataceae**

Small, delicate agaric, with dark-brown cap and stem, white or grey gills; in small trooping groups on soil on fire sites.

Dimensions cap 2 - 5 cm dia; stem 2 - 5 cm tall x 0.1 - 0.3 cm dia.
Cap dark brown or blackish-brown, more pallid when dry; convex and often slightly umbilicate, margin faintly striate, smooth. Flesh greyish-white and thin.
Gills at first white, becoming pallid grey, adnate, broad, close or somewhat distant. **Spores** hyaline, smooth, ellipsoid, non-amyloid, droplets, 5-7 x 3-5 µm. Basidia 4-spored. Cystidia absent.
Stem concolorous with cap, slightly hairy at the base, otherwise smooth. Ring absent. Flesh greyish, cartilaginous and hollow.
Odour faint, of meal. **Taste** not distinctive.
Chemical tests none.
Occurrence autumn to spring; infrequent or rare. (Forest of Dean, Gloucestershire)
■ Inedible.
Species differs microscopically from *T. anthracophila* which has spherical spores.

Tephrocybe palustris (Peck) Donk   **Tricholomataceae**

Small, delicate, agaric with brown cap and stem, white or grey gills; in small trooping groups in *Sphagnum* bogs.

Dimensions cap 1 - 2 cm dia; stem 2 - 3 cm tall x 0.1 - 0.2 cm dia.
Cap umber-brown, sometimes darker at centre; at first convex, then flattened or slightly umbilicate, transparently striate two-thirds to the centre, smooth. Flesh buff and thin.
Gills at first white becoming pallid grey, deeply adnate-emarginate, broad, close or somewhat distant. **Spores** hyaline, smooth, ellipsoid, non-amyloid, occasionally with droplets, 6-9 x 3-5 µm. Basidia 4-spored. Cystidia absent.
Stem concolorous with cap, slightly pruinose when young, otherwise smooth. Ring absent. Flesh greyish, fragile and hollow.
Odour faint, of meal. **Taste** not distinctive.
Chemical tests none.
Occurrence spring to autumn; common.
■ Inedible.

Macrocystidia cucumis (Pers.: Fr.) Joss. **Tricholomataceae**
= *Naucoria cucumis* (Pers.: Fr.) Kummer

Smallish, dark-brown agaric, with whitish gills and smelling of fish; in trooping groups on soil but typically associated with rotting wood fragments in broad-leaf and mixed woods.

Dimensions cap 2 - 7 cm dia.; stem 3 - 7 cm tall x 0.3 - 0.5 cm dia.
Cap reddish-brown or darker when damp, hygrophanous, becoming ochraceous (from the margin) when dry; at first conical or campanulate, becoming expanded and flatly umbonate, densely downy. Flesh dark brown, thin. Cap cystidia very large, lanceolate.
Gills at first whitish becoming more ochraceous at maturity, adnexed with decurrent tooth, fairly broad, crowded. **Spores** reddish, smooth, ellipsoid, non-amyloid, 7-9 x 3.5-4.5 µm. Basidia 4-spored. Cystidia very large, lanceolate.
Stem concolorous with cap, or blackish-brown, but more ochraceous at apex, cylindrical, densely downy. Ring absent. Flesh brown, cartilaginous, tough, full or hollow.
Odour strong, of fish oil or cucumber. **Taste** mild, of fish oil.
Chemical tests none.
Occurrence autumn; infrequent.
■ Inedible.

Marasmius alliaceus (Jacq.: Fr.) Fr.   **Tricholomataceae**

Thin, brownish agaric, with pale gills and long, slender, rigid stem, smelling distinctively of garlic; on leaf litter, buried twigs and other plant debris, typically in small trooping groups in mixed woods, favouring beech, and on chalky soils.

Dimensions cap 1 - 4 cm dia; stem 4 - 20 cm tall x 0.1 - 0.3 cm dia.
Cap pallid or clay-brown; convex or campanulate, smooth but striate or sulcate two-thirds towards the centre. Flesh whitish and thin.
Gills whitish or pallid grey, adnexed or free, narrow, distant.
Spores hyaline, smooth, ovoid, non-amyloid, droplets, 7-10 x 6-8 µm. Basidia 4-spored. Cystidia more or less clavate.
Stem concolorous with cap, faintly pruinose at the apex, becoming much darker towards base, finely downy, slender, rooting. Ring absent. Flesh brownish, thin and tough.
Odour strong, of garlic. **Taste** not distinctive.
Chemical tests none.
Occurrence autumn; infrequent.
■ Inedible.

Marasmius androsaceus L.: Fr.   **Tricholomataceae**
= *Androsaceus androsaceus* (L.: Fr.) Rea   Horsehair Fungus

Minute, delicate, brownish agaric, with pinkish gills and tough horsehair-like stem; trooping and often carpeting large areas on pine needles and other coniferous litter, also on dead heather.

Dimensions cap 0.5 - 1 cm dia; stem 2 - 6 cm tall x 0.05 - 0.1 cm dia.
Cap clay-brown with pinkish tinge, more reddish-brown at centre; convex, sometimes slightly depressed, at the centre, membraneous, radially wrinkled. Flesh whitish and very thin.
Gills buff, adnate or decurrent, broad, distant. **Spores** hyaline, smooth, ellipsoid, non-amyloid, droplets, 6-8 x 3-4 µm. Basidia 4-spored. Cystidia absent.
Stem black, shiny, very slender. Ring absent. Flesh brownish and tough.
Odour not distinctive. **Taste** not distinctive.
Chemical tests none.
Occurrence spring to late autumn; common.
■ Inedible.

Marasmius bulliardii Quél.   Tricholomataceae

Very small, delicate, whitish agaric, with parachute-like cap and slender tough stem; typically in trooping groups on surface roots and dead twigs in mixed woods, often favouring woodland paths.

Dimensions cap 0.3 - 1 cm dia; stem 2 - 4 cm tall x 0.02 - 0.05 cm dia.
Cap whitish and typically dark brown at centre; at first convex then umbilicate, radially grooved with a crenulate margin. Flesh white and very thin.
Gills whitish, free, with distinct collar, broad, very distant.
Spores hyaline, smooth, ellipsoid, non-amyloid, droplets, 7-10 x 3-5 µm. Basidia 4-spored. Gill edge cystidia absent but clavate or vesicular cells present with small warty projections.
Stem concolorous with cap at apex, becoming darker brown towards base, very slender, shiny. Ring absent. Flesh brownish, thin and tough.
Odour not distinctive. **Taste** not distinctive.
Chemical tests none.
Occurrence late summer to autumn; infrequent.
■ Inedible.
Distinguished from *M. rotula* by overall size and smaller number of gills (12-16)

Marasmius buxi Fr.   Tricholomataceae

Minute, delicate, light-brown agaric, with darker stem; specifically on fallen leaves of box.

Dimensions cap 0.2 - 0.5 cm dia; stem 0.5 - 1.5 cm tall x 0.05 - 0.1 cm dia,
Cap pallid brown, more reddish-brown at centre; convex, sometimes slightly depressed at the centre, membraneous, covered with white downy hairs. Flesh whitish and very thin.
Gills white, adnate or decurrent, broad, distant. **Spores** hyaline, smooth, cylindrical-ellipsoid, non-amyloid, droplets, 8-10 x 2.5-3.5 µm. Basidia 1- or 2-spored. Gill edge cystidia fusiform.
Stem purplish-brown, very slender, finely pruinose when young, then smooth. Ring absent. Flesh brownish and tough.
Odour not distinctive. **Taste** not distinctive.
Chemical tests none.
Occurrence spring to autumn; infrequent.
■ Inedible.

Marasmius epiphyllus (Pers.: Fr.) Fr.   Tricholomataceae
= *Androsaceus epiphyllus* (Pers.: Fr.) Pat.

Very small, delicate, whitish agaric, with long wiry stem; typically in large trooping groups on leaf blades, petioles and twigs, in mixed woods.

Dimensions cap 0.3 - 1 cm dia; stem 0.5 - 3 cm tall x 0.05 - 0.1 cm dia.
Cap whitish or pallid cream; at first convex, then flattened and often depressed at centre, membraneous, radially wrinkled. Flesh whitish and very thin.
Gills whitish, decurrent, narrow, very distant and branched.
Spores hyaline, smooth, fusiform-clavate, non-amyloid, 10-14 x 3-4 µm. Basidia 4-spored. Gill edge cystidia cylindrical or fusiform.
Stem concolorous with cap at apex, becoming more reddish-brown towards base, very slender filiform, shiny, rooting. Ring absent. Flesh brownish, thin and tough.
Odour not distinctive. **Taste** not distinctive.
Chemical tests none.
Occurrence late summer to autumn; infrequent.
■ Inedible.

Marasmius hudsonii (Pers: Fr.) Fr.   **Tricholomataceae**
= *Marasmius pilosus* (Huds.) Quél.

Minute, delicate, whitish agaric, with tough, dark, horsehair-like stem; trooping on damp fallen leaves of holly.

Dimensions cap 0.2 - 0.6 cm dia; stem 1 - 4 cm tall x 0.03 - 0.05 cm dia.
Cap cream with flesh-pink tinge, more reddish-brown at centre; convex, membraneous, radially wrinkled, covered with erect bristly reddish-brown hairs. Flesh whitish and very thin.
Gills white, adnate or decurrent, broad, distant, sometimes rudimentary. **Spores** hyaline, smooth, ellipsoid, non-amyloid, 10-14.5 x 4.5-6.5 µm. Basidia 3- or 4-spored. Cystidia fusiform.
Stem reddish-brown, shiny, very slender, tough. Ring absent. Flesh brownish and tough.
Odour not distinctive. **Taste** not distinctive.
Chemical tests none.
Occurrence autumn to spring; rare. (Holt Forest, Dorset)
■ Inedible.

Marasmius oreades (Bolt.: Fr.) Fr.   **Tricholomataceae**
Fairy Ring Mushroom

Smallish, pale tan, fleshy agaric, with blunt umbo and tough rooting stem; typically in rings, on soil in short grass, often favouring garden lawns.

Dimensions cap 2 - 5 cm dia; stem 2 - 10 cm tall x 0.3 - 0.5 cm dia.
Cap tan, hygrophanous, drying to buff but retaining tan tinge at centre; at first convex, then flattened and broadly umbonate, smooth but striate at the margin. Flesh whitish-buff, thick at the centre, otherwise thin.
Gills whitish becoming ochraceous-cream, adnexed or free, fairly broad, distant. **Spores** hyaline, smooth, ellipsoid, non-amyloid, occasional droplets, 7-10 x 4-6 µm. Basidia 4-spored. Cystidia absent.
Stem concolorous with cap, smooth or finely scurfy, slender, more or less equal, whitish downy at the base and slightly rooting, stiff. Ring absent. Flesh whitish buff and tough.
Odour slight, woody. **Taste** slight, peppery.
Chemical tests none.
Occurrence spring to autumn; common.
☐ Edible. **Caution** Care must be taken to avoid confusion with the dangerously poisonous species *Clitocybe dealbata* and *Clitocybe rivulosa*, both of which are of similar size and habit.

Marasmius quercophilus Pouz.   **Tricholomataceae**
= *Marasmius splachnoides* (Horn.: Fr.) Fr.

Very small, delicate, whitish-pink agaric with short tough brown stem; typically in small trooping groups on leaves of oak.

Dimensions cap 0.3 - 1 cm dia; stem 1 - 2.5 cm tall x 0.01 - 0.03 cm dia.
Cap whitish-buff, often darker pink at centre; at first convex, then flattened and often depressed at centre, membraneous, radially wrinkled, surface finely roughened. Flesh concolorous and very thin.
Gills whitish-pink, adnate, narrow, distant. **Spores** hyaline, smooth, elongated-ellipsoid with 1 large droplet, non-amyloid, 7.5-9 x 4.5-5.5 µm. Basidia 4-spored. Gill edge cystidia broadly clavate or pear-shaped with short blunt projections.
Stem reddish-brown, pallid at the apex, very slender, smooth. Ring absent. Flesh concolorous, thin and tough.
Odour not distinctive. **Taste** not distinctive.
Chemical tests none.
Occurrence autumn; infrequent.
■ Inedible.

Marasmiellus ramealis (Bull.:Fr.) Sing.   **Tricholomataceae**
= *Marasmius ramealis* (Bull.: Fr.) Fr.

Very small, delicate, whitish or cream agaric, with short tough stem; typically in large trooping groups on twigs, favouring bramble but also on other broad-leaf and conifer debris.

Dimensions cap 0.3 - 1.5 cm dia; stem 0.3 - 2 cm tall x 0.05 - 0.1 cm dia.
Cap whitish-buff, often darker at centre; at first convex, then flattened and often depressed at centre, membraneous, radially wrinkled, surface finely roughened. Flesh concolorous and very thin.
Gills whitish-pink, adnate, narrow, distant. **Spores** hyaline, smooth, elongated-ellipsoid with 1 large droplet, non-amyloid, 8-10 x 3-4.5 µm. Basidia 2- or 4-spored. Gill edge cystidia broadly clavate with short, blunt projections.
Stem concolorous with cap at apex, becoming darker towards curved base, very slender, with white scurfy patches, sometimes tapering downwards. Ring absent. Flesh concolorous, thin and tough.
Odour not distinctive. **Taste** not distinctive.
Chemical tests none.
Occurrence late spring to autumn; common.
■ Inedible.

Marasmius rotula (Scop.: Fr.) Fr.   **Tricholomataceae**

Very small, delicate, whitish agaric, with parachute-like cap and slender tough stem; typically in trooping groups on surface roots and dead twigs in mixed woods, often favouring woodland paths.

Dimensions cap 0.5 - 1.5 cm dia; stem 2 - 7 cm tall x 0.05 - 0.15 cm dia.
Cap whitish and typically dark brown at centre; at first convex then umbilicate, radially grooved with a crenulate margin. Flesh white and very thin.
Gills whitish, free with distinct collar, broad, distant. **Spores** hyaline, smooth, ellipsoid, non-amyloid, droplets, 7-10 x 3-5 µm. Basidia 4-spored. Gill edge cystidia clavate or vesicular with small warty projections.
Stem concolorous with cap at the apex, becoming darker brown towards base, very slender, shiny. Ring absent. Flesh brownish, thin and tough.
Odour not distinctive. **Taste** not distinctive.
Chemical tests none.
Occurrence late summer to autumn; common.
■ Inedible.
Distinguished from *M. bulliardii* by overall size and larger number of gills (16-22).

Megacollybia platyphylla (Pers.: Fr.) Kotl. & Pouz.
= *Collybia platyphylla* (Pers.: Fr.) Kummer
= *Oudemansiella platyphylla* (Pers.: Fr.) Moser
= *Tricholomopsis platyphylla* (Pers.: Fr.) Sing.   **Tricholomataceae**

Medium to large fleshy agaric, with grey-brown fibrillose cap, whitish gills and stem; gregarious or tufted, with rotting broad-leaf and coniferous stumps often attached to buried wood by mycelial cords.

Dimensions cap 4 - 12 cm dia; stem 6 - 15 cm tall x 1.5 - 2.5 cm dia.
Cap whitish, densely covered with grey-brown fibrils; at first campanulate, becoming expanded convex, sometimes slightly umbonate, darkening with age, splitting radially. Flesh white, fragile and thin.
Gills white becoming cream, adnexed, very broad, crowded.
Spores hyaline, smooth, sub-spherical or broadly ellipsoid, non-amyloid, droplets, 6-10 x 5-8 µm. Basidia 4-spored. Gill edge cystidia clavate. Gill face cystidia absent.
Stem whitish, with darker fine fibrils (less dense than on cap), more or less equal, thickened at base, with long mycelial cords. Ring absent. Flesh white, tough and full, becoming stuffed or hollow.
Odour earthy. **Taste** not distinctive.
Chemical tests none.
Occurrence late spring to early autumn; infrequent but locally common.
■ Inedible.

Melanoleuca cinerascens Reid   Tricholomataceae

Medium or large agaric, with pale grey, broad, umbonate cap and white gills; solitary or scattered on soil in woodland and parks.

Dimensions cap 4 - 8 cm dia; stem 4 - 8 cm tall x 0.5 - 1 cm dia.
Cap pallid, greyish with darker brown-tinged centre; at first flattened convex, becoming umbonate-depressed, smooth. Flesh white, thickish and soft.
Gills white, slightly decurrent, broad, crowded. **Spores** hyaline, minutely warty, ellipsoid, amyloid, droplets, 6.5-9.5 x 5-6 µm. Basidia 4-spored. Gill edge cystidia fusiform.
Stem pallid grey with silvery effect, tapering slightly upwards from more or less bulbous base. Ring absent. Flesh white, tinged brown towards the base, soft and full.
Odour not distinctive. **Taste** not distinctive.
Chemical tests none.
Occurrence autumn; infrequent.
■ Inedible.

Melanoleuca cognata (Fr.) Konrad & Maubl. Tricholomataceae
= *Tricholoma cognatum* (Fr.) Gill.

Medium or large agaric, with pale brown, broad, umbonate cap and white gills; solitary or scattered on soil in coniferous woods.

Dimensions cap 4 - 10 cm dia; stem 5 - 12 cm tall x 1 - 1.5 cm dia
Cap ochraceous-brown, drying more greyish-brown; at first flattened convex, becoming umbonate-depressed, smooth, shiny. Flesh whitish or cream, thickish and soft.
Gills cream, emarginate, broad, crowded. **Spores** hyaline, minutely warty, ellipsoid, amyloid, droplets, 9-10 x 5.5-6 µm. Basidia 4- or 5-spored. Gill edge cystidia lanceolate and encrusted at the apex.
Stem pallid with brownish fibrils, tapering slightly upwards from more or less bulbous base. Ring absent. Flesh cream, soft and full.
Odour faint, of meal. **Taste** not distinctive.
Chemical tests none.
Occurrence spring to autumn; infrequent.
□ Edible.

Melanoleuca melaleuca (Pers.: Fr.) Murr. Tricholomataceae
= *Tricholoma melaleucum* (Pers.: Fr.) Kummer

Medium or large agaric, with dark brown cap, drying paler, broadly umbonate with white gills; solitary or scattered on soil amongst grass in open broad-leaf woodland and on lawns near trees.

Dimensions cap 3 - 8 cm dia; stem 4 - 7 cm tall x 0.8 - 1.4 cm dia.
Cap dark brown, hygrophanous, drying buff; at first flattened convex, becoming umbonate-depressed, smooth. Flesh white, tinged ochraceous, thickish and soft.
Gills white, emarginate, broad, crowded. **Spores** hyaline, minutely warty, ellipsoid, amyloid, droplets, 7-8.5 x 5-5.5 µm. Basidia 4-spored. Gill edge cystidia harpoon-shaped.
Stem pallid with dark brownish-grey fibrils, tapering slightly upwards from more or less bulbous base. Ring absent. Flesh brownish, becoming darker towards the base, soft and full.
Odour not distinctive. **Taste** not distinctive.
Chemical tests none.
Occurrence late summer to autumn; common.
■ Inedible.

Melanoleuca strictipes (Karst.) Schaeff. Tricholomataceae
= *Melanoleuca evenosa* (Sacc.) Konrad & Maubl.

Medium or large, often tall, agaric, with cream cap, broadly umbonate, with pale cream gills; solitary or scattered on soil amongst grass in open broad-leaf woodland or grassland near trees.

Dimensions cap 4 - 10 cm dia; stem 8 - 14 cm tall x 0.8 - 1.2 cm dia.
Cap cream, sometimes tinged brown at centre; at first convex, then expanded with broad umbo, smooth. Flesh white, thickish and soft.
Gills pallid cream, sometimes with pink tinge, emarginate, broad, crowded. **Spores** hyaline, minutely warty, ellipsoid, amyloid, droplets, 8-9 x 4.5-5 µm. Basidia 4-spored. Gill edge cystidia lanceolate and encrusted at apex.
Stem concolorous with cap, fibrillose, tapering slightly upwards from more or less bulbous base. Ring absent. Flesh white, soft, stuffed or full.
Odour strong, of meal. **Taste** not distinctive.
Chemical tests none.
Occurrence spring to autumn; rare. (Soudley, Forest of Dean, Gloucestershire)
□ Edible.

Melanoleuca stridula (Fr.) Metr.   Tricholomataceae

Small or medium-sized agaric, with blackish-brown, broadly umbonate cap and white gills; solitary or scattered on soil amongst grass near coniferous trees.

Dimensions cap 2 - 6 cm dia; stem 5 - 6 cm tall x 0.5 - 0.8 cm dia.
Cap dark grey-brown, hygrophanous, drying more pallid; convex, becoming umbonate shield-shaped, smooth. Flesh pallid, tinged brown, thickish and soft.
Gills white, adnate, broad, crowded. **Spores** hyaline, minutely warty, broadly ellipsoid, amyloid, droplets, 7-9 x 5-6 µm. Basidia 1-, 2- or 4-spored. Cystidia absent.
Stem pallid, greyish with darker fibrils, tapering upwards from stoutly bulbous base. Ring absent. Flesh pallid, becoming darker brown towards base, soft and full.
Odour not distinctive. **Taste** not distinctive.
Chemical tests none.
Occurrence autumn; infrequent.
■ Inedible.

Micromphale brassicolens (Romagn.) P Orton
Tricholomataceae

Small, delicate agaric, with pale brown cap and gills, smelling distinctively foul; in trooping groups on leaf litter and other rotting debris specifically of beech.

Dimensions cap 1 - 2 cm dia; stem 1.5 - 2.5 cm tall x 0.1 - 0.2 cm dia.
Cap brownish, darker towards centre, more tan at margin; convex, becoming expanded, flattened or slightly depressed with striate margin when damp, otherwise smooth. Flesh concolorous and thin.
Gills at first white, becoming pallid dirty-brown, adnate, irregular, somewhat distant. **Spores** hyaline, smooth, ellipsoid, non-amyloid, 5-7 x 3-3.5 µm. Basidia 2- or 4-spored. Cystidia absent.
Stem darker brown than cap, smooth or with slight fibrous flecking, more or less equal. Ring absent. Flesh pallid brown and full.
Odour of rotting cabbage. **Taste** not distinctive.
Chemical tests none.
Occurrence autumn; rare. (Bridgham, near Thetford, Norfolk)
■ Inedible.

Micromphale foetidum (Sow.: Fr.) Sing. Tricholomataceae

Small agaric with brown cap, whitish gills and dark slender stem; more or less solitary on rotting wood and bark.

Dimensions cap 1.5 - 3 cm dia; stem 1 - 2.5 cm tall x 0.2 - 0.3 cm dia.
Cap dull reddish-brown; convex, becoming expanded, flattened with somewhat sulcate margin, smooth. Flesh pallid and thin.
Gills pallid-brown, adnate, distant. **Spores** hyaline, smooth, narrowly ellipsoid, non-amyloid, with droplets, 8-10 x 3-4 µm. Basidia 4-spored. Gill edge cystidia more or less fusiform; gill face cystidia absent.
Stem blackish-brown, downy, more or less equal or tapering slightly at base, sometimes compressed. Ring absent. Flesh brownish, fragile, full.
Odour foul, of rotting cabbage. **Taste** not distinctive.
Chemical tests none.
Occurrence summer to autumn; infrequent.
■ Inedible.

Micromphale inodorum (Pat.) Svrcek Tricholomataceae

Small agaric with brown cap, whitish gills and dark slender stem; more or less solitary on rotting wood and bark.

Dimensions cap 1 - 4.5 cm dia; stem 2 - 4 cm tall x 0.2 - 0.4 cm dia.
Cap dull brown; convex, becoming expanded, flattened with slightly sulcate margin, smooth. Flesh pallid and thin.
Gills pallid brown, adnate, distant. **Spores** hyaline, smooth, narrowly ellipsoid, non-amyloid, with droplets, 6-10 x 3.5-4.5 µm. Basidia 4-spored. Gill edge cystidia fusiform, knobbly, some forked; gill face cystidia absent.
Stem concolorous with cap, but slightly darker, with white floccose patches, more or less equal or tapering slightly at base, sometimes compressed. Ring absent. Flesh brownish, fragile and full .
Odour not distinctive. **Taste** not distinctive.
Chemical tests none.
Occurrence autumn; infrequent.
■ Inedible.
Only member of the genus which does not possess an unpleasant smell.

Mycena abramsii Murr. Tricholomataceae
= *Mycena praecox* Vel.

Small, pale grey-brown conical agaric with white gills; solitary or in small trooping groups on rotting stumps and other wood, sometimes buried, of broad-leaf trees, favouring beech.

Dimensions cap 1 - 2.5 cm dia; stem 3 - 6 cm tall x 0.1 - 0.3 cm dia.
Cap pallid greyish-brown or beige but also darker brown, more pallid at the margin; at first campanulate becoming expanded conical, striate-sulcate 3/4 to centre when damp. Flesh pallid grey and thin.
Gills whitish, adnexed with decurrent tooth, broad, close.
Spores hyaline, smooth, cylindrical-ellipsoid, amyloid, with droplets, 7.5-11 x 4.5-5.5 µm. Basidia 4-spored. Gill cystidia fusiform or thinly clavate, thin-walled, sometimes slightly forked at the apices.
Stem concolorous with cap, darker towards the base, smooth or finely pruinose, slender, more or less equal, downy at the base. Ring absent. Flesh pallid greyish, fragile and hollow.
Odour faint, of radish. **Taste** not distinctive.
Chemical tests none.
Occurrence summer to autumn; infrequent.
■ Inedible.
Amongst the earliest seasonal appearances of *Mycena* species.

Mycena acicula (Schaeff.: Fr.) Kummer   **Tricholomataceae**

Very small orange agaric with white gills and lemon-yellow stem; solitary or in small tufts on buried wood and other plant debris.

Dimensions cap 0.3 - 1 cm dia; stem 2 - 5 cm tall x 0.05 - 0.1 cm dia.
Cap bright orange, reddish at the centre, more pallid yellow towards the margin; at first hemispherical becoming convex-campanulate, striate-sulcate with wavy margin. Flesh yellowish and very thin.
Gills whitish-yellow, adnexed, broad, close. **Spores** hyaline, smooth, fusiform or cylindrical, non-amyloid, without droplets, 8-12 x 2.5-4 µm. Basidia 4-spored. Gill cystidia fusiform, thin-walled, sometimes with yellowish apical excrescence.
Stem pallid lemon-yellow, more pallid towards the base, smooth or finely pruinose, slender, more or less equal, downy at the base. Ring absent. Flesh yellowish, fragile and hollow.
Odour not distinctive. **Taste** not distinctive.
Chemical tests none.
Occurrence summer to autumn; infrequent.
■ Inedible.

Mycena aetites (Fr.) Quél.   **Tricholomataceae**
= *Mycena miserior* Huijsman
= *Mycena umbellifera* (Schaeff.) Quél.

Small, pale grey-brown agaric with radially wrinkled and umbonate cap and white gills, solitary or scattered on soil or on rotting wood.

Dimensions cap 0.8 - 2 cm dia; stem 2 - 6 cm tall x 0.1 - 0.2 cm dia.
Cap pallid grey-brown, darker at the centre; at first conical becoming broadly campanulate, sulcate when damp, striate when dry. Flesh whitish and thin.
Gills pallid grey, edges slightly more pallid and floccose, adnate, broad, close. **Spores** hyaline, smooth, ellipsoid, amyloid, without droplets, 6-10 x 4-6.5 µm. Basidia 2- or 4-spored. Gill cystidia fusiform, thin-walled and smooth.
Stem concolorous with cap, more pallid towards the apex, smooth, slender, more or less equal, downy at the base and slightly rooting. Ring absent. Flesh whitish and firm.
Odour faint, of radish. **Taste** not distinctive.
Chemical tests none.
Occurrence late summer to autumn; infrequent.
■ Inedible.

Mycena alcalina (Fr.) Kummer   **Tricholomataceae**

Smallish-pale grey-brown agaric with radially wrinkled and umbonate cap and white gills, smelling distinctively of ammonia; in small troops typically on or by conifer stumps.

Dimensions cap 1 - 3 cm dia; stem 2 - 6 cm tall x 0.1 - 0.3 cm dia.
Cap pallid grey-brown; at first conical becoming broadly campanulate, sulcate when damp, striate when dry. Flesh whitish and thin.
Gills pallid grey with lighter edges, adnate, broad, fairly distant.
Spores hyaline, smooth, elongated-ellipsoid, amyloid, with droplets, 8-11 x 4.5-6 µm. Basidia 4-spored. Gill cystidia fusiform, thin-walled and smooth.
Stem concolorous with cap, smooth, slender, more or less equal, downy at the base and slightly rooting. Ring absent. Flesh whitish and firm.
Odour strong, of ammonia. **Taste** not distinctive.
Chemical tests none.
Occurrence late summer to autumn; common.
■ Inedible.

Mycena amicta (Fr.) Quél.   Tricholomataceae
= *Mycena calorhiza* Bres.

Small grey and white agaric with radially striate, bluntly conical cap, woolly-downy stem and greyish gills; scattered or trooping amongst litter and on buried cones, typically in coniferous woodlands but also under broad-leaf trees.

Dimensions cap 1 - 2 cm dia; stem 2 - 6 cm tall x 0.1 - 0.3 cm dia.
Cap creamy-grey with detachable whitish gelatinous skin; at first conical becoming broadly campanulate, translucently striate, minutely downy. Flesh pallid ochraceous-grey and thin.
Gills white becoming pallid grey, adnate, broad, close.
Spores hyaline, smooth, ellipsoid, amyloid, with droplets, 6-11 x 4-5 μm. Basidia 4-spored. Gill edge cystidia cylindrical-fusiform, thin-walled and smooth. Gill face cystidia absent.
Stem concolorous with cap, densely pruinose towards the apex, white woolly towards base, more or less equal, slightly rooting. Ring absent. Flesh pallid brown and firm.
Odour of radish. **Taste** not distinctive.
Chemical tests none.
Occurrence summer to autumn; frequent.
■ Inedible.

Mycena ascendens (Lasch) Geest.   Tricholomataceae
= *Mycena tenerrima* (Berk.) Sacc.

Minute whitish agaric with bell-shaped cap; in troops on bark of living or dead broad-leaf or coniferous trees, and on other woody debris.

Dimensions cap 0.25 - 0.75 cm dia; stem 1 - 2.5 cm tall x 05 - 0.08cm dia.
Cap white; at first convex becoming flattened, striate and decorated with minute shining granules. Flesh white and very thin.
Gills white, adnexed or free, crowded. **Spores** hyaline, smooth, sub-spherical or ellipsoid, amyloid, 8-10 x 5-7 μm. Basidia 2-spored. Gill cystidia clavate or flask-shaped, warty and thin-walled.
Stem concolorous with cap, smooth, very slender, more or less equal, downy at the base which may possess a very small disc. Ring absent. Flesh white, fragile and hollow.
Odour not distinctive. **Taste** not distinctive.
Chemical tests none.
Occurrence summer to autumn; infrequent.
■ Inedible.

Mycena atroalba (Bull.: Fr.) S F Gray   Tricholomataceae

Small agaric with a dark, blackish-brown, conical cap, white gills and dark stem; solitary or in small trooping groups on needle litter or moss in coniferous woods, also less frequently with broad-leaf trees.

Dimensions cap 1 - 2 cm dia; stem 3 - 7 cm tall x 0.05 - 0.1 cm dia.
Cap dark blackish-brown, faintly pruinose with a downy appearance; at first conical becoming campanulate, margin sulcate. Flesh greyish- brown and thin.
Gills whitish or pallid grey when old, bristly with cystidia (under lens), adnexed-adnate, broad, close. **Spores** hyaline, smooth, ellipsoid, amyloid, with droplets, 10-13 x 5-6 μm. Basidia 4-spored. Gill cystidia fusiform, projecting (up to 100 μm), thin-walled and smooth.
Stem concolorous with cap, smooth, slender, more or less equal, faintly pruinose at the apex, coarsely downy at the base. Ring absent. Flesh greyish, cartilaginous and hollow.
Odour not distinctive. **Taste** not distinctive.
Chemical tests none.
Occurrence early to late autumn; infrequent.
■ Inedible.

Mycena avenacea (Fr.) Quél.　　Tricholomataceae
= *Mycena olivaceomarginata* (Mass.) Mass.

Small, greyish-brown agaric with broadly campanulate cap and pale greyish gills; in small troops on soil in lawns and other mown grasslands.

Dimensions cap 1 - 3 cm dia; stem 2 - 4 cm tall x 0.1 - 0.2 cm dia
Cap pallid grey-brown sometimes with olivaceous tinge; at first conical becoming more flattened, umbonate, sulcate when damp, striate when dry. Flesh whitish and thin.
Gills greyish-white sometimes with yellowish tinge to the edge, adnexed, narrow, crowded. **Spores** hyaline, smooth, ellipsoid, amyloid, with droplets, 9-12 x 5-6 µm. Basidia 4-spored. Gill cystidia clavate with finger-like processes of varying lengths, thin-walled.
Stem concolorous with cap, smooth, slender, more or less equal, downy white at base. Ring absent. Flesh whitish, fragile and hollow.
Odour faint, of radish. **Taste** not distinctive.
Chemical tests none.
Occurrence late summer to autumn; frequent.
■ Inedible.

Mycena capillaripes Peck　　Tricholomataceae

Small, delicate agaric with dark brown bell-shaped radially grooved cap, whitish gills with brown edges and brown stem; in small troops on needle litter in spruce woods.

Dimensions cap 0.5 - 2.5 cm dia; stem 3 - 6 cm tall x 0.05 - 0.2 cm dia.
Cap brown, more pallid brown towards margin, blackish-brown towards centre; at first conical becoming campanulate, translucently striate and sulcate. Flesh greyish-brown and thin.
Gills greyish-white with reddish-brown margin, adnate with decurrent tooth, broad, close. **Spores** hyaline, smooth, ellipsoid, amyloid, with droplets, 7-12.5 x 4-5.5 µm. Basidia 3- or 4-spored. Gill cystidia fusiform or clavate; sometimes forked, thin-walled, smooth.
Stem reddish-brown, darker at the base, smooth or faintly pruinose at the apex, very slender, more or less equal. Ring absent. Flesh greyish-brown, somewhat stiff.
Odour faint, of ammonia, or radish. **Taste** not distinctive.
Chemical tests none.
Occurrence late summer to autumn; frequent.
■ Inedible.

Mycena cinerella (Karst.) Karst.　　Tricholomataceae

Small, delicate agaric with grey, convex, striate cap, whitish gills and greyish stem; in small troops on twigs and litter mainly in coniferous but also broad-leaf woods.

Dimensions cap 0.5 - 1.5 cm dia; stem 2 - 5 cm tall x 0.05 - 0.1 cm dia.
Cap ash-grey or with brown tinge; hemispherical then convex, becoming expanded, transparently striate almost to the centre. Flesh pallid, watery and thin.
Gills whitish-grey, adnate with decurrent tooth, broad, close. **Spores** hyaline, smooth, ellipsoid, amyloid, with droplets, 7-10 x 4-5.5 µm. Basidia 4-spored. Gill edge cystidia clavate with warty outgrowths; gill face cystidia clavate, smooth.
Stem greyish-brown, smooth or faintly pruinose when young, very slender, often bent, more or less equal. Ring absent. Flesh pallid, fragile and hollow.
Odour slight, of meal. **Taste** not distinctive.
Chemical tests none.
Occurrence late summer to autumn; infrequent.
■ Inedible.

Mycena crocata (Schrad.: Fr.) Kummer   Tricholomataceae

Small, delicate agaric with brown bell-shaped radially grooved cap, whitish gills and brown stem, producing orange juice when damaged; in small troops on twigs and litter in broad-leaf woods, favouring beech.

Dimensions cap 1 - 3 cm dia; stem 6 - 10 cm tall x 0.1 - 0.2 cm dia.
Cap brown with olivaceous tinge, more pallid brown towards the margin; at first conical becoming campanulate, silky-striate. Flesh greyish-brown, thin, with orange juice.
Gills whitish, becoming yellowed with orange-brown spotting, adnate with decurrent tooth, broad, close. **Spores** hyaline, smooth, ellipsoid, amyloid, with droplets, 8.5-10.5 x 4.5-6.5 µm. Basidia 2- or 4-spored. Gill edge cystidia clavate with warty outgrowths; gill face cystidia clavate, smooth, with thicker walls.
Stem yellowish-brown, more reddish at base, smooth or faintly pruinose at the apex, very slender, often bent, more or less equal. Ring absent. Flesh greyish-brown, fragile, hollow, with orange juice.
Odour not distinctive. **Taste** not distinctive.
Chemical tests none.
Occurrence late summer to autumn; infrequent.
■ Inedible.

Mycena epipterygia (Scop.: Fr.) S F Gray. Tricholomataceae

Small, delicate agaric with yellowish-brown bell-shaped radially grooved cap, white gills and yellowish-green stem; in small troops on grasses, mosses and other debris in damp places both in and out of woodlands.

Dimensions cap 1 - 2 cm dia; stem 4 - 8 cm tall x 0.1 - 0.2 cm dia.
Cap pallid tan with yellowish tinge, darker brown towards the centre; at first conical, becoming campanulate, cuticle sticky and peelable, striate-sulcate and margin faintly denticulate. Flesh whitish and thin.
Gills whitish or with pink tinge, more or less adnate with decurrent tooth, with peelable slightly sticky margin, narrow, fairly distant.
Spores hyaline, smooth, ellipsoid, amyloid, with droplets, 8-10 x 4-5 µm. Gill cystidia clavate with irregularly warty outgrowths.
Stem yellowish, smooth, sticky, slender, more or less equal. Ring absent. Flesh whitish, fragile and hollow.
Odour faint, rancid or of meal. **Taste** not distinctive.
Chemical tests none.
Occurrence late summer to autumn; common.
■ Inedible.
Note: cap colour, shape of spores and cystidia, and habitat are important when distinguishing varieties of *M. epipterygia*.

Mycena epipterygia (Peck) Geest.
var Splendidipes Peck
= *Mycena splendidipes* Peck   Tricholomataceae

Small, delicate, slimy agaric with yellowish-brown bell-shaped radially grooved cap, white gills and yellowish-green stem; in small troops on needle litter and twigs in coniferous woods.

Dimensions cap 1 - 3 cm dia; stem 3 - 7 cm tall x 0.1 - 0.3 cm dia.
Cap pallid olive-brown, darker brown towards the centre, at first hemispherical, becoming convex, cuticle viscid and peelable, striate-sulcate and margin irregular. Flesh whitish and thin.
Gills whitish or cream with pink tinge at maturity, broadly adnate with decurrent tooth, with peelable slightly sticky margin, broad, fairly distant. **Spores** hyaline, smooth, broadly ellipsoid, amyloid, with droplets, 8-11 x 6-8 µm. Basidia 4-spored. Gill edge cystidia clavate with few irregularly warty outgrowths.
Stem yellowish-green, smooth, viscid, slender, more or less equal. Ring absent. Flesh whitish.
Odour faint, rancid or mealy. **Taste** not distinctive.
Chemical tests none.
Occurrence late summer to autumn; common.
■ Inedible
Note: cap colour, shape of spores and cystidia, and habitat are important when distinguishing varieties of *M. epipterygia*.

Mycena galericulata (Scop.: Fr.) S F Gray   **Tricholomataceae**
= *Mycena rugosa* (Fr.) Quél.

Small or medium, greyish-brown agaric with bell-shaped radially grooved cap and pinkish gills; in clusters on stumps and wood, favouring broad-leaf trees, particularly oak, but also on conifer stumps, etc.

Dimensions cap 2 - 8 cm dia; stem 5 - 10 cm tall x 0.2 - 0.4 cm dia.
Cap greyish-brown with more pallid margin; at first conical, becoming campanulate and broadly umbonate, striate-sulcate 1/2 way to centre. Flesh whitish and thin.
Gills at first white, becoming flesh-pink, adnate with decurrent tooth, broad, fairly distant. **Spores** hyaline, smooth, ellipsoid, amyloid, without droplets, 9-12 x 6-8 µm. Basidia 2-spored. Gill edge cystidia clavate, with long slender processes, thin-walled.
Stem concolorous with cap but more pallid near the apex, smooth, slender, more or less equal, white hairy at the base and somewhat rooting. Ring absent. Flesh whitish, tough, hollow.
Odour of meal, sometimes rancid. **Taste** not distinctive.
Chemical tests none.
Occurrence throughout the year but mainly summer and autumn; common.
☐ Edible.
Note: the only commonly occuring *Mycena* with a fruiting body growing to these dimensions.

Mycena galopus (Pers.: Fr.) Kummer   **Tricholomataceae**
= *Mycena annae* Benedix

Small, pale grey-brown agaric with bell-shaped radially grooved cap and whitish-grey gills; in small troops on leaf litter and amongst grasses in mixed woods and hedgerows; oozing white juice when cut.

Dimensions cap 1 - 2 cm dia; stem 5 - 10 cm tall x 0.2 - 0.3 cm dia.
Cap greyish-brown with darker centre; at first conical becoming campanulate, markedly striate-sulcate, finely pruinose when young, otherwise smooth. Flesh whitish and thin.
Gills pallid grey, adnate, broad, fairly distant. **Spores** hyaline, smooth, ellipsoid, amyloid, with droplets, 10-13 x 5-6 µm. Basidia 2- or 4-spored. Gill edge cystidia fusiform, thin-walled, conspicuous; gill face cystidia sparse.
Stem greyish, smooth, slender, more or less equal, coarsely whitish-woolly at base and slightly rooting. Ring absent. Flesh whitish, exuding white juice when cut, fragile and hollow.
Odour faint, of radish. **Taste** not distinctive.
Chemical tests none.
Occurrence summer to autumn; common.
■ Inedible.
Note: *M. galopoda* which appears in some reference books is a misspelling of this species.

Mycena galopus var candida J Lange **Tricholomataceae**
= *Mycena galopus var alba* Rea

Small, pure-white agaric with bell-shaped radially grooved cap; in small troops on leaf litter in mixed woods and hedgerows; oozing white juice when cut.

Dimensions cap 1 - 2 cm dia; stem 5 - 10 cm tall x 0.2 - 0.3 cm dia.
Cap wholly white; at first conical becoming campanulate, markedly striate-sulcate. Flesh white and thin.
Gills white, adnate, fairly distant. **Spores** hyaline, smooth, ellipsoid, amyloid, with droplets, 10-13 x 5-6 µm. Basidia 2- or 4-spored. Gill cystidia fusiform, thin-walled, conspicuous.
Stem concolorous with cap, smooth, slender, more or less equal, downy or woolly at the base and slightly rooting, exuding white juice when cut. Ring absent. Flesh white, fragile and hollow.
Odour not distinctive. **Taste** not distinctive.
Chemical tests none.
Occurrence summer to autumn; common.
■ Inedible.

Mycena haematopus (Pers.: Fr.) Kummer **Tricholomataceae**

Smallish grey-brown agaric with bell-shaped radially grooved cap and pinkish gills; in clusters, or tufted caespitose, on stumps and logs of broad-leaf trees; oozing blood-red juice when cut.

Dimensions cap 2 - 4 cm dia; stem 4 - 10 cm tall x 0.2 - 0.3 cm dia.
Cap greyish-brown when damp, hygrophanous, drying more pallid buff; at first conical, becoming campanulate, striate-sulcate from the margin when damp. Flesh blood-red and thin.
Gills at first white, becoming pallid pink with darker edge, adnate, fairly distant. **Spores** hyaline, smooth, ellipsoid, amyloid, with droplets, 7-10 x 5-6 µm. Basidia 2- or 4-spored. Gill cystidia flask-shaped with elongated apex, thin-walled, smooth.
Stem pinkish-grey, at first finely downy, becoming smooth, slender, more or less equal, woolly at the base. Ring absent. Flesh blood-red, exuding blood-red juice when cut, firm and hollow.
Odour not distinctive. **Taste** not distinctive
Chemical tests none.
Occurrence late summer to autumn; common.
■ Inedible.

Mycena inclinata (Fr.) Quél.   **Tricholomataceae**
= *Mycena galericulata var calopus* (Fr.) Karst.

Smallish bay-brown agaric with bell-shaped radially grooved cap, pinkish gills and brown stem; in dense caespitose tufts, on stumps of oak and, occasionally, other broad-leaf trees.

Dimensions cap 2 - 3.5 cm dia; stem 5 - 10 cm tall x 0.2 - 0.4 cm dia.
Cap bay-brown with darker centre; at first conical, becoming campanulate, striate-sulcate with overhanging, finely scalloped, margin. Flesh whitish and thin.
Gills at first whitish, becoming flesh pink, adnate, fairly broad, close or somewhat distant. **Spores** hyaline, smooth, ellipsoid, amyloid, with droplets, 8-11 x 5.5-6.5 µm. Basidia 4-spored. Gill edge cystidia clavate with irregular hair-like apical processes, thin-walled; gill face cystidia absent.
Stem whitish at the apex becoming bay-brown towards the base, smooth, slender, more or less equal, downy or woolly at the base and slightly rooting. Ring absent. Flesh whitish, firm and hollow.
Odour of meal or rancid. **Taste** not distinctive.
Chemical tests none.
Occurrence summer to autumn; frequent.
■ Inedible.

Mycena leucogala (Cke.) Sacc.   **Tricholomataceae**
= *Mycena galopus var nigra* Rea

Small, dark-grey agaric with bell-shaped radially grooved cap and white gills; in small troops on soil in mixed woods, often on old fire sites.

Dimensions cap 1 - 2 cm dia; stem 5 - 10 cm tall x 0.2 - 0.3 cm dia.
Cap dark or blackish-grey; at first conical, becoming campanulate, deeply striate-sulcate. Flesh whitish and thin.
Gills at first whitish then grey, adnate, fairly distant. **Spores** hyaline, smooth, ellipsoid, amyloid, with droplets, 10-13 x 5-6 µm. Basidia 4-spored. Gill cystidia fusiform, thin-walled and smooth.
Stem concolorous with cap, smooth, slender, more or less equal, downy or woolly at the base and slightly rooting, exuding white juice when cut. Ring absent. Flesh whitish, firm and hollow.
Odour faint, of radish. **Taste** not distinctive.
Chemical tests none.
Occurrence summer to autumn; infrequent.
■ Inedible.

Mycena metata (Fr.) Kummer   **Tricholomataceae**
= *Mycena phyllogena* (Pers.) Sing.

Small, delicate, pale flesh-coloured agaric with conical radially grooved cap, white gills and long very slender stem; in small troops on needles and other debris in coniferous woods.

Dimensions cap 1 - 2 cm dia; stem 4 - 8 cm tall x 0.1 - 0.2 cm dia.
Cap beige or pallid flesh, centre tinged darker; at first conical becoming expanded-campanulate, translucently striate-sulcate almost to the centre when moist. Flesh pallid, watery and thin.
Gills white or pallid beige, adnexed or with decurrent tooth, broad, close. **Spores** hyaline, smooth, broadly ellipsoid, amyloid, with droplets, 7-9.5 x 3.5-4.5 µm. Basidia 2- or 4-spored. Gill cystidia clavate or fusiform, with warty apical processes, thin-walled.
Stem concolorous with cap, smooth or finely pruinose at the apex, very slender, more or less equal, downy at the base. Ring absent. Flesh pallid and watery, fragile and hollow.
Odour faint, of iodoform. **Taste** not distinctive or slightly antiseptic.
Chemical tests none.
Occurrence late summer to autumn; frequent.
■ Inedible.

Mycena oortiana Hora   **Tricholomataceae**
= *Mycena arcangeliana var oortiana* Kühn.

Small, pale grey-brown agaric with broadly conical cap and pinkish gills; in small troops on stumps and branches of broad-leaf trees.

Dimensions cap 1 - 4 cm dia; stem 2 - 4 cm tall x 0.1 - 0.2 cm dia.
Cap pallid grey-brown, sometimes with olivaceous tinge; at first conical, becoming more flattened, umbonate, striate when damp, the margin becoming irregular. Flesh whitish and thin.
Gills white, becoming pinkish tinged, adnexed, narrow, crowded.
Spores hyaline, smooth, ellipsoid, amyloid, with droplets, 7-8 x 4.5-5 µm. Basidia 4-spored. Gill cystidia vesicular, thin-walled and warty.
Stem greyish, smooth, shiny, slender, more or less equal, downy white at the base. Ring absent. Flesh whitish, fragile and hollow.
Odour strong, of iodoform. **Taste** not distinctive.
Chemical tests none.
Occurrence late summer to autumn; infrequent.
■ Inedible.

Mycena pearsoniana Dennis ex Sing.   **Tricholomataceae**
= *Mycena pseudopura* (Cke.) Sacc.

Small, pinkish-brown agaric with convex or flattish faintly striate cap; in small troops on needle litter of spruce.

Dimensions cap 1 - 2.5 cm dia; stem 3 - 5 cm tall x 0.2 - 0.3 cm dia.
Cap pinkish-brown with violaceous or lilaceous tinges, hygrophanous, drying more pallid buff; at first convex, becoming flattened and some-times rather reflexed at the margin, striate 1/2 to the centre when damp. Flesh brownish and thin at the margin.
Gills pallid with violaceous tinge, adnate with decurrent tooth, fairly broad and crowded. **Spores** hyaline, smooth, broadly ellipsoid, non-amyloid, with droplets 5-7 x 3.5-4.5 µm. Basidia 4-spored. Gill cystidia cylindrical or slightly clavate, thin-walled and smooth.
Stem pallid buff, smooth, slender, more or less equal, with fibrils at the base and slightly rooting. Ring absent. Flesh brownish, firm and hollow.
Odour distinct, of radish. **Taste** not distinctive.
Chemical tests none.
Occurrence ; rare. (Longleat Park, Wiltshire)
■ Inedible.
Note: a partly decomposed specimen, supporting a Phycomycete on the cap, has been included, inadvertently, in the photograph.

Mycena pelianthina (Fr.) Quél.   **Tricholomataceae**

Smallish, violet or buff agaric with flattened , radially wrinkled bell-shaped cap on stoutish stem; in small troops on calcareous soil amongst leaf litter of broad-leaf trees, favouring beech.

Dimensions cap 2 - 4 cm dia; stem 5 - 6 cm tall x 0.4 - 0.8 cm dia.
Cap brownish-buff with violaceous tinge, hygrophanous, drying more pallid; at first conical-campanulate becoming flattened or broadly umbonate, finely sulcate. Flesh whitish and thin at the margin.
Gills violaceous-brown with darker margin, adnate, broad, close.
Spores hyaline, smooth, ellipsoid, amyloid, with droplets, 5-7 x 2.5-3 μm. Basidia 4-spored. Gill edge cystidia clavate or fusiform, thin-walled, smooth, abundant; gill face cystidia fusiform.
Stem brownish-violet, faintly grooved, stout, more or less equal, fibrous at the base. Ring absent. Flesh whitish, firm and hollow.
Odour faint, of radish. **Taste** not distinctive.
Chemical tests none.
Occurrence summer to autumn; infrequent.
■ Inedible.

Mycena polygramma (Bull.: Fr.) S F Gray
Tricholomataceae

Small, pale grey-brown agaric with broadly conical cap, pinkish gills; in small troops on stumps and branches of broad-leaf and occasionally coniferous trees.

Dimensions cap 1 - 4 cm dia; stem 2 - 4 cm tall x 0.1 - 0.2 cm dia.
Cap pallid grey-brown, sometimes with olivaceous tinge; at first conical, becoming more flattened, umbonate, striate at the margin when damp, the margin becoming irregular. Flesh whitish and thin.
Gills white becoming pinkish tinged, adnexed, narrow, crowded.
Spores hyaline, smooth, ellipsoid, amyloid, occasionally with droplets, 8-10 x 5.5-7.5 μm. Basidia 4-spored. Gill edge cystidia vesicular, thin-walled, warty; gill face cystidia absent.
Stem greyish, smooth or grooved, shiny, slender, more or less equal, downy white at the base. Ring absent. Flesh whitish, firm and hollow.
Odour of radish. **Taste** not distinctive.
Chemical tests none.
Occurrence late summer to autumn; infrequent.
■ Inedible.

Mycena pseudocorticola Kühn.   **Tricholomataceae**

Very small bluish-grey agaric with conical cap and whitish gills solitary or more typically in trooping groups amongst moss on trunks of broad-leaf trees, favouring ash, elm and sycamore.

Dimensions cap 0.3 - 1 cm dia; stem 1 - 2 cm tall x 0.03 - 0.1 cm dia.
Cap bluish-grey with darker centre; at first conical, becoming campanulate, translucently striate 3/4 to centre when damp, often finely pruinose, margin slightly denticulate. Flesh pallid and thin.
Gills whitish or pallid buff, adnate, broad, distant. **Spores** hyaline, smooth, sub-spherical, amyloid, with droplets, 10-13 x 9-12 μm. Basidia 2-spored. Gill edge cystidia clavate with finger-like outgrowths; gill face cystidia absent.
Stem bluish grey, pruinose, slender, more or less equal, downy at base. Ring absent. Flesh pallid, fragile and hollow.
Odour not distinctive. **Taste** not distinctive.
Chemical tests none.
Occurrence late autumn and through mild winters; infrequent.
■ Inedible.

Mycena pura (Pers.: Fr.) Kummer   **Tricholomataceae**

Smallish, lilac-pink agaric, with broadly convex radially grooved cap and pale lilac gills; in small troops on soil amongst leaf litter of broad-leaf and coniferous trees.

Dimensions cap 2 - 5 cm dia; stem 5 - 10 cm tall x 0.4 - 0.8 cm dia.
Cap pallid lilac or pink when damp, drying more buff; at first convex, becoming flattened with broad umbo, striate-sulcate when damp. Flesh whitish and moderately thin.
Gills pallid lilaceous-grey, adnexed with decurrent tooth, broad, fairly distant. **Spores** hyaline, smooth, ellipsoid, amyloid, with droplets, 6-8 x 3.5-4 μm. Basidia 4-spored. Gill cystidia flask-shaped, thin-walled, smooth.
Stem concolorous with cap, smooth, stout, fairly stiff, more or less equal, downy at base. Ring absent. Flesh whitish, firm and hollow.
Odour faint, of radish. **Taste** not distinctive.
Chemical tests none.
Occurrence summer to autumn; common.
■ Inedible.

Mycena rorida (Scop.: Fr.) Quél.   **Tricholomataceae**

Very small whitish agaric with flattened rather wavy cap; in small troops on leaf litter and other plant debris in mixed woods.

Dimensions cap 0.5 - 1.5 cm dia; stem 1 - 3 cm tall x 0.1 - 0.2 cm dia.
Cap whitish; at first convex, becoming flattened, typically with a small umbo, smooth but scurfy when dry, striate, denticulate at the margin. Flesh whitish and very thin.
Gills concolorous with cap, adnate or decurrent, fairly distant.
Spores hyaline, smooth, ellipsoid, amyloid, with droplets, 9 -12 x 3.5-5 μm. Basidia 2-, 3- or 4-spored. Gill edge cystidia cylindrical, thin-walled, smooth; gill face cystidia absent.
Stem concolorous with cap, smooth with sticky sheath, slender, more or less equal. Ring absent. Flesh white, tough and hollow.
Odour not distinctive. **Taste** not distinctive.
Chemical tests none.
Occurrence summer to autumn; infrequent.
■ Inedible.

Mycena rosea (Bull.) Gramberg   **Tricholomataceae**

Small or medium, whitish or pink agaric, with broadly convex radially grooved cap and pale pink gills; in small troops on soil, typically calcareous, amongst leaf litter in broad-leaf or mixed woods.

Dimensions cap 2 - 6 cm dia; stem 5 - 10 cm tall x 0.4 - 1 cm dia.
Cap pink when damp, drying whitish (see inset); at first convex, becoming flattened with broad umbo, striate-sulcate when damp. Flesh white and thin.
Gills pallid pink, adnexed with decurrent tooth, broad, close.
Spores hyaline, smooth, broadly ellipsoid, amyloid, with droplets, 6.5-7.5 x 4-5 μm. Basidia 4-spored. Gill cystidia fusiform or clavate, smooth, thin-walled.
Stem white, smooth, longitudinally fibrillose, more or less equal, downy at base. Ring absent. Flesh white, firm and hollow.
Odour faint, of radish. **Taste** not distinctive.
Chemical tests none.
Occurrence summer to autumn; infrequent.
■ Inedible.

Mycena stylobates (Pers.: Fr.) Kummer **Tricholomataceae**
= *Mycena dilatata* (Fr.:Fr.) Gill.

Very small whitish agaric with bell-shaped, minutely bristly cap and basal disc in stem; in small troops on leaves and other debris of broad-leaf and coniferous trees.

Dimensions cap 0.3 - 1 cm dia; stem 2 - 4 cm tall x 0.05 - 0.1 cm dia.

Cap whitish or pallid beige; at first campanulate becoming more flattened, umbonate, striate when damp, with scattered bristles (hand lens). Flesh whitish and thin.
Gills white, becoming greyish tinged, adnexed or free, narrow, close. **Spores** hyaline, smooth, cylindrical-ellipsoid, amyloid, with droplets, 7-11 x 3.5-5 µm. Basidia 4-spored. Gill edge cystidia cylindrical or narrowly clavate, with finger-like processes, and thin-walled; gill face cystidia absent.
Stem greyish or hyaline, smooth, slender, more or less equal, downy white at base and attached to substrate by a small disc. Ring absent. Flesh whitish, fragile and hollow.
Odour not distinctive. **Taste** not distinctive.
Chemical tests none.
Occurrence late summer to autumn; infrequent.
■ Inedible.

Mycena vitilis (Fr.) Quél.   **Tricholomataceae**

Small, pale beige agaric with broadly conical cap, whitish gills; in small troops on branches and detached bark of broad-leaf trees.

Dimensions cap 1 - 2 cm dia; stem 4 - 9 cm tall x 0.1 - 0.3 cm dia.
Cap beige, sometimes with greyish tinge, more pallid at margin; at first conical becoming more flattened, umbonate, translucently striate and slightly viscid when damp. Flesh whitish and thin.
Gills whitish, sometimes pinkish tinged with age, adnexed-adnate, broad, margin finely serrated, crowded. **Spores** hyaline, smooth, ellipsoid, amyloid, with droplets, 9-12 x 5.5-7.5 µm. Basidia 4-spored. Gill edge cystidia narrowly clavate, with finger-like processes and thin-walled; gill face cystidia absent.
Stem brown with pinkish tinge, more pallid towards the apex, finely pruinose or smooth, slender, more or less equal. Ring absent. Flesh pallid grey, stiff and hollow.
Odour not distinctive. **Taste** not distinctive.
Chemical tests none.
Occurrence summer to autumn; common.
■ Inedible.

Myxomphalia maura (Fr.) Hora   **Tricholomataceae**
= *Mycena maura* (Fr.) Kühn.
= *Omphalia maura* (Fr.) Gill.

Small agaric with dark brown depressed cap and white decurrent gills; trooping on soil on fire sites, often where conifers have grown.

Dimensions cap 1 - 3 cm dia; stem 2 - 4 cm tall x 0.2 - 0.4 cm dia.
Cap dark brown with greyish tinge; at first more or less convex, becoming depressed, margin incurved, shiny, striate. Flesh grey and thin.
Gills white or pallid grey, decurrent, close. **Spores** hyaline, smooth, broadly ellipsoid, amyloid, with droplets, 5-6.5 x 3.5-4.5 µm. Basidia 4-spored. Gill edge cystidia cylindrical or ventricose, smooth; gill face cystidia sparse.
Stem concolorous with cap or more pallid, smooth, equal, slender. Ring absent. Flesh grey, thin, full or stuffed.
Odour not distinctive. **Taste** not distinctive.
Chemical tests none.
Occurrence late summer to autumn; infrequent.
■ Inedible.

Nyctalis asterophora Fr.   Tricholomataceae
= *Asterophora lycoperdoides* (Bull.) Ditmar

Small pale or brownish agaric, parasitic in caespitose tufts on old specimens of *Russula nigricans.*

Dimensions cap 1 - 4 cm dia; stem 1 - 2 cm tall x 0.05 - 0.06 cm dia.
Cap whitish, becoming brown; at first convex or sub-spherical, becoming more expanded, powdery with chlamydospores. Flesh pallid grey-brown and very thin.
Gills grey-brown, thick, almost rudimentary. **Spores** chlamydospores pallid-brown, star-shaped with long conical spines, 12-15 μm.; basidiospores - white, smooth, ellipsoid, non-amyloid, 5-5.5 x 3.5-4 μm. Basidia 4-spored. Cystidia absent.
Stem concolorous with cap then blackening, short, slender. Ring absent. Flesh brownish, thin and full.
Odour unpleasant (probably from the rotting host tissue). **Taste** not distinctive.
Chemical tests none.
Occurrence late summer to autumn; rare.
■ Inedible.
Very similar in appearance to *N. parasitica* which is distinguished by the sparse presence of smooth chlamydospores on the cap surface.

Nyctalis parasitica (Bull.: Fr.) Fr.   Tricholomataceae
= *Asterophora parasitica* (Bull.: Fr.) Sing.

Small, pale agaric, parasitic in caespitose tufts on old specimens of *Lactarius* and *Russula.*

Dimensions cap 0.5 - 1.5 cm dia; stem 1 - 3 cm tall x 0.1 - 0.3 cm dia.
Cap whitish, with brown or lilac tinges; at first convex or sub-spherical becoming more expanded, silky. Flesh pallid grey-brown and very thin.
Gills at first grey-brown, becoming whitish, floury with developing chlamydospores. **Spores** chlamydospores pallid brown, smooth, ellipsoid, 2 large droplets, 14-15 x 9-11 μm.; basidiospores generally absent but, if present, white, smooth, ellipsoid, non-amyloid, 5-5.5 x 3-4 μm.
Stem concolorous with cap, slender, often twisted. Ring absent. Flesh brownish, thin and full.
Odour not distinctive. **Taste** not distinctive.
Chemical tests none.
Occurrence summer to autumn; infrequent.
■ Inedible.
Very similar in appearance to *N. asterophora* which is distinguished by the presence of abundant spiny chlamydospores on the cap surface.

Omphalina ericetorum (Fr.: Fr.) J Lange
= *Omphalina umbellifera* (L.: Fr.) Quél.   Tricholomataceae

Small yellowish-buff agaric with funnel-shaped cap and broad decurrent pale gills; in troops on soil amongst grass and moss on heathland, favouring peaty soils.

Dimensions cap 0.5 - 2 cm dia; stem 1 - 2 cm tall x 0.2 - 0.4 cm dia.
Cap yellowish-buff with olivaceous tinge; at first more or less convex, becoming depressed, infundibuliform or umbilicate, margin incurved, smooth, striate or sulcate at the margin. Flesh whitish-ochre and thin.
Gills pallid, creamy-yellow, decurrent, broad, distant. **Spores** hyaline, smooth, ellipsoid, non-amyloid, with droplets, 8-9 x 4.5-5.5 μm. Basidia 4-spored. Cystidia absent.
Stem concolorous, equal, finely downy and darker at the base. Ring absent. Flesh whitish-ochre, thin and full.
Odour not distinctive. **Taste** not distinctive.
Chemical tests none.
Occurrence spring to autumn; common.
■ Inedible.

Phaeotellus griseopallida (Desm.) Kühn. & Lamoure
= *Omphalina griseopallida* (Desm.) Quél.
= *Leptoglossum griseopallidum* (Desm.) Moser **Tricholomataceae**

Small brownish agaric with funnel-shaped cap and broad decurrent brown gills; in troops on soil amongst grass and moss on open heathland, favouring sandy soils.

Dimensions cap 0.5 - 1.5 cm dia; stem 1 - 1.5 cm tall x 0.2 - 0.3 cm dia.
Cap dark greyish-brown, drying more pallid; at first more or less convex, becoming depressed or umbilicate, margin incurved, smooth, sulcate at the margin.. Flesh greyish-brown and thin.
Gills concolorous but becoming darker at maturity, decurrent, thick, broad, sometimes forked, distant. **Spores** hyaline, smooth, broadly ellipsoid, non-amyloid, droplets absent, 9-12 x 5-8 µm. Basidia 2- or, occasionally, 4-spored. Cystidia absent.
Stem concolorous, sometimes slightly eccentric, stoutish, equal, finely downy at base. Ring absent. Flesh greyish-brown, thin, and full becoming hollow.
Odour not distinctive. **Taste** not distinctive.
Chemical tests none.
Occurrence late summer to autumn; infrequent or rare. (Thetford Warren, Norfolk)
■ Inedible.

Gerronema postii (Fr.) Sing. **Tricholomataceae**

Small orange agaric with funnel-shaped cap and broad decurrent gills; in troops on soil amongst moss on peaty soils favouring *Sphagnum.*

Dimensions cap 0.5 - 5 cm dia; stem 2 - 5 cm tall x 0.1 - 0.3 cm dia.
Cap orange; at first more or less convex, becoming infundibuliform or umbilicate, with incurved margin, smooth, striate or sulcate at the margin. Flesh concolorous, thin.
Gills yellow, decurrent, broad, sometimes forked, distant.
Spores hyaline, smooth, ellipsoid, non-amyloid, with droplets, 6-10 x 4-5.5 µm. Basidia 4-spored. Cystidia absent.
Stem concolorous with cap but slightly more pallid, equal and slender. Ring absent. Flesh yellow, thin and full.
Odour not distinctive. **Taste** not distinctive.
Chemical tests none.
Occurrence spring to autumn; rare. (Lullingstone, Kent)
■ Inedible.

Omphalina pyxidata (Bull.: Fr.) Quél. **Tricholomataceae**

Small tan-brown agaric with funnel-shaped cap and broad decurrent pale gills; in troops on soil amongst grass and moss, typically on sandy soils, less commonly calcareous.

Dimensions cap 0.5 - 1.5 cm dia; stem 2 - 5 cm tall x 0.1 - 0.2 cm dia.
Cap tan; at first more or less convex, becoming infundibuliform or umbilicate, with incurved margin, smooth, translucently striate almost to the centre. Flesh concolorous and thin.
Gills pallid, decurrent, broad, occasionally forked, distant.
Spores hyaline, smooth, ellipsoid, non-amyloid, with droplets, 6.5-9 x 3.5-6 µm. Basidia 4-spored. Cystidia-like gill edge cells cylindrical or knobbly.
Stem concolorous with cap but slightly more pallid, equal, slender. Ring absent. Flesh yellow, thin and hollow.
Odour not distinctive. **Taste** not distinctive.
Chemical tests none.
Occurrence summer to autumn; infrequent.
■ Inedible.

Arrhenia acerosa (Fr.) Kühn.   **Tricholomataceae**
= *Leptoglossum acerosum* (Fr.) Moser
= *Pleurotus acerosus* (Fr.) Konr. & Maubl.

Small pale agaric with irregularly funnel-shaped and lobed cap, ecentric stem and grey-brown gills; solitary or in small trooping groups on soil amongst mossy grass or debris, often associated with buried rotten wood.

Dimensions cap 0.8 - 2.5 cm dia; stem 0.3 - 1 cm tall x 0.3 - 0.5 cm dia.
Cap at first whitish and button-shaped becoming pallid grey-brown and irregularly way-infundibuliform, at first finely downy then minutely and radially fibrillose. Flesh greyish-brown, watery, thin.
Gills pallid grey or grey-brown, adnexed, broad, close.
Spores hyaline, smooth, broadly ellipsoid or pip-shaped, non-amyloid, 6.5-8.5 x 3.5-5.0 µm. Basidia 4-spored. Cystidia absent.
Stem white, finely woolly or downy, lateral or eccentic, rudimentary. Ring absent. Flesh greyish-brown, firm and full.
Odour not distinctive. **Taste** not distinctive.
Chemical tests none.
Occurrence autumn to early winter; infrequent or rare. (Campus, Leicester University)
■ Inedible.

Arrhenia rickenii (Fr.) Watl.   **Tricholomataceae**
= *Leptoglossum rickenii* (Sing.: Hora) Sing.

Pale greyish-brown agaric with funnel-shaped cap and broad whitish decurrent gills; in troops on soil amongst moss, often on or near walls, also on bare ground of woodland embankments.

Dimensions cap 0.5 - 2 cm dia; stem 1.5 - 3.5 cm tall x 0.1 - 0.3 cm dia.
Cap pallid brown with greyish or ochraceous tinges; at first more or less convex, becoming infundibuliform or umbilicate, with undulating margin, smooth, striate or sulcate at the margin in older specimens. Flesh concolorous and thin.
Gills pallid, decurrent, broad, forked and interveined, distant.
Spores hyaline, smooth, ellipsoid, non-amyloid, with droplets, 6-10 x 3.5-5.5 µm. Basidia 4-spored. Cystidia absent.
Stem concolorous with cap but slightly more pallid, equal, slender, sometimes eccentric. Ring absent. Flesh yellow, thin and full.
Odour not distinctive. **Taste** not distinctive.
Chemical tests none.
Occurrence autumn to winter; infrequent.
■ Inedible.

Oudemansiella mucida (Schrad.: Fr.) Höhn.
= *Armillaria mucida* (Schrad.: Fr.) Kummer   **Tricholomataceae**
Porcelain Agaric

Whitish, very slimy and semi-translucent, medium-sized agaric; caespitose on trunks and branches of beech, often high up.

Dimensions cap 2 - 8 cm dia; stem 3 - 10 cm tall x 0.3 - 1 cm dia.
Cap pallid grey at first, becoming white with faint ochraceous tinge; at first convex, becoming flattened, very viscid. Flesh white and thin.
Gills white, becoming slightly ochraceous, adnate, broad, distant.
Spores hyaline, smooth, sub-spherical, non-amyloid, 15-20 x 15-19 µm. Basidia 4-spored. Gill cystidia sparse, fusiform or somewhat ventricose.
Stem concolorous with cap, slender, more or less equal, faintly striate above ring, slightly scaly below, viscid. Ring white, membraneous, superior. Flesh white and stuffed.
Odour not distinctive. **Taste** not distinctive.
Chemical tests none.
Occurrence summer to winter; common.
□ Edible.

Oudemansiella radicata (Relh.: Fr.) Sing.
= *Collybia radicata* (Relh.: Fr.) Quél.   **Tricholomataceae**

Medium-sized agaric, yellow to olive-brown, somewhat slimy caps with white gills and long, tough rooting stems; typically solitary or trooping, attached to roots and buried wood of broad-leaf trees, favouring beech but sometimes appearing as if growing on soil.

Dimensions cap 3 - 10 cm dia; stem 8 - 20 cm tall x 0.5 - 1 cm dia.
Cap pallid ochraceous or olive-brown; convex or campanulate with broad umbo, sulcate, viscid but drying shiny. Flesh pallid, otherwise concolorous and thin.
Gills white, adnexed, thick, broad, distant. **Spores** hyaline, smooth, ellipsoid, non-amyloid, with droplets, 12-16 x 10-12 µm. Basidia 3- or 4-spored. Gill edge cystidia fusiform; gill face cystidia clavate, thick-walled.
Stem white at the apex, becoming tinged cap colour below, slender, tapering upwards, deeply rooting. Ring absent. Flesh white, very firm and stuffed.
Odour not distinctive. **Taste** not distinctive.
Chemical tests none.
Occurrence summer to winter; common.
☐ Edible but poor.

Panellus mitis (Pers: Fr.) Sing.   **Tricholomataceae**
= *Pleurotus mitis* (Pers.: Fr.) Quél.
= *Urosporellina mitis* (Pers.: Fr.) Kreisel

Kidney- or fan-shaped cap, white, with buff gills beneath and with lateral sometimes rudimentary stem; on twigs, branches and logs of coniferous trees.

Dimensions cap 0.5 - 2.5 cm dia; stem 0.5 - 1 cm tall x 0.3 - 0.5 cm dia.
Cap white, with pinkish tinge in older specimens; fan-shaped or reniform, smooth with peelable skin. Flesh white, thin and rubbery.
Gills white or cream, adnate, narrow, crowded with gelatinous edge.
Spores hyaline, smooth, ellipsoid, amyloid, with droplets, 3.5-6 x 1-1.5 µm. Basidia 4- spored. Cystidia absent.
Stem concolorous with cap, lateral (may be rudimentary), finely fibrillose or floccose-granular when dry. Ring absent. Flesh white, firm and full.
Odour not distinctive. **Taste** not distinctive.
Chemical tests none.
Occurrence late summer to autumn; rare. (Macclesfield Forest, Cheshire)
■ Inedible.

Panellus serotinus (Schrad.: Fr.) Kühn. **Tricholomataceae**
= *Pleurotus serotinus* (Schrad.: Fr.) Kummer
= *Sarcomyxa serotina* (Schrad.: Fr) Karst.

Kidney-shaped, greenish cap with yellowish gills and with lateral sometimes rudimentary stem; on fallen trunks and branches of broad-leaf trees, favouring beech and birch.

Dimensions cap 3 - 10 cm dia; stem 1 - 2.5 cm tall x 0.8 - 1.5 cm dia.
Cap olivaceous-green, sometimes with ochraceous tinge; reniform, viscid or sticky in wet weather. Flesh white, with gelatinous layer beneath cap cuticle.
Gills at first cream then yellow, adnate, narrow, forked, crowded.
Spores hyaline, smooth, cylindrical, curved, amyloid, with droplets, 4-7 x 1-2 µm. Basidia 4-spored. Gill cystidia clavate or fusiform.
Stem yellowish, lateral (may be rudimentary), stout, covered with minute brown scurfy scales. Ring absent. Flesh white, firm and full.
Odour not distinctive. **Taste** not distinctive.
Chemical tests none.
Occurrence late summer to winter; infrequent or rare. (Alfred's Tower near Stourhead, Wiltshire)
■ Inedible.

Panellus stipticus (Bull.: Fr.) Karst.   **Tricholomataceae**
= *Pleurotus stipticus* (Bull.: Fr.) Fr.

Kidney-shaped, pale yellow cap, with buff gills and with lateral sometimes rudimentary stem; clustered on stumps, fallen trunks and
branches of broad-leaf trees, favouring oak.

Dimensions cap 1 - 4 cm dia; stem 0.5 - 2 cm tall x 0.2 - 0.5 cm dia.
Cap pallid ochraceous-brown; reniform, minutely scurfy. Flesh white
or pallid yellow, thin and tough.
Gills pinkish-buff, adnate or decurrent, broad, some forked crowded.
Spores hyaline, smooth, ellipsoid, amyloid, droplets absent 3-6 x
2-3 µm. Basidia 4-spored. Gill edge cystidia clavate or fusiform,
some knobbly; gill face cystidia absent.
Stem concolorous with cap but more pallid, lateral (may be rudi-
mentary), smooth, tapering towards the base. Ring absent. Flesh
white, firm and full.
Odour not distinctive. **Taste** not distinctive.
Chemical tests none.
Occurrence throughout the year; infrequent.
■ ✛ Inedible, possibly poisonous.

Rhodotus palmatus (Bull.: Fr.) Maire   **Tricholomataceae**
= *Pleurotus palmatus* (Bull.: Fr.) Quél.

Medium-sized pale pink agaric with wrinkled cap and interveined
gills; solitary or in small tufts on stumps, trunks and felled
timber of elm.

Dimensions cap 3 - 8 cm dia; stem 3 - 6 cm tall x 1 - 1.5 cm dia.
Cap peach or pallid orange-pink; at first convex, becoming flattened
with inrolled margin, surface sulcate. Flesh pallid pink and firm.
Gills pallid, concolorous with cap, adnate, narrow, close with inter-
connecting veins. **Spores** pink, finely warty, sub-spherical, non-
amyloid, 5-8 µm. Basidia 4-spored. Gill cystidia large, flask-shaped
with elongated apex.
Stem pinkish, decoated with white fibres, sometimes curved and
eccentric, more or less equal. Ring absent. Flesh white, with pinkish
tinge, firm, full or narrowly hollow.
Odour not distinctive. **Taste** bitter.
Chemical tests none.
Occurrence autumn to winter; infrequent.
■ Inedible.

Rickenella fibula (Bull.: Fr.) Raith.   **Tricholomataceae**
= *Mycena fibula* (Bull.: Fr.) Kühn
= *Gerronema fibula* (Bull.: Fr.) Sing.
= *Omphalia fibula* (Bull.: Fr.) Kummer

Very small, delicate, bright orange agaric with flattened grooved cap;
in small troops amongst damp grass or moss.

Dimensions cap 0.5 - 1 cm dia; stem 2 - 4 cm tall x 0.1 - 0.2 cm dia.
Cap yellowish-orange, slightly darker at the centre; at first convex
becoming flattened and slightly depressed, striate or crenate. Flesh
orange and very thin.
Gills pallid yellow, decurrent, broad, distant. **Spores** hyaline, smooth,
elongate-ellipsoid, non-amyloid, with droplets, 4-5 x 2-2.5 µm. Basidia
4-spored. Gill cystidia fusiform and thin-walled.
Stem concolorous with cap, finely downy, slender, more or less equal.
Ring absent. Flesh orange, fragile and hollow.
Odour not distinctive. **Taste** not distinctive.
Chemical tests none.
Occurrence summer to autumn; common.
■ Inedible.

Rickenella swartzii (Fr.) Kujp.   Tricholomataceae
= *Mycena swartzii* (Fr.: Fr.) A H Smith
= *Gerronema setipes* (Fr.: Fr.) Sing.
= *Omphalia swartzii* (Fr.) Quél.

Very small, delicate agaric with flattened or funnel-shaped cream cap, dark at the centre, and white deeply decurrent gills; in scattered trooping groups in lawns and other grasslands, typically with moss.

Dimensions cap 0.5 - 1 cm dia; stem 2 - 4 cm tall x 0.1 - 0.2 cm dia.
Cap ochraceous-cream or pallid brown with darker centre; at first convex, becoming flattened and finally depressed, striate when damp. Flesh cream, very thin.
Gills pallid, deeply decurrent, broad, distant. **Spores** hyaline, smooth, elongated-ellipsoid, non-amyloid, some with droplets, 4-5 x 2-2.5µm. Basidia 4-spored. Gill cystidia flask-shaped with elongated apex and thin-walled.
Stem violaceous at apex, otherwise pallid ochraceous, finely downy, slender, more or less equal. Ring absent. Flesh cream, fragile and hollow.
Odour not distinctive. **Taste** not distinctive.
Chemical tests none.
Occurrence summer to winter; frequent.
■ Inedible.

Strobilurus esculentus (Wulf.:Fr.) Sing.   Tricholomataceae
= *Pseudohiatula esculenta* (Wulf.: Fr.) Sing.

Small, delicate agaric with brown conical cap and whitish gills; solitary or in small groups, attached to buried or partly buried spruce cones. Note: this species never occurs on pine cones.

Dimensions cap 1 - 2 cm dia; stem 2 - 7 cm tall x 0.1 - 0.2 cm dia.
Cap dull brown; at first conical or convex, becoming flattened, smooth. Flesh white and thin.
Gills whitish-grey, adnexed, crowded. **Spores** hyaline, smooth, ellipsoid, non-amyloid, droplets absent, 5-7 x 2-4 µm. Basidia 4-spored. Gill edge cystidia bluntly fusiform and sometimes with slight heads, sparsely encrusted; gill face cystidia bluntly fusiform, heavily encrusted at the apex.
Stem pallid, concolorous with cap, finely silky-pruinose, slender, tapering slightly upwards, typically deeply rooting. Ring absent. Flesh white, fragile and hollow.
Odour not distinctive. **Taste** not distinctive.
Chemical tests none.
Occurrence autumn to spring; locally common.
■ Inedible.

Strobilurus stephanocystis (Hora) Sing.
= *Pseudohiatula stephanocystis* Kühn & Romagn. ex Hora
                 Tricholomataceae

Small, delicate agaric with tawny-brown cap and whitish gills; solitary or in small groups attached to buried or partly buried pine cones.

Dimensions cap 1 - 2 cm dia; stem 2 - 7 cm tall x 0.1 - 0.2 cm dia.
Cap brown or tawny; at first convex, becoming flattened, smooth. Flesh white and thin.
Gills whitish-cream, adnexed, crowded. **Spores** hyaline, smooth, ellipsoid, non-amyloid, droplets absent, 6-10 x 3-4 µm. Basidia 4-spored. Gill edge and gill face cystidia broadly clavate, thick-walled, encrusted at apex.
Stem pallid, concolorous with cap, finely silky-pruinose, slender, tapering upwards, typically deeply rooting. Ring absent. Flesh white, fragile and hollow.
Odour not distinctive. **Taste** not distinctive.
Chemical tests none.
Occurrence autumn to spring; infrequent.
■ Inedible.

Strobilurus tenacellus (Pers.: Fr.) Sing. **Tricholomataceae**
= *Pseudohiatula tenacella* Metrod
= *Collybia tenacella* (Pers.: Fr.) Kummer

Small, delicate agaric with tawny cap and whitish gills; solitary or in small groups attached to buried or partly buried pine cones.

Dimensions cap 1 - 2.5 cm dia; stem 2 - 7 cm tall x 0.1 - 0.2 cm dia.
Cap tawny; at first convex, becoming flattened, smooth, slightly striate when damp. Flesh white and thin.
Gills whitish-cream, adnexed, crowded. **Spores** hyaline, smooth, ellipsoid, non-amyloid, droplets absent, 4.5-6.5 x 2.5-3.5 μm. Basidia 4-spored. Gill edge and gill face cystidia fusiform-pointed, thick-walled, with some encrustation at apex.
Stem pallid concolorous with cap, finely silky-pruinose, slender, tapering upwards, typically deeply rooting. Ring absent. Flesh white, fragile and hollow.
Odour not distinctive. **Taste** not distinctive.
Chemical tests none.
Occurrence autumn to spring; frequent.
■ Inedible.

Tricholoma album (Schaeff.: Fr.) Quél. **Tricholomataceae**

Large fleshy agaric, whitish throughout; scattered or in small trooping groups, on soil in broad-leaf and coniferous woods.

Dimensions cap 5 - 10 cm dia; stem 6 - 9 cm tall x 0.8 - 1.5 cm dia.
Cap white; at first convex, becoming flattened, sometimes with slight umbo, smooth, dry. Flesh white and thick.
Gills white, emarginate, fairly broad, crowded. **Spores** hyaline, smooth, ellipsoid, non-amyloid, with droplets, 3-4.5 x 7-8 μm. Basidia 4-spored. Cystidia absent.
Stem white, smooth, fibrous lined, more or less equal or tapering slightly upwards. Ring absent. Flesh white, stuffed or full.
Odour of meal. **Taste** not distinctive.
Chemical tests none.
Occurrence late summer to autumn; rare. (Mendip Forest, Somerset)
✣ Poisonous.

Tricholoma argyraceum (Bull.: Fr.) Quél.
= *Tricholoma chrysites* (Fr.) Gill. **Tricholomataceae**

Medium-sized fleshy agaric with silver-grey scaly cap and white gills; solitary or in trooping, sometimes tufted, groups, on soil in broad-leaf or coniferous woods, favouring both pine and beech.

Dimensions cap 4 - 8 cm dia; stem 4 - 8 cm tall x 0.6 - 1.2 cm dia.
Cap pallid, silvery-grey or grey-brown, breaking up into felty scales at maturity, showing white flesh beneath; at first convex, becoming flattened, sometimes with slight umbo, margin irregular, wavy, sometimes remaining slightly inrolled. Flesh white and thick.
Gills white, yellowing slightly with age, emarginate, fairly crowded.
Spores hyaline, smooth, ellipsoid, non-amyloid, with droplets, 5-6 x 3-4 μm. Basidia 4-spored. Cystidia absent.
Stem white or flushed cap colour with small dark scales, more or less equal or tapering slightly upwards. Ring absent. Flesh white or with greyish tinges, stuffed or full.
Odour of meal. **Taste** of meal.
Chemical tests none.
Occurrence summer to autumn; common.
■ Inedible.
Similar to *T. scalpturatum* which has a somewhat darker base colour in the cap.

Tricholoma atrosquamosum (Chev.) Sacc.
= *Tricholoma terreum var atrosquamosum* (Chev.) Mass.
Tricholomataceae

Large fleshy agaric, pale grey with dense dark grey scales on cap, greyish gills, aromatic; scattered or in small trooping groups, on calcareous soil in broad-leaf and coniferous woods.

Dimensions cap 4 - 10 cm dia; stem 3 - 8 cm tall x 1 - 2 cm dia.
Cap pallid grey or clay, densely covered with dark grey pointed scales; at first convex, becoming flattened, sometimes with slight umbo. Flesh pallid grey and thin.
Gills white or greyish-white sometimes with darker edge, emarginate, fairly distant. **Spores** hyaline, smooth, ellipsoid, non-amyloid, with droplets, 5-8 x 3.5-5 μm. Basidia 4-spored. Cystidia absent.
Stem white with dark fibrous scales, more or less equal or tapering slightly upwards. Ring absent. Flesh pallid grey, stuffed or full.
Odour aromatic or of pepper. **Taste** not distinctive.
Chemical tests none.
Occurrence late summer to autumn; rare. (Lyme Park, Cheshire)
☐ Edible.

Tricholoma auratum (Fr.) Gill.   **Tricholomataceae**
= *Tricholoma equestre* (L.: Fr.) Kummer

Large fleshy agaric, yellowish throughout; scattered or in small trooping groups, on soil in coniferous woods, typically on sandy or acid soils.

Dimensions cap 7 - 12 cm dia; stem 4 - 6 cm tall x 0.8 - 1.2 cm dia.
Cap pallid greenish-sulphur-yellow with small adpressed scales towards the centre; at first convex, becoming flattened, sometimes with slight umbo, slightly sticky. Flesh white, yellowish immediately below the cap cuticle and thin.
Gills greenish-sulphur-yellow, emarginate or adnexed, broad, more or less crowded. **Spores** hyaline, smooth, broadly ellipsoid, non-amyloid, droplets absent, 6-8 x 4-5 μm. Basidia 4-spored. Cystidia absent.
Stem concolorous, smooth or slightly scaly, more or less equal or tapering slightly upwards. Ring absent. Flesh white with yellow tinge, stuffed or full.
Odour of meal. **Taste** of meal.
Chemical tests none.
Occurrence autumn; rare. (Thetford Warren, Norfolk)
✢ Poisonous.

Tricholoma cingulatum (Fr.) Jacobasch **Tricholomataceae**
= *Tricholoma ramentacea* (Bull.: Fr.) Kummer

Medium-sized fleshy agaric, pale felty grey, with ring on stem; typically in trooping groups, on soil with broad-leaf trees with willow.

Dimensions cap 3.5 - 6 cm dia; stem 5 - 8 cm tall x 0.8 - 1.2 cm dia.
Cap pallid grey with fine dark felty scales; at first convex, becoming flattened, sometimes with slight umbo and inrolled margin. Flesh white and thin.
Gills white, emarginate or adnexed, crowded. **Spores** hyaline, smooth, ellipsoid, non-amyloid, with droplets, 4-5.5 x 2.5-3.5 μm. Basidia 4-spored. Cystidia absent.
Stem whitish, smooth, more or less equal or tapering slightly upwards. Ring white, woolly, superior. Flesh white, stuffed or full.
Odour of meal. **Taste** of meal.
Chemical tests none.
Occurrence summer to autumn; infrequent or rare. (Clifton Maybank woods, Somerset)
☐ Edible.

Tricholoma columbetta (Fr.) Kummer   Tricholomataceae

Large fleshy agaric, whitish throughout; scattered or in small trooping groups, on soil in broad-leaf woods.

Dimensions cap 5 - 10 cm dia; stem 4 - 10 cm tall x 1 - 2 cm dia.
Cap white, sometimes bluish-green, spotted with age; at first convex, becoming flattened, sometimes with slight umbo, silky, dry. Flesh white and moderately thick.
Gills white, emarginate, crowded. **Spores** hyaline, smooth, broadly ellipsoid, non-amyloid, with droplets, 5-6 x 3.5-4.5 µm. Basidia 4-spored. Cystidia absent.
Stem white, smooth, fibrillose, more or less equal or tapering slightly downwards, sometimes blue, spotted at the base. Ring absent. Flesh white, stuffed or full.
Odour not distinctive. **Taste** not distinctive.
Chemical tests none.
Occurrence summer to autumn; infrequent.
☐ Edible.
The blue-green spotting is rarely seen and perhaps misleading. Care should be exercised not to confuse with poisonous species including *T. album*.

Tricholoma focale (Fr.) Ricken   Tricholomataceae

Medium-sized fleshy agaric, reddish-orange with cream gills and distinctively banded stem; scattered or in small trooping groups, on soil with pines in coastal locations.

Dimensions cap 5 - 10 cm dia; stem 6 - 8 cm tall x 1 - 1.5 cm dia.
Cap reddish-brown or orange tinged, with radiating scaly fibrils; at first convex, becoming flattened, sometimes with slight umbo, slightly sticky. Flesh white and thick.
Gills cream, becoming more brown or brown-spotted with age, emarginate, crowded. **Spores** hyaline, smooth, broadly ellipsoid, non-amyloid, 3-4.5 x 2.5-3 µm. Basidia 4-spored. Cystidia absent.
Stem pallid, creamy-brown girdled with scaly bands concolorous with cap, more or less equal then tapered at the base. Ring concolorous with cap, woolly, superior. Flesh white, stuffed or full.
Odour of meal. **Taste** of meal.
Chemical tests none.
Occurrence late summer to autumn; very localised.
■ Inedible.

Tricholoma fulvum (DC.: Fr.) Sacc.   Tricholomataceae
= *Tricholoma flavobrunneum* (Fr.) Kummer

Large fleshy agaric, reddish-brown with yellowish gills; scattered or in small trooping groups, on soil in broad-leaf and mixed woods, favouring birch and spruce.

Dimensions cap 5 - 10 cm dia; stem 3 - 7 cm tall x 0.8 - 1.4 cm dia.
Cap reddish-brown with fine radiating fibrils; at first convex, becoming flattened, sometimes with slight umbo, smooth, slightly sticky. Flesh white and thick.
Gills yellow, becoming brown spotted with age, emarginate, crowded.
Spores hyaline, smooth, broadly ellipsoid, non-amyloid, with droplets, 5-7 x 3-4.5 µm. Basidia 4-spored. Cystidia absent.
Stem concolorous with cap, smooth, fibrous lined, more or less equal, slender, slightly sticky. Ring absent. Flesh yellowish, stuffed or full.
Odour of meal. **Taste** of meal or bitterish.
Chemical tests none.
Occurrence late summer to autumn; common.
☐ Edible.

Tricholoma imbricatum (Fr.: Fr.) Kummer
Tricholomataceae

Large fleshy agaric, dark reddish-brown with cream gills; scattered or in small trooping groups on sandy soil in coniferous woods with pine.

Dimensions cap 5 - 10 cm dia; stem 8 - 12 cm tall x 1 - 2 cm dia.
Cap dark reddish-brown with fine radiating scaly fibrils; at first hemispherical, becoming campanulate or bluntly umbonate, margin inrolled for a long time, smooth. Flesh white and thin.
Gills creamy white, becoming brown with pinkish tinge, emarginate, edges slightly crenated, crowded. **Spores** hyaline, smooth, broadly ellipsoid, non-amyloid, with droplets, 5.5-7.5 x 3.5-5.5 µm. Basidia 4-spored. Cystidia absent.
Stem whitish at the apex, otherwise pallid, concolorous with cap, becoming brown spotted where bruised, smooth, fibrous lined, more or less equal or tapering at the base. Ring absent. Flesh whitish, stuffed or full.
Odour not distinctive. **Taste** bitter.
Chemical tests none.
Occurrence late summer to autumn; localised.
■ Inedible.

Tricholoma lascivum (Fr.) Gill.   **Tricholomataceae**

Largish fleshy agaric, whitish tan with whitish gills; scattered or in small trooping groups on soil in broad-leaf woods.

Dimensions cap 4 - 7 cm dia; stem 7 - 11 cm tall x 1 - 1.5 cm dia.
Cap dirty-white with tan staining; at first convex, becoming flattened or slightly depressed, smooth, silky, dry. Flesh white and thick.
Gills white, becoming cream with age, adnexed, fairly broad, crowded. **Spores** hyaline, smooth, ellipsoid, non-amyloid, with droplets, 6-7 x 3.5-4 µm. Basidia 4-spored. Cystidia absent.
Stem white with tan staining, smooth or slightly farinose towards the apex, more or less equal. Ring absent. Flesh white, stuffed or full.
Odour sweetish, of coal gas. **Taste** mild then bitterish.
Chemical tests none.
Occurrence late summer to autumn; infrequent.
✚ Probably poisonous.

Tricholoma pessundatum (Fr.) Quél. **Tricholomataceae**

Medium-sized fleshy agaric, dark brown, spotted, slimy; scattered or in small trooping groups on soil with conifers, favouring pine and spruce.

Dimensions cap 5 - 9 cm dia; stem 4 - 10 cm tall x 1.5 - 3 cm dia.
Cap reddish-brown with darker brown spots; at first convex, becoming expanded, smooth, very viscid. Flesh white and medium.
Gills white, becoming spotted rust with age, adnexed, fairly broad, crowded. **Spores** hyaline, smooth, ellipsoid, non-amyloid, with droplets, 4-6 x 2.5-3 µm. Basidia 4-spored. Cystidia absent.
Stem whitish with brown staining, smooth, more or less equal or somewhat clavate. Ring absent. Flesh white, stuffed or full.
Odour strong, of meal. **Taste** of meal.
Chemical tests none.
Occurrence late summer to autumn; localised in Scottish Highlands and other montane regions of Europe.
✚ Probably poisonous.

Tricholoma portentosum (Fr.) Quél.   **Tricholomataceae**

Large fleshy agaric, greyish with radiating darker streaks, white gills; scattered or in large trooping groups on soil in coniferous woods.

Dimensions cap 5 - 10 cm dia; stem 4 - 10 cm tall x 1 - 2 cm dia.
Cap pallid grey with fine, dark radiating streaks, sometimes with olivaceous or vinaceous tinges; at first convex, becoming flattened, with broad umbo, smooth, dry. Flesh white, becoming yellowish with age and thin.
Gills white becoming yellowish, emarginate, distant. **Spores** hyaline, smooth, sub-spherical or broadly ellipsoid, non-amyloid, droplets absent, 5-6 x 3.5-5 µm. Basidia 2- or 4- spored. Cystidia absent.
Stem white with yellow tinge below, smooth, tapering slightly upwards, sometimes curved. Ring absent. Flesh white, becoming yellowish with age, stuffed or full.
Odour of meal. **Taste** of meal.
Chemical tests none.
Occurrence late summer to autumn; infrequent.
☐ Edible.

Tricholoma saponaceum (Fr.) Kummer **Tricholomataceae**

Large fleshy agaric, grey-brown with whitish gills bearing characteristic reddish stains; scattered or in small trooping groups on soil in broad-leaf and coniferous woods.

Dimensions cap 5 - 10 cm dia; stem 5 - 10 cm tall x 1 - 3 cm dia.
Cap greyish-brown, darker at the centre; at first convex, becoming flattened, with broad umbo, smooth, dry. Flesh white and thick.
Gills white, becoming reddish spotted, emarginate, fairly distant.
Spores hyaline, smooth, ellipsoid, non-amyloid, occasionally with droplets, 5-6 x 3-4 µm. Basidia 4-spored. Cystidia absent.
Stem white with reddish tinge, smooth, fairly stout or bulbous, tapering into a rooting base. Ring absent. Flesh white, stuffed or full.
Odour of soap, or sour. **Taste** slight, bitter.
Chemical tests none.
Occurrence late summer to autumn; common.
■ Inedible.

Tricholoma scalpturatum (Fr.) Quél.   **Tricholomataceae**

Large fleshy agaric, with grey-brown cap decorated with darker scales and with pale gills; often in extensive trooping groups on soil in broad-leaf and mixed woods.

Dimensions cap 3 - 8 cm dia; stem 3 - 6 cm tall x 0.5 - 1 cm dia.
Cap pallid, grey-brown with darker brown radially-arranged fibrous scales; at first convex, becoming flattened, sometimes with slight umbo. Flesh white, yellowing with age, thin and soft.
Gills pallid grey or with yellowish tinge, emarginate or adnexed, edges irregular, broad, close. **Spores** hyaline, smooth, ellipsoid, non-amyloid, with droplets, 5-7 x 3-4 µm. Basidia 4-spored. Cystidia absent.
Stem white or with greyish tinge, silky smooth or finely fibrous, more or less equal or slightly thickened at the base. Ring absent or with faint zone. Flesh white, soft, stuffed or full.
Odour faint, of meal. **Taste** of meal or rancid.
Chemical tests none.
Occurrence summer to autumn; frequent.
✛ Poisonous.
Similar to *T. argyraceum* which has a very pale silvery base colour in the cap.

Tricholoma sciodes (Sécr.) Martin   **Tricholomataceae**

Medium-sized fleshy agaric, pale, streaked radially with very fine dark, slightly scaly fibrils, greyish-white gills; solitary or in small trooping groups on calcareous soil in broad-leaf woods, favouring beech.

Dimensions cap 3 - 7 cm dia; stem 5 - 8 cm tall x 0.8 - 1.8 cm dia.
Cap pallid, with dark somewhat scaly streaks; at first conical, then convex, and finally flattened, with broad umbo. Flesh pallid grey, thin and soft.
Gills pallid, becoming more greyish, edges crenate, emarginate, broad, crowded. **Spores** hyaline, smooth, broadly ellipsoid, non-amyloid, with droplets, 6-8 x 4.5-5.5 µm. Basidia 4-spored. Cystidia-like cells cylindrical or clavate, in groups.
Stem greyish-white, smooth or finely fibrous, equal or tapering slightly upwards. Ring absent. Flesh whitish-grey, soft, stuffed or full.
Odour not distinctive. **Taste** hot after being chewed (a minute at least).
Chemical tests none.
Occurrence late summer to autumn; frequent.
■ Inedible.

Tricholoma sejunctum (Sow.: Fr.) Quél. **Tricholomataceae**

Large fleshy agaric, whitish with fine tan streaks; scattered or in small trooping groups on soil in coniferous and mixed woods.

Dimensions cap 4 - 10 cm dia; stem 5 - 8 cm tall x 1 - 3 cm dia.
Cap pallid, yellowish with fine brownish or tan radiating fibrils; at first convex, becoming flattened, with broad umbo, slightly sticky. Flesh white and thick.
Gills whitish ochre, emarginate, very broad, crowded.
Spores hyaline, smooth, broadly ellipsoid, non-amyloid, with droplets, 6-8 x 4-6 µm. Basidia 4-spored. Cystidia-like cells present.
Stem concolorous but more pallid, smooth, more or less equal or tapering slightly upwards. Ring absent. Flesh white, stuffed or full.
Odour of meal. **Taste** of meal or bitter.
Chemical tests none.
Occurrence late summer to autumn; infrequent.
■ Inedible.

Tricholoma stans (Fr.) Sacc.   **Tricholomataceae**

Largish fleshy agaric, with reddish-brown spotted, slightly greasy cap and whitish gills; solitary or in small trooping groups on soil specifically with pines.

Dimensions cap 4 - 9 cm dia; stem 5 - 8 cm tall x 1 - 1.5 cm dia.
Cap dull reddish-brown, with more pallid margin and characteristic brown spotting, smooth, greasy in damp conditions; at first hemispherical becoming flattened or somewhat depressed around the blunt umbo. Flesh white or with slight reddish tinge, thin and soft.
Gills whitish or with reddish tinges, emarginate or adnexed, close.
Spores hyaline, smooth, ellipsoid, non-amyloid, with droplets, 4.5-6 x 3.5-4 µm. Basidia 4-spored. Cystidia absent.
Stem reddish-brown, finely fibrillose, more or less equal or somewhat tapered towards the base. Ring, faint zone. Flesh white, soft, stuffed or full.
Odour of meal when cut. **Taste** of meal.
Chemical tests none.
Occurrence summer to autumn; rare, limited to Scottish Highlands and other montane areas of Europe.
■ Inedible.

Tricholoma sulphureum (Bull.: Fr.) Kummer
= *Tricholoma bufonium* (Pers.: Fr.) Gill.   **Tricholomataceae**

Largish fleshy agaric, yellowish-green throughout; scattered or in small trooping groups, on soil in broad-leaf and, more rarely, coniferous woods.

Dimensions cap 3 - 8 cm dia; stem 5 - 10 cm tall x 0.6 - 1 cm dia.
Cap sulphur-yellow with brownish or olivaceous tinges; at first convex, becoming flattened, sometimes with slight umbo, at first silky, becoming smooth, dry. Flesh sulphur-yellow and thin.
Gills concolorous, emarginate, fairly broad, distant. **Spores** hyaline, smooth, broadly ellipsoid, non-amyloid, with droplets, 9-12 x 5-6 µm. Basidia 4-spored. Cystidia absent.
Stem concolorous with cap, fibrous lined, more or less equal, often curved, fairly slender. Ring absent. Flesh sulphur-yellow, stuffed or full.
Odour strong, of coal gas or tar. **Taste** not distinctive.
Chemical tests none.
Occurrence late summer to autumn; infrequent.
■ Inedible.

Tricholoma terreum (Schaeff.: Fr.) Kummer
Tricholomataceae

Largish fleshy agaric, felty grey with greyish gills; often in extensive trooping groups on calcareous soil in coniferous and, occasionally, broad-leaf woods.

Dimensions cap 4 - 7 cm dia; stem 3 - 8 cm tall x 1 - 1.5 cm dia.
Cap pallid or dark grey, finely felty or downy; at first convex, becoming flattened, sometimes with slight umbo. Flesh pallid grey, thin and soft.
Gills pallid grey or with yellowish tinge, emarginate or adnexed, distant. **Spores** hyaline, smooth, broadly ellipsoid, non-amyloid, with droplets, 6-7.5 x 3.5-4.5 µm. Basidia 4-spored. Cystidia absent.
Stem white, silky-smooth or finely fibrous, more or less equal. Ring absent. Flesh white, soft, stuffed or full.
Odour not distinctive. **Taste** not distinctive.
Chemical tests none.
Occurrence summer to autumn; infrequent.
□ Edible.

Tricholoma ustaloides Romagn.   **Tricholomataceae**

Largish fleshy agaric, chestnut-brown with whitish gills, greasy; in loose trooping groups or clustered on soil in broad-leaf woods, favouring beech.

Dimensions cap 4 - 9 cm dia; stem 5 - 10 cm tall x 1 - 1.5 cm dia.
Cap chestnut-brown, more pallid at the striate and somewhat inrolled margin; at first more or less hemispherical, becoming convex and slightly umbonate, viscid when damp. Flesh white, thin and firm.
Gills white or with brown spots, emarginate or adnexed, crowded. **Spores** hyaline, smooth, sub-spherical, non-amyloid, with droplets, 6-7 x 4-5 µm. Basidia 4-spored. Cystidia absent.
Stem pallid, decorated with rust-brown speckled fibrils below ring zone, more or less equal. Ring ephemeral cortinal zone, sub-apical. Flesh white, fibrous, stuffed or full.
Odour of meal. **Taste** bitter.
Chemical tests none.
Occurrence late summer to autumn; infrequent.
■ Inedible.

Tricholomopsis rutilans (Schaeff.: Fr.) Sing.
= *Tricholoma rutilans* (Schaeff.: Fr.) Kummer.   **Tricholomataceae**
Plums and Custard

Medium to large, fleshy agaric, cap with plum fibrils on custard-yellow backround, gills egg-yellow; solitary or in small caespitose tufts on or close to rotting conifer stumps.

Dimensions cap 4 - 12 cm dia; stem 3.5 - 5.5 cm tall x 1 - 1.5 cm dia.
Cap egg-yellow, densely covered with reddish-purple fibrillose scales typically in tufts, thinning towards the margin; at first convex becoming broadly umbonate. Flesh pallid yellow and thin.
Gills egg-yellow, adnexed, very broad, crowded. **Spores** hyaline, smooth, broadly ellipsoid, non-amyloid, with droplets, 7-8 x 5-6 µm. Basidia 4-spored. Gill edge cystidia clavate.
Stem yellow and covered with reddish-purple fibrillose scales, less densely than on cap, more or less equal. Ring absent. Flesh pallid yellow, tough, stuffed or full.
Odour of rotting wood. **Taste** not distinctive.
Chemical tests none.
Occurrence summer to autumn; common.
☐ Edible.

Amanita aspera (Fr.) S F Gray   **Amanitaceae**

Small or medium straw, then dull brown, agaric with yellow pointed scales, white gills and whitish scaly stem with ring and volval remains at the base; solitary or scattered on soil in broad-leaf woods.

Dimensions cap 5 - 9 cm dia; stem 5 - 9 cm tall x 1.5 - 2.5 cm dia.
Cap at first straw, becoming dull brown; convex, becoming ex-panded, sulcate at the margin, decorated with pyramidal yellow warty patches of volva. Flesh white and thin.
Gills white, adnate, crowded. **Spores** hyaline, smooth, broadly ellipsoid, non-amyloid, 8-9 x 6-6.5 µm. Basidia 4-spored. Cystidia not distinctive.
Stem pallid buff, with yellowish belted remnants of veil; incon-spicuous bulb encased in yellowish volval remnants. Ring yellowish. Flesh white, firm, hollow or stuffed.
Odour not distinctive. **Taste** not distinctive.
Chemical test none.
Occurrence late summer to autumn; infrequent.
■ Inedible.

Amanita ceciliae (Berk. & Br.) Bes.   **Amanitaceae**
= *Amanita inaurata* Sécr.
= *Amanita strangulata* (Fr.) Sacc.

Large, greyish-brown, fleshy agaric with a scaly stem, white gills, no ring, volval bag; solitary or scattered on soil in broad-leaf or mixed woods.

Dimensions cap 7 - 12 cm dia; stem 8 - 13 cm tall x 1.5 - 2 cm dia.
Cap pallid grey-brown; convex becoming more expanded, sulcate at the margin, decorated with large, coarse, dingy grey patches of volva. Flesh white and firm. Velum with many sphaerocysts.
Gills white, free, crowded. **Spores** hyaline, smooth, spherical, non-amyloid, 10-13 µm. Basidia 4-spored. Cystidia not distinctive.
Stem pallid, greyish-brown with white shaggy horizontal bands of veil; inconspicuous bulb encased in a volval bag which breaks away leaving oblique ridges. Ring absent. Flesh white, firm, hollow or stuffed.
Odour not distinctive. **Taste** not distinctive.
Chemical test none.
Occurrence summer to autumn; infrequent.
■ Inedible.

Amanita citrina (Schaeff.) S F Gray
= *A. mappa* (Batsch) Quél.

Amanitaceae
False Death Cap

Medium to large, whitish-lemony, fleshy agaric with white gills, cap scales, ring, bulbous base; on soil, solitary or scattered, in woods generally but favouring beech.

Dimensions cap 4-10 cm dia; stem 6-8 cm tall x 0.8-1.2 cm dia.
Cap pallid lemon or ivory, slightly deeper towards the centre; at first sub-spherical, becoming expanded-convex to flattened, more or less smooth and dry, decorated with coarse whitish velar patches discolouring ochraceous-brown. Flesh white and firm.
Gills whitish, adnexed, crowded. **Spores** hyaline, smooth, sub-spherical, amyloid, 9.5-7.5 µm. Basidia 4-spored. Cystidia not distinctive.
Stem whitish, smooth but lined above ring; basal bulb with volval rim creating a gutter around the stem. Ring white, membraneous and fairly firm, superior. Flesh white, stuffed, becoming hollow.
Odour of raw potato. **Taste** unpalatable.
Chemical tests none.
Occurrence summer to autumn; common.
■ Inedible.
Extreme caution when tasting because of possible confusion with dangerous species. (See also *A. citrina var alba*.)

Amanita citrina var **alba** (Gill.) Gilb.   **Amanitaceae**

Medium to large, white, fleshy agaric with cap scales, similar to *A. citrina*; solitary or scattered, on soil, in woods generally but favouring beech.
This species is a form of *A. citrina*, occurring infrequently in similar locations and season, and distinguished by being white throughout.

Amanita crocea (Quél.) Kühn & Romagn.   **Amanitaceae**
= *Amanita vaginata var crocea* Quél.
= *Amanitopsis crocea* (Quél.) Gilb.

Medium to large, fleshy agaric with yellowish-orange cap, cream gills, no ring, volval bag; solitary or scattered on soil with broad-leaf trees, favouring beech and birch, often in open or grassy places.

Dimensions cap 4 - 10 cm dia; stem 10 - 15 cm tall x 1 - 2 cm dia.
Cap bright yellowish-orange with apricot tinge at centre; at first sub-spherical, or hemi-spherical becoming expanded-convex to flattened, and often reflexing at the margin in older specimens; smooth, without obvious velar remnants. Flesh white or tinged orange below cap cuticle, medium.
Gills cream, free or adnexed, crowded. **Spores** hyaline, smooth, sub-spherical, non-amyloid, 10-12 x 9-10 µm. Basidia 4-spored. Cystidia not distinctive.
Stem pallid, concolorous with cap, covered in silky tufts, narrowing slightly towards the apex, base non-bulbous with white volval bag tinged cap colour on the inner surface. Ring absent. Flesh white, firm, stuffed or full.
Odour sweet. **Taste** sweetish with nutty tones.
Chemical tests none.
Occurrence summer to autumn; infrequent.
□ Edible. Possible confusion with old washed-out *A. muscaria*.

Amanita echinocephala (Vitt.) Quél. Amanitaceae
= *Aspidella echinocephala* (Vitt.) Gilb.

White, fleshy agaric with warty-scaly cap, thin ring, stout rooting stem and warty volval sheath; solitary or scattered on calcareous soil under broad-leaf trees.

Dimensions cap 6 - 20 cm dia; stem 6 - 14 cm tall x 2 - 3 cm dia.
Cap pure white or tinged pallid brown; at first sub-spherical, becoming convex and finally irregularly expanded, covered when young in conical white warts. Flesh white or with green tinge and firm.
Gills white, free, crowded. **Spores** hyaline, smooth, ellipsoid, amyloid, 9.5-11.5 x 6.5-8 µm. Basidia 4-spored. Cystidia not distinctive.
Stem white, smooth towards the apex, lower part sheathed in warty volval remains, base tapering and deeply rooted in soil. Ring thin and superior. Flesh white, firm, stuffed or full.
Odour unpleasant. **Taste** unpleasant
Chemical test none
Occurrence summer to autumn; rare. (Blenheim, Oxfordshire)
✣ Poisonous.

Amanita echinocephala (Vitt.) Quél. (immature)
Amanitaceae

The illustration depicts the species in its young state showing the typical characteristics which account for its title. The spiny warts which clothe the immature fruiting body are largely dispersed as the cap expands.

Amanita eliae Quél. Amanitaceae

Medium to large, pale cream agaric with cap patches, white gills, ring on the elongated stem ; solitary or scattered on soil in broad-leaf woods.

Dimensions cap 5 - 10 cm dia; stem 6 - 12 cm tall x 1.5 - 2.5 cm dia.
Cap whitish-cream with pink tinge; at first convex, becoming expanded-convex or flattened, smooth and slightly greasy in damp conditions, decorated with whitish velar patches. Flesh white, firm and somewhat thin.
Gills white, adnexed or with slight decurrent tooth, crowded.
Spores hyaline, smooth, broadly ellipsoid, amyloid, 9 -15 x 6.5-8.5 µm. Basidia 4-spored. Cystidia not distinctive.
Stem white, smooth, typically elongated and sunk deep into the substrate, basal bulb somewhat slender and pointed with sparse volval sheathing. Ring white, ephemeral and superior. Flesh white, firm and full.
Odour not distinctive. **Taste** not distinctive.
Chemical tests none.
Occurrence summer to autumn; frequent, or common locally.
■ Inedible.

Amanita excelsa (Fr.) Kummer   **Amanitaceae**
= *A. spissa* (Fr.) Kummer

Medium to large, brown fleshy agaric with cap patches, white gills, ring, basal bulb; solitary or scattered on soil in mixed woods.

Dimensions cap 5 - 10 cm dia; stem 6 - 12 cm tall x 1.5 - 2.5 cm dia.
Cap greyish-brown, variable; at first convex, becoming expanded-convex or flattened, smooth and slightly greasy in damp conditions, decorated with delicate, non-persistent, whitish or whitish-grey velar patches which eventually wash off or fall away. Flesh white, firm and medium.
Gills white, adnexed or with a slight decurrent 'tooth', crowded.
Spores hyaline, smooth, broadly ellipsoid, amyloid, 9-10 x 8-9 μm. Basidia 4-spored. Cystidia not distinctive.
Stem white, smooth, lined above the ring and covered in small scales below; basal bulb not obviously guttered and sometimes deeply buried, no volval sheath generally apparent. Ring large, white and fairly firm, superior. Flesh white, firm and full.
Odour faint, unpleasant. **Taste** not distinctive.
Chemical tests flesh purple with sulphuric acid.
Occurrence summer to autumn; frequent, or common locally.
☐ Allegedly edible but worthless.
Note: *A. pantherina* differs in the pure white colour of the velar remains and a volval 'ridge' at the stem base.

Amanita fulva (Schaeff.) Sing.   **Amanitaceae**
= *Amanitopsis vaginata var fulva* Schaeff.

Medium-sized agaric, distinctive tawny cap and white gills, no ring, usually without cap patches but with volval bag; solitary or scattered on soil in mixed woodlands favouring birch.

Dimensions cap 4 - 9 cm dia; stem 7 - 12 cm tall x 0.8 - 1.2 cm dia.
Cap tan or orange-brown with sulcate margin, occasionally with brownish velar patches; at first ovoid, becoming expanded-convex and flattened with a slight umbo. Flesh white and brittle.
Gills white, free, crowded. **Spores** hyaline, smooth, spherical, non-amyloid, 9-11 μm. Basidia 4-spored. Cystidia not distinctive.
Stem whitish or pallid cap colour, narrower towards the apex and arising from white volval bag. Ring absent. Flesh white, firm but brittle, hollow in mature specimens.
Odour not distinctive. **Taste** not distinctive.
Chemical tests none.
Occurrence summer to autumn; very common.
☐ Edible. Care must be taken to avoid confusion with poisonous members of the *Amanita* genus.

Amanita gemmata (Fr.) Bertillon   **Amanitaceae**
= *Amanitopsis adnata* (W G Smith) Sacc.

Medium-sized fleshy agaric with pale yellowish cap, white gills, thin membraneous ring, volval bag; solitary or scattered on soil with coniferous trees.

Dimensions cap 5 - 7 cm dia; stem 7 - 10 cm tall x 1 - 1.4 cm dia.
Cap pallid yellow, more ochraceous at the centre; at first convex, becoming expanded-convex or flattened; smooth with striate margin, decorated with white velar patches. Flesh white with yellowish tinge beneath cap cuticle, thin, firm.
Gills white, adnexed or free, crowded. **Spores** hyaline, smooth, sub-spherical or broadly ellipsoid, non-amyloid, 8.5-9 x 7-7.5 μm. Basidia 4-spored. Cystidia not distinctive.
Stem white or flushed pallid yellow, finely woolly, bulbous at the base surrounded by a short, thin, white volval bag. Ring white, thin, ragged, tending to be inferior. Flesh white or tinged yellowish, firm, stuffed or full.
Odour not distinctive. **Taste DO NOT ATTEMPT TO TASTE ANY PART OF THE SPECIMEN**.
Chemical tests none.
Occurrence summer to autumn; rare.
✚ Lethally poisonous.

Amanita muscaria (L.: Fr.) Hook.   **Amanitaceae**
Fly Agaric

Large fleshy agaric, cap red with white patches, white gills, ring, bulbous base; solitary or scattered, on poor and sandy soils, favouring birch woods but also pines.

Dimensions cap 8 - 10 cm dia; stem 8 - 18 cm tall x 1 - 2 cm dia.
Cap deep scarlet, fading to orange-red or orange-yellow in older specimens; at first sub-spherical, becoming expanded-convex or flattened, smooth, sometimes with a distinct sheen, decorated with pure-white warty velar remnants, readily washed off by rain. Flesh white, or tinged cap colour beneath the cuticle, firm.
Gills white, free, crowded. **Spores** hyaline, smooth, ellipsoid, non-amyloid, 9.5-10.5 x 7-8 μm. Basidia 4-spored. Cystidia not distinctive.
Stem white or ivory, smooth or slightly sculptured, often with velar fragments in several warty rings immediately above the basal bulb. Ring white or tinged yellow, membraneous and pendant with double margin, superior or sub-apical. Flesh white, firm, and stuffed in part.
Odour not distinctive. **Taste** not distinctive.
Chemical tests none.
Occurrence late summer to autumn; common.
✚ Dangerously poisonous, hallucinogenic, but generally non-fatal.

Amanita pantherina (DC.: Fr.) Krombh.   **Amanitaceae**

Large fleshy agaric with brownish cap, pure white patches, white stem, ring, bulbous base with distinct margin; solitary or scattered on soil in broad-leaf, occasionally coniferous, woods, favouring beech.

Dimensions cap 5 - 10 cm dia; stem 8 - 12 cm tall x 1 - 1.5 cm dia.
Cap brown with ochraceous tinge; at first convex, becoming expanded-convex and then flattened, decorated with pure-white velar remnants. Flesh white, sometimes tinged cap colour below cuticle and becoming brownish with age, medium.
Gills white, free, crowded. **Spores** hyaline, smooth, broadly ellipsoid, non-amyloid, 8-12 x 7-8 μm. Basidia 4-spored. Cystidia not distinctive.
Stem white, tapering upwards, with belt-like velar remnants below the ring, basal bulb sheathed by small white volval sac which forms a distinct gutter. Ring white, pendulous, thin, non-striate and superior. Flesh white, firm and full.
Odour not distinctive. **Taste DO NOT ATTEMPT TO TASTE ANY PART OF THE SPECIMEN**
Chemical tests flesh vinaceous with phenol.
Occurrence summer to autumn; infrequent.
✚ Poisonous, perhaps lethal.

Amanita phalloides (Vaill.: Fr.) Link   **Amanitaceae**
Death Cap

Large fleshy agaric with greenish-olive cap, whitish stem, ring, no patches but distinctive volval sheath; solitary or scattered on soil, favouring mixed light broad-leaf woods, usually with oak.

Dimensions cap 6 - 12 cm dia; stem 7 - 12 cm tall x 1 - 1.5 cm dia.
Cap greenish, with yellow or olivaceous tinges, with radiating silky fibrils suggesting a 'bloom' when dry, sometimes pallid dirty-white; at first convex, becoming expanded or flattened, slightly viscid when damp, velar remains absent. Flesh white, tinged cap colour below cuticle, medium. **Gills** white, free, crowded. **Spores** hyaline, smooth, sub-spherical or broadly ellipsoid, amyloid, 8-10 x 7-8 μm. Basidia 4-spored. Cystidia not distinctive.
Stem white or tinged cap colour, tapering upwards, sometimes sculpted into faint bands, basal bulb sheathed by large, loose, ragged, white volval sac. Ring white, pendulous, thin and superior. Flesh white, firm and solid.
Odour sweetish or sickly. **Taste DO NOT ATTEMPT TO TASTE ANY PART OF THE SPECIMEN.**
Chemical tests gills pallid lilaceous with sulphuric acid.
Occurrence summer to autumn; infrequent but more common in some localities and during some years.
✚ Lethally poisonous even in very small quantities.

Amanita rubescens ([Pers.] Fr.) S F Gray   **Amanitaceae**
The Blusher

Large, rosy-brown, fleshy agaric with cap patches, white gills, bulbous base and ring on stem, bruising reddish; solitary or scattered on soil in woods generally.

Dimensions cap 6 - 15 cm dia; stem 6 - 15 cm tall x 1 - 2.5 cm dia.
Cap dull reddish-brown or buff, sometimes spotted, decorated with warty velar remnants from pallid clay to reddish; at first sub-spherical, becoming convex and finally flattened. Flesh white, becoming brownish-pink where cut or damaged, moderately firm (spongy in older specimens).
Gills white, spotted brownish-pink with age and where bruised, free, crowded. **Spores** hyaline, smooth, ellipsoid, amyloid, 8-9 x 5-6 μm. Basidia 4-spored. Cystidia not distinctive.
Stem white or pallid clay above the ring, flushed cap colour below, often with patterning, bruising reddish-brown; basal bulb may retain small rows of volval patches but without gutter. Ring membraneous, lax, typically striated, superior. Flesh at first white, then spotted brownish pink with age and where bruised; firm becoming spongy, stuffed.
Odour not distinctive. **Taste** mild, then somewhat astringent.
Chemical tests none. **Occurrence** summer to autumn; common.
✧ Poisonous when raw; edible only when cooked by thorough boiling.

Amanita rubescens var **annulosulphurea** Gill.
Amanitaceae

Large, rosy-brown, fleshy agaric with cap patches, very similar to *A. rubescens* but with distinctive yellow coloration on the apex of the stalk and the ring; solitary or scattered on soil in woods and on heaths. Characteristics otherwise as for *A. rubescens*. Infrequent or rare. (Thetford Warren, Suffolk)

Amanita strobiliformis (Vitt.) Quél.   **Amanitaceae**
= *Amanita solitaria* (Bull.: Fr.) Mérat

Large, white, conspicuously shaggy, fleshy agaric with ring and volval bag; typically solitary, on soil favouring, but not limited to, calcareous soils, in or near broad-leaf woods.

Dimensions cap 6 - 20 cm dia; stem 6 - 10 cm tall x 1 - 2 cm dia.
Cap pure-white; convex becoming more expanded, covered with dense flat shaggy scales which overhang the margin. Flesh white and firm.
Gills white, free, crowded. **Spores** hyaline, smooth, broadly ellipsoid, amyloid, 10,0-12.5 x 8-10 μm. Basidia 4-spored. Cystidia not distinctive.
Stem white, shaggy, slightly bulbous and rooting at the base with remains of volval bag. Ring large, pendulous, farinose, thin and superior. Flesh white, cream towards the base, firm, partly stuffed, becoming hollow.
Odour faint, pleasant. **Taste** slight, pleasant.
Chemical tests none.
Occurrence summer to autumn; rare. (Bridgham near Thetford, Norfolk)
☐ Edible.
Note: beware confusion with *A. virosa* and *A. phalloides var alba*.

Amanita vaginata (Bull.: Fr.) Vitt.   **Amanitaceae**
= *Amanitopsis vaginata* (Bull.:Fr.) Roze

Medium-sized, greyish, fleshy agaric with volval bag but no ring or cap patches; solitary or scattered on soil in broad-leaf woods and on heaths.

Dimensions cap 5-9 cm dia; stem 13-20 cm tall x 1.5-2 cm dia.
Cap greyish-brown; ovoid, expanding to almost flat, with a slight umbo and with sulcate margin; otherwise smooth, without velar remains. Flesh white and firm.
Gills white, adnexed, crowded. **Spores** hyaline, smooth, spherical, non-amyloid, 9-12 µm. Basidia 4-spored. Cystidia not distinctive.
Stem white or tinged cap colour, tapering upwards, arising from large volval bag. Ring absent. Flesh white, firm, becoming hollow.
Odour not distinctive. **Taste** not distinctive.
Chemical tests none.
Occurrence summer to autumn; infrequent.
☐ Edible. Best avoided because of possible confusion with dangerous species.

Amanita virosa (Fr.) Bertillon   **Amanitaceae**
Destroying Angel

Large, white, fleshy agaric with a shaggy stalk and volval bag; solitary or scattered on soil in broad-leaf or mixed woods.

Dimensions cap 5-12 cm dia; stem 9-12 cm tall x 1-1.5 cm dia.
Cap pure-white; at first conical, becoming campanulate and finally irregularly expanded, but often retaining a flat umbo, viscid. Flesh white and firm.
Gills white, free, crowded. **Spores** hyaline, smooth, spherical, amyloid, 8-10 µm. Basidia 4-spored. Cystidia not distinctive.
Stem white, shaggy or fibrous, typically curved, arising from large bag-like volva often buried deep in the soil. Ring large, fragile and typically ruptured, superior. Flesh white, firm, stuffed or full.
Odour sickly sweet. **Taste DO NOT ATTEMPT TO TASTE ANY PART OF THE SPECIMEN.**
Chemical test flesh immediately yellow with KOH.
Occurrence summer to autumn; infrequent but more common in Scotland.
✚ Lethally poisonous even in very small quantities.

Limacella glioioderma (Fr.) Maire   **Amanitaceae**

Small agaric with greasy brownish cap, white gills and pinkish woolly stem; solitary or in small groups on soil in broad-leaf woods.

Dimensions cap 2 - 6 cm dia; stem 2 - 6 cm tall x 0.4 - 0.8 cm dia.
Cap reddish-brown, slightly darker at the centre, becoming more pallid; at first hemispherical then shallowly convex, greasy, more or less smooth. Flesh pallid, pinkish-brown, thin.
Gills whitish-buff, free, distant. **Spores** hyaline, smooth, spherical, 4-8 µm. Basidia 4-spored. Cystidia not distinctive.
Stem whitish or pallid pink, more or less equal, with concentric floccose belts below the ring. Ring concolorous, floccose, ascendant, median. Flesh pallid pinkish-brown, stuffed or hollow.
Odour of meal. **Taste** not distinctive.
Chemical tests none.
Occurrence summer to autumn; infrequent.
■ Inedible.

Pluteus atromarginatus (Konrad) Kühn.   **Pluteaceae**
= *Pluteus nigrofloccosus* (Schultz.) Favre
= *Pluteus tricuspidatus* Vel.

Medium or large agaric with dark brown scaly-velvety cap and pink, dark-edged gills; solitary or in small more or less clustered groups, on rotting stumps and other woody debris of coniferous trees.

Dimensions cap 5 - 15 cm dia; stem 5 - 15 cm tall x 0.5 - 1.5 cm dia.
Cap sepia or dark umber-brown; convex becoming more flattened, decorated with fine pointed almost downy scales. Flesh whitish tinged brown, firm and thin. Cap skin fibrous.
Gills at first whitish, becoming pink with darker edges, broad, crowded, free. **Spores** pink, smooth, broadly ellipsoid, 6-7.5 x 4.5-5.5 µm. Basidia 4-spored. Gill face cystidia thick-walled with hooked ends.
Stem pallid background covered with dark brown fibres, more or less equal or tapering slightly upwards, smooth. Ring absent. Flesh whitish, tinged brown, firm, full or stuffed.
Odour not distinctive. **Taste** not distinctive.
Chemical tests none.
Occurrence spring to autumn; infrequent or rare. (Thetford Forest, Norfolk)
☐ Edible.

Pluteus cervinus (Schaeff.: Fr.) Kummer   **Pluteaceae**

Medium or large agaric with dark brown radially streaked cap and pink gills; solitary or in small more or less clustered groups, on rotting stumps and other woody debris of broad-leaf and, very occasionally, coniferous trees.

Dimensions cap 4 - 12 cm dia; stem 7 - 10 cm tall x 0.5 - 1.5 cm dia.
Cap sepia or dark umber-brown with darker radial streaks; convex, becoming more flattened, more or less smooth. Flesh whitish, firm, moderate. Cap skin fibrous.
Gills at first whitish, becoming dull pink, broad, crowded, free.
Spores pink, smooth, broadly ellipsoid, 7-8 x 5-6 µm. Basidia 4-spored. Gill face cystidia thick-walled with hooked ends.
Stem whitish with darker fibres, more or less equal or slightly swollen at the base, smooth. Ring absent. Flesh whitish, firm, full or stuffed.
Odour not distinctive. **Taste** not distinctive.
Chemical tests none.
Occurrence spring to winter; common.
☐ Edible.

Pluteus galeroides P Orton   **Pluteaceae**

Smallish agaric with yellow cap and yellowish-pink gills; in groups, on rotting wood, favouring elm and beach.

Dimensions cap 2 - 4 cm dia; stem 2 - 4 cm tall x 0.2 - 0.4 cm dia.
Cap at first chrome-yellow becoming yellowish-brown and darker towards centre; convex becoming more flattened, sulcate at the margin, otherwise smooth, not viscid. Flesh yellowish, thin.
Gills at first ochraceous becoming tinged pink at maturity, broad, free, close. **Spores** pink, smooth, sub-spherical, 5-8 x 5-6 µm. Basidia 4-spored. Gill face cystidia flask-shaped or fusiform.
Stem yellowish-ochre, slender, more or less equal, smooth. Ring absent. Flesh whitish, firm, full or stuffed.
Odour not distinctive. **Taste** not distinctive.
Chemical tests none.
Occurrence late summer to autumn; infrequent.
■ Inedible.

Pluteus griseopus P Orton   **Pluteaceae**

Smallish agaric with dark brown cap and pinkish gills; solitary or in small more or less clustered groups on woody debris of broad-leaf trees.

Dimensions cap 1.5 - 4 cm dia; stem 1.5 - 5 cm tall x 0.2 - 0.5 cm dia.
Cap dark umber-brown with greyish tinge; sometimes more pallid at the margin, convex becoming more flattened, slightly wrinkled at the centre, otherwise smooth, suede-like. Flesh brown and thin. Cap skin cellular.
Gills pink, becoming brownish-pink at maturity, broad, free, crowded. **Spores** pink, smooth, sub-spherical, 6-8 x 5-6.5 μm. Basidia 4-spored. Gill face cystidia fusiform or flask-shaped.
Stem pallid, greyish, slender, more or less equal, silk fibrous or somewhat pruinose. Ring absent. Flesh whitish, firm, full or stuffed.
Odour not distinctive. **Taste** not distinctive.
Chemical tests none.
Occurrence late summer to autumn; infrequent.
■ Inedible.

Pluteus luteovirens Rea   **Pluteaceae**

Smallish agaric with mustard-yellow cap, pink gills and whitish stem; solitary or in small groups on rotting stumps and trunks or broad-leaf trees, favouring elm.

Dimensions cap 2 - 4 cm dia; stem 3 - 6 cm tall x 0.3 - 0.4 cm dia.
Cap mustard-yellow or with ochraceous-brown tinge, convex becoming more flattened, matt. Flesh whitish, firm and thin. Cap skin cellular.
Gills at first whitish, becoming pink, broad, crowded, free.
Spores pink, smooth, broadly ellipsoid or sub-spherical, 5-6 x 4.5-5 μm. Basidia 4-spored. Gill face cystidia flask-shaped with elongated necks.
Stem white or with faint yellow tinge at the base, more or less equal or tapering slightly upwards, smooth. Ring absent. Flesh whitish, tinged yellowish in the base, firm, full or stuffed.
Odour not distinctive. **Taste** not distinctive.
Chemical tests none.
Occurrence late summer to autumn; infrequent.
■ Inedible.

Pluteus lutescens (Fr.) Bres.   **Pluteaceae**
= *Pluteus nanus var lutescens* (Fr.) Karst.
= *Pluteus romellii* (Britz.) Sacc.

Medium-sized agaric with brown and yellow cap and yellowish-pink gills; solitary or in small more or less clustered groups on woody debris and wood chips of broad-leaf trees, favouring beech.

Dimensions cap 1.5 - 5 cm dia; stem 1.5 - 7 cm tall x 0.2 - 0.6 cm dia.
Cap cinnamon-brown, becoming more yellowish at the margin; convex becoming more flattened, faintly striate at the margin and sometimes slightly sulcate at the centre, otherwise smooth, not viscid. Flesh yellowish and thin. Cap skin cellular.
Gills at first chrome-yellow, becoming pink tinged at maturity, broad, crowded, free. **Spores** pink, smooth, sub-spherical, 6.5-7 x 5.5-6 μm. Basidia 4-spored. Gill face cystidia clavate or fusiform, thin-walled.
Stem yellowish, slender, more or less equal, smooth. Ring absent. Flesh whitish, firm, full or stuffed.
Odour not distinctive. **Taste** not distinctive.
Chemical tests none.
Occurrence summer to autumn; infrequent.
□ Edible.

Pluteus phlebophorus (Ditmar: Fr.) Kummer   **Pluteaceae**

Small or medium-sized agaric with brown, veined cap and pink gills; solitary or in small more or less tufted groups, on rotting stumps and other woody debris of broad-leaf trees.

Dimensions cap 2.5 - 6 cm dia; stem 2.5 - 9 cm tall x 0.4 - 1 cm dia.
Cap cinnamon or umber-brown; convex becoming more flattened, covered with fine network of raised veins. Flesh white with brown tinge and thin. Cap skin cellular.
Gills at first whitish, becoming pink, broad, crowded, free.
Spores pink, smooth, sub-spherical or broadly ellipsoid, 5-7 x 4.5-6 µm. Basidia 4-spored. Gill face cystidia flask-shaped with long projecting apex.
Stem pallid background finely lined and grooved, more or less equal or tapering slightly upward. Ring absent. Flesh white with brown tinges, firm, full or stuffed.
Odour not distinctive. **Taste** not distinctive.
Chemical tests none.
Occurrence summer to autumn; infrequent.
■ Inedible.

Pluteus rimulosus Kühn. & Romagn.   **Pluteaceae**

Smallish agaric with dark brown cap and pink gills; solitary or in small more or less tufted groups, near rotting stumps and on other woody debris of broad-leaf trees.

Dimensions cap 3.5 - 6 cm dia; stem 3 - 5 cm tall x 0.4 - 0.7 cm dia.
Cap dark brown, downy-pruinose or matt; convex becoming more flattened, smooth but faintly sulcate when damp, cuticle tearing radially when mature revealing white flesh. Flesh whitish and moderate. Cap skin cellular.
Gills at first whitish becoming pink, crowded, free. **Spores** pink, smooth, broadly ellipsoid, 6.5-9 x 5-6 µm. Basidia 4-spored. Gill face cystidia flask-shaped.
Stem whitish tinged cap colour at the base, more or less equal or slightly swollen at the base, smooth. Ring absent. Flesh white, firm, full or stuffed.
Odour not distinctive. **Taste** not distinctive.
Chemical tests none.
Occurrence late summer to autumn; infrequent.
■ Inedible.

Pluteus salicinus (Pers.: Fr.) Kummer   **Pluteaceae**

Medium-sized agaric with dark greyish cap and pink gills; solitary or in small more or less tufted groups, on rotting stumps and other woody debris of broad-leaf trees.

Dimensions cap 2 - 5 cm dia; stem 3 - 5 cm tall x 0.2 - 0.6 cm dia.
Cap grey with blue or green tinges, darker at the disc, more or less radially streaked; convex, becoming more flattened, smooth but faintly striate when damp. Flesh white with greyish tinge, moderate. Cap skin fibrous.
Gills at first whitish, becoming pink, broad, crowded, free.
Spores pink, smooth, broadly ellipsoid, 8-9 x 6-7 µm. Basidia 4-spored. Gill face cystidia fusiform with hooked ends.
Stem whitish tinged cap colour at the base, more or less equal or slightly swollen at the base, smooth. Ring absent. Flesh white with greyish tinge, firm, full or stuffed.
Odour not distinctive. **Taste** not distinctive.
Chemical tests none.
Occurrence spring to autumn; frequent.
□ Edible.

Pluteus semibulbosus (Lasch apud Fr.) Gill.   Pluteaceae
= *Pluteus plautus* (Weinm.) Gill.

Medium-sized, pale-greyish agaric with pink gills and bulbous base to the stem; solitary or in small groups on rotting wood of broad-leaf trees.

Dimensions cap 3 - 6 cm dia; stem 3 - 6 cm tall x 0.3 - 0.5 cm dia.
Cap pallid grey or darker, with brown tinge at the centre; flattened with incurved margin becoming unevenly expanded, finely scaly. Flesh whitish and thin. Cap skin fibrous-hyphal.
Gills at first whitish becoming pink, broad, crowded, free.
Spores pink, smooth, sub-spherical or broadly ellipsoid, 6-8 x 5-7 µm. Basidia 4-spored. Gill edge cystidia clavate; gill face cystidia flask-shaped.
Stem white with brownish tinges towards the base, slender, tapering upwards from distinctly bulbous base, striate. Ring absent. Flesh whitish and firm.
Odour not distinctive. **Taste** not distinctive.
Chemical tests none.
Occurrence late summer to autumn; infrequent.
■ Inedible.

Pluteus umbrosus (Pers.: Fr.) Kummer   Pluteaceae

Medium-sized, brown-veined, agaric with pink gills and whitish stem; solitary or in small groups on rotting wood of broad-leaf trees, favouring elm.

Dimensions cap 3 - 9 cm dia; stem 3 - 9 cm tall x 0.4 - 1.2 cm dia.
Cap mid-brown, finely scaly and radially sulcate, the ridges decorated with darker brown scales; convex becoming expanded. Flesh whitish and medium. Cap skin fibrous-hyphal.
Gills at first whitish, becoming pink with brown flaky edges, broad, crowded, free. **Spores** pink, smooth, sub-spherical or broadly ellipsoid, 5-7 x 4-5 µm. Basidia 4-spored. Gill face cystidia flask-shaped, thin-walled.
Stem white, decorated with minute brown scales towards the base, more or less equal. Ring absent. Flesh whitish and firm.
Odour slight, of garlic. **Taste** not distinctive.
Chemical tests none.
Occurrence late summer to autumn; infrequent.
☐ Edible.

Volvariella bombycina (Schaeff.: Fr.) Sing.   Pluteaceae
= *Volvaria bombycina* (Schaeff.: Fr.) Kummer

Medium to very large agaric with whitish silky-fibrous cap, pink gills and the base of the stem encased in a volval bag; solitary or in small tufts, on rotted wood including fissures and knot holes of sickly or dead broad-leaf trees.

Dimensions cap 8 - 20 cm dia; stem 8 - 15 cm tall x 1 - 3.5 cm dia.
Cap ivory-white; at first ovoid, becoming broadly campanulate, covered with long silky yellowish fibres which overhang the margin. Flesh cream, fragile and thick.
Gills at first white, becoming flesh pink at maturity, free, broad, crowded. **Spores** pink, smooth, ellipsoid, non-amyloid, 7-10 x 5-6.5 µm. Basidia 4-spored. Cystidia not significant.
Stem concolorous with cap, smooth, tapering upwards, the base enclosed in a volval sheath at first whitish discolouring brown. Ring absent. Flesh white, fibrous, tough and full.
Odour strong, pleasantly fungoid. **Taste** not distinctive.
Chemical tests none.
Occurrence summer to early autumn; rare. (Corse Lawns, near Gloucester)
☐ Edible and excellent.

Volvariella gloiocephala (DC.: Fr.) Sing.   **Pluteaceae**
= *Volvariella speciosa* (Fr.: Fr.) Sing.
= *Volvaria speciosa* (Fr.: Fr.) Kummer

Medium to large agaric with whitish cap, greasy when damp, pink gills and the base of the stem encased in a volval bag; solitary or scattered, on dung heaps, rotted straw, manured ground, compost heaps.

Dimensions cap 5 - 10 cm dia; stem 5- 9 cm tall x 1 - 1.5 cm dia.
Cap white with greyish-brown tinge at the centre; at first ovoid, becoming convex and finally more or less expanded, greasy or viscid when damp, otherwise smooth. Flesh white, firm and moderate.
Gills at first white, becoming flesh pink at maturity, free, broad, crowded. **Spores** pink, smooth, ellipsoid, non-amyloid, 13-18 x 8-10 μm. Basidia 4-spored. Cystidia not significant.
Stem white, smooth, tapering upwards, the base enclosed in a white or greyish volval sheath. Ring absent. Flesh white and firm.
Odour earthy. **Taste** not distinctive.
Chemical tests none.
Occurrence summer to early autumn; infrequent.
☐ Edible.
Care must be taken to avoid possible confusion with dangerously poisonous members of the *Amanita* genus.

Volvariella murinella (Quél.) Moser ex Courtec. **Pluteaceae**

Smallish, delicate, often tall agaric with pale greyish-brown cap, pink gills and the base of the stem encased in a volval bag; solitary or scattered, in pastures, on forest rides and other grasslands.

Dimensions cap 2 - 4 cm dia; stem 4 - 8 cm tall x 1 - 1.5 cm dia.
Cap pallid grey with brown tinge; at first ovoid, becoming convex and finally more or less expanded, smooth, striate at the margin. Flesh white, fragile and thin.
Gills at first white, becoming flesh pink at maturity, free, broad, crowded. **Spores** pink, smooth, ellipsoid, non-amyloid, 6-9 x 3.5-4.5 μm. Basidia 4-spored. Cystidia not significant.
Stem whitish, smooth, tapering upwards, the base enclosed in a white or greyish volval sheath. Ring absent. Flesh white and fragile.
Odour faint, of pelargonium. **Taste** not distinctive.
Chemical tests none.
Occurrence summer to early autumn; rare. (Bridgham, near Thetford, Norfolk)
■ Inedible.

Volvariella surrecta (Knapp) Sing.   **Pluteaceae**
= *Volvariella loveiana* (Berk.) Gill.

Small or medium-sized agaric with whitish cap, pink gills and the base of the stem encased in a volval bag; solitary or in small groups, parasitic on caps of decaying *Clitocybe nebularis* and, more rarely, other large agarics.

Dimensions cap 4 - 7 cm dia; stem 4 - 8 cm tall x 1 - 1.5 cm dia.
Cap whitish, becoming tinged brown with age; at first convex, becoming flattened, finely woolly-downy. Flesh white and thin.
Gills at first white, becoming flesh pink at maturity, free, broad and crowded. **Spores** pink, smooth, ellipsoid, non-amyloid, 5-7 x 3-5 μm. Basidia 4-spored. Gill cystidia flask-shaped.
Stem whitish, smooth, tapering upwards, the base enclosed in a whitish lobed volval sheath. Ring absent. Flesh white, stuffed or full.
Odour not distinctive. **Taste** not distinctive.
Chemical tests none.
Occurrence autumn to early winter; rare. (Near Newbury, Berkshire)
■ Inedible.

Clitopilus prunulus (Scop.: Fr.) Kummer **Entolomataceae**
= *Paxillopsis prunulus* (Scop.: Fr.) J Lange

Whitish cream, medium-sized agaric with pinkish gills, on comparatively short stem and with a distinctive mealy smell; scattered in troops on soil amongst grass, typically in the vicinity of trees.

Dimensions cap 3 - 12 cm dia; stem 1.5 - 2.5 cm tall x 0.4 - 1.2 cm dia.
Cap ivory-white or cream; at first convex, becoming irregularly wavy and depressed, surface textured like chamois leather. Flesh white, thin and firm.
Gills white becoming pinkish, decurrent, fairly broad, crowded.
Spores pink, with 6-8 longitudinal ribs (may be difficult to observe), ellipsoid, non-amyloid, 8-13 x 5-7 μm. Basidia 4-spored.
Stem concolorous with cap, slightly eccentric, more or less equal, slender, finely downy. Ring absent. Flesh white and firm.
Odour strong, of meal. **Taste** strong, of meal.
Chemical tests none.
Occurrence late summer to autumn; frequent.
□ Edible.

Entoloma aprile (Britz.) Sacc. **Entolomataceae**
= *Rhodophyllus aprilis* (Britz.) Romagn.

Medium-sized agaric with greyish-brown cap and stem, pinkish gills; solitary or in small trooping groups on soil under trees and shrubs.

Dimensions cap 3 - 6 cm dia; stem 4 - 7 cm tall x 0.5 - 1 cm dia.
Cap creamy-buff, with yellowish or greyish tinges, hygrophanous, drying more pallid; campanulate, becoming expanded with a distinct blunt umbo and wavy margin, not greasy. Flesh pallid grey and firm.
Gills at first whitish, becoming pallid pink at maturity, adnate, crowded. **Spores** pink, smooth, sub-spherical, regularly angular, 7-8 sided, non-amyloid, 9-12 x 8-10 μm. Basidia 4-spored. Cystidia not distinctive.
Stem greyish, drying more pallid, stout, more or less equal, fibrous. Ring absent. Flesh greyish, stuffed or narrowly hollow.
Odour faint, of meal. **Taste** faint, of meal.
Chemical tests none.
Occurrence spring to early summer; infrequent.
■ Inedible.

Entoloma cetratum (Fr.: Fr.) Moser **Entolomataceae**
= *Nolanea cetrata* (Fr.: Fr.) Kummer
= *Rhodophyllus cetratus* (Fr.: Fr.) Quél.

Small agaric with tan cap and stem, flesh-brown gills; solitary or in small trooping groups on soil with conifers, favouring spruce, often amongst mosses.

Dimensions cap 1 - 4 cm dia; stem 4 - 8 cm tall x 0.2 - 0.4 cm dia.
Cap tan or ochraceous-brown; campanulate becoming expanded, translucently striate, otherwise smooth. Flesh brown, thin and fragile.
Gills at first buff, becoming tan with pinkish tinge at maturity, emarginate or free, crowded. **Spores** pink, smooth, sub-spherical, regularly angular, non-amyloid, 9-12 x 6.5-8 μm. Basidia frequently 2-spored. Cystidia not distinctive.
Stem pallid tan, slender, more or less equal, silky fibrillose. Ring absent. Flesh concolorous, stuffed or narrowly hollow.
Odour not distinctive. **Taste** not distinctive.
Chemical tests none.
Occurrence summer to autumn; frequent.
■ Inedible.

Entoloma clypeatum (L.: Fr.) Kummer   Entolomataceae
= *Rhodophyllus clypeatus* (L.: Fr.) Quél.

Medium-sized agaric with brownish-grey cap and pinkish gills, stem white with darker streaks, comparatively short; solitary or in small trooping groups on soil under trees and bushes of the *Rosaceae* including hawthorn, cherry and rose.

Dimensions cap 3 - 10 cm dia; stem 3 - 5 cm tall x 0.8 - 1.5 cm dia.
Cap brownish-grey, with darker radiating fibrillose streaks and sometimes with a yellowish tinge showing through, hygrophanous, drying more pallid; convex becoming expanded with wavy margin, broadly umbonate, finally somewhat depressed. Flesh white, tinged cap colour, thin.
Gills at first pallid grey becoming flesh pink at maturity, adnate, crowded. **Spores** pink, smooth, sub-spherical, angular, non-amyloid, 8-11.5 x 7.5-9 µm. Basidia 4-spored. Cystidia not distinctive.
Stem white, with silky fibrillose streaks tinged cap colour, fairly stout, sometimes compressed, more or less equal. Ring absent. Flesh white, firm and hollow.
Odour strong, of meal. **Taste** of meal.
Chemical tests none.
Occurrence spring to early summer; frequent.
■ Inedible.

Entoloma conferendum (Britz.) Noord.   Entolomataceae
= *Nolanea staurospora* Bres.
= *Rhodophyllus staurosporus* (Bres.) J Lange

Small agaric with brownish cap, pinkish gills and pale stem; solitary or in small trooping groups on soil in pastures, parks and open woodland.

Dimensions cap 1 - 3 cm dia; stem 2 - 6 cm tall x 0.1 - 0.3 cm dia.
Cap reddish-brown, hygrophanous, drying more pallid; conical becoming bluntly expanded, striate when damp, otherwise smooth. Flesh concolorous and thin.
Gills at first whitish, becoming pink at maturity, adnexed or emarginate, close. **Spores** pink, smooth, cuboid or irregularly star-shaped, non-amyloid, 9-11 x 6-9 µm. Basidia 4-spored. Gill edge cystidia absent.
Stem pallid, concolorous with cap, slender, more or less equal, fibrillose. Ring absent. Flesh concolorous, fragile.
Odour of meal. **Taste** not distinctive.
Chemical tests none.
Occurrence spring to autumn; common.
■ Inedible.

Entoloma corvinum (Kühn.) Noord.   Entolomataceae
= *Leptonia corvina* (Kühn.) P Orton
= *Rhodophyllus corvinus* Kühn.

Small agaric with blackish fibrous dimpled cap, pink gills and greyish stem; in trooping groups on soil amongst grass in pastures.

Dimensions cap 1 - 3 cm dia; stem 3 - 6 cm tall x 0.2 - 0.4 cm dia.
Cap blackish; at first convex, becoming flattened and slightly umbilicate, finely radially fibrous. Flesh greyish and thin.
Gills pallid, becoming pink at maturity, adnate, fairly distant.
Spores hyaline (pink in the mass), smooth, broadly ellipsoid, angular, 8.5-12 x 6.5-7.5 µm. Basidia 4-spored. Gill edges with many sterile cells.
Stem greyish, more pallid with age, slender, more or less equal. Ring absent. Flesh greyish and fragile.
Odour not distinctive. **Taste** not distinctive.
Chemical tests none.
Occurrence summer to autumn; common.
■ Inedible.

Entoloma griseocyaneum (Fr.: Fr.) Kummer
= *Rhodophyllus griseocyaneus* (Fr.: Fr.) Quél.   **Entolomataceae**
= *Agaricus griseocyaneus* Fr.: Fr.

Small agaric with greyish scaly cap, pink gills and blue-grey stem; in trooping groups on soil amongst grass in pastures.

Dimensions cap 2 - 3 cm dia; stem 6 - 8 cm tall x 0.3 - 0.7 cm dia.
Cap pallid grey with lilaceous tinge; at first campanulate, becoming more convex, fibrous, scaly or flaky. Flesh greyish and thin.
Gills pallid, becoming pink at maturity, adnate, fairly distant.
Spores pink, smooth, broadly ellipsoid, angular, 10-12 x 7-8 µm. Basidia 4-spored. Cystidia not distinctive.
Stem pallid blue-grey, slender, more or less equal, silky, whitish fibrous at base. Ring absent. Flesh greyish and fibrous, hollow.
Odour not distinctive. **Taste** not distinctive.
Chemical tests none.
Occurrence summer to autumn; infrequent.
■ Inedible.

Entoloma icterinum (Fr.) Moser   **Entolomataceae**

Small agaric with silky olive-yellow cap and stem, pinkish-yellow gills, distinctive fruity smell; solitary or in small trooping groups on soil in grass under conifers, less frequently in open grassland.

Dimensions cap 1 - 3 cm dia; stem 2 - 6 cm tall x 0.2 - 0.3 cm dia.
Cap olivaceous-yellow with brown tinge, silky when dry; campanulate, becoming expanded umbonate, sometimes with a distinct papilla, margin slightly striate. Flesh pallid cap colour, thin and fragile.
Gills at first whitish, becoming pallid ochraceous-pink at maturity, adnate, close. **Spores** pink, smooth, angular, 5-6 sided, non-amyloid, 10-12 x 6-8 µm. Basidia 4-spored. Cystidia not distinctive.
Stem concolorous with cap, slender, more or less equal. Ring absent. Flesh concolorous with cap, hollow.
Odour of fruit (reminiscent of acid drops). **Taste** faint, of meal.
Chemical tests none.
Occurrence summer to autumn; infrequent.
■ Inedible.

Entoloma incanum (Fr.: Fr.) Hesler   **Entolomataceae**
= *Leptonia incana* (Fr.: Fr.) Gill.
= *Rhodophyllus incanus* (Fr.: Fr.) Kühn. & Romagn.

Small agaric with yellow-brown dimpled cap, whitish gills and distinctive green stem; solitary or in small trooping groups on calcareous soil in exposed grassland.

Dimensions cap 1 - 3 cm dia; stem 2 - 4 cm tall x 0.1 - 0.3 cm dia.
Cap olivaceous-brown with yellowish tinge, darker at the centre; convex, becoming umbilicate, margin striate. Flesh pallid greenish, thin and fragile.
Gills at first whitish-green, becoming pallid buff at maturity, adnate, somewhat distant. **Spores** pink, smooth, angular, non-amyloid, 11-14 x 8-9 µm. Basidia 4-spored. Cystidia not distinctive.
Stem bright grass-green, slender, more or less equal. Ring absent. Flesh concolorous with cap, hollow.
Odour strong, of mice, or of burnt corn. **Taste** not distinctive.
Chemical tests none.
Occurrence late summer to autumn; infrequent.
✣ Poisonous.

Entoloma infula (Fr.) Noord.   Entolomataceae
= *Rhodophyllus infulus* (Fr.) Quél.

Small agaric with brown papillate cap, pinkish gills and brown stem; solitary or in small trooping groups on calcareous soil in exposed poor grassland, often at montane altitudes.

Dimensions cap 1 - 3 cm dia; stem 2 - 8 cm tall x 0.1 - 0.3 cm dia.
Cap sepia-brown, sometimes with yellowish tinge, hygrophanous, grey-brown when dry; conical, becoming campanulate and typically papillate, margin striate. Flesh pallid brown, thin and fragile.
Gills at first whitish, becoming more pink at maturity, narrowly adnate or more ventricose, crowded. **Spores** pink, smooth, angular, non-amyloid, 7-9.5 x 6-7 μm. Basidia 4-spored. Cystidia not distinctive.
Stem brown, smooth, slender, more or less equal or narrowing upwards. Ring absent. Flesh concolorous with cap, cartilaginous.
Odour not distinctive. **Taste** not distinctive.
Chemical tests none.
Occurrence late summer to autumn; rare. (Mendip Hills, Somerset)
■ Inedible.

Entoloma lampropus (Fr.) Hesler   Entolomataceae

Small agaric with black, flattish, scaly cap, pinkish gills and blackish stem; solitary or in small trooping groups on soil in short grass.

Dimensions cap 1.5 - 2.5 cm dia; stem 3 - 4 cm tall x 0.3 - 0.4 cm dia.
Cap very dark greyish-brown with bluish tinge, almost wholly black at the centre; conical, becoming flattened and somewhat umbilicate, radially fibrous and scaly. Flesh brownish and thin.
Gills pallid then pinkish, tinged brown at maturity, adnexed or emarginate, close. **Spores** pink, smooth, ellipsoid, angular, non-amyloid, 7-10.5 x 5.5-6 μm. Basidia 4-spored. Cystidia not distinctive.
Stem blackish with blue or violaceous tinges, smooth, more or less equal. Ring absent. Flesh concolorous, fibrillose.
Odour not distinctive. **Taste** not distinctive.
Chemical tests none.
Occurrence late summer to autumn; locally frequent.
■ Inedible.

Entoloma lazulinum (Fr.) Noord.   Entolomataceae
= *Entoloma chalybaeum* (Fr. Fr.) Nord *var lazulinum* (Fr.) Noord.
= *Leptonia lazulina* (Fr.) Quél.
= *Rhodophyllus lazulinus* (Fr.) Quél.

Small agaric with blue-black flatly umbonate cap and stem, blue or pale brownish gills; solitary or in small trooping groups on soil in pastures, parks and open woodland.

Dimensions cap 1.5 - 2.5 cm dia; stem 3 - 4 cm tall x 0.1 - 0.2 cm dia.
Cap blue-black, almost wholly black at the centre, then dark brown; conical, becoming bluntly umbonate or umbilicate, striate at the margin when damp, otherwise smooth. Flesh dark blue and thin.
Gills at first blue, becoming pinkish-brown at maturity, adnexed or emarginate, somewhat distant. **Spores** pink, smooth, ellipsoid, angular, non-amyloid, 10-12 x 6.5-8 μm. Basidia 4-spored. Cystidia not distinctive.
Stem blue-black with violet tinge, smooth, slender, more or less equal. Ring absent. Flesh concolorous, fibrillose.
Odour not distinctive. **Taste** not distinctive.
Chemical tests none.
Occurrence late summer to autumn; locally frequent.
■ Inedible.

Entoloma lucidum (P Orton) Moser   **Entolomataceae**
= *Nolanea lucida* P Orton

Small agaric with brownish shining umbonate cap, pinkish gills and grey stem; solitary or in small trooping groups on soil in pastures, parks and open woodland.

Dimensions cap 1 - 4 cm dia; stem 2 - 6 cm tall x 0.2 - 0.5 cm dia.
Cap sepia-brown, hygrophanous, drying more pallid; conical, becoming bluntly expanded-umbonate or papillate, striate at the margin when damp, otherwise smooth with distinctive sheen. Flesh concolorous and thin.
Gills at first whitish, becoming pink at maturity, adnexed or emarginate, somewhat distant. **Spores** pink, smooth, sub-spherical or broadly ellipsoid, angular, non-amyloid, 7.5-10 x 5.5-7.5 µm. Basidia 4-spored. Cystidia not distinctive.
Stem pallid brownish-grey with cap, pruinose at the apex, slender, more or less equal, fibrillose. Ring absent. Flesh concolorous.
Odour of meal. **Taste** not distinctive.
Chemical tests none.
Occurrence summer to autumn; infrequent.
■ Inedible.

Entoloma nidorosum (Fr.) Quél.   **Entolomataceae**
= *Rhodophyllus nidorosus* (Fr.) Quél.

Small to medium agaric with creamy-grey cap, pinkish gills and white stem; solitary or in small trooping groups on soil in broad-leaf woods.

Dimensions cap 3 - 7 cm dia; stem 5 - 10 cm tall x 0.3 - 1.5 cm dia.
Cap creamy-grey; convex, becoming expanded with wavy margin, slightly umbonate, finally somewhat depressed. Flesh white and watery.
Gills at first whitish, becoming flesh pink at maturity, adnate, crowded. **Spores** pink, smooth, sub-spherical, angular, non-amyloid, 8-11 x 7-8 µm. Basidia 4-spored. Cystidia not distinctive.
Stem white or tinged cap colour, slender, more or less equal, fibrillose. Ring absent. Flesh white, watery.
Odour strong, of ammonia or nitric acid when fresh, then not distinctive. **Taste** not distinctive.
Chemical tests none.
Occurrence summer to autumn; infrequent.
■ Inedible.

Entoloma ortonii Arnould & Noord.   **Entolomataceae**
= *Entoloma farinolens* (P Orton) Moser
= *Nolanea farinolens* P Orton

Small agaric with dark brown flattish cap, pinkish gills and pale stem; solitary or in small trooping groups on soil in shaded grassy places in damp woodland.

Dimensions cap 1.5 - 3 cm dia; stem 2 - 6 cm tall x 0.15 - 0.3 cm dia.
Cap dark sepia-brown, blackish-brown at the centre; convex, becoming flattened or slightly depressed, finely striate at the margin when damp otherwise smooth, cuticle tough. Flesh concolorous and thin.
Gills at first whitish, becoming brownish-pink at maturity, adnexed or emarginate, somewhat distant. **Spores** pink, smooth, sub-spherical or broadly ellipsoid, angular, non-amyloid, 6.5-10 x 5.5-7.5 µm. Basidia 4-spored. Cystidia not distinctive.
Stem pallid, concolorous with cap, sometimes pruinose at the apex, slender, more or less equal, fibrillose. Ring absent. Flesh concolorous, cartilaginous, stuffed.
Odour strong, of meal. **Taste** strong, of meal.
Chemical tests none.
Occurrence summer to autumn; infrequent.
■ Inedible.

Entoloma papillatum (Bres.) Dennis   Entomolataceae
= *Nolanea mammosa subsp. papillata* (Bres.) Konrad & Maubl.

Small agaric with brownish-grey cap and stem, pinkish-grey gills;
solitary or in small trooping groups on soil in lawns and other short
grass.

Dimensions cap 1.5 - 4 cm dia; stem 4 - 8 cm tall x 0.1 - 0.3 cm dia.
Cap brown, hygrophanous, drying more greyish; at first conical,
becoming campanulate and then expanded-convex with a sharp low
umbo, striate when damp, otherwise smooth and shining-silky. Flesh
concolorous and thin.
Gills at first pallid, becoming greyish-pink at maturity, adnexed or
emarginate, close. **Spores** pink, smooth, sub-spherical, angular, non-
amyloid, 9-12 x 7-8 µm. Basidia 4-spored. Cystidia not distinctive.
Stem concolorous with cap, slender, more or less equal, fibrillose.
Ring absent. Flesh concolorous.
Odour of fish oil or cucumber. **Taste** not distinctive.
Chemical tests none.
Occurrence spring to autumn; common.
■ Inedible.

Entoloma porphyrophaeum (Fr.) Karst.   Entomolataceae
= *Phodophyllus porphyrophaeus* (Fr.) J Lange

Small to medium agaric with pale grey-brown cap, pinkish gills and
grey-brown fibrous stem; solitary or in small trooping groups in lawns
and on other grassy places.

Dimensions cap 3 - 9 cm dia; stem 4 - 8 cm tall x 0.5 - 1 cm dia.
Cap grey-brown, sometimes with purplish-brown tinge at the
margin, hygrophanous, drying more pallid; convex-campanulate,
becoming expanded but sharply umbonate, silky-fibrous, margin
sometimes reflexed and split in older specimens. Flesh whitish and
thin.
Gills at first whitish, becoming greyish-flesh pink at maturity, adnate,
somewhat distant. **Spores** pink, smooth, broadly ellipsoid, angular,
non-amyloid, 10-12 x 6.5-8.5 µm. Basidia 4-spored. Gill edge cystidia
flask-shaped, clavate or fusiform.
Stem concolorous, slender, more or less equal, fibrillose, sometimes
laterally compressed, white-downy at the base. Ring absent. Flesh
whitish and fibrous.
Odour not distinctive. **Taste** not distinctive.
Chemical tests none.
Occurrence spring to autumn; infrequent.
■ Inedible.

Entoloma saepium (Noullet & Dassier) Richon & Roze
= *Rhodophyllus saepius* (Noullet & Dassier) Romagn.
                                        Entomolataceae

Medium-sized agaric with pale greyish-straw cap, pinkish gills and
whitish stem; solitary or in small trooping groups on soil under trees
and shrubs of the *Rosaceae* family.

Dimensions cap 3 - 10 cm dia; stem 4 - 10 cm tall x 0.5 - 1.5 cm dia.
Cap pallid grey with straw tinge, not hygrophanous and slightly
greasy when young; conical-convex with wavy margin. Flesh white
and firm.
Gills at first whitish, becoming pallid pink at maturity, adnate,
crowded. **Spores** pink, smooth, sub-spherical, irregularly angular, 6-7 sided, non-
amyloid, 8.5-10 x 7.5-9 µm. Basidia 4-spored. Cystidia not distinctive.
Stem white or tinged cap colour, sometimes with reddish tone to the
fibres, more or less equal or slightly clavate at the base, fibrillose.
Ring absent. Flesh white, pithy or narrowly hollow.
Odour faint, of meal. **Taste** faint, of meal.
Chemical tests none.
Occurrence spring to early summer; infrequent.
■ Inedible.
Distinguished from *E. saundersii* mainly by odour and slightly smaller
spore size.

Entoloma saundersii (Fr.) Sacc. **Entomolataceae**
= *Rhodophyllus saundersii* (Fr.) Romagn.

Medium-sized agaric with creamy-buff cap, pinkish gills and white stem; solitary or in small trooping groups on soil under trees and shrubs of the *Rosaceae* family.

Dimensions cap 3 - 12 cm dia; stem 2 - 5 cm tall x 1 - 2.4 cm dia.
Cap pallid creamy-buff, with yellowish or greyish tinges, not hygrophanous and slightly greasy when young; campanulate, becoming expanded with a distinct blunt umbo and wavy margin. Flesh white and firm.
Gills at first whitish, becoming pallid pink at maturity, adnate, crowded. **Spores** pink, smooth, sub-spherical, regularly angular, 7-8 sided, non-amyloid, 8.5-12 x 8-10.5 µm. Basidia 4-spored. Cystidia not distinctive.
Stem white or tinged cap colour, stout, more or less equal, fibrillose. Ring absent. Flesh white, pithy or narrowly hollow.
Odour unpleasant, rancid. **Taste** faint, of meal.
Chemical tests none.
Occurrence spring to summer; infrequent.
■ Inedible.
Distinguished from *E. saepium* mainly by the smell of older specimens and slightly larger spore size.

Entoloma sericeum (Bull.) Quél. **Entolomataceae**
= *Rhodophyllus sericeus* (Bull.) Quél.

Small agaric with dark brownish-grey cap and stem, pale gills; solitary or in small trooping groups on soil in grass and open moorland, less often in open woodland.

Dimensions cap 2 - 4 cm dia; stem 2 - 7 cm tall x 0.15 - 0.5 cm dia.
Cap greyish-brown, darker at the centre; campanulate, becoming expanded with small umbo, silky-fibrous and often distinctly striate. Flesh pallid concolorous and medium.
Gills at first pallid grey, becoming pink tinged at maturity, adnexed or emarginate with decurrent tooth, close. **Spores** pink, smooth, sub-spherical, 5-8 angular, non-amyloid, 7.5-10 x 6.5-8 µm. Basidia 4-spored. Cystidia not distinctive.
Stem concolorous with cap, more pallid at the apex and base, silky fibrillose, more or less equal, base slightly swollen. Ring absent. Flesh pallid concolorous and stuffed.
Odour strong, of meal. **Taste** of meal.
Chemical tests none.
Occurrence summer to autumn; more common in northern regions.
■ Inedible.

Entoloma serrulatum (Fr.: Fr.) Hesler **Entolomataceae**
= *Leptonia serrulata* (Fr.: Fr.) Kummer
= *Rhodophyllus serrulatus* (Fr.: Fr.) Quél.

Small agaric with blackish cap and stem, pale gills edged with blue; solitary or in small trooping groups on soil (acidic or calcareous) in pastures and open woodland.

Dimensions cap 1.5 - 3.5 cm dia; stem 4 - 7 cm tall x 0.15 - 0.3 cm dia.
Cap blue-black, with brown tinge when old; convex, becoming umbilicate, radially fibrillose, striate when damp, particularly with age. Flesh pallid concolorous and thin.
Gills at first blue-grey, becoming pink tinged at maturity, edges blue-black and minutely flaky, adnexed or emarginate with decurrent tooth, somewhat distant. **Spores** pink, smooth, broadly ellipsoid, angular, non-amyloid, 7-11 x 5-7.5 µm. Basidia 4-spored. Cystidia not distinctive.
Stem concolorous with cap, blackish punctate at the apex, slender, more or less equal, fibrillose, white-downy at the base. Ring absent. Flesh pallid concolorous and stuffed.
Odour not distinctive. **Taste** not distinctive.
Chemical tests none.
Occurrence summer to autumn; more common in northern regions.
■ Inedible.

Entoloma strigosissimum (Rea) Noord.   **Entolomataceae**
= *Nolanea strigosissima* Rea
= *Rhodophyllus strigosissimus* (Rea) Horak.

Very small agaric with dark brown intensely hairy cap and stem, pinkish-brown gills; solitary or scattered, on rotting wood.

Dimensions cap 0.5 - 1 cm dia; stem 2 - 4 cm tall x 0.05 - 0.1 cm dia.
Cap dark brown, decorated thickly with greyish hairs; conical, becoming bluntly expanded-umbonate. Flesh concolorous and thin. Hairs 400-600 µm, septate.
Gills at first whitish, becoming pinkish-brown at maturity, adnexed or emarginate, close. **Spores** pink, smooth, ellipsoid, angular, non-amyloid, 14-19 x 8-9 µm. Basidia 4-spored. Gill edge cystidia large cylindrical or fusiform.
Stem concolorous with cap and similarly hairy, slender, more or less equal, somewhat elastic. Ring absent. Flesh concolorous.
Odour not distinctive. **Taste** not distinctive.
Chemical tests none.
Occurrence summer to autumn; infrequent.
■ Inedible.

Entoloma subradiatum (Kühn. & Romagn.) Moser
= *Rhodophyllus subradiatus* Kühn. & Romagn.   **Entolomataceae**

Small agaric with brownish cap, pinkish gills and pale stem; solitary or in small trooping groups on soil in pastures, parks and open woodland.

Dimensions cap 1.5 - 3 cm dia; stem 2 - 7 cm tall x 0.3 - 0.8 cm dia.
Cap grey-brown; conical, becoming bluntly expanded, striate when damp otherwise smooth. Flesh concolorous and thin.
Gills at first whitish, then with brown tinge and becoming pink at maturity, adnexed or emarginate, close. **Spores** pink, smooth, sub-spherical, angular, non-amyloid, 9-11.5 x 6-8 µm. Basidia 4-spored. Cystidia not distinctive.
Stem whitish, slender, more or less equal, fibrillose. Ring absent. Flesh concolorous.
Odour of meal when fresh. **Taste** not distinctive.
Chemical tests none.
Occurrence summer to autumn; rare. (Forest of Dean, Gloucestershire)
■ Inedible.

Entoloma vernum Lund.   **Entolomataceae**
= *Nolanea verna* (Lund.) Kotl. & Pouz.
= *Rhodophyllus cucullatus* Favre

Small agaric with grey-brown cap and stem, brown gills; solitary or in small trooping groups in grasslands, often in the vicinity of conifers.

Dimensions cap 2 - 4 cm dia; stem 2 - 7 cm tall x 0.2 - 0.6 cm dia.
Cap greyish-tan or darker; conical with a sharpish umbo, silky smooth. Flesh brown, thin and fragile.
Gills tan-brown, emarginate or free, broad, close or more or less distant. **Spores** pink, smooth, sub-spherical, regularly angular, non-amyloid, 8-12 x 7-9.5 µm. Basidia 4-spored. Cystidia not distinctive.
Stem pallid tan, slender, more or less equal, apex pruinose, otherwise silky fibrillose. Ring absent. Flesh concolorous, pithy or narrowly hollow.
Odour not distinctive. **Taste** not distinctive.
Chemical tests none.
Occurrence mainly in spring but also summer to autumn; infrequent.
■ Inedible.

Chamaemyces fracidus (Fr.) Donk   Agaricaceae
= *Drosella fracida* (Fr.) Sing.
= *Lepiota irrorata* Quél.

Medium-sized, fleshy, yellow agaric with white gills and stem with ring; solitary or scattered on soil in mixed woods and pastures.

Dimensions cap 2.5 - 10 cm dia; stem 3 - 4 cm tall x 0.7 - 1 cm dia.
Cap yellowish or straw; convex, smooth but when fresh exuding watery droplets which dry into brown discolorations. Flesh white and firm.
Gills white, becoming cream with age, free, crowded.
Spores hyaline, smooth, ellipsoid, non-amyloid, 4.5-5 x 4 µm. Basidia 4-spored. Gill edge and gill face cystidia abundant; clavate or fusiform.
Stem white, smooth above ring; covered with small brown scales below, also exuding droplets. Ring concolorous with stem, membraneous, superior, ephemeral. Flesh white, firm and full.
Odour unpleasant, perhaps of radish. **Taste** not distinctive.
Chemical tests none.
Occurrence summer to autumn; infrequent or rare. (Noxon Park, Forest of Dean)
■ Inedible.

Cystoderma amianthinum ([Scop.] Fr.) Fayod
= *Lepiota amianthina* ([Scop.] Fr.) Karst.   Agaricaceae

Small to medium, fleshy agaric with ochre-yellow cap, white or cream gills, and coarsely granular stem with ring; solitary, scattered, or in small tufted groups on soil amongst short grass in coniferous woods and on heaths.

Dimensions cap 2 - 5 cm dia; stem 3 - 5 cm tall x 0.4 - 0.8 cm dia.
Cap ochraceous or yellowish-tan; at first campanulate, becoming expanded-convex or flattened, radially wavy with age, granular. Flesh pallid yellow and thin.
Gills white, becoming creamy-yellow, adnate, crowded.
Spores hyaline, smooth, ellipsoid, amyloid, 4-6 x 3-4 µm. Basidia 4-spored. Cystidia not distinctive.
Stem concolorous with cap, more or less equal, smooth above ring, coarsely granular below. Ring concolorous with stem or slightly darker tan, upwardly directed, median or superior, persistent. Flesh dirty-yellow, firm and stuffed.
Odour unpleasant, of mould. **Taste** not distinctive.
Chemical tests brown coloration on cap with KOH solution.
Occurrence summer to autumn; common.
□ Edible.

Cystoderma carcharias (Pers.) Fayod   Agaricaceae

Small to medium, fleshy agaric with whitish tan cap, white gills, and coarsely granular stem with ring; solitary, scattered, or in small tufted groups on soil amongst short grass in coniferous woods and on heaths.

Dimensions cap 2 - 6 cm dia; stem 4 - 7 cm tall x 0.4 - 0.8 cm dia.
Cap off-white, tinged tan towards the centre; typically slightly umbonate, wavy, granular, sometimes with delicately fringed margin. Flesh white or pallid yellow and thin.
Gills white, adnate, crowded. **Spores** hyaline, smooth, sub-spherical, amyloid, 4-5.5 x 3-4 µm. Basidia 4-spored. Cystidia not distinctive.
Stem off white, more or less equal, smooth above ring, coarsely granular below. Ring white, upwardly directed, median or superior, persistent. Flesh white, firm and stuffed.
Odour unpleasant, of mould. **Taste** not distinctive.
Chemical tests brown coloration on cap with KOH solution.
Occurrence late summer to autumn; rare. (Priddy Forest near Wells, Somerset)
■ Inedible.

Cystolepiota aspera (Pers.: Fr.) Knudsen   **Agaricaceae**
= *Lepiota fresii* (Lasch) Quél.
= *Lepiota acutesquamosa* (Weinm.: Fr.) Gill.

Medium-sized fleshy agaric with dark brown scaly cap, white gills, and stem with ring and bulbous base; solitary or scattered on soil in broad-leaf woods.

Dimensions cap 5 - 10 cm dia; stem 3 - 5 cm tall x 0.5 - 1 cm dia.
Cap dark brown at the centre, breaking up into brownish-black pointed scales revealing pallid flesh beneath; at first sub-spherical, becoming flattened-convex. Flesh white, occasionally turning yellowish with age, thin.
Gills white, free (remote), crowded, forked. **Spores** hyaline, smooth, ellipsoid or narrowly fusiform, dextrinoid, 6-8 x 3-4 μm. Basidia 4-spored. Gill edge cystidia stalked, sub-sperical, thin-walled. Gill face cystidia absent.
Stem pallid, smooth, or with scattered dark brown scales below, more or less equal, with bulbous base. Ring whitish, broad, spreading and often adhering to the cap margin, slightly woolly, superior, persistent. Flesh white, firm and hollow.
Odour strong, unpleasant. **Taste** not distinctive.
Chemical tests none.
Occurrence late summer to autumn; infrequent.

Lepiota brunneoincarnata Chod. & Martin   **Agaricaceae**

Small umbonate agaric with cap bearing brownish scales on a white background and somewhat scaly stem, no ring, smelling fruity; solitary or scattered on soil in grass close to woodland edges.

Dimensions cap 2 - 4 cm dia; stem 2 - 4 cm tall x 0.3 - 0.5 cm dia.
Cap flesh-brown with purplish tinge at the centre, breaking up (apart from centre) into flattened scales and revealing whitish flesh beneath; conico-convex and broadly umbonate. Flesh white or with vinaceous tinge, woolly, not reddening where bruised, thin.
Gills whitish, free, crowded. **Spores** hyaline, smooth, ellipsoid or narrowly ovoid, dextrinoid, 7-9 x 4.5-5 μm. Basidia 4-spored.
Stem white but with zone of scales concolorous with cap below a faint ring-like zone, more or less equal. Ring absent. Flesh white or with vinaceous tinge, woolly-stuffed, not reddening where bruised.
Odour faint, of fruit. **Taste DO NOT ATTEMPT TO TASTE ANY PART OF THE SPECIMEN.**
Chemical tests none
Occurrence late summer and autumn; infrequent.
✚ Lethally poisonous.

Cystolepiota bucknallii (Berk. & Br.) Sing. & Clc.
= Lepiota bucknallii (Berk. & Br.) Sacc.
= Cystoderma bucknallii (Berk. & Br.) Sing.   **Agaricaceae**

Small agaric with violet tinge on the pale cap and stem, yellowish gills, no ring; in scattered groups on damp nitrogen rich soil with broad-leaf trees.

Dimensions cap 2 - 5 cm dia; stem 2 - 6 cm tall x 0.3 - 0.5 cm dia.
Cap whitish, with violaceous tinge; conico-convex with umbo, mealy. Flesh concolorous, thin and fragile. **Cap** skin broken up into large rounded cells.
Gills creamy-yellow, free, crowded. **Spores** hyaline, smooth, ellipsoid or sub-fusiform, weakly dextrinoid, 7.5-10 x 1-3.5 μm. Basidia 4-spored. Cystidia not distinctive.
Stem pallid, becoming violaceous particularly when bruised, slender, farinose, more or less equal. Ring absent. Flesh concolorous, fragile and hollow.
Odour strong, of gas tar or sulphur. **Taste** not distinctive.
Chemical tests none.
Occurrence late summer to autumn
■ ⚜ Inedible, possible poisonous

Lepiota clypeolaria (Bull.: Fr.) Kummer   Agaricaceae

Small to medium, umbonate agaric with cap bearing brownish wool-ly scales on a white background and stem with ring; solitary or scat-tered on soil in coniferous and broad-leaf woods.

Dimensions cap 4 - 8 cm dia; stem 5 - 10 cm tall x 0.5 - 0.8 cm dia.
Cap reddish or ochraceous-brown breaking up into flattened woolly scales and revealing whitish flesh beneath; conico-convex and broadly umbonate. Flesh white and woolly.
Gills white or pallid yellow, free, crowded. **Spores** hyaline, smooth, fusiform, dextrinoid, 13-15 x 5-6 µm. Basidia 4-spored. Cystidia not distinctive.
Stem white, very woolly, tapering slightly upwards. Ring whitish, narrow, woolly, superior, ephemeral. Flesh white, woolly-stuffed.
Odour not distinctive. **Taste** not distinctive.
Chemical tests none.
Occurrence late summer to autumn; infrequent.
■ Inedible.

Lepiota cristata (Fr.) Kummer   Agaricaceae

Small, delicate agaric with whitish scaly cap, tan at the centre, white gills and stem with ring; solitary or scattered on soil in woods and gardens.

Dimensions cap 2 - 5 cm dia; stem 2 - 3.5 cm tall x 0.3 - 0.4 cm dia.
Cap reddish-brown, breaking up into scales other than at the centre, revealing whitish flesh beneath; irregularly campanulate or umbonate. Flesh white and thin.
Gills white becoming darker with age, free, crowded.
Spores hyaline, smooth, bullet-shaped, dextrinoid, 6-7.5 x 3-3.5 µm. Basidia 4-spored. Cystidia not distinctive.
Stem white, tinged buff towards the base, more or less smooth, equal. Ring white, membraneous, superior, ephemeral. Flesh white and stuffed.
Odour unpleasant. **Taste** mild.
Chemical tests none.
Occurrence summer to autumn; common
✣ Poisonous.

Macrolepiota excoriata (Schaeff.: Fr.) Wasser   Agaricaceae
= *Leucocoprinus excoriatus* (Schaeff.: Fr.) Sing.
= *Lepiota excoriata* (Schaeff.: Fr.) Kummer

Large fleshy agaric with white cap, gills and stem with ring; in scattered groups on soil in pastures.

Dimensions cap 6 - 10 cm dia; stem 4 - 6 cm tall x 0.8 - 1 cm dia.
Cap white with fine adpressed pallid buff scales; at first bun-shaped, becoming expanded-convex and slightly umbonate, margin charac-teristically frayed. Flesh white, thick and soft.
Gills white, free (remote), crowded. **Spores** hyaline, smooth, ellipsoid, large germ pore, dextrinoid, 12-15 x 8-9 µm. Basidia 4-spored. Cystidia not distinctive.
Stem white, more or less equal but with slightly swollen base. Ring white, woolly, narrow, 2-tiered superior, persistent but movable. Flesh white, soft and hollow.
Odour not distinctive. **Taste** not distinctive.
Chemical tests none.
Occurrence summer to autumn; rare. (Somerset Levels near Glastonbury)
☐ Edible.

Lepiota felina (Pers.: Fr.) Karst.   Agaricaceae

Small umbonate agaric with blackish scaly cap on a pale background and whitish stem with ring; solitary or scattered on soil in coniferous woods.

Dimensions cap 2 -3 cm dia; stem 3 - 5 cm tall x 0.2 - 0.3 cm dia.
Cap blackish, breaking up into minute pointed scales and revealing whitish flesh beneath; conical or convex and more or less umbonate. Flesh white, becoming slightly brown with age, thin and fragile.
Gills white or pallid yellow, free, crowded. **Spores** hyaline, smooth, ellipsoid, dextrinoid, 6.5-7.5 x 3.5-4 µm. Basidia 4-spored. Cystidia not distinctive.
Stem whitish, silky fibrillose, decorated with a few scales below ring. Ring whitish, narrow, superior, ephemeral. Flesh white, becoming slightly brown with age, fragile and stuffed.
Odour unpleasant, of mould. **Taste** not distinctive.
Chemical tests none.
Occurrence late summer to autumn; infrequent.
■ Inedible.

Lepiota ignivolvata Bousset & Joss.   Agaricaceae

Medium to large fleshy agaric with yellowish scaly cap, white gills, ring on the stem and slightly bulbous orange-tinged base; solitary or scattered on soil in coniferous and broad-leaf woods.

Dimensions cap 4 - 10 cm dia; stem 6 - 12 cm tall x 0.6 - 1.5 cm dia
Cap reddish-brown, breaking up, other than at the centre, into tiny ochraceous scales revealing cream flesh beneath, more thinly dispersed towards the margin; at first conico-convex becoming expanded. Flesh white, fleshy and soft.
Gills white or cream, free, crowded. **Spores** hyaline, smooth, fusiform, dextrinoid, 11-13 x 6 µm. Basidia 4-spored. Cystidia not distinctive.
Stem whitish, more or less smooth, with slightly bulbous base, bright orange at the upper margin. Ring white, with traces of bright orange coloration on the underside, fragile, median. Flesh whitish, soft. hollow or stuffed.
Odour strong, rancid. **Taste** foul.
Chemical tests none.
Occurrence late summer to autumn; rare. (Halden Forest near Exeter, Devon)
■ Inedible.

Leucoagaricus leucothites (Vitt.) Wasser   Agaricaceae
= *Lepiota leucothites* (Vitt.) P Orton

Medium-sized fleshy white agaric with silky cap, white gills, ring on the stem and slightly bulbous base; solitary or scattered amongst grass, often on roadside verges or in gardens.

Dimensions cap 5 - 8 cm dia; stem 5 - 8 cm tall x 1 - 2 cm dia.
Cap white, later with faint buff tinge, smooth, silky, at first convex, becoming expanded and flattened. Flesh white, thick and fairly soft.
Gills white, tinged pallid buff with age, free, crowded.
Spores hyaline, smooth, fusiform, small germ pore, dextrinoid, 7-9 x 4.5-5 µm. Basidia 4-spored. Cystidia not distinctive.
Stem whitish, smooth, with slightly bulbous base. Ring white, narrow, superior. Flesh whitish, soft. hollow or stuffed.
Odour not distinctive. **Taste** not distinctive.
Chemical tests none.
Occurrence late summer to autumn; rare. (Near Petworth, Sussex)
■ Inedible.

Macrolepiota mastoidea (Fr.) Sing.   Agaricaceae
= *Lepiota umbonata* (Schum.) Schroet.
= *Lepiota mastoidea* (Fr.) Kummer

Large fleshy agaric with creamy-yellow scaly cap, white gills, tall stem with ring and slightly swollen base; solitary or scattered on soil in open mixed woodland and grassy places adjacent to woodland.

Dimensions cap 8 - 12 cm dia; stem 8 - 10 cm tall x 0.8 - 1.5 cm dia.
Cap ochraceous-cream, breaking up into minute granular scales; at first sub-spherical, becoming expanded-convex with distinct umbo. Flesh white and soft.
Gills white or pallid cream, free (remote), crowded. **Spores** hyaline, smooth, ellipsoid, large germ pore, dextrinoid, 12-15 x 8-9 µm. Basidia 4-spored. Cystidia not distinctive.
Stem white with small densely crowded pallid scales, more or less equal, base slightly swollen. Ring white, thick, movable, superior, persistent. Flesh white, firm and hollow.
Odour not distinctive. **Taste** not distinctive.
Chemical tests none.
Occurrence late summer to autumn; infrequent.
☐ Edible and good.

Lepiota ochraceofulva P Orton   Agaricaeae

Smallish agaric with creamy-brown scaly cap, white gills and stoutish stem with ring and slightly swollen base; solitary or scattered on soil in open mixed woodland and grassy places adjacent to woodland.

Dimensions cap 2 - 6 cm dia; stem 4 - 9 cm tall x 0.5 - 1 cm dia.
Cap brown or rust, breaking up into scales revealing cream flesh beneath; at first sub-spherical, becoming expanded-convex with slight umbo. Flesh white and soft.
Gills white or pallid cream, free (remote), crowded. **Spores** hyaline, smooth, ellipsoid, dextrinoid, 6-9 x 3.5-4.5 µm. Basidia 4-spored. Cystidia not distinctive.
Stem ochraceous-brown, smooth, more or less equal, base slightly swollen. Ring membraneous, zone-like, superior. Flesh white, firm and hollow.
Odour not distinctive. **Taste** not distinctive.
Chemical tests none.
Occurrence late summer to autumn; infrequent.
■ Inedible.

Macrolepiota procera (Scop.: Fr.) Sing.   Agaricaceae
= *Leucocoprinus procerus* (Scop.: Fr.) Pat.   Field Parasol
= *Lepiota procera* (Scop.: Fr.) S F Gray

Large, distinctive, pale brownish agaric with scaly cap, white gills, and pale brownish stem with ring; solitary, scattered or clustered on soil in open grassy places and in mixed woods.

Dimensions cap 10 - 25 cm dia; stem 15 - 30 cm tall x 1 - 1.5 cm dia.
Cap pallid brown decorated with broad darker brown scales; at first spherical or ovoid, becoming expanded-convex and finally flattened with a slight broad umbo. Flesh white, soft and thin.
Gills white, free (remote), crowded. **Spores** hyaline, smooth, ellipsoid, large germ pore, dextrinoid, 15-20 x 10-13 µm. Basidia 4-spored. Cystidia not distinctive.
Stem distinctive grey-brown with banded markings on a whitish background, woody, bulbous at the base and tapering slightly upwards. Ring white above and brown below, large, double, superior, movable. Flesh tough, hollow or loosely stuffed.
Odour not distinctive. **Taste** pleasant, sweet.
Chemical tests none.
Occurrence late summer to autumn; infrequent, locally common.
☐ Edible and excellent.

Macrolepiota rhacodes (Vitt.) Sing.   Agaricaceae
= *Lepiota rhacodes* (Vitt.) Quél.   Shaggy Parasol

Large, fleshy agaric with shaggy, pale grey-brown cap and whitish stem with ring and obliquely bulbous base; solitary or scattered on soil in woods generally, often with conifers.

Dimensions cap 5 - 15 cm dia; stem 10 - 15 cm tall x 1 - 1.5 cm dia.
Cap pallid buff or grey-brown, decorated with darkish brown, broad, slightly reflexed, fibrous shaggy scales; at first bun-shaped, becoming broadly umbonate or flattened. Flesh soft, thin, flushing carmine red where cut.
Gills white, bruising reddish, free (remote), crowded.
Spores hyaline, smooth, ellipsoid, large germ pore, dextrinoid, 8-12 x 6-7 µm. Basidia 4-spored. Cystidia not distinctive.
Stem whitish, tinged pinkish-brown, tapering slightly upwards, base slightly bulbous, eccentric. Ring concolorous with stem, spreading, double, felty, superior, movable . Flesh white, flushing carmine red where cut, firm and hollow.
Odour strong, aromatic. **Taste** not distinctive.
Chemical tests none.
Occurrence early summer to autumn; frequent.
☐ Edible. Note: causes an allergic reaction when eaten by some individuals.

Lepiota rhacodes var bohemica (Wichansky) Pil.
= *Lepiota bohemica* Wichansky
= *Macrolepiota rhacodes var hortensis* Pil.   **Agaricaceae**

Large, fleshy agaric, cap shaggy brown on a white background, stem with ring, large bulbous base, no volva; solitary or in small groups on soil in gardens and on compost heaps.

Dimensions cap 10 - 15 cm dia; stem 6 - 7 cm tall x 1.8 - 2.4 cm dia.
Cap whitish, decorated with reddish-brown, large, angular scales; at first bun-shaped, becoming expanded convex. Flesh white, becoming reddish when cut (less strongly than in the type species), soft and thin.
Gills white or cream, becoming dirty buff with age, free (remote), crowded. **Spores** hyaline, smooth, ellipsoid, large germ pore, dextrinoid, 10-13 x 7.5-9.5 µm. Basidia 4-spored. Cystidia not distinctive.
Stem whitish, discolouring brownish below, more or less equal but with large bulbous (sub-spherical) base. Ring whitish, thick, double, superior, movable. Flesh white, becoming reddish where cut, firm and hollow.
Odour pleasant. **Taste** pleasant.
Chemical tests none.
Occurrence summer to autumn; infrequent.
☐ Edible. Note: causes an allergic reaction when eaten by some individuals.

Cystolepiota sistrata (Fr.) Sing.: Bon. & Bellu   Agaricaceae
= *Lepiota seminuda* (Lasch) Kummer
= *Lepiota sistrata* (Fr.) Quél.
= *Cystoderma seminudum* (Lasch) Sing.

Small delicate agaric with white cap, white gills and slender stem; scattered or in tufted groups on soil in short grass amongst, or close to, coniferous and broad-leaf trees.

Dimensions cap 0.5 - 1.5 cm dia; stem 1.5 - 2.5 cm tall x 0.1 - 0.2 cm dia.
Cap white with faint buff tinge; conico-convex with distinct umbo, distinctively pruinose. Flesh white, fragile.
Gills white, free, crowded. **Spores** hyaline, smooth, ellipsoid, non-dextrinoid, without germ pore, 3-4 x 2-2.5 µm. Basidia 4-spored. Cystidia not distinctive.
Stem white, slender and fragile, more or less equal. Ring white, pruinose, superior, very ephemeral and sometimes absent. Flesh white or purple-pinkish, fragile and stuffed.
Odour not distinctive. **Taste** not distinctive.
Chemical tests none.
Occurrence late summer autumn; infrequent.
■ Inedible.

Leucocoprinus badhamii Berk. & Br.   **Agaricaceae**

Medium-sized agaric with pale grey or blackened fibrous cap, white gills and stem with fragile ring, parts reddening where damaged; scattered on soil with broad-leaf trees.

Dimensions cap 2 - 8 cm dia; stem 4 - 8 cm tall x 0.6 - 1.5 cm dia.
Cap pallid greyish-brown, decorated with fine knots of blackish-brown fibres, surface blackening in patches; at first conical, becoming expanded, margin sulcate. Flesh white, reddening, delicate and thin.
Gills white, free, crowded. **Spores** hyaline, smooth, broadly ellipsoid or almond-shaped, small germ pore, 5-7 x 3.5-4.5 um. Basidia 4-spored. Gill edge cystidia with red contents and with appendage.
Stem white, then dark brown below the ring, finely fibrous-floccose, more or less equal. Ring white, fragile, ascending, superior. Flesh white but reacting as in cap, fragile, stuffed or hollow.
Odour not distinctive. **Taste** not distinctive.
Chemical tests gills green with NH_3.
Occurrence late summer to autumn; infrequent.
■ Inedible.

Leucocoprinus brebissonii (Godey) Locq.   **Agaricaceae**

Small whitish cap with dark scales, white stem with fragile ring; scattered on soil in broad-leaf woods and in greenhouses.

Dimensions cap 2 - 3 cm dia; stem 4 - 6 cm tall x 0.2 - 0.4 cm dia.
Cap white decorated with dark brown small scales, uniform at the centre; at first conical, becoming expanded, margin sulcate. Flesh white, delicate and thin.
Gills white, free, crowded. **Spores** hyaline, smooth, broadly ellipsoid or almond-shaped, small germ pore, 9-12.5 x 5.5-7 um. Basidia 4-spored. Cystidia not distinctive.
Stem white, smooth, more or less equal, base slightly swollen. Ring white, fragile, ephemeral, superior. Flesh white, fragile, stuffed or hollow.
Odour not distinctive. **Taste** not distinctive.
Chemical tests none.
Occurrence late summer to autumn; infrequent.
■ Inedible.

Leucocoprinus cepestipes (Sow: Fr.) Pat. **Agaricaceae**
= *Lepiota rorulenta* (Paniz.) Barl.

Whitish cap with small, pale brown scales, white stem with fragile ring; scattered or clustered on soil of flowerpots in greenhouses.

Dimensions cap 2 - 4 cm dia; stem 5 - 6 cm tall x 0.3 - 0.4 cm dia.
Cap whitish, decorated with pallid brownish-grey small scales, uniform at the centre; at first conical, becoming expanded, margin grooved. Flesh white, delicate and thin.
Gills white, brownish-grey when dry, free, crowded. **Spores** hyaline, smooth, broadly ellipsoid or almond-shaped, small germ pore, 8-10 x 6-7 um. Basidia 4-spored. Cystidia not distinctive.
Stem white, yellowish-brown where bruised, smooth, more or less equal, base downy and slightly swollen. Ring white, fragile, ephemeral, superior. Flesh white, fragile, stuffed or hollow.
Odour not distinctive. **Taste** not distinctive.
Chemical tests none.
Occurrence late summer to autumn; infrequent.
■ Inedible.

Leucocoprinus jubilaei Joss.
Agaricaceae

Whitish cap with fine amethyst-brown fibres, white stem with fragile ring; scattered or clustered on soil in mixed woods.

Dimensions cap 2 - 4.5 cm dia; stem 5 - 6 cm tall x 0.4 - 0.6 cm dia.
Cap whitish or ivory decorated with fine amethyst-brown fibres towards centre; at first conical becoming expanded, margin sulcate. Flesh white, delicate, thin, discolouring reddish-brown where bruised.
Gills white, discolouring brownish, free, crowded. **Spores** hyaline, smooth, broadly ellipsoid or almond-shaped, germ pore, 6-8 x 3.5-4.5 µm. Basidia 4-spored. Cystidia not distinctive.
Stem white, yellowish-brown where bruised, smooth, more or less equal. Ring white, fragile, ephemeral, superior. Flesh white, reacting as in cap, fragile, stuffed or hollow.
Odour not distinctive. **Taste** not distinctive.
Chemical tests gills green with NH$_3$.
Occurrence late summer and autumn; infrequent.
■ Inedible.

Agaricus arvensis Schaeff.
Agaricaceae
= *Psalliota arvensis* (Schaeff.) Kummer   Horse Mushroom

Large, sometimes massive, agaric with white cap and ringed stem, pink or chocolate gills; in trooping groups, often forming rings, on soil amongst grass in pastures and other open places.

Dimensions cap 8 - 20 cm dia; stem 8 - 10 cm tall x 2 - 3 cm dia.
Cap white or cream, yellowing slightly with age or on bruising; at first ovoid, becoming convex and later expanded, smooth or finely scaly. Flesh white, firm and thick.
Gills at first white, becoming pink, then chocolate-brown or blackish, free, crowded. **Spores** purple-brown, smooth, ellipsoid, 7-8 x 4.5-5 µm. Basidia 4-spored. Gill edge cystidia sub-spherical or ovoid, thin-walled.
Stem white or cream, slightly clavate, smooth or finely scaly below the ring. Ring white or cream, double membrane splitting in cog-wheel shape below, pendulous, superior. Flesh white, pithy, stuffed, becoming hollow.
Odour strong, of aniseed. **Taste** not distinctive.
Chemical tests positive reaction with Schaeffer's test.
Occurrence late summer to autumn; frequent.
☐ Edible and good.

Agaricus augustus Fr.
Agaricaceae
= *Psalliota augusta* (Fr.) Quél.

Tall, sometimes massive, agaric with brownish scaly cap, pale stem with ring, pale or chocolate gills; in trooping groups on soil with broad-leaf or coniferous trees.

Dimensions cap 10 - 20 cm dia; stem 10 - 20 cm tall x 2 - 4 cm dia.
Cap chestnut, soon breaking up into fibrous scales in more or less concentric rings against a yellow-tinged background; at first obtusely ovoid, becoming convex and later expanded. Flesh whitish, firm and thick.
Gills at first pallid pink, becoming chocolate-brown or blackish at maturity. **Spores** chocolate-brown, smooth, ellipsoid, 7-10 x 4.5-5.5 µm. Basidia 4-spored. Gill edge cystidia formed of sub-spherical units arranged in chains.
Stem whitish, bruising faintly yellow, tapering slightly upwards, finely scaly below the ring, penetrating deeply into the substrate. Ring white, broad, superior. Flesh whitish, tinged pink in the stem base, stuffed, sometimes becoming hollow.
Odour strong, of bitter almonds. **Taste** not distinctive.
Chemical tests positive reaction with Schaeffer's test.
Occurrence late summer to autumn; infrequent.
☐ Edible and good.

Agaricus bisporus (J Lange) Imbach   Agaricaceae
= *Psalliota brunnescens* Peck

Medium to large agaric with greyish-brown cap and stem, pink or chocolate gills and ring on stem; in trooping groups, often tufted, on compost, manure and manured soil.

Dimensions cap 5 - 15 cm dia; stem 3.5 - 5.5 cm tall x 0.8 - 1.4 cm dia.
Cap greyish-brown, with darker radiating fibres; at first hemispherical becoming convex and later expanded, smooth or finely scaly with age. Flesh white, reddish on cutting, firm and thick.
Gills at first white, becoming dull pink, then chocolate-brown or blackish, free, crowded. **Spores** chocolate-brown, smooth, broadly ellipsoid, 6-9 x 4.5-7 µm. Basidia 2-spored. Gill edge cystidia elongated clavate, thin-walled.
Stem white, more or less equal, smooth or with some flaking below the ring. Ring white, single, superior, persistent. Flesh white, reacting as in cap, stuffed becoming hollow.
Odour strong, of mushroom. **Taste** not distinctive.
Chemical tests no reaction with Schaeffer's test.
Occurrence early summer to autumn; infrequent.
☐ Edible and good.
The only member of the genus with 2-spored basidia

Agaricus bisporus var hortensis (J Lange) Pil.
Agaricaceae

Medium to large agaric with whitish cap and stem, pink or chocolate gills and ring on stem; in trooping groups, often tufted, cultivated on compost or as a naturalised escape.

Dimensions cap 5 - 15 cm dia; stem 3.5 - 5.5 cm tall x 0.8 - 1.4 cm dia.
Cap at first pure-white becoming brown-tinged at the centre with age; at first hemispherical becoming convex and later expanded, smooth or finely scaly with age. Flesh white, firm and thick.
Gills at first white, becoming dull pink, then chocolate-brown or blackish, free, crowded. **Spores** chocolate-brown, smooth, broadly ellipsoid or sub-spherical, 5-8 x 4.5-6 µm. Basidia 2-spored. Gill edge cystidia clavate, thin-walled.
Stem white, more or less equal, smooth or with some flaking below the ring. Ring white, single, superior, persistent. Flesh white, stuffed becoming hollow.
Odour less distinctive than in the wild form. Taste not distinctive.
Chemical tests no reaction with Schaeffer's test.
Occurrence throughout the year in culture, otherwise summer and autumn; infrequent in the wild.
☐ Edible and good.

Agaricus bitorquis (Quél.) Sacc.   Agaricaceae
= *Agaricus edulis* (Vitt.) Möller & Schaeff.
= *Agaricus campestris subsp. bitorquis* (Quél.) Konrad & Maubl.
= *Psalliota rodmanii* (Peck) Kauffm.

Medium to large agaric with white cap, stem with 2 rings and pink or chocolate gills; in trooping groups, often tufted, on manure, also favouring sandy soil, often by roadsides.

Dimensions cap 4 - 10 cm dia; stem 3 - 6 cm tall x 1.5 - 2 cm dia.
Cap whitish with faint ochraceous tinge; at first convex becoming flattened-convex, then expanded, smooth or finely flaky. Flesh white, with pink tinge on cutting, firm and thick.
Gills at first dull pink, becoming clay and chocolate-brown or blackish at maturity, free, crowded. **Spores** chocolate-brown, smooth, sub-spherical, 4-6.5 x 4-5 µm. Basidia 4-spored. Gill edge cystidia clavate, thin-walled.
Stem white, more or less equal but tapered at the base, smooth, silky. Ring white, double; upper rigid, spreading, striate above; lower thin, erect, reminiscent of a volva; both persistent. Flesh white, tinged pink where cut and full.
Odour not distinctive. **Taste** not distinctive.
Chemical tests no reaction with Schaeffer's test.
Occurrence early summer to autumn; infrequent or rare. (Near Fordingbridge, New Forest, Hampshire)
☐ Edible and good.

Agaricus campestris L.: Fr.
= *Psalliota campestris* (L.: Fr.) Quél.

Agaricaceae
Field Mushroom

Medium to large agaric with creamy-white cap and stem, deep pink or chocolate gills and ring on stem; in trooping groups on soil amongst grass in pastures.

Dimensions cap 3 - 10 cm dia; stem 3 - 10 cm tall x 1 - 1.8 cm dia.
Cap white with cream or yellow tinges; sub-spherical, eventually becoming convex and expanded, smooth or finely scaly with age. Flesh white, unchanging or with pink tinge on cutting, firm and thick.
Gills at first deep pink, becoming chocolate-brown or blackish at maturity, free, crowded. **Spores** chocolate-brown, smooth, ellipsoid, 7-8 x 4-5 µm. Basidia 4-spored. Gill edge cystidia absent.
Stem white, more or less equal, smooth above the ring, woolly and scaly below. Ring white, single, fragile, superior, ephemeral. Flesh white, reacting as in cap and full.
Odour not distinctive. **Taste** not distinctive.
Chemical tests no reaction with Schaeffer's test.
Occurrence summer to autumn; frequent.
☐ Edible and good.

Agaricus comtulus Fr.
= *Psalliota comtula* (Fr.) Quél.

Agaricaceae

Small agaric with white or cream-tinged cap, pure pink or brown gills and ring on stem; solitary or scattered, on soil in pastures and mown grass.

Dimensions cap 2 - 5 cm dia; stem 3 - 5 cm tall x 0.4 - 0.6 cm dia.
Cap white, with cream tinge particularly towards the centre with age, not yellowing; at first convex, becoming expanded, more or less smooth. Flesh white, firm and fairly thin.
Gills clear pink from the outset and then brown at maturity, free, crowded. **Spores** brown, smooth, broadly ellipsoid, 4.5-5.5 x 3-3.5 µm. Basidia 4-spored. Gill edge cystidia absent.
Stem white or tinged yellow at base, more or less equal, smooth, slightly bulbous at base. Ring white, single, superior. Flesh white, full or stuffed.
Odour of aniseed. **Taste** not distinctive.
Chemical tests no reaction with Schaeffer's test.
Occurrence late summer to autumn; infrequent.
☐ Edible.

Agaricus excellens (Miller) Miller
= *Psalliota excellens* Miller

Agaricaceae

Large agaric with white or yellow-tinged cap, greyish-pink gills and ring on stem; in trooping groups on soil in open broad-leaf and coniferous woodlands, favouring spruce.

Dimensions cap 10 - 15 cm dia; stem 10 - 14 cm tall x 2 - 3.5 cm dia.
Cap white, with yellow tinge particularly towards the centre with age; at first convex, becoming expanded, breaking up into very small, fine, fibrous scales. Flesh white, unchanging or with pink tinge on cutting, firm and thick.
Gills at first whitish, becoming pink and then chocolate-brown or blackish at maturity, free, crowded. **Spores** chocolate-brown, smooth, broadly ellipsoid, 9-12 x 5-7 µm. Basidia 4-spored. Gill edge cystidia sub-spherical to broadly clavate, thin-walled.
Stem white, stout, tapering slightly upwards, smooth above the ring, scaly below. Ring white, single, thick, superior, ephemeral. Flesh white, reacting as in cap and full.
Odour not distinctive. **Taste** not distinctive.
Chemical tests positive reaction with Schaeffer's test.
Occurrence summer to autumn; frequent.
☐ Edible and good.

Agaricus fuscofibrillosus (Miller) Pil.   **Agaricaceae**

Medium-sized agaric with brown scaly cap, dull pink or chocolate gills and ring on stem; in trooping groups, sometimes tufted, on soil in woods or, less typically, in open grassland.

Dimensions cap 4 - 9 cm dia; stem 5 - 7 cm tall x 0.8 - 1.5 cm dia.
Cap dull brown, densely fibrous and, in exposed situations, somewhat adpressed-scaly against a pallid background; at first subspherical, becoming convex and expanded. Flesh white, reddening slightly where cut, firm and thinnish.
Gills at first pinkish-fawn, becoming chocolate-brown or blackish at maturity, free, crowded. **Spores** chocolate-brown, smooth, ellipsoid, 5-7.5 x 3.5-4.5 µm. Basidia 4-spored. Gill edge cystidia ovoid or balloon-like, thin-walled.
Stem white, then tinged brown, more or less equal, fairly slender, smooth. Ring white, single, pendulous, thin, superior. Flesh white, reacting as in cap and full.
Odour not distinctive. **Taste** not distinctive.
Chemical tests no reaction with Schaeffer's test.
Occurrence summer to autumn; infrequent.
☐ Edible.

Agaricus haemorrhoidarius Kchb. & Schulz.
                                              Agaricaceae

Medium to large agaric with brown scaly cap, dull pink or chocolate gills, stem with ring and bulbous base, flesh reddening where damaged; in trooping groups on soil with broad-leaf trees.

Dimensions cap 8 - 12 cm dia; stem 8 - 12 cm tall x 1.5 - 2.5 cm dia.
Cap dull umber-brown, breaking up into adpressed scales against a slightly more pallid background; at first sub-spherical, becoming convex and expanded. Flesh white, reddening strongly where cut, firm and thinnish.
Gills at first pinkish-fawn with white flaky edges, becoming chocolate-brown or blackish at maturity, free, crowded.
Spores chocolate-brown, smooth, ovoid, 4.5-6.5 x 3-5 µm. Basidia 4-spored. Gill edge cystidia clavate, thin-walled.
Stem white, then tinged brown, more or less equal but with broad bulbous base, finely scaly below ring, otherwise smooth. Ring white, single, pendulous, superior. Flesh white, reacting as in cap and full.
Odour not distinctive. **Taste** not distinctive.
Chemical tests no reaction with Schaeffer's test.
Occurrence late summer to autumn; infrequent.
☐ Edible.

Agaricus impudicus (Rea) Pil.   **Agaricaceae**
= *Agaricus variegans* Miller

Medium or large agaric with dull brown scaly cap, pinkish or dark brown gills with white edges, upturned ring on stem; in trooping groups on soil in coniferous woods.

Dimensions cap 5 - 10 cm dia; stem 6 - 10 cm tall x 0.8 - 1.2 cm dia.
Cap dull brown, soon breaking up into adpressed fibrous scales against a buff-brown background; at first convex becoming expanded-convex, or somewhat flattened at the centre. Flesh white, becoming slightly reddened where cut, firm and medium.
Gills at first pallid pink, becoming darker brown with whitish but not flaky edges, free, crowded. **Spores** chocolate-brown, smooth, broadly ellipsoid, 5-6 x 3-3.5 µm. Basidia 4-spored. Gill edge cystidia utriform, thin-walled.
Stem at first whitish, becoming more brown with age, more or less equal, with slightly bulbous base. Ring whitish or discoloured brown, boot-like, superior. Flesh white, browning in stem base, full or narrowly hollow.
Odour not distinctive. **Taste** not distinctive.
Chemical tests no reaction with Schaeffer's test.
Occurrence late summer to autumn; frequent.
☐ Edible

Agaricus langei (Miller) Miller   Agaricaceae
= *Psalliota langei* Miller

Medium to large agaric with brown scaly cap, dull pink or chocolate gills and ring on stem; in trooping groups, sometimes tufted, on soil in coniferous or mixed woods.

Dimensions cap 4 - 12 cm dia; stem 3 - 12 cm tall x 1.5 - 3 cm dia.
Cap tawny or rust-brown, soon breaking up into dense fibrous adpressed scales against a more pallid background; at first sub-spherical, becoming convex and expanded. Flesh white, bright red where cut, firm and thick.
Gills at first pinkish-fawn, becoming chocolate-brown or blackish at maturity, free, crowded. **Spores** purple-brown, smooth, ellipsoid, 7-9 x 4-5 µm. Basidia 4-spored. Gill edge cystidia ovoid or broadly clavate, thin-walled.
Stem white, tinged pink, bruising reddish, more or less equal, stout, smooth above the ring, finely woolly-scaly below. Ring white, single, thick, superior. Flesh white, reacting as in cap and full.
Odour not distinctive. **Taste** not distinctive.
Chemical tests no reaction with Schaeffer's test.
Occurrence early summer to early autumn; infrequent.
☐ Edible.

Agaricus lanipes (Miller & Schaeff.) Sing.   Agaricaceae

Medium to large agaric with brown scaly cap, dull pink or chocolate gills and ring on stem; in trooping groups, sometimes tufted, on soil in open mixed woods.

Dimensions cap 5 - 10 cm dia; stem 3 - 10 cm tall x 1.5 - 3 cm dia.
Cap chocolate-brown, soon breaking up into broad scales against a pallid background; at first sub-spherical, becoming convex and expanded. Flesh white, more or less unchanging, firm and thickish.
Gills at first pinkish, becoming chocolate-brown or blackish at maturity, free, crowded. **Spores** purple-brown, smooth, ellipsoid, 5.5-6.5 x 4.5 µm. Basidia 4-spored. Gill edge cystidia broadly clavate, thin-walled and tufted.
Stem white, tinged pink, bruising reddish, more or less equal, stout, smooth above the ring, brown woolly-scaly below. Ring white, single, thick, superior. Flesh white, yellow in base in old specimens.
Odour not distinctive. **Taste** not distinctive.
Chemical tests no reaction with Schaeffer's test.
Occurrence early summer to early autumn; infrequent.
☐ Edible.

Agaricus macrosporus (Miller & Schaeff.) Pil.
                                          Agaricaceae
= *Psalliota arvensis subsp. macrospora* Mller & Schaeff.

Large or massive agaric with creamy-white cap, greyish or chocolate gills and ring on stem; in rings, on soil amongst grass in pastures.

Dimensions cap 8 - 30 cm dia; stem 5 - 12 cm tall x 2.5 - 4 cm dia.
Cap cream or ochraceous, splitting into woolly or scaly patches against a white background; at first sub-spherical, becoming broadly convex. Flesh white, unchanging, firm and thick.
Gills at first pallid, becoming fawn and then chocolate-brown or blackish at maturity, free, crowded. **Spores** chocolate-brown, smooth, ellipsoid, 8-12 x 5.5-6.5 µm. Basidia 4-spored. Gill edge cystidia utriform, thin-walled.
Stem white, massively fusiform, smooth above the ring, coarsely floccular below. Ring white, single, thick but fragile sometimes with denticulate margin, superior. Flesh white, sometimes pinkish towards the base on cutting, full.
Odour strong, of aniseed, in young and mature specimens, more rarely ammoniacal. **Taste** not distinctive.
Chemical tests variable reaction with Schaeffer's test.
Occurrence early summer to early autumn; infrequent.
☐ Edible and good.

Agaricus praeclavesquamosus Freeman
= *Agaricus placomyces* Peck   **Agaricaceae**
= *Psalliota meleagris* Schaeff.

Medium to large agaric with greyish-brown finely scaly cap, pink or chocolate gills and stem with ring and bulbus base; in trooping groups on soil, typically amongst grass, in mixed woods.

Dimensions cap 5 - 10 cm dia; stem 6 - 9 cm tall x 1 - 1.2 cm dia.
Cap greyish-brown, soon breaking up into fine, dense greyish-brown scales against a whitish background; at first sub-spherical or ovoid, becoming expanded-convex. Flesh white, with yellowish and then with brown tinge where cut, firm and thick.
Gills pallid pink, eventually becoming chocolate-brown or blackish at maturity, free, crowded. **Spores** chocolate-brown, smooth, broadly ovoid, 4.5-6 x 3-4 µm. Basidia 4-spored. Gill edge cystidia sub-spherical or pear-shaped, thin-walled.
Stem white, with yellowish tinge, becoming brown, tapering slightly upwards from bulbous base, smooth above the ring, coarsely flaky below. Ring white, single, large, membraneous and pendulous, superior. Flesh white, reacting as in cap, more yellow in stem base, firm and full.
Odour faint, of phenol. **Taste** not distinctive.
Chemical tests no reaction with Schaeffer's test.
Occurrence early summer to early autumn; infrequent.
✵ Poisonous.

Agaricus semotus Fr.   **Agaricaceae**
= *Agaricus comtulus var amethystinus* (Quél.) Konrad & Maubl.
= *Psalliota amethystina* (Quél.) J Lange

Small or medium agaric with yellowish-brown, or lilac-tinged cap, pinkish or chocolate gills, white stem with ring and bulbous base; solitary or in small trooping groups on soil typically in grass in or close to woodland.

Dimensions cap 2 - 4 cm dia; stem 3 - 6 cm tall x 0.4 - 0.8 cm dia.
Cap at first whitish, becoming radially fibrous ochraceous or brown with lilaceous or vinaceous tinges, sometimes slightly scaly at the centre; at first sub-spherical or bluntly ovoid, becoming broadly convex and flattened. Flesh white, unchanging, firm and rather thin.
Gills at first pallid pinkish-grey, becoming greyish-brown at maturity, free, crowded. **Spores** brown, smooth, ovoid, 4-5 x 3-3.5 µm. Basidia 4-spored. Gill edge cystidia broadly clavate, thin-walled, hyaline or stained brownish, prolific.
Stem whitish, but yellowing at the base, more or less equal with bulbous base, smooth, silky. Ring white, double, pendulous, superior. Flesh white, yellow in the base, narrowly hollow or full.
Odour not distinctive. **Taste** not distinctive.
Chemical tests positive reaction with Schaeffer's test.
Occurrence late summer to autumn; infrequent.
■ ✵ Inedible, possibly poisonous.

Agaricus silvaticus Schaeff.   **Agaricaceae**

Medium or large agaric with brownish fibrous scaly cap, pinkish or dark brown gills and ring on stem; in trooping groups on soil in coniferous woods, favouring spruce.

Dimensions cap 5 - 10 cm dia; stem 5 - 8 cm tall x 1 - 1.2 cm dia.
Cap ochraceous-brown, soon breaking up into small, adpressed, fibrous scales against a pallid background; at first convex, becoming expanded-convex. Flesh white, becoming red then brown where cut, firm and thinnish.
Gills at first pallid rose-pink, becoming reddish and finally darker brown, edges white and flaky, free, crowded. **Spores** chocolate-brown, smooth, ovoid, 4.5-6 x 3-3.5 µm. Basidia 4-spored. Gill edge cystidia clavate, thin-walled.
Stem whitish, more or less equal, with slightly bulbous base, brownish woolly-scaly below the ring. Ring dirty-brown, single, large, thick, superior. Flesh white, reacting as in cap, full or narrowly hollow.
Odour not distinctive. **Taste** not distinctive.
Chemical tests no reaction with Schaeffer's test.
Occurrence late summer to autumn; infrequent or rare.
☐ Edible and good.

Agaricus silvicola (Vitt.) Peck

Agaricaceae
Wood Mushroom

Medium or large agaric with creamy white cap, pinkish or chocolate gills, stem with ring and bulbous base; in trooping groups on soil in broad-leaf and coniferous woods.

Dimensions cap 5 - 10 cm dia; stem 5 - 8 cm tall x 1 - 1.5 cm dia.
Cap cream, bruising ochraceous and generally yellowing with age; at first sub-spherical or ovoid becoming broadly convex and flattened. Flesh white, unchanging, firm and thinnish.
Gills at first pallid pinkish-grey, becoming chocolate-brown or blackish at maturity, free, crowded. **Spores** chocolate-brown, smooth, ovoid, 5-6 x 3-4 µm. Basidia 4-spored. Gill edge cystidia sub-spherical or spherical, thin-walled, prolific.
Stem concolorous and reacting as cap, more or less equal, with bulbous base, smooth, silky. Ring white above, pallid dirty-brown below, single, large, pendulous, superior. Flesh white, sometimes with pinkish tinges, full or narrowly hollow.
Odour strong, of aniseed. **Taste** not distinctive.
Chemical tests positive reaction with Schaeffer's test.
Occurrence late summer to autumn; infrequent.
☐ Edible and good.

Agaricus subperonatus (J Lange) Sing.   **Agaricaceae**

Medium or large agaric with rust-brown scaly cap, pinkish or dark brown gills and ring on stem with distinctive belted zones below; solitary or in trooping groups on soil in open woodlands and pastures.

Dimensions cap 7 - 12 cm dia; stem 6 - 8 cm tall x 1.5 - 3 cm dia.
Cap reddish-brown, soon breaking up into large adpressed scales against a pallid background; margin draped with velar remnants; at first convex, becoming expanded-convex. Flesh white, becoming flesh pink where cut, firm and thick.
Gills at first pallid pink, becoming darker brown, free, crowded.
Spores chocolate-brown, smooth, sub-spherical, 5.5-7.5 x 4-5.5 µm. Basidia 4-spored. Gill edge cystidia clavate, thin-walled.
Stem whitish, tinged brown, more or less equal or somewhat clavate, several brownish scaly-belted zones below the ring. Ring whitish or discoloured brown, large, thick, superior. Flesh white, reacting as in cap, full or narrowly hollow.
Odour not distinctive or of fruit. **Taste** not distinctive.
Chemical tests no reaction with Schaeffer's test.
Occurrence late summer to autumn; rare. (Mendip Hills, Somerset)
☐ Edible

Agaricus vaporarius (Vitt.) Capelli   **Agaricaceae**

Medium or large agaric with brownish scaly cap, pinkish or dark brown gills, and thick ring on stem with distinctive belted zones below; in trooping groups on soil in open woodlands and pastures.

Dimensions cap 10 - 15 cm dia; stem 6 - 12 cm tall x 2.5 - 5 cm dia
Cap dull brown, soon breaking up into large adpressed scales against a pallid background; at first convex, becoming expanded-convex, sometimes whilst semi-submerged in the soil. Flesh white, becoming slightly reddened where cut, firm and thick.
Gills at first pallid pink, becoming darker brown, free, crowded.
Spores chocolate-brown, smooth, sub-spherical, 6-7 x 5-6 µm. Basidia 4-spored. Gill edge cystidia narrowly clavate, thin-walled.
Stem whitish, stout more or less equal, with tapered base buried deep in the soil, several brownish scaly-belted zones below the ring. Ring whitish or discoloured brown, large, thick, superior. Flesh white, reacting as in cap, full or narrowly hollow.
Odour not distinctive. **Taste** not distinctive.
Chemical tests no reaction with Schaeffer's test.
Occurrence late summer to autumn; rare. (Arne Peninsula, Dorset)
☐ Edible

Agaricus xanthodermus Genevier   **Agaricaceae**
= *Psalliota xanthoderma* (Genevier) Richon & Roze   Yellow Stainer

Large agaric with white cap, pinkish or greyish-brown gills and ring on stem, distinctive colour change in exteme stem base; in trooping groups on soil amongst grass in woods, pastures and parks.

Dimensions cap 5 - 15 cm dia; stem 5 - 15 cm tall x 1 - 2 cm dia.
Cap at first white, with minute greyish scales on ageing, bruising yellowish; at first sub-spherical, becoming broadly convex and flattened. Flesh white, unchanging, thick and firm.
Gills at first whitish, then pallid pink, becoming grey-brown at maturity, free, crowded. **Spores** purple-brown, smooth, broadly ellipsoid, 5-6.5 x 3.5 µm. Gill edge cystidia sub-spherical or broadly clavate, thin-walled, prolific.
Stem concolorous and reacting as cap, more or less equal, with somewhat bulbous base, smooth, silky. Ring white, single but with deceptively thick edge, large, spreading, superior. Flesh white, unchanging except in extreme base which turns immediately chrome-yellow on cutting, stuffed or full.
Odour faint, of ink or phenol. **Taste** not distinctive.
Chemical tests no reaction with Schaeffer's test.
Occurrence late summer to autumn; infrequent.
✛ Poisonous.

Coprinus atramentarius (Bull.: Fr.) Fr.   **Coprinaceae**
Common Ink Cap

Medium-sized greyish-brownish conical agaric, later blackening; generally in tufts, in fields, gardens and waste ground, near broad-leaf tree stumps or from buried wood, sometimes pushing up through tarmac paths, less frequently near bases of living trees.

Dimensions cap 3 - 7 cm tall x variable dia; stem 7 - 14 cm tall x 1 - 1.5 cm dia.
Cap dirty greyish-brown or brown, more rust at the centre with scurfy brown scales (velar remnants); at first ovoid, then bell-shaped, broadly conical and finally reflexed at the margin, sulcate, often splitting. Flesh white in young specimens but soon discolouring, fairly thin in relation to size, fragile, auto-digesting.
Gills at first white, rapidly becoming grey-brown and finally black, free, crowded. **Spores** date-brown, smooth, ellipsoid, germ pore, 8-11 x 5-6 µm. Basidia 4-spored. Cystidia not distinctive.
Stem white (later stained with products of auto-digestion), smooth. Ring-like zone near the base. Flesh white, hollow and medium.
Odour not distinctive. **Taste** not distinctive.
Chemical tests none.
Occurrence early summer to autumn; common.
☐ Edible with caution. Adverse symptoms when combined with alcohol.

Coprinus auricomus Pat.   **Coprinaceae**

Small, delicate, yellowish-brown, parasol-like agaric with blackening gills; solitary or scattered on soil, typically amongst leaf litter, also on burnt ground.

Dimensions cap 1 - 2 cm dia; stem 6 - 9 cm tall x 0.2 - 0.3 cm dia.
Cap greyish-brown, decorated with minute yellowish-brown downy hairs particularly at the centre, finely sulcate, the ridges sometimes forked; at first ovoid and then expanded, without velar remnants. Flesh thin, fragile, and barely auto-digesting.
Gills white becoming grey, adnate, crowded. **Spores** dark brown (black in the mass), smooth, ellipsoid, germ pore, 12-14 x 6-7 µm. Basidia 4-spored. Cystidia not distinctive.
Stem white, smooth, very slender, more or less equal. Ring absent. Flesh thin and fragile.
Odour not distinctive. **Taste** not distinctive.
Chemical tests none.
Occurrence summer to autumn; infrequent.
■ Inedible.

Coprinus cinereofloccosus P Orton   Coprinaceae

Small, very delicate ovoid, then conical, agaric with blackening gills; solitary or scattered on soil in short grass on lawns.

Dimensions cap 1 - 2 cm tall; stem 2 - 10 cm tall x 0.2 - 0.3 cm dia.
Cap greyish, darker and more ochraceous at the centre, finely sulcate; at first ovoid and then expanded, at first covered with fine whitish mealy flakes. Flesh thin, fragile and rapidly auto-digesting.
Gills translucent, somewhat floccose becoming black, adnate, crowded. **Spores** dark brown (black in the mass), smooth, ellipsoid or almond-shaped, germ pore, 11-15 x 5.5-7 µm. Basidia 4-spored. Gill cystidia flask-shaped or pear-shaped.
Stem white, pruinose, very slender, more or less equal, without basal sclerotium. Ring absent. Flesh thin and fragile.
Odour not distinctive. **Taste** not distinctive.
Chemical tests none.
Occurrence early summer to early autumn; infrequent.
■ Inedible.

Coprinus cinereus (Schaeff.: Fr.) S F Gray   Coprinaceae

Small dark grey, ovoid or conical agaric with blackening gills, at first covered woolly white; solitary or caespitose tufted, on rotting or manured straw.

Dimensions cap 1.5 - 4 cm tall x variable dia; stem 2 - 10 cm tall x 0.5 - 0.8 cm dia.
Cap at first dirty-white with velar covering which flakes away to reveal dark grey surface; at first ovoid and then conical and expanded, sulcate to the centre and splitting with reflexed margin when mature. Flesh thin, fragile and auto-digesting.
Gills cream, becoming vinaceous-brown and finally black, adnate, crowded. **Spores** dark brown (black in the mass), smooth, ellipsoid, germ pore, 9-12 x 6-7 µm. Basidia 4-spored. Cystidia not distinctive.
Stem white, pruinose at the apex, woolly-fibrous below, at first stoutish, then more slender, swollen at base and then somewhat rooting. Ring absent. Flesh thin and fragile.
Odour not distinctive. **Taste** not distinctive.
Chemical tests none.
Occurrence early summer to early autumn; common.
■ Inedible.

Coprinus comatus (Müll.: Fr.) S F Gray   Coprinaceae
Shaggy Ink Cap

Tall agaric with white, conical, shaggy cap, blackening; in scattered trooping groups on soil in short grass.

Dimensions cap 5 - 15 cm tall x variable dia; stem 10 - 30 cm tall x 1.5 - 2.5 cm dia.
Cap white; at first ovoid-cylindrical, pressed to the stalk at the margin, becoming campanulate-conical; decorated with feathery, overlapping, partly reflexed velar remnants, white when young or stained brownish. Flesh white, soon discolouring, fleshy in relation to cap size, fragile, auto-digesting.
Gills white at first, rapidly becoming pinkish, grey-brown, and finally black, free, crowded. **Spores** date-brown, smooth, almond-shaped, germ pore, 10-13 x 6.5-8 µm. Basidia 4-spored. Cystidia not distinctive.
Stem white, smooth, often very tall, slightly swollen at base and sometimes rooting. Ring white, thin, loose and often slipping down stem towards base. Flesh white, medium, hollow and fragile.
Odour slightly acidic. **Taste** not distinctive.
Chemical tests none.
Occurrence late summer to autumn; common.
□ Edible and excellent.

Coprinus congregatus Bull.: Fr.   Coprinaceae

Small, yellowish, conical agaric with blackening gills; caespitose tufted, on dung, rich decaying vegetable matter and rotting straw.

Dimensions cap 0.5 - 2 cm tall x variable dia; stem 2 - 8 cm tall x 0.1 - 0.4 cm dia.
Cap pallid ochraceous-buff, with greyish tinge at the margin and usually slightly darker buff or fulvous towards the centre; at first ovoid and then bell-shaped or conical, deeply sulcate to the centre. Flesh thin, fragile and not auto-digesting. Cap cystidia thin-walled, sub-cylindrical, 48-106 x 7.5-17 µm.
Gills cream, becoming vinaceous-brown and finally black, adnate, crowded. **Spores** dark brown (black in the mass), smooth, ellipsoid or almond-shaped, 12-14 x 6-7 µm. Gill face cystidia large and distinctive, gill edge cystidia shorter.
Stem white, smooth, slender, often slightly rooting. Ring absent. Flesh thin and fragile.
Odour not distinctive. **Taste** not distinctive.
Chemical tests none.
Occurrence summer; rare. (On rotting straw bales on Mendip Hills near Wells, Somerset)
■ Inedible.

Coprinus cothurnatus Godey apud Gill.   Coprinaceae

Small, whitish, then dark grey, ovoid or conical agaric with blackening gills; solitary or caespitose tufted, on rotting or manured straw.

Dimensions cap 0.7 - 2 cm tall x variable dia; stem 4 - 10 cm tall x 0.2 - 0.6 cm dia.
Cap at first with dense, white, floccose, velar covering which flakes away to reveal a pallid surface becoming dark grey; at first ovoid, and then conical and expanded, sulcate to the centre and splitting with reflexed margin when mature. Flesh thin, fragile and auto-digesting.
Gills cream, becoming vinaceous-brown and finally black, adnate, crowded. **Spores** dark brown (black in the mass), smooth, ellipsoid, germ pore, 11-15 x 7-9 µm. Basidia 4-spored. Cystidia not distinctive.
Stem whitish, or with pink tinge, floccose below, slender. Ring absent. Flesh thin and fragile.
Odour not distinctive. **Taste** not distinctive.
Chemical tests none.
Occurrence early summer to early autumn; infrequent.
■ Inedible.

Coprinus disseminatus (Pers.: Fr.) S F Gray
= *Psathyrella disseminata* (Pers.: Fr.) Quél.   Coprinaceae

Small, pale, conical agaric with blackening gills; massed distinctively on stumps of broad-leaf trees and spreading to adjacent soil, often in very large groups.

Dimensions cap 0.5 - 1.5 cm tall x variable dia; stem 1.5 - 4 cm tall x 0.1 - 0.3 cm dia.
Cap pallid buff or clay with greyish tinge, and usually slightly darker buff or fulvous towards the centre; ovoid, then campanulate-conical, deeply sulcate to the centre. Flesh thin, fragile, not auto-digesting. Cap cystidia thin-walled, bluntly cylindrical, 75-200 x 20-30 µm.
Gills white, then greyish-brown and finally black, adnate, crowded.
Spores dark brown (black in the mass), smooth, ellipsoid or almond-shaped, 7-9.5 x 4-5 µm. Basidia 4-spored. Gill cystidia cylindrical or flask-shaped.
Stem white or slightly buff near base, which is minutely downy, often curved. Ring absent. Flesh whitish and fragile.
Odour not distinctive. **Taste** not distinctive.
Chemical tests none.
Occurrence spring to autumn; frequent.
□ Edible but worthless.

Coprinus domesticus (Bolt.: Fr.) S F Gray   Coprinaceae

Tall, pale buff, conical agaric, at first scurfy, then smooth, blackening; solitary or in small trooping or tufted groups on logs and stumps of broad-leaf trees.

Dimensions cap 1 - 3 cm tall x variable dia; stem 4 - 15 cm tall x 0.2 - 1 cm dia.
Cap pallid buff, darker and more tawny-ochraceous towards the centre; ovoid then campanulate-conical, finally reflexed and split, becoming sulcate from the margin; scaly-pruinose with velar remnants, becoming smooth. Flesh whitish at first, thin and auto-digesting.
Gills whitish, then purplish-brown and finally black, adnexed, crowded. **Spores** dark brown (black in the mass), smooth, cylindrical, ellipsoid or reniform, with germ pore, 7.5-10 x 4-5 µm. Basidia 4-spored. Cystidia not distinctive.
Stem white, with buff tinge near the swollen, slightly ridged base arising from distinctive orange mycelial mass; smooth. Ring absent. Flesh whitish and fairly firm.
Odour not distinctive. **Taste** not distinctive.
Chemical tests none.
Occurrence spring to summer; infrequent.
■ Inedible.
Mycelial mass often visible but not in this photograph.

Coprinus extinctorius (Bull.) Fr.   Coprinaceae

Tall, pale brownish agaric, with conical cap, blackening; solitary or in small groups, typically caespitose tufted, on stumps or wounds of broad-leaf trees.

Dimensions cap 1.5 - 2.5 cm tall x variable dia; stem 3 - 11 cm tall x 0.4 - 0.6 cm dia.
Cap pallid ochraceous-buff, darker and more tawny towards the centre; at first ovoid, then bell-shaped or conical, finally flattened with strongly reflexed margin, deeply sulcate as far as the centre, covered with small, pointed, fibrous scales, becoming smoother. Flesh whitish at first, thin, fragile and auto-digesting.
Gills white, then greyish-brown and finally black, adnexed, crowded.
Spores dark brown (black in the mass), smooth, pip- or almond-shaped, germ pore, 8-10 x 6.5-7.5 µm. Basidia 4-spored. Cystidia not distinctive.
Stem white, slender, more or less equal or tapering slightly upwards, at first pruinose at apex, then smooth, downy at base. Ring absent. Flesh whitish and fragile.
Odour not distinctive. **Taste** not distinctive.
Chemical tests none.
Occurrence summer to autumn; infrequent.
■ Inedible.

Coprinus hemerobius Fr.   Coprinaceae

Delicate, pale brownish agaric with egg-shaped cap, greying; solitary or in small trooping groups on damp soil in damp grassy places.

Dimensions cap 1 - 1.5 cm tall x variable dia; stem 3 - 6 cm tall x 0.1 - 0.2 cm dia.
Cap ochraceous or pallid leather, becoming grey at maturity; at first ovoid and then campanulate and finally flattened, deeply sulcate as far as the centre, otherwise more or less smooth, without velar remnants. Flesh pallid, thin, fragile and barely auto-digesting.
Gills buff, then greyish-brown, adnexed or free, close. **Spores** dark brown, smooth, ellipsoid or almond-shaped, germ pore, 11.5-12.5 x 6-8 µm. Basidia 4-spored. Cystidia not distinctive.
Stem whitish-cream, very slender, more or less equal, smooth. Ring absent. Flesh pallid and fragile.
Odour not distinctive. **Taste** not distinctive.
Chemical tests none.
Occurrence late summer to autumn; frequent.
■ Inedible.

Coprinus heterosetulosus (Locq.) Watl.   Coprinaceae

Small, delicate, yellowish-brown agaric with egg-shaped cap; solitary or in small trooping groups on dung.

Dimensions cap 0.4 - 0.7 cm tall x variable dia; stem 1 - 3 cm tall x 0.3 - 0.1 cm dia.
Cap ochraceous, with umber brown centre, becoming greyish, then black at maturity; at first ovoid and then campanulate, densely pruinose at the centre. Flesh pallid, thin, fragile, auto-digesting rapidly. Cap cystidia very long, 30-140 μm, lying between shorter thick-walled, brown sclerocystidia.
Gills buff, then blackish, adnexed or free, close. **Spores** dark brown, smooth, ellipsoid or almond-shaped, germ pore, 8-12.5 x 4.5-6.5 μm. Basidia 4-spored. Gill edge cystidia more or less rounded, gill face cystidia absent.
Stem whitish, tinged brown at the base, more or less equal, smooth. Ring absent. Flesh pallid and fragile.
Odour not distinctive. **Taste** not distinctive.
Chemical tests none.
Occurrence summer to autumn; frequent.
■ Inedible.

Coprinus impatiens (Fr.) Quél.   Coprinaceae
= *Psathyrella impatiens* (Fr.) Kühn.

Delicate, brownish agaric with convex cap, barely blackening; solitary or in small trooping groups on soil amongst litter with broad-leaf trees, favouring beech.

Dimensions cap 1 - 3 cm tall x variable dia; stem 7 - 10 cm tall x 0.2 - 0.4 cm dia.
Cap pallid buff, tawny or cinnamon towards the centre, drying more pallid; at first ovoid, then conical-convex, and finally flattened, markedly sulcate as far as the centre. Flesh whitish, thin, fragile and barely auto-digesting. Cap surface cystidia thin-walled, 70-120 x 8-14 μm.
Gills buff, then greyish-brown, adnexed or free, distant.
Spores dark brown, smooth, ellipsoid or almond-shaped, germ pore, 9-12 x 5-6 μm. Basidia 4-spored. Cystidia not distinctive.
Stem whitish, very slender, more or less equal, at first minutely pruinose, then smooth, silky. Ring absent. Flesh whitish and fragile.
Odour not distinctive. **Taste** not distinctive.
Chemical tests none.
Occurrence late summer to autumn; infrequent or rare.
(MOD property near Farnham, Surrey)
■ Inedible.

Coprinus lagopides Karst.   Coprinaceae

Tallish, fragile, grey agaric covered with whitish scurf, blackening; scattered or solitary on soil or charred wood.

Dimensions cap 2 - 6 cm tall x variable dia; stem 3 - 11 cm tall x 0.3 - 1.2 cm dia.
Cap greyish, surface covered with whitish or whitish-grey fibrils which flake off; at first ovoid or cylindrical, becoming convex and then almost flat, finely sulcate-striate almost to the centre, the margin splitting and finally reflexed. Flesh thin, fragile, barely auto-digesting. Velar hyphae ending in long cell-like chains.
Gills at first white, rapidly becoming dark vinaceous, then black, adnexed to almost free, close. **Spores** black with slight violet tinge, ellipsoid or sub-spherical, smooth, germ pore, 6-9 x 5-7 μm. Basidia 4-spored. Cystidia not distinctive.
Stem white, at first minutely downy, then smooth, white-woolly at base. Ring absent. Flesh hollow and fragile.
Odour not distinctive. **Taste** not distinctive.
Chemical tests none.
Occurrence autumn to winter; frequent.
■ Inedible.
Note: *C. lagopides* is readily distinguished from *C. lagopus* by its habitat.

Coprinus lagopus (Fr.) Fr.   Coprinaceae

Tallish, fragile, grey agaric covered with whitish mealy remnants, blackening; solitary or scattered on soil amongst leaf litter in well-shaded woods, occasionally in field margins.

Dimensions cap 2 - 4 cm tall x variable dia; stem 6 - 13 cm tall x 0.2 - 0.3 cm dia.
Cap greyish, covered with delicate whitish or whitish-grey fibrils which slowly flake off; at first ovoid or cylindrical, becoming convex and then almost flat, sulcate-striate almost to the centre, the margin splitting and finally reflexed. Flesh thin, fragile and barely auto-digesting.
Gills at first white, rapidly becoming black, adnexed or free, crowded. **Spores** black with slight violet tinge, ellipsoid or almond-shaped, smooth, germ pore, 11-13.5 x 6-7 μm. Basidia 4-spored. Cystidia not distinctive.
Stem white, at first minutely downy, then smooth, swollen at base. Ring absent. Flesh hollow and fragile.
Odour not distinctive. **Taste** not distinctive.
Chemical tests none.
Occurrence summer to autumn; infrequent.
☐ Edible but worthless.

Coprinus leiocephalus P Orton   Coprinaceae
= *Coprinus plicatilis var microsporus* Kühn.

Delicate, brownish agaric with egg-shaped cap, greying; solitary or in small trooping groups on damp soil amongst litter in woods generally.

Dimensions cap 0.5 - 1.5 cm tall x variable dia; stem 3 - 9 cm tall x 0.1 - 0.2 cm dia.
Cap hazel or reddish-brown with ochraceous tinge, becoming grey at maturity, darker at the centre; at first ovoid, then conical-convex and finally flattened, markedly sulcate as far as the centre, otherwise more or less smooth, without velar remnants. Flesh pallid, thin, fragile and barely auto-digesting.
Gills buff, then greyish-brown, adnexed or free, close. **Spores** dark brown, smooth, ellipsoid or almond-shaped, germ pore, 8-11 x 5.5-8.5 μm. Basidia 4-spored. Cystidia not distinctive.
Stem whitish-cream becoming tinged brown, very slender, more or less equal, smooth. Ring absent. Flesh pallid and fragile.
Odour not distinctive. **Taste** not distinctive.
Chemical tests none.
Occurrence late summer to autumn; frequent.
■ Inedible.

Coprinus micaceus (Bull.: Fr.) Fr.   Coprinaceae

Yellowish-tan agaric with mica-like particles on the cap, eventually blackening; in dense clusters, sometimes extensive, on or around the stumps of broad-leaf trees, less frequently arising from buried wood.

Dimensions cap 1 - 4 cm tall x variable dia; stem 4 - 10 cm tall x 0.2 - 0.5 cm dia.
Cap ochre-brown, becoming cinnamon towards the centre; at first ovoid, then bell-shaped and expanded-conical, surface deeply sulcate almost to the centre and splitting at margin, covered at first with pallid, mica-like, glistening velar particles which eventually disperse. Flesh pallid, thin except at centre, auto-digesting.
Gills white, becoming date-brown and eventually black, adnexed or free, close. **Spores** dark-brown, smooth, mitriform, 7-10 x 4.5-6 μm. Basidia 4-spored. Cystidia not distinctive.
Stem white with buff tinge below, at first finely downy, becoming smooth. Ring absent. Flesh pallid, hollow or loosely stuffed.
Odour not distinctive. **Taste** not distinctive.
Chemical tests velar remnants pinkish with KOH or NH₃.
Occurrence early summer to winter; very common.
☐ Edible.
Note: *C. truncorum* does not appear downy in the young stem.

Coprinus miser (Karst.) Karst.   Coprinaceae
= *Coprinus subtilis* (Fr.) Quél.

Very small, delicate brownish agaric with egg-shaped cap; solitary or in small trooping groups on cow dung.

Dimensions cap 0.1 - 0.3 cm tall x variable dia; stem 0.3 - 0.6 cm tall x 0.05 - 0.1 cm dia.
Cap brownish-grey, more reddish at the centre, becoming grey at maturity; at first ovoid, then conical-convex and finally flattened, faintly sulcate as far as the centre, otherwise more or less smooth, without velar remnants. Flesh pallid, thin, fragile and barely auto-digesting.
Gills buff, then greyish-brown, adnexed or free, close. **Spores** dark brown, smooth, more or less heart-shaped, germ pore, 7-9 x 4-5 µm. Basidia 4-spored. Cystidia not distinctive.
Stem whitish, diminutive, more or less equal, smooth. Ring absent. Flesh pallid and fragile.
Odour not distinctive. **Taste** not distinctive.
Chemical tests none.
Occurrence summer to autumn; frequent.
■ Inedible.

Coprinus niveus (Pers.: Fr.) Fr.   Coprinaceae

Fragile pale agaric with chalk-white mealy covering, margin splitting, blackening; solitary or in small groups on cow or horse dung and manured straw.

Dimensions cap 1.5 - 3 cm tall x variable dia; stem 3 - 9 cm tall x 0.4 - 0.7 cm dia.
Cap pallid whitish, covered with chalk-white, coarsely pruinose velar remnants; at first ovoid or conical, becoming expanded-campanulate, smooth or faintly sulcate, the margin splitting and finally reflexed. Flesh thin, fragile and auto-digesting.
Gills at first white, rapidly becoming black, adnate or free, close. **Spores** black, ellipsoid, smooth, germ pore, 15-19 x 11-13 µm. Basidia 4-spored. Gill face cystidia more or less clavate, abundant. Smooth or faintly sulcate.
Stem white, at first finely pruinose, becoming smooth and more or less silky, downy at base. Ring absent. Flesh hollow and fragile.
Odour not distinctive. **Taste** not distinctive.
Chemical tests none.
Occurrence summer to autumn; infrequent.
■ Inedible.

Coprinus patouillardii Quél.   Coprinaceae
= *Coprinus cordisporus* Gibbs.

Very small, delicate, pale translucent agaric; in small groups on manure.

Dimensions cap 0.15 - 0.7 cm tall x variable dia; stem 1 - 3 cm tall x 0.05 - 0.15 cm dia.
Cap buff, more cinnamon at centre and later with greyish tinge towards the margin with mealy velar remnants which flake away; at first ovoid or cylindrical, then convex, surface sulcate and smooth. Flesh very thin and fragile, rapidly auto-digesting.
Gills pallid clay, soon grey-brown and finally blackish, adnexed or free, somewhat distant. **Spores** black, smooth, narrowly ellipsoid, 6.5-11.5 x 3-5 µm. Basidia 4-spored. Gill edge cystidia tapered, gill face cystidia absent.
Stem semi-translucent whitish, very slender. Ring absent. Flesh brittle and very delicate.
Odour faint, rancid. **Taste** not distinctive.
Chemical tests none.
Occurrence summer to autumn; common.
■ Inedible.

Coprinus pellucidus Karst.   **Coprinaceae**

Very small, delicate, pale greyish agaric; loosely tufted on rotting and manured straw.

Dimensions cap 0.15 - 0.7 cm tall x variable dia; stem 1 - 3 cm tall x 0.05 - 0.15 cm dia.
Cap buff, more cinnamon at centre and later with greyish tinge towards the margin; at first ovoid or cylindrical, then convex, sulcate, very finely downy. Flesh very thin and fragile, rapidly auto-digesting. Cap cystidia tapered, 20-125 x 2-12 µm.
Gills pallid clay, soon grey-brown and finally blackish, adnexed or free, crowded. **Spores** black, smooth, narrowly ellipsoid, 6.5-11.5 x 3-5 µm. Basidia 4-spored. Gill edge cystidia utriform or tapered; gill face cystidia absent.
Stem semi-translucent whitish, slender. Ring absent. Flesh brittle and very delicate.
Odour faint, rancid. **Taste** not distinctive.
Chemical tests none.
Occurrence summer to autumn; common.
■ Inedible.

Coprinus picaceus (Bull.: Fr.) S F Gray   **Coprinaceae**

Tall, dark, greyish-brown agaric, covered with striking white patches, blackening; solitary or scattered, on soil (often calcareous), with beech.

Dimensions cap 5 - 8 cm tall x variable dia; stem 9 - 30 cm tall x 0.6 - 1.5 cm dia.
Cap dark greyish-brown or sepia, covered with whitish velar patches which slowly flake away; at first ovoid or cylindrical, becoming convex and then broadly campanulate, the margin finally slightly reflexed. Flesh thin, fragile and auto-digesting.
Gills at first white, becoming clay-pink and then rapidly black, adnexed or free, crowded. **Spores** black, ellipsoid, smooth, germ pore, 13-17 x 10-12 µm. Basidia 4-spored. Cystidia not distinctive.
Stem white, covered with downy or woolly hairs at first, swollen at base. Ring absent. Flesh hollow and fragile.
Odour strong, unpleasant. **Taste** not distinctive.
Chemical tests none.
Occurrence late summer to autumn; infrequent.
■ Inedible.

Coprinus plicatilis (Curt.: Fr.) Fr.   **Coprinaceae**

Small, delicate, buff, parasol-like agaric; solitary or scattered on soil in lawns and other short grass.

Dimensions cap 0.5 - 1.5 cm tall x variable dia; stem 3 - 7 cm tall x 0.1 - 0.2 cm dia.
Cap buff, more cinnamon at the centre and later with grey tinge at the margin; at first ovoid or cylindrical, then campanulate and finally shallowly convex like a parasol, wholly flat, or slightly depressed at the centre, surface deeply sulcate, otherwise smooth, without velar remnants. Flesh very thin and fragile; withering with age, no obvious auto-digestion.
Gills pallid clay, sometimes with pinkish tinge, soon grey and finally black, somewhat distant, separated from the stem apex by a distinct collar. **Spores** black, smooth, almond-shaped, 10-13 x 8.5-10.5 µm. Basidia 4-spored. Cystidia not distinctive.
Stem semi-translucent white or buff, particularly near the base more or less equal, slender, with slight basal bulb. Ring absent. Flesh brittle and very delicate.
Odour not distinctive. **Taste** not distinctive.
Chemical tests none.
Occurrence spring to autumn; common.
□ Edible but worthless.

Coprinus pseudoradiatus (Kühn. & Joss.) Watl.
Coprinaceae

Small, delicate, egg-shaped agaric; solitary or in groups on horse dung.

Dimensions cap 0.3 - 0.5 cm tall x variable dia; stem 2 - 5 cm tall x 0.05 - 0.1 cm dia.
Cap at first whitish, shaggy with velar remnants which fall away, then soot-grey; at first ovoid or cylindrical becoming campanulate-expanded, sulcate. Flesh very thin and fragile; limited auto-digestion.
Gills pallid clay, sometimes with pinkish tinge, soon grey and finally black, distant, free. **Spores** black, smooth, ellipsoid, 7-9 x 4-5.5 μm. Basidia 4-spored. Cystidia not distinctive.
Stem hyaline, minutely pruinose or woolly. Ring absent. Flesh brittle and very delicate.
Odour not distinctive. **Taste** not distinctive.
Chemical tests none.
Occurrence spring to autumn; frequent.
■ Inedible.
Note: Taller than *C. radiatus* with pruinose hyaline stem and smaller spores.

Coprinus radiatus (Bolt.: Fr.) S F Gray　　　**Coprinaceae**

Small, delicate, egg-shaped agaric; solitary or in groups, on horse dung.

Dimensions cap 0.2 - 0.8 cm tall x variable dia; stem 2 - 3 cm tall x 0.05 - 0.1 cm dia.
Cap at first whitish, shaggy with velar remnants which fall away, then soot-grey; at first ovoid or cylindrical becoming campanulate-expanded, sulcate. Flesh very thin and fragile; limited auto-digestion.
Gills pallid clay, sometimes with pinkish tinge, soon grey and finally black, distant, free. **Spores** black, smooth, ellipsoid, 11-12.5 x 6.5-7.5 μm. Basidia 4-spored. Cystidia not distinctive.
Stem white or buff. Ring absent. Flesh brittle and very delicate.
Odour not distinctive. **Taste** not distinctive.
Chemical tests none.
Occurrence spring to autumn; common.
■ Inedible.

Coprinus stercoreus (Bull.) Fr.　　　**Coprinaceae**

Small, delicate, greyish agaric with white mealy covering when young, blackening; solitary or in small groups on dung and manured straw.

Dimensions cap 0.25 - 1.25 cm tall x variable dia; stem 1 - 4 cm tall x 0.1 - 0.2 cm dia.
Cap greyish, with brown tinge, with whitish velar covering which flakes off revealing radial sulcation; at first ovoid, then convex and finally flattened with split and reflexed margin. Flesh thin, fragile and auto-digesting.
Gills whitish, soon dark brown and finally black, adnate or free, fairly distant. **Spores** black, smooth, elongated-ellipsoid or reniform, germ pore, 7-8 x 3.5-4 μm. Basidia 4-spored. Cystidia not distinctive.
Stem at first slightly pruinose or scaly, then smooth, cottony at base. Ring absent. Flesh brittle and very delicate.
Odour not distinctive. **Taste** not distinctive.
Chemical tests none.
Occurrence spring to autumn; infrequent.
■ Inedible.

Coprinus truncorum (Schaeff.) Fr.   Coprinaceae

Yellowish-tan agaric with mica-like particles on the cap, eventually blackening; in dense tufts on or around stumps of trees, less frequently arising from buried wood.

Dimensions cap 1 - 3 cm tall x variable dia; stem 5 - 9 cm tall x 0.2 - 0.5 cm dia.
Cap ochre-brown, becoming cinnamon towards centre; at first ovoid, then bell-shaped and expanded-conical, surface deeply sulcate almost to centre and splitting at margin, covered at first with pallid, mica-like, glistening velar particles which eventually flake off. Flesh pallid, thin except at margin, limited auto-digestion.
Gills white, becoming purplish-brown and eventually black, adnexed or free, close. **Spores** dark brown, smooth, ellipsoid or almond-shaped, germ pore, 6.5-8.5 x 5-6.5 µm. Basidia 4-spored. Cystidia not distinctive.
Stem white with buff tinge in the lower part, at first finely downy, becoming smooth. Ring absent. Flesh pallid, hollow or loosely stuffed.
Odour not distinctive. **Taste** not distinctive.
Chemical tests velar remnants pinkish with KOH or NH_3.
Occurrence early summer to winter; infrequent.
■ Inedible.
Note: *C. micaceus* differs in spore characteristics and in the wholly smooth young stem.

Lacrymaria velutina (Pers.: Fr.) Konr. & Maubl.Coprinaceae
= *Hypholoma velutinum* (Pers.: Fr.) Kummer
= *Psathyrella lacrymabunda* (Fr.) Moser

Medium-sized agaric, yellowish-brown cap with ragged margin, dark brown gills and ring-zone; solitary or tufted, on soil amongst grass by paths in woods, and on roadside verges.

Dimensions cap 2 - 10 cm dia; stem 4 - 8 cm tall x 0.5 - 1 cm dia.
Cap ochraceous-tan, becoming darker brown; convex and broadly umbonate becoming expanded, at first woolly fibrous then smooth, margin ragged with velar remnants. Flesh ochraceous-brown, rather soft and thick at centre.
Gills at first dark clay, becoming purplish-brown and more or less black at maturity with pallid edges, 'weeping' when damp, adnate or adnexed, close. **Spores** black, warty, lemon-shaped, with truncated germ pore, 8-12 x 5-7 µm. Basidia 4-spored. Gill edge cystidia filamentous, some more or less capitate; gill face cystidia absent.
Stem whitish above, flushed cap colour and fibrous-scaly below ring-zone, more or less equal. Ring zone-like, fibrous-scaly. Flesh ochraceous-brown, fibrous, hollow.
Odour not distinctive. **Taste** slight, bitter.
Chemical tests none.
Occurrence summer to autumn; frequent.
☐ Edible but poor.

Panaeolus ater (J Lange) Kühn. & Romagn. ex Bon
= *Panaeolus fimicola* (Fr.) Quél. s. Ricken   Coprinaceae

Small brown agaric, drying paler from margin to centre, with mottled gills becoming dark brown; in trooping groups, on soil in lawns, in parks and other grassy situations.

Dimensions cap 1.5 - 4.5 cm dia; stem 3 - 8 cm tall x 0.2 - 0.5 cm dia.
Cap darkish brown when damp, hygrophanous, drying pallid tan or buff; hemispherical, becoming expanded-convex with poorly defined umbo, smooth. Flesh brown and thin.
Gills at first pallid grey, becoming mottled darker brown with pallid margin and finally black, adnate, crowded. **Spores** black, smooth, lemon-shaped, oblique germ pore, 10-14 x 7-8 µm. Basidia 4-spored. Gill edge cystidia flask-shaped; gill face cystidia generally more or less clavate.
Stem pallid above, otherwise concolorous with cap, white-downy at base, equal, slender. Ring absent. Flesh brown and fragile.
Odour not distinctive. **Taste** not distinctive.
Chemical tests none.
Occurrence spring to autumn; common.
■ Inedible.

Panaeolus fimicola (Fr.) Quél. **Coprinaceae**
= *Agaricus fimicola* Fr.

Small greyish agaric with mottled gills, becoming blackish; in trooping groups on soil in lawns, parks and other grassy situations.

Dimensions cap 1.5 - 3.5 cm dia; stem 6 - 8 cm tall x 0.1 - 0.2 cm dia.
Cap greyish, with sepia or hazel tinges, hygrophanous, drying pallid grey; hemispherical or somewhat umbonate, smooth. Flesh pallid and thin.
Gills at first pallid olivaceous, becoming mottled darker and finally black with white edge, adnate, crowded. **Spores** blackish-grey, smooth, ellipsoid or lemon-shaped, germ pore, 10-14 x 7-8 µm. Basidia 4-spored. Gill edge cystidia fusiform, typically with long necks; gill face cystidia absent.
Stem clay, becoming darker brownish towards base, minutely pruinose at the apex, equal, very slender. Ring absent. Flesh dirty ochraceous-buff and fragile.
Odour not distinctive. **Taste** not distinctive.
Chemical tests none.
Occurrence summer to autumn; common.
■ Inedible.

Panaeolina foenisecii (Pers.: Fr.) Maire **Coprinaceae**
= *Panaeolus foenisecii* (Pers.: Fr.) Schroet.
= *Psilocybe foenisecii* (Pers.: Fr.) Quél.

Small brown agaric, drying paler from centre to margin, with mottled gills becoming dark brown; in trooping groups on soil in lawns, parks and other grassy situations.

Dimensions cap 1 - 2 cm dia; stem 4 - 7 cm tall x 0.2 - 0.3 cm dia.
Cap dull brown when damp, more reddish-brown at centre, hygrophanous, drying pallid clay; campanulate or convex, becoming expanded-convex, smooth. Flesh buff-brown and thin.
Gills at first pallid brown, becoming mottled darker and finally chocolate brown, adnate, crowded. **Spores** blackish-brown, warty, ellipsoid or lemon-shaped, large germ pore, 12-15 x 7-8.5 µm. Basidia 4-spored. Gill edge cystidia cylindrical to fusiform or flask-shaped; gill face cystidia absent.
Stem more pallid than cap, otherwise concolorous, equal, slender. Ring absent. Flesh buff-brown and fragile.
Odour not distinctive. **Taste** not distinctive.
Chemical tests none.
Occurrence summer to autumn; common.
■ Inedible.

Panaeolus papilionaceus (Bull.: Fr.) Quél. **Coprinaceae**

Small whitish agaric, with mottled gills becoming black; in trooping groups in fields and on dung.

Dimensions cap 2 - 3 cm dia; stem 5 - 10 cm tall x 0.4 - 0.8 cm dia.
Cap whitish-grey, with ochraceous tinge towards centre, hemi-spherical or more flattened-convex, dry, smooth or finely crazed, margin regularly dentate. Flesh whitish and thin.
Gills at first greyish, becoming mottled black and finally black, sometimes with a whitish margin, broad, adnate, crowded.
Spores black, smooth, lemon-shaped or ellipsoid, germ pore, 11-16 x 9-12 µm. Basidia 4-spored. Gill edge cystidia cylindrical; gill face cystidia absent.
Stem concolorous with cap, equal or tapering slightly upwards, slender. Ring absent. Flesh whitish and brittle.
Odour not distinctive. **Taste** not distinctive.
Chemical tests none.
Occurrence summer to autumn; infrequent.
■ Inedible.

Panaeolina rickenii Hora   **Coprinaceae**

Small dark brown agaric, drying paler, with mottled gills becoming blackish with white edge; in trooping groups on soil in lawns, parks and other grassy situations.

Dimensions cap 0.5 - 2 cm dia; stem 5 - 10 cm tall x 0.1 - 0.2 cm dia.
Cap dark brown when damp, hygrophanous, drying reddish-brown or bay; elongated-campanulate, barely expanding, smooth, margin producing small watery droplets. Flesh clay or vinaceous-buff and thin.
Gills at first pallid greyish-brown, becoming mottled darker and finally black with a white edge producing droplets, adnate, crowded.
Spores blackish-purple, smooth, germ pore, 13-16 x 8-9 µm. Basidia 4-spored. Gill edge cystidia cylindrical; gill face cystidia absent.
Stem vinaceous-buff at the apex, darkening towards base, equal but slightly bulbous at the base, very slender. Ring absent. Flesh buff-brown and fragile.
Odour not distinctive. **Taste** not distinctive.
Chemical tests none.
Occurrence summer to autumn; common.
■ Inedible.

Panaeolus semiovatus (Sow.: Fr.) Lund.   **Coprinaceae**
= *Panaeolus separatus* (L.: Fr.) Gill.
= *Anellaria separata* (L.: Fr.) Karst.

Smallish cream agaric, drying distinctively shiny, with mottled gills becoming dark brown; in trooping groups on dung.

Dimensions cap 2 - 6 cm dia; stem 5 - 10 cm tall x 0.4 - 0.8 cm dia.
Cap pallid clay, with yellow tinge towards the centre; hemispherical or ovoid, smooth, sticky when damp, drying shiny and slightly crazed and wrinkled. Flesh whitish and thin.
Gills at first whitish, becoming mottled darker brown and finally black, sometimes with a pallid margin, adnate, crowded.
Spores black, smooth, ellipsoid, germ pore, 16-20 x 9-12 µm. Basidia 4-spored. Gill edge cystidia variable but mainly flask-shaped; gill face cystidia mainly ovoid.
Stem concolorous with cap, equal or slightly thickened at base, slender. Ring median-superior, white, fragile, persistent. Flesh white, tinged yellowish and brittle.
Odour not distinctive. **Taste** not distinctive.
Chemical tests none.
Occurrence spring to winter; locally frequent otherwise infrequent.
■ Inedible.

Panaeolus sphinctrinus (Fr.) Quél.   **Coprinaceae**
= *Panaeolus campanulatus var sphinctrinus* (Fr.) Quél.

Small greyish agaric, drying paler, with mottled gills becoming black; in trooping groups on soil in pastureland and other grassy situations, also on rotting straw, favouring proximity to dung.

Dimensions cap 2 - 4 cm dia; stem 6-12 cm tall x 0.2 - 0.3 cm dia.
Cap darkish grey when damp, hygrophanous, drying pallid grey with ochraceous tinge at centre; broadly conical or campanulate, smooth but with velar remnants overhanging the margin as fine regularly spaced 'teeth'. Flesh pallid grey-brown and thin.
Gills at first pallid grey, becoming mottled darker grey and finally black, broad, adnate, crowded. **Spores** black, smooth, lemon-shaped, germ pore, 14-18 x 10-11.5 µm. Basidia 4-spored. Gill edge cystidia cylindrical; gill face cystidia absent.
Stem pallid grey, minutely floury, long, equal, slender. Ring absent. Flesh pallid grey-brown and fragile.
Odour not distinctive. **Taste** not distinctive.
Chemical tests none.
Occurrence spring to autumn; common.
■ Inedible.

Panaeolus subbalteatus (Berk. & Br.) Sacc. Coprinaceae

Small brown agaric, drying paler with characteristically darker marginal zone, mottled gills becoming black; in trooping groups on manured ground and compost heaps.

Dimensions cap 2 - 6 cm dia; stem 6 - 9 cm tall x 0.3 - 0.5 cm dia.
Cap darkish brown when damp, hygrophanous, drying pallid tan or buff from centre outwards; at first convex, becoming expanded with poorly defined umbo, smooth. Flesh brown and thin.
Gills at first pallid tan, becoming mottled darker brown and finally black, fairly broad, adnate, crowded. **Spores** black, smooth, ellipsoid or lemon-shaped, large germ pore, 12-14 x 7.5-8.5 µm. Basidia 4-spored. Gill edge cystidia flask-shaped; gill face cystidia absent.
Stem pallid above otherwise concolorous with cap, silky fibrous, equal, slender. Ring absent. Flesh brown and fragile.
Odour not distinctive. **Taste** not distinctive.
Chemical tests none.
Occurrence summer to autumn; infrequent.
■ Inedible.

Psathyrella ammophila (Dur. & Lév.) P Orton
Coprinaceae

Small agaric with clay-brown cap, brownish stem, and dark brown gills; solitary or scattered on sandy soil with marram grass (*Ammophila*).

Dimensions cap 1 - 2.5 cm dia; stem 3 - 5 cm tall x 0.1 - 0.3 cm dia.
Cap pallid clay or brownish, not hygrophanous; conical-convex or campanulate, smooth with microscopic hairs. Flesh pallid and thin.
Gills at first pallid, becoming dark brown or black, adnate or adnexed, fairly broad, crowded. **Spores** dark brown, smooth, ellipsoid, large germ pore, 10-11 x 6-7 µm. Basidia 4-spored. Gill edge cystidia flask-shaped, thin-walled at the neck.
Stem pallid brown, smooth, more or less equal, slender, deeply rooting. Ring absent. Flesh pallid, stuffed or hollow.
Odour not distinctive. **Taste** not distinctive.
Chemical tests none.
Occurrence summer to autumn; localised.
■ Inedible.

Psathyrella artemisiae (Pass.) Konrad & Maubl.
Coprinaceae

= *Psathyrella squamosa* (Karst.) Moser apud Gams
= *Drosophila squamosa* (Karst.) Kühn. & Romagn.

Small fragile agaric with whitish cottony cap, white flaky stem, and white or dark brown gills; solitary or scattered on soil in beech-woods.

Dimensions cap 2.5 - 3.5 cm dia; stem 3 - 6 cm tall x 0.3 - 0.5 cm dia.
Cap ochraceous with brown tinge, drying cream, decorated with white woolly fibrils, particularly towards the margin; conical convex. Flesh pallid and thin.
Gills at first pallid cream, becoming violaceous-brown, adnate or adnexed, crowded. **Spores** purplish-brown, smooth, ellipsoid, large germ pore, 8.5-9.5 x 4.5-5 µm. Basidia 4-spored. Gill cystidia narrowly flask-shaped, thick-walled, tinged yellow.
Stem white, flaky fibrillose, more or less equal, slender. Ring white, flaky fibrous, soon disappearing. Flesh pallid, fragile and hollow.
Odour not distinctive. **Taste** not distinctive.
Chemical tests none.
Occurrence summer to autumn; infrequent.
■ Inedible.

Psathyrella atomata (Fr.) Quél.   Coprinaceae

Small fragile agaric with mainly ash-grey cap, almost translucent stem, and greyish or black gills; solitary scattered on soil in grasslands.

Dimensions cap 0.5 - 2 cm dia; stem 6 - 10 cm tall x 0.05 - 0.15 cm dia.
Cap pallid ochraceous with brown tinge, drying pallid ash-grey and slightly glistening; conical convex, smooth but finely striate or sulcate towards the margin. Flesh pallid and thin.
Gills at first pallid grey, becoming black, adnate or adnexed, crowded. **Spores** dark brown, smooth, ellipsoid, large germ pore, 14-17 x 7-8 μm. Basidia 4-spored. Gill cystidia flask-shaped, thin-walled, occasionally with small encrustations.
Stem pallid, almost translucent, smooth, more or less equal, very slender. Ring absent. Flesh pallid, fragile and hollow.
Odour not distinctive. **Taste** not distinctive.
Chemical tests none.
Occurrence summer to autumn; frequent.
■ Inedible.

Psathyrella candolleana (Fr.) Maire   Coprinaceae

Smallish agaric with pale yellowish or cream cap and stem, greyish or chocolate-brown gills; typically tufted on soil close to broad-leaf trees, also on stumps and other woody debris.

Dimensions cap 2 - 6 cm dia; stem 4 - 8 cm tall x 0.4 - 0.8 cm dia.
Cap pallid ochraceous with brown tinge, drying cream or almost white; at first campanulate, becoming expanded, more or less flattened, smooth but typically with ragged velar remnants at the margin. Flesh white and thin.
Gills at first pallid grey with lilaceous tinge, becoming chocolate brown, adnate or adnexed, crowded. **Spores** dark brown, smooth, ellipsoid, germ pore, 6-8 x 3.5-4.5 μm. Basidia 4-spored. Gill edge cystidia cylindrical, thin-walled; gill face cystidia absent.
Stem white, more or less equal, smooth. Ring absent. Flesh white, fragile and hollow.
Odour not distinctive. **Taste** not distinctive.
Chemical tests none.
Occurrence summer to autumn; frequent.
■ Inedible.

Psathyrella clivensis (Berk. & Br.) P Orton   Coprinaceae

Smallish agaric with brown or cream cap, whitish stem, and clay then purplish gills; solitary or grouped on calcareous soil in grasslands.

Dimensions cap 1.5 - 3 cm dia; stem 3 - 4 cm tall x 0.15 - 0.3 cm dia.
Cap date-brown, hygrophanous, drying ochraceous-cream; hemispherical becoming conical-convex, barely striate towards the margin otherwise smooth, shiny when damp. Flesh white and thin.
Gills at first pallid, becoming coffee, then purplish-black, adnate or emarginate, close. **Spores** dark brown, smooth, ellipsoid, germ pore, 8-11 x 4.5-5.5 μm. Basidia 4-spored. Gill edge cystidia scarce, more or less fusiform, thin-walled; gill face cystidia utriform.
Stem white or tinged brown at base, pruinose at apex, otherwise silky, striate, more or less equal, slender. Ring absent. Flesh white, fragile and hollow.
Odour not distinctive. **Taste** not distinctive.
Chemical tests none.
Occurrence spring to autumn; rare. (Hope Woods, Gloucestershire)
■ Inedible.

Psathyrella coronata (Fr.) Moser **Coprinaceae**

Smallish agaric with whitish scaly cap and stem, and clay then purplish gills; solitary or grouped on soil in grasslands.

Dimensions cap 1.5 - 3 cm dia; stem 3 - 4 cm tall x 0.15 - 0.3 cm dia.
Cap whitish, more ochraceous at centre, decorated with soft ochraceous flakes; conical-convex, barely striate towards margin which bears velar remnants. Flesh white and thin.
Gills at first pallid, becoming purplish-black, adnate or adnexed, crowded. **Spores** dark brown, smooth, ellipsoid, germ pore, 8-9.5 x 4-4.5 µm. Basidia 4-spored. Gill cystidia flask-shaped, thin-walled.
Stem white or tinged yellow, finely floccose, more or less equal, slender. Ring absent. Flesh white, fragile and hollow.
Odour not distinctive. **Taste** not distinctive.
Chemical tests none.
Occurrence summer to autumn; infrequent.
■ Inedible.

Psathyrella corrugis (Pers.: Fr.) Konrad & Maubl.
Coprinaceae

Small agaric with dull brown, bell-shaped, wrinkled cap, greyish-brown gills and slender silvery mottled stem; in trooping groups or tufted on soil amongst plant debris in open mixed woodland.

Dimensions cap 2 - 3.5 cm dia; stem 5 - 13 cm tall x 0.1 - 0.3 cm dia.
Cap dull greyish-brown; at first conical, becoming campanulate, markedly radially sulcate. Flesh brownish-buff and thin.
Gills at first pallid greyish-brown with lilaceous tinge, becoming blackish-brown, adnate or adnexed, crowded. **Spores** black, smooth, ellipsoid, germ pore, 11-13 x 6.5-7 µm. Basidia 4-spored. Gill cystidia tapering, awl-shaped.
Stem white, more or less equal, slender, smooth, mottled silvery, downy at base, slightly rooting. Ring absent. Flesh white, fragile and hollow.
Odour not distinctive. **Taste** not distinctive.
Chemical tests none.
Occurrence summer to autumn; infrequent.
■ Inedible.

Psathyrella cotonea (Quél.) Konrad & Maubl. **Coprinaceae**

Medium-sized agaric with striking whitish scaly cap, and stem, and pallid then grey-brown gills; tufted on or near stumps and felled trunks of broad-leaf trees favouring beech.

Dimensions cap 4 - 8 cm dia; stem 5 - 12 cm tall x 0.6 - 0.8 cm dia.
Cap whitish, tinged ochraceous with age, decorated with scaly fibres, at first pallid grey, becoming darker grey-brown; at first ovoid or convex, expanding and becoming slightly umbonate. Flesh white, soft and medium.
Gills at first pallid, becoming greyish-brown, edges sometimes weeping, adnate or adnexed, crowded. **Spores** dark brown, smooth, narrowly ellipsoid, germ pore, 7-9.5 x 3-4.5 µm. Basidia 4-spored. Gill cystidia flask-shaped, thin-walled.
Stem concolorous with cap but tinged yellow at base, finely floccose, more or less equal, slender. Ring absent. Flesh white, soft, stuffed or hollow, arising from distinctive yellow mycelium.
Odour not distinctive. **Taste** not distinctive.
Chemical tests none.
Occurrence late summer to autumn; infrequent.
■ Inedible.

Psathyrella hydrophila (Bull.) Maire　　　**Coprinaceae**

= *Psathyrella piluliformis* (Bull.: Fr) P Orton
= *Hypholoma hydrophilum* (Bull.) Quél.

Small agaric with brownish-tan cap, silvery mottled stem, clay or chocolate-brown gills; densely tufted on stumps and woody debris in broad-leaf woods.

Dimensions cap 2 - 3 cm dia; stem 4 - 10 cm tall x 0.3 - 0.6 cm dia.
Cap tan or date-brown, drying more pallid brown; at first convex, becoming expanded, smooth but young cap with velar remnants at the margin. Flesh whitish and thin.
Gills at first clay, becoming chocolate-brown at maturity, adnate or adnexed, fairly crowded. **Spores** dark brown, smooth, ellipsoid, germ pore, 4.5-7 x 3-4 µm. Basidia 4-spored. Gill cystidia fusiform, thin-walled.
Stem white, more or less equal, smooth, mottled silver. Ring absent. Flesh white, fibrous and hollow.
Odour not distinctive. **Taste** bitter.
Chemical tests none.
Occurrence summer to autumn; common.
■ Inedible.

Psathyrella impexa (Romagn.) Gall.　　　**Coprinaceae**

Small fragile agaric with brownish cottony cap, white flaky stem, and white or dark brown gills; solitary or scattered on soil in grass with broad-leaf and coniferous trees.

Dimensions cap 2.5 - 3.5 cm dia; stem 3 - 6 cm tall x 0.3 - 0.5 cm dia.
Cap ochraceous-brown, with pinkish tinge, drying pinkish-cream, decorated with white woolly fibrils, particularly towards the margin; conical convex. Flesh pallid and thin.
Gills at first pallid cream, becoming violaceous-brown, adnate or adnexed, crowded. **Spores** purplish-brown, smooth, ellipsoid, large germ pore, 8.5-9.5 x 4.5-5 µm. Basidia 4-spored. Gill cystidia broadly flask-shaped, thick-walled, tinged yellowish.
Stem white, flaky-fibrillose, more or less equal, slender. Ring white, flaky fibrous, soon disappearing. Flesh pallid, fragile and hollow.
Odour not distinctive. **Taste** not distinctive.
Chemical tests none.
Occurrence late summer to autumn; infrequent.
■ Inedible.
Distinguished from *P. artemisiae* by the pinkish tinge to the cap and the more conventionally shaped gill cystidia.

Psathyrella leucotephra (Berk. & Br.) P Orton **Coprinaceae**

Smallish agaric with white or brown tinged cap and stem, greyish-brown gills; typically tufted or in small groups, on soil in broad-leaf woods.

Dimensions cap 3 - 7 cm dia; stem 6 - 10 cm tall x 0.5 - 1.2 cm dia.
Cap whitish or with pallid brown tinge towards centre; at first campanulate, becoming expanded, more or less flattened, smooth or finely sulcate but typically with ragged velar remnants at margin. Flesh white and thin.
Gills at first whitish, becoming grey-buff, then blackish-brown with white edges, adnate or adnexed, crowded. **Spores** blackish-brown, smooth, ellipsoid, germ pore, 8-9 x 5-6 µm. Basidia 4-spored. Gill edge cystidia cylindrical, thin-walled; gill face cystidia absent.
Stem white, more or less equal, smooth. Ring superior, pendulous, fragile. Flesh white, fragile and hollow.
Odour not distinctive. **Taste** not distinctive.
Chemical tests none.
Occurrence late summer to autumn; infrequent.
■ Inedible.

Psathyrella marcescibilis (Britz.) Sing.   Coprinaceae
= *Hypholoma marcescibilis* (Britz.) Sacc.

Smallish agaric with clay brown or pale cap, silvery mottled stem, grey or chocolate-brown gills with pale edges; in trooping groups on soil, usually calcareous in gardens, pastures and damp shady places.

Dimensions cap 1.5 - 5 cm dia; stem 3 - 10 cm tall x 0.2 - 0.5 cm dia.
Cap grey-brown, drying whitish-clay; at first conical, becoming more campanulate, young cap finely woolly-fibrous with velar remnants, later more smooth but typically with appendicular remnants at margin, striate to halfway. Flesh whitish and thin.
Gills at first grey, becoming purplish-brown and more or less black at maturity but with whitish edges, almost free or adnexed, crowded.
Spores dark brown, smooth, ellipsoid, germ pore, 12-16 x 5-7 µm. Basidia 4-spored. Gill edge cystidia swollen at base with short neck, thin-walled; gill face cystidia absent.
Stem white, more or less equal, smooth or finely fibrous. Ring absent. Flesh white, hollow and fragile.
Odour not distinctive. **Taste** bitter.
Chemical tests none.
Occurrence spring to autumn; infrequent.
■ Inedible.

Psathyrella multipedata (Peck) A H Smith   Coprinaceae

Small agaric with clay-brown cap, silvery mottled stem, clay or chocolate-brown gills; densely tufted on soil in open broad-leaf woods.

Dimensions cap 1 - 3 cm dia; stem 7 - 12 cm tall x 0.2 - 0.4 cm dia.
Cap dull clay-brown, drying more pallid; conical-convex, radially striate to halfway. Flesh whitish and thin.
Gills at first clay, becoming purplish-brown at maturity, adnate or adnexed, fairly crowded. **Spores** dark brown, smooth, ellipsoid, germ pore, 6.5-10 x 3.5-4.5 µm. Basidia 4-spored. Gill edge cystidia narrowly fusiform with swollen base, thin-walled.
Stem white, flushed cap colour towards base, more or less equal, smooth, mottled silver. Ring absent. Flesh white, fragile and hollow.
Odour not distinctive. **Taste** bitter.
Chemical tests none.
Occurrence summer to autumn; rare. (Whitley Common, near Guildford, Surrey)
■ Inedible.

Psathyrella obtusata (Fr.) A H Smith   Coprinaceae
= *Drosophila obtusata* (Fr.) Kühn. & Romagn.

Small agaric with dull brown or pale cap, silvery mottled stem, pinkish or chocolate-brown gills; in small trooping groups on soil with plant debris in broad-leaf woods, also in grass of field borders close to trees.

Dimensions cap 1.5 - 4 cm dia; stem 3 - 7 cm tall x 0.2 - 0.4 cm dia.
Cap dull brown, with tan tinge at centre, drying cream; at first bluntly conical, becoming expanded-convex, smooth, margin striate when wet, finely sulcate when dry. Flesh whitish and thin.
Gills at first buff, becoming chocolate-brown at maturity, adnate or adnexed, fairly crowded. **Spores** dark brown, smooth, ellipsoid, 7.5-100 x 4.5-5.5 µm. Basidia 4-spored. Gill cystidia bluntly fusiform, thin-walled.
Stem white, more or less equal, smooth, silky. Ring absent. Flesh white, fibrous and hollow.
Odour not distinctive. **Taste** bitter.
Chemical tests none.
Occurrence spring to autumn; infrequent.
■ Inedible.

Psathyrella pseudogracilis (Romagn.) Moser
Coprinaceae

Small agaric with grey or pale cap, brown gills with a grey tinge and silky white stem; in small trooping groups on soil and woody debris in broad-leaf and coniferous woods and scrub.

Dimensions cap 2 - 4 cm dia; stem 8 - 12 cm tall x 0.1 - 0.3 cm dia.
Cap grey, with ochraceous tinge, more pallid with pinkish or vinaceous tinge when dry; at first conical, becoming convex, smooth. Flesh buff and thin.
Gills at first grey-brown then blackish-brown, adnate or adnexed, crowded. **Spores** dark brown (purplish-brown in the mass), smooth, ellipsoid or lemon-shaped, germ pore, 11.5-16.5 x 5.5-7.5 µm. Basidia 4-spored. Gill cystidia with blunt thickish tips; profuse on gill edges.
Stem whitish, more or less equal, smooth, silvery. Ring absent. Flesh whitish, stiff and hollow.
Odour not distinctive. **Taste** not distinctive.
Chemical tests none.
Occurrence late summer to autumn; infrequent.
■ Inedible.

Psathyrella sarcocephala (Fr.: Fr.) Sing.   **Coprinaceae**
= *Psathyrella spadicea* ss. J Lange

Smallish agaric with rich brown cap drying almost white, silvery white stem, clay or dark brown gills; typically tufted on or beside stumps of broad-leaf trees.

Dimensions cap 3 - 7 cm dia; stem 4 - 6 cm tall x 0.4 - 0.6 cm dia.
Cap date- or chocolate-brown, strongly hygrophanous, drying whitish-clay; at first convex, becoming expanded, smooth, non-striate. Flesh pallid concolorous and thin.
Gills at first clay, becoming dark brown with greyish tinge at maturity, adnate or adnexed, fairly crowded. **Spores** dark brown, smooth, ellipsoid, fairly large germ pore, 6-10 x 3.5-5.5 µm. Basidia 4-spored. Gill cystidia thick-walled with apical encrustations.
Stem white, more or less equal, smooth, silvery. Ring absent. Flesh white, fibrous and hollow.
Odour not distinctive. **Taste** not distinctive.
Chemical tests none.
Occurrence autumn; frequent.
■ Inedible.

Psathyrella spadiceogrisea (Fr.) Maire   **Coprinaceae**

Smallish agaric with dull brown cap drying almost white, silvery mottled stem, clay or chocolate-brown gills; typically solitary on soil in broad-leaf, and less frequently, coniferous woods.

Dimensions cap 3 - 6 cm dia; stem 6 - 10 cm tall x 0.3 - 0.6 cm dia.
Cap date-brown, strongly hygrophanous, drying whitish-clay; at first conical, becoming expanded but often with distinct umbo, striate when damp, otherwise smooth. Flesh whitish and thin.
Gills at first clay, becoming chocolate-brown with violaceous tinge at maturity, adnate or adnexed, fairly crowded. **Spores** blackish-brown, smooth, ellipsoid, fairly large germ pore, 6-10 x 3.5-5.5 µm. Basidia 4-spored. Gill edge cystidia balloon-shaped and distinct stem, apices sometimes divided, gill face cystidia bottle-shaped, thin-walled.
Stem white, more or less equal, smooth, mottled silvery. Ring absent. Flesh white, fibrous and hollow.
Odour not distinctive. **Taste** bitter.
Chemical tests none.
Occurrence spring to autumn; frequent.
■ Inedible.
One of the more distinctive spring species of *Psathyrella*.

Psathyrella subnuda (Karst.) A H Smith   **Coprinaceae**

Smallish agaric with brown cap, brown gills with a violet tinge and silky white stem; in small trooping groups on soil and woody debris in broad-leaf and coniferous woods and scrub.

Dimensions cap 2 - 3 cm dia; stem 6 - 8 cm tall x 0.2 - 0.4 cm dia.
Cap brown, more pallid when dry; at first conical, becoming convex, smooth or finely striate to halfway from margin. Flesh buff or greyish-brown and thin.
Gills at first brown with violaceous tinge, then blackish-brown, adnate or adnexed, crowded. **Spores** dark reddish-brown (purplish-brown in the mass), smooth, ellipsoid or lemon-shaped, germ pore, 7.5-9.5 x 4.5-5 µm. Basidia 4-spored. Gill cystidia narrowly clavate, thin-walled.
Stem whitish, more or less equal, smooth, silvery. Ring absent. Flesh whitish, fragile and hollow.
Odour not distinctive. **Taste** not distinctive.
Chemical tests none.
Occurrence spring to summer; infrequent.
■ Inedible.

Psathyrella vernalis (J Lange) Moser   **Coprinaceae**

Smallish agaric with yellow-brown or whitish cap, beige-brown gills and silky white stem; in small trooping groups on soil and woody debris in broad-leaf woods.

Dimensions cap 2 - 4 cm dia; stem 4 - 6 cm tall x 0.2 - 0.4 cm dia.
Cap ochraceous, whitish when dry; at first conical, becoming convex, smooth. Flesh pallid and thin.
Gills at first cream, then beige-brown, adnate or adnexed, crowded. **Spores** pallid brown, smooth, ellipsoid or lemon-shaped, germ pore, 7.5-9.5 x 4-5 µm. Basidia 4-spored. Gill cystidia flask-shaped.
Stem whitish or with yellow tinge, more or less equal, smooth, silvery. Ring absent. Flesh whitish, fragile and hollow.
Odour not distinctive. **Taste** not distinctive.
Chemical tests none.
Occurrence spring to summer; infrequent.
■ Inedible.

Agrocybe arvalis (Fr.) Sing.   **Bolbitiaceae**
= *Agrocybe tuberosa* (Henn.) Sing.

Smallish agaric with pale whitish-yellow cap, light brown gills, no ring on stem; solitary, scattered or in loose trooping groups on soil in pastures.

Dimensions cap 1.5 - 3 cm dia; stem 4 - 10 cm tall x 0.4 - 0.7 cm dia.
Cap pallid ochraceous with brown tinge; flattened convex becoming expanded, smooth, slightly greasy when damp, sulcate when dry. Flesh white, firm and thin.
Gills at first pallid buff, becoming reddish-brown at maturity, adnate or adnexed, close. **Spores** hazel-brown, smooth, ellipsoid or slightly allantoid, largish germ pore, 9-12 x 5-6 µm. Basidia 4-spored. Gill edge cystidia cylindrical or flask-shaped; gill face cystidia clavate or with 2-5 finger-like processes.
Stem whitish, smooth, pruinose, more or less equal, deeply rooting and arising from a black sclerotium buried in the substrate. Ring absent. Flesh whitish, fibrous, and full other than narrow central canal.
Odour not distinctive. **Taste** not distinctive.
Occurrence spring to summer; infrequent.
■ Inedible.

Agrocybe cylindracea (DC.: Fr.) Maire   **Bolbitiaceae**
= *Pholiota aegerita* (Brig.) Quél.

Medium-sized agaric with whitish cap, light brown gills, ring on stem; in tufts, on base of trunks or on soil in vicinity of broad-leaf trees, favouring willow and poplar.

Dimensions cap 4 - 10 cm dia; stem 5 - 10 cm tall x 1 - 1.5 cm dia.
Cap whitish or pallid buff; flattened-convex, becoming expanded and slightly depressed, at first smooth, becoming crazed near centre with age. Flesh white, firm and moderate.
Gills at first cream, turning hazel-brown at maturity, adnate or slightly decurrent, crowded. **Spores** hazel-brown, smooth, ellipsoid, minute germ pore, 8.5-10.5 x 5-6 µm. Basidia 4-spored. Gill edge cystidia utriform; gill face cystidia rarely seen.
Stem whitish smooth but pruinose at apex, more or less equal. Ring white, becoming brown with spores, spreading, superior, persistent. Flesh whitish, becoming tinged brown in stem base, fibrous and full.
Odour not distinctive. **Taste** nutty.
Chemical tests none
Occurrence throughout the year; infrequent.
☐ Edible.

Agrocybe gibberosa (Fr.) Fayod   **Bolbitiaceae**
= *Agarious gibberosus* Fr.
= *Pholiota gibberosa* (Fr.) Sacc.

Small agaric with buff or wine-tinged cap, buff or purplish gills, fragile ring on stem; in trooping groups on soil and woody debris.

Dimensions cap 1 - 3.5 cm dia; stem 3 - 4 cm tall x 0.4 - 0.5 cm dia.
Cap pallid buff, becoming more vinaceous-brown, particularly at the margin; flattened-convex becoming expanded, smooth or faintly striate at the margin. Flesh white, firm and thin.
Gills at first cream, then vinaceous-buff and purplish-brown at maturity, edges white, adnate or almost free, close. **Spores** purplish-brown, smooth, ellipsoid, minute germ pore, 7.5-9 x 5-6 µm. Basidia 4-spored. Gill cystidia utriform or flask-shaped.
Stem whitish, striate at the apex, fibrillose below, more or less equal or slightly swollen at the base. Ring white, becoming brown with spores, membraneous, superior, fragile and soon fragmenting. Flesh whitish, becoming tinged ochraceous in the stem base, fibrous, full.
Odour faint, of meal. **Taste** of meal, then unpleasant.
Chemical tests none.
Occurence spring; infrequent.
■ Inedible.

Agrocybe molesta (Lasch) Sing.   **Bolbitiaceae**
= *Agrocybe dura* (Bolt. Fr.) Sing.
= *Pholiota dura* (Bolt. Fr.) Kummer

Smallish agaric with whitish cap, light brown gills, ring on stem; typically in trooping groups on soil in pastures or amongst grass at roadsides.

Dimensions cap 3 - 7 cm dia; stem 5 - 8 cm tall x 0.3 - 0.7 cm dia.
Cap whitish with yellow tinge; flattened-convex, becoming expanded and slightly umbonate, smooth. Flesh white, firm and moderate.
Gills at first pallid clay, becoming darker brown at maturity, adnate, crowded. **Spores** hazel-brown, smooth, ellipsoid, prominent germ pore, 8.5-10.5 x 5-6 µm. Basidia 4-spored. Gill edge cystidia utriform or flask-shaped; gill face cystidia rarely seen.
Stem whitish, smooth but pruinose at apex, more or less equal. Ring white, becoming brown with spores, spreading, superior, ephemeral. Flesh whitish, fibrous and full.
Odour not distinctive. **Taste** slight, bitter.
Chemical tests none.
Occurrence spring to summer; infrequent.
☐ Edible but poor.

Agrocybe paludosa (J Lange) Kühn. & Romagn.
Bolbitiaceae

Small agaric with pale grey-brown cap, light brown gills becoming dirty-brown, ring on stem; in trooping groups, often rings, on soil in marshy fields or places liable to periodical flooding.

Dimensions cap 1.5 - 4 cm dia; stem 5 - 9 cm tall x 0.2 - 0.4 cm dia.
Cap pallid tawny with greyish tinge, hygrophanous, drying cream; convex, becoming expanded sometimes with faint umbo, smooth, greasy when damp, wrinkled when dry. Flesh pallid, firm and very thin.
Gills at first pallid buff, becoming dirty-brown at maturity, adnate or adnexed, close. **Spores** tobacco-brown, smooth, broadly ellipsoid, large germ pore, 8-11 x 5-7 µm. Basidia 4-spored. Gill edge cystidia clavate or utriform; gill face cystidia utriform or flask-shaped.
Stem whitish, smooth, silky, more or less equal. Ring whitish, membraneous, ascendant, sub-apical. Flesh whitish, firm and hollow.
Odour of meal. **Taste** not distinctive.
Chemical tests none.
Occurrence spring to summer; infrequent.
■ Inedible.
Easily distinguished from *A. sphaleromorpha* which has a bulbous base to stem; both species have ascendant rings.

Agrocybe pediades (Pers.: Fr.) Fayod   **Bolbitiaceae**

Smallish agaric with pale yellow or whitish cap, light brown gills, no ring on stem; solitary, scattered or in loose trooping groups on soil, often sandy in grassy places.

Dimensions cap 1.5 - 5 cm dia; stem 2 - 4 cm tall x 0.3 - 0.6 cm dia.
Cap pallid ochraceous, drying almost white; hemispherical or convex becoming expanded, smooth. Flesh white, firm and thin.
Gills at first cream, turning hazel-brown at maturity, adnate, crowded. **Spores** hazel-brown, smooth, ellipsoid, distinct germ pore, 10-12 x 6-7 µm. Basidia 4-spored. Gill edge cystidia, flask-shaped; gill face cystidia rarely seen.
Stem whitish, smooth but pruinose at apex, more or less equal. Ring absent or faint cortinal zone. Flesh whitish, becoming tinged brown in stem base, fibrous and full.
Odour not distinctive. **Taste** not distinctive.
Chemical tests none.
Occurrence summer to autumn; infrequent.
■ Inedible.

Agrocybe praecox (Pers.: Fr.) Fayod   **Bolbitiaceae**
= *Pholiota praecox* (Pers.: Fr.) Kummer

Medium or large agaric with cream cap, light brown gills, ring on stem; tufted or scattered on soil amongst grass, sometimes on rotting straw, etc. in thickets and woodland edges.

Dimensions cap 3 - 9 cm dia; stem 4 - 10 cm tall x 0.4 - 1.5 cm dia.
Cap pallid tan or cream, sometimes more brownish-ochre, hygrophanous, drying almost white; convex, becoming expanded, smooth or sometimes wrinkled and greasy when young, becoming slightly crazed and sulcate with age. Flesh whitish, firm and moderate.
Gills pallid buff, dirty-brown at maturity, adnate or adnexed, crowded. **Spores** hazel-brown, smooth, ellipsoid, prominent germ pore, 8.5-10 x 5-6 µm. Basidia 4-spored. Gill edge cystidia utriform or broadly fusiform; gill face cystidia flask-shaped or fusiform.
Stem whitish, discoloured brown with age, pruinose at the apex, more or less equal but with bulbous base. Ring white, becoming brown with spores, fibrous, spreading and often remaining partly attached to the cap margin, superior. Flesh buff, becoming tinged brown with age, fibrous and full.
Odour faint, of meal. **Taste** of nuts.
Chemical tests none.
Occurrence summer to autumn; infrequent.
□ Edible.

Agrocybe semiorbicularis (Bull.: Fr.) Fayod   Bolbitiaceae

Small agaric with yellow or whitish greasy cap, light brown gills, no ring on stem; solitary, scattered or in loose trooping groups, on soil or in grass.

Dimensions cap 1 - 3 cm dia; stem 3 - 4 cm tall x 1 - 2.5 cm dia.
Cap ochraceous, drying almost white; hemispherical or slightly expanded, smooth, greasy. Flesh white, firm and thin.
Gills at first cream, turning coffee-brown at maturity, adnate or slightly decurrent, crowded. **Spores** rust-brown, smooth, ellipsoid, indistinct germ pore, 10-14 x 8-11 μm. Basidia 4-spored. Gill edge cystidia flask-shaped; gill face cystidia rarely seen.
Stem pallid yellow or whitish, smooth, more or less equal. Ring absent. Flesh whitish, becoming tinged brown in stem base, fibrous and full.
Odour not distinctive. **Taste** not distinctive.
Chemical tests none.
Occurrence summer to autumn; frequent.
■ Inedible.

Agrocybe sphaleromorpha (Bull.: Fr.) Fayod   Bolbitiaceae

Smallish agaric with creamy-yellow cap, light brown gills, ring on stem; solitary, scattered or in loose trooping groups on soil in grass.

Dimensions cap 2 - 4 cm dia; stem 5 - 9 cm tall x 0.3 - 0.5 cm dia.
Cap pallid ochraceous, with brown tinge, cream at the margin; flattened-convex becoming expanded, smooth, slightly greasy when damp. Flesh white, firm and thin.
Gills at first pallid buff, becoming more brown at maturity, adnate or adnexed, close. **Spores** hazel-brown, smooth, ellipsoid or slightly allantoid, largish germ pore, 8.5-12 x 6-8 μm. Basidia 4-spored. Gill edge cystidia utriform; gill face cystidia flask-shaped or fusiform.
Stem whitish or tinged cap colour, smooth, more or less equal, with bulbous base. Ring white, membraneous, ascendant, sub-apical. Flesh whitish, fibrous and full other than narrow central canal.
Odour not distinctive. **Taste** not distinctive.
Chemical tests none.
Occurrence spring and summer; infrequent.
■ Inedible.
Easily distinguished from *A. paludosa* which has no bulbous base to stem; both species have ascendant rings.

Bolbitius lacteus J Lange   Bolbitiaceae

Delicate, pale yellowish, sticky agaric; scattered or in small tufts on soil in rich or manured grasslands, rotting manured straw.

Dimensions cap 0.3 - 1 cm dia; stem 5 - 6 cm tall x 0.1 - 0.3 cm dia.
Cap whitish, with chrome-yellow tinge, sticky when damp, otherwise dry, peelable; at first convex or campanulate, becoming flattened, margin at first striate, then more markedly sulcate. Flesh whitish and fragile.
Gills at first pallid straw, becoming rust or cinnamon, free, crowded. **Spores** rust or cinnamon, smooth, ellipsoid, 10.5-11.5 x 6-6.5 μm. Basidia 4-spored. Gill edge cystidia flask-shaped; gill face cystidia rarely seen.
Stem whitish yellow, pruinose, slender, fragile. Ring none. Flesh delicate and hollow.
Odour not distinctive. **Taste** not distinctive.
Chemical tests none.
Occurrence early summer to early autumn; rare.
■ Inedible.

Bolbitius reticulatus (Pers.: Fr.) Ricken   Bolbitiaceae
= *Pluteolus aleuriatus* (Fr.: Fr.) Karst.

Small or medium-sized agaric with creamy-grey slimy cap and pink gills; solitary or in small more or less tufted groups on rotting stumps and other woody debris of broad-leaf trees.

Dimensions cap 1.5 - 5 cm dia; stem 2.4 - 4 cm tall x 0.2 - 0.4 cm dia.
Cap pallid greyish-cream with lilaceous tinge; convex becoming more flattened, striate, viscid. Flesh whitish and thin.
Gills at first whitish, becoming pink and finally cinnamon, free, crowded. **Spores** pink, smooth, narrowly ellipsoid, 9-10.5 x 4-5 μm. Basidia 4-spored. Gill edge cystidia fusiform or utriform; gill face absent; all with hooked ends.
Stem whitish, more or less equal or slightly swollen at base, smooth. Ring absent. Flesh whitish, firm, full or stuffed.
Odour not distinctive. **Taste** not distinctive.
Chemical tests none.
Occurrence late summer to autumn; infrequent.
■ Inedible.

Bolbitius titubans (Bull.: Fr.) Fr.   Bolbitiaceae

Delicate, lemon-yellow, sticky agaric; scattered or in small clusters on soil in rich or manured grasslands and rotting manured straw.

Dimensions cap 1 - 4 cm dia; stem 3 - 10 cm tall x 0.2 - 0.4 cm dia.
Cap at first lemon-yellow, fading, margin more pallid and tinged greyish with age; at first ovoid, becoming campanulate, sulcate to the centre, viscid when damp otherwise smooth and shiny, peelable. Flesh thin and easily splitting radially.
Gills at first pallid yellow, becoming rust or cinnamon, adnate, crowded. **Spores** rust or cinnamon, smooth, ellipsoid, 13-15 x 7-9 μm. Basidia 4-spored. Gill edge cystidia flask-shaped; gill face cystidia rarely seen.
Stem whitish, wholly pruinose, downy at base. Ring absent. Flesh delicate and hollow.
Odour not distinctive. **Taste** not distinctive.
Chemical tests none.
Occurrence summer to autumn; frequent.
■ Inedible.
Note: Distinguished from *B. vitellinus* by colour, pruinosity of stipe and spore size.

Bolbitius vitellinus (Pers.: Fr.) Fr.   Bolbitiaceae

Delicate, chrome-yellow, sticky agaric; scattered or in small clusters on soil in rich or manured grasslands, and rotting manured straw.

Dimensions cap 1 - 4 cm dia; stem 3 - 10 cm tall x 0.2 - 0.4 cm dia.
Cap at first chrome-yellow, fading, margin more pallid and tinged greyish with age; at first ovoid becoming campanulate, sulcate at the margin, viscid when damp otherwise smooth and shiny, peelable. Flesh thin and easily splitting radially.
Gills at first pallid yellow, becoming rust or cinnamon, adnate, crowded. **Spores** rust or cinnamon, smooth, ellipsoid, 12-13 x 6-7 μm. Basidia 4-spored. Gill edge cystidia flask-shaped; gill face cystidia rarely seen.
Stem whitish, apex granular, downy at base. Ring absent. Flesh delicate and hollow.
Odour not distinctive. **Taste** not distinctive.
Chemical tests none.
Occurrence summer to autumn; infrequent.
■ Inedible.

Conocybe aporos Kits. v. War.   Bolbitiaceae
= *Pholiotina aporos* (Kits. v. War.) Clc.

Small fragile agaric, hazel or yellowish-brown with pale ring on the stem; solitary or in small trooping groups on soil in coniferous and broad-leaf woods, favouring clay or calcareous soils.

Dimensions cap 1 - 4 cm dia; stem 2 - 6 cm tall x 0.1 - 0.3 cm dia.
Cap hazel-brown, more ochraceous-brown towards margin; at first convex or campanulate, becoming expanded and sharply umbonate, margin finely sulcate when damp, slightly sticky. Flesh ochraceous brown, thin and fragile.
Gills concolorous, edges finely serrated and more pallid, adnexed, crowded. **Spores** ochraceous-brown, smooth, ellipsoid, germ pore absent, 7-10 x 4-5.5 µm. Basidia 4-spored. Gill edge cystidia skittle-shaped with small heads; gill face cystidia rarely seen.
Stem concolorous with cap, at first slightly pruinose at apex, other-wise silky, fragile. Ring pallid, membraneous, folded, median. Flesh concolorous and fragile.
Odour not distinctive. **Taste** not distinctive.
Chemical tests none.
Occurrence spring to autumn; frequent.
■ Inedible.

Conocybe arrhenii (Fr.) Kits. v. War.   Bolbitiaceae
= *Pholiotina arrhenii* (Fr.) Sing.

Small fragile agaric, reddish-brown with pale ring on the stem; solitary or in small trooping groups, on soil in parks, open woods, field borders and by path sides.

Dimensions cap 1 - 3 cm dia; stem 2 - 6 cm tall x 0.1 - 0.3 cm dia.
Cap reddish-brown, hygrophanous, ochraceous-brown when dry; at first convex or campanulate, becoming expanded and sometimes umbonate, margin finely striate when damp. Flesh ochraceous-brown, thin and fragile.
Gills concolorous, becoming tinged reddish at maturity, adnexed, crowded. **Spores** ochraceous-brown, smooth, ellipsoid, germ pore indistinct, 7-10 x 4-5.5 µm. Basidia 4-spored. Gill edge cystidia skittle-shaped with small heads; gill face cystidia rarely seen.
Stem concolorous, at first slightly pruinose at apex, otherwise silky, fragile. Ring pallid, membraneous, folded, median. Flesh concolorous and fragile.
Odour not distinctive. **Taste** not distinctive.
Chemical tests none.
Occurrence spring to autumn; frequent.
■ Inedible.

Conocybe brunneola (Kühn.) Kühn. & Romagn.   Bolbitiaceae

Small, fragile agaric, yellowish-brown, slender stem without ring; solitary or in small trooping groups on soil in grassy places, often with moss.

Dimensions cap 1 - 1.5 cm dia; stem 3 - 7 cm tall x 05 - 0.1 cm dia.
Cap ochraceous-brown, hygrophanous, drying more pallid; convex or bluntly campanulate becoming expanded and sometimes umbonate, margin finely striate when damp. Flesh ochraceous-brown, thin and fragile.
Gills concolorous becoming tinged reddish at maturity, adnexed, crowded. **Spores** ochraceous-brown, smooth, ellipsoid, germ pore indistinct, 6-8.5 x 3-4.5 µm. Basidia 4-spored. Gill edge cystidia skittle-shaped with small heads; gill face cystidia rarely seen.
Stem concolorous, finely pruinose, slender, more or less equal but with small basal bulb. Ring absent. Flesh concolorous and fragile.
Odour not distinctive. **Taste** not distinctive.
Chemical tests none.
Occurrence summer to autumn; frequent.
■ Inedible.

Conocybe kuehneriana Sing.   **Bolbitiaceae**

Small fragile agaric, yellowish-brown, drying paler, with yellowish gills; solitary or in small trooping groups on soil on grassy pathsides and clearings in woods.

Dimensions cap 1 - 2.5 cm dia; stem 3 - 8 cm tall x 0.05 - 0.15 cm dia.
Cap sienna, or ochraceous-brown when damp, hygrophanous, drying whitish buff; at first convex or campanulate, becoming expanded-conical, faintly downy, margin striate when damp. Flesh ochraceous-buff, thin and fragile.
Gills at first clay becoming ochraceous or sienna at maturity, adnate, crowded. **Spores** ochraceous, smooth, ellipsoid, large germ pore, 9.5-14.5 x 6-8 µm. Basidia 4-spored. Gill edge cystidia skittle-shaped with small heads; gill face cystidia rarely seen.
Stem whitish above, concolorous with cap below, filiform and slightly swollen at the base, fragile. Ring absent. Flesh concolorous and fragile.
Odour not distinctive. **Taste** not distinctive.
Chemical tests none.
Occurrence summer; infrequent.
■ Inedible.

Conocybe pubescens (Gill.) Kühn.   **Bolbitiaceae**

Small, fragile agaric, ochre-brown, drying paler, with yellowish or reddish-brown gills; solitary or in small trooping groups on well-manured soil or dung in or near coniferous woods.

Dimensions cap 1 - 2 cm dia; stem 3 - 8 cm tall x 0.1 - 0.3 cm dia.
Cap tawny or ochraceous-brown when damp, hygrophanous, drying more pallid ochraceous; always conical or campanulate, margin striate to halfway when damp, downy when young. Flesh ochraceous-buff, thin and fragile.
Gills at first ochraceous, becoming more reddish-brown at maturity, adnate, close. **Spores** ochraceous, smooth, ellipsoid, large germ pore, 16-20 x 9-10 µm. Basidia 4-spored. Gill edge cystidia skittle-shaped with small heads; gill face cystidia rarely seen.
Stem concolorous with cap, striate when young, with white fibrils at base, fragile. Ring absent. Flesh concolorous but darker in the stem base and fragile.
Odour not distinctive. **Taste** not distinctive.
Chemical tests none.
Occurrence spring to autumn; infrequent.
■ Inedible.

Conocybe siennophylla (Berk. & Br.) Sing. **Bolbitiaceae**
= *Conocybe siliginea var ochracea* Kühn.
= *Conocybe ochracea* (Kühn.) Sing.

Small, fragile agaric, tawny-brown, drying paler, with yellowish gills; solitary or in small trooping groups on soil in lawns and other grassy situations.

Dimensions cap 1 - 2 cm dia; stem 3 - 6 cm tall x 0.1 - 0.3 cm dia.
Cap tawny, or ochraceous-brown when damp, hygrophanous, drying whitish; at first convex or campanulate, becoming expanded and sometimes bluntly umbonate, margin striate to halfway when damp. Flesh ochraceous buff, thin and fragile.
Gills at first clay, becoming ochraceous at maturity, adnexed or free, crowded. **Spores** ochraceous, smooth, ellipsoid, germ pore, 8.5-14 x 5-8.5 µm. Basidia 4-spored. Gill edge cystidia skittle-shaped with small heads; gill face cystidia rarely seen.
Stem pallid above, concolorous with cap below and with white fibrils at the base, fragile. Ring absent. Flesh concolorous and fragile.
Odour not distinctive. **Taste** not distinctive.
Chemical tests none.
Occurrence summer; common.
■ Inedible.

Conocybe siliginea (Fr. : Fr.) Kühn.   Bolbitiaceae

Small fragile agaric, creamy-yellow with brownish-yellow gills and no ring on stem; solitary or scattered on soil in grasslands.

Dimensions cap 0.7 - 1.5 cm dia; stem 3 - 7 cm tall x 0.05 - 0.15 cm dia.
Cap cream, more ochraceous towards centre; at first convex-hemispherical or campanulate, barely hygrophanous, smooth, not obviously sulcate. Flesh pallid, thin and fragile.
Gills ochraceous-brown, adnexed, crowded. **Spores** ochraceous-brown, smooth, ellipsoid, germ pore, 13-17 x 7.5-8.5 µm. Basidia 2-spored. Gill edge cystidia skittle-shaped with distinctive heads; gill face cystidia rarely seen.
Stem pallid, slightly silky-fibrous with hair-like cystidia, fragile. Ring absent. Flesh concolorous and fragile.
Odour not distinctive. **Taste** not distinctive.
Chemical tests none.
Occurrence summer to autumn; frequent.
■ Inedible.

Conocybe tenera (Schaeff.: Fr.) Kühn.   Bolbitiaceae

Small fragile agaric, brownish-yellow with slightly darker stem and no ring; scattered or trooping on soil in grass, at woodland edges and in parks.

Dimensions cap 1 - 2 cm dia; stem 3 - 7 cm tall x 0.05 - 0.15 cm dia.
Cap ochraceous-brown, glistening slightly when dry, sulcate almost to centre; at first campanulate then more convex, barely hygrophanous, smooth. Flesh pallid, thin and fragile.
Gills ochraceous-brown, adnexed, crowded. **Spores** ochraceous-brown, smooth, ellipsoid, germ pore, 8.5-14.5 x 5-8 µm. Basidia 4-spored. Gill edge cystidia skittle-shaped with distinctive heads; gill face cystidia rarely seen.
Stem darker than cap, particularly below, pruinose throughout with capitate cystidia, fragile. Ring absent. Flesh concolorous and fragile.
Odour not distinctive. **Taste** not distinctive.
Chemical tests none.
Occurrence summer to autumn; frequent.
■ Inedible.

Flammulaster carpohiloides (Kühn.) Watl.
Strophariaceae

Very small, delicate, creamy agaric with bluntly conical, densely granular cap and stem, pale brown gills; on twigs, leaves and other litter of broad-leaf trees.

Dimensions cap 1 - 1.5 cm dia; stem 1 - 2.5 cm tall x 0.05 - 0.2 cm dia.
Cap cream with ochraceous or brown tinges; at first conical-convex then more flattened, densely pruinose or micaceous, with chains of sphaeroidal cells. Flesh concolorous and thin.
Gills cream, becoming pallid brown, adnexed-adnate, close.
Spores pallid brown, smooth, broadly ellipsoid or lemon-shaped, non-amyloid, 7.5-10.5 x 5-6 µm. Basidia 4-spored. Gill edge cystidia flask-shaped with elongated apex; gill face cystidia absent.
Stem concolorous with cap, granular-fibrous, slender, more or less equal. Ring absent. Flesh concolorous and somewhat cartilaginous.
Odour not distinctive. **Taste** not distinctive.
Chemical tests none.
Occurrence early summer to autumn; infrequent.
■ Inedible.

Flammulaster granulosa (J Lange) Watl. **Strophariaceae**

Very small, delicate, brown agaric with bluntly conical granular cap and stem, pale brown gills; scattered, trooping, on soil often amongst mosses and short grass by trees.

Dimensions cap 0.5 - 1.2 cm dia; stem 1 - 2.5 cm tall x 0.05 - 0.2 cm dia.
Cap brown with cinnamon or sepia tinges; at first conical-convex then more flattened, densely and darker pruinose or micaceous with chains of sphaeroidal cells. Flesh concolorous and thin.
Gills more pallid brown, adnexed-adnate, close. **Spores** pallid brown, smooth, ellipsoid or almond-shaped, non-amyloid, 7.5-10.5 x 4-5.5 µm. Basidia 4-spored. Gill edge cystidia shaped as a series of vesicles diminishing towards the apex; gill face cystidia absent.
Stem concolorous with cap but more pallid at the apex, darker below, granular-fibrous, slender, more or less equal. Ring absent. Flesh concolorous and somewhat cartilaginous.
Odour not distinctive. **Taste** not distinctive.
Chemical tests none.
Occurrence early summer to autumn; infrequent.
■ Inedible.

Hypholoma fasciculare (Huds.: Fr.) Kummer
= *Geophila fascicularis* (Huds.: Fr.) Quél.   **Strophariaceae**
= *Naematoloma fasciculare* (Huds.: Fr.) Karst.

Medium-sized agaric, with yellowish-tan cap and stem, sulphur or blackish-brown gills, faint ring zone; densely caespitose, or tufted, on stumps of broad-leaf and coniferous trees.

Dimensions cap 2 - 7 cm dia; stem 4 - 10 cm tall x 0.5 - 1 cm dia.
Cap sulphur-yellow, tan towards centre; convex or slightly umbonate, expanded with age, smooth but with velar remnants attached to margin. Flesh sulphur-yellow, firm and moderate.
Gills at first sulphur-yellow, becoming more olivaceous and finally blackish-brown with purple tinge, adnate, crowded. **Spores** purplish-brown, smooth, ellipsoid, germ pore, 6-8 x 4-4.5 µm. Basidia 4-spored. Gill edge cystidia fusiform-ventricose, thin-walled; gill face cystidia ovoid or ventricose with elongated apex.
Stem more or less concolorous with cap, but darker brown towards the base, more or less equal, typically curved, smooth. Ring zone-like, faint, coloured at maturity with spores, superior. Flesh sulphur-yellow, brown in stem base, fibrous, more or less full.
Odour not distinctive. **Taste** bitter.
Chemical tests none.
Occurrence throughout the year but mainly summer and autumn; common.
■ Inedible.

Hypholoma marginatum (Pers.: Fr.) Schroet.
= *Hypholoma dispersum* (Fr.) Quél.   **Strophariaceae**

Smallish agaric with tan cap, stem with silvery mottled appearance and yellow or olive-brown gills, no ring zone; in small trooping groups on needles or rotting wood with coniferous trees.

Dimensions cap 1.5 - 4 cm dia; stem 3 - 7cm tall x 0.2 - 0.5 cm dia.
Cap tan, with more pallid margin; convex or campanulate, becoming expanded and broadly umbonate, smooth but with whitish velar remnants near the cap margin. Flesh whitish, firm and thin.
Gills at first pallid yellow, then olivaceous-brown, adnate or emarginate, crowded. **Spores** chocolate-brown, smooth, ellipsoid, germ pore, 7-9.5 x 4-5 µm. Basidia 4-spored. Gill edge cystidia fusiform or finely ventricose, thin-walled; gill face cystidia ovoid with pointed apex.
Stem pallid but bruising darker, covered with silky-scaly fibres, slender, more or less equal. Ring absent. Flesh pallid brown, darker in the stem base, fibrous, more or less full.
Odour not distinctive. **Taste** mild.
Chemical tests Gill face cystidia yellow with KOH.
Occurrence late summer to autumn; frequent.
■ Inedible.

Hypholoma sublateritium (Fr.) Quél.   Strophariaceae
= *Naematoloma sublateritium* (Fr.) Karst.

Small or medium agaric, with reddish-brown cap and stem on a yellow
background, and yellow or olive-brown gills, faint ring zone; densely
caespitose, or clustered, on stumps and debris of broad-leaf trees.

Dimensions cap 3 - 10 cm dia; stem 5 - 15 cm tall x 0.5 - 1.2 cm dia.
Cap reddish-brown, colour deepening towards centre, more ochra-
ceous at margin; convex, becoming more expanded with age,
smooth but with dark velar remnants attached near cap margin. Flesh
pallid yellowish, firm and moderate.
Gills pallid yellow, then olivaceous-brown, adnate, crowded.
Spores chocolate-brown, smooth, ellipsoid, germ pore, 6-7 x 3-4.5 µm.
Basidia 4-spored. Gill edge cystidia fusiform-ventricose with blunt
apex, thin-walled; gill face cystidia ovoid with pointed apex.
Stem pallid yellowish above but more concolorous with cap towards
base, more or less equal, typically curved, smooth. Ring zone-like,
faint, superior. Flesh pallid yellowish, brown in stem base, fibrous,
more or less full.
Odour not distinctive. **Taste** mild or slightly astringent.
Chemical tests spores yellowish-brown in KOH.
Occurrence late summer to autumn; infrequent.
■ Inedible.

Hypholoma udum (Pers.: Fr.) Kühn.   Strophariaceae
= *Naematoloma udum* (Pers.: Fr.) Karst.

Smallish agaric with tawny cap, long slender stem and pale lilac
tinged gills, no ring zone; solitary or in small trooping groups
amongst moss and plant debris on damp heaths and moors.

Dimensions cap 1 - 2 cm dia; stem 4 - 10 cm tall x 0.2 - 0.3 cm dia.
Cap tan, with more pallid margin; convex or campanulate, becoming
more expanded, smooth but with faint velar remnants attached near
cap margin. Flesh pallid brown, firm and thin.
Gills at first pallid, becoming greyish-lilac with olivaceous tinge and
finally violaceous-brown, adnate or emarginate, crowded.
Spores chocolate-brown, smooth, ellipsoid, germ pore, 13-18 x
6-7 µm. Basidia 4-spored. Gill edge cystidia fusiform-ventricose,
thin-walled; gill face cystidia ovoid with pointed apex.
Stem pallid at apex but darker brown below, covered with silky,
scaly fibres giving a mottled appearance, very slender, more or less
equal. Ring absent. Flesh pallid brown, darker in the stem base,
fibrous, more or less full.
Odour not distinctive. **Taste** slight, astringent.
Chemical tests spores brown in KOH.
Occurrence late summer to autumn; frequent.
■ Inedible.

Kuehneromyces mutabilis (Schaeff.: Fr.) Sing. & A H Smith
= *Pholiota mutabilis*(Schaeff.: Fr.) Kummer
= *Galerina mutabilis* (Schaeff.: Fr.) P Orton   Strophariaceae

Medium-sized agaric, with bright tan cap, drying paler from the centre,
yellowish or reddish-brown gills, ring on stem; caespitose or densely
clustered on stumps and logs of broad-leaf trees, favouring birch.

Dimensions cap 3 - 6 cm dia; stem 3 - 8 cm tall x 0.5 - 1 cm dia.
Cap tan when damp, drying pallid ochraceous, distinctively two-tone
from centre; convex, becoming flattened with a blunt umbo, smooth,
not greasy or viscid. Flesh pallid cap colour and thin.
Gills pallid ochraceous, becoming cinnamon at maturity, adnate,
crowded. **Spores** reddish-ochre, smooth, ellipsoid or almond-shaped,
germ pore, 6-7.5 x 4-5 µm. Basidia 4-spored. Gill edge cystidia
fusiform-ventricose, thin-walled; gill face cystidia absent.
Stem pallid tan above ring, darker tan below, shading to almost black
at base, slender, scaly, more or less equal, smooth. Ring brown, torn,
sub-apical. Flesh concolorous with cap above, becoming darker tan
towards base, stuffed.
Odour not distinctive. **Taste** not distinctive.
Chemical tests none.
Occurrence spring to winter, but mainly in autumn; common.
☐ Edible and good.

Phaeomarasmius erinaceus (Fr.) Kühn. **Strophariaceae**
= *Phaeomarasmius aridus* (Pers.) Sing.

Small, densely scaly agaric with yellowish-tan cap and darker brown stem; solitary or in small trooping groups on twigs, small branches and other woody debris of broad-leaf trees favouring willow in damp places.

Dimensions cap 1 - 1.5 cm dia; stem 1 - 2 cm tall x 0.2 - 0.3 cm dia.
Cap ochraceous-tan with reddish tinge, more pallid at margin; convex, becoming flattened or umbilicate, densely covered with small pointed scales and with fringed margin when young. Flesh concolorous and thin.
Gills ochraceous, becoming rust, adnate, fairly distant. **Spores** rust, smooth, ellipsoid or lemon-shaped, 9-13 x 6-9 μm. Basidia 1-, 2- or 4-spored. Gill edge cystidia more or less cylindrical; gill faced cystidia absent.
Stem concolorous with cap but darker, coarsely scaly as in cap, tapering upwards and sometimes curved. Ring absent. Flesh concolorous and full.
Odour not distinctive. **Taste** not distinctive.
Chemical tests none.
Occurrence throughout the year but mainly late summer and autumn; infrequent or rare. (Haldon Forest, near Exeter, Devon)
■ Inedible.

Pholiota adiposa (Fr.) Kummer   **Strophariaceae**
= *Dryophila adiposa* (Fr.) Quél.

Medium to large agaric, with golden-yellow slimy cap covered in rust scales, yellow gills becoming rust and ring on stem; caespitose or in dense clusters specifically on beech, stumps and at base of living trees.

Dimensions cap 5 - 12 cm dia; stem 2 - 5 cm tall x 0.5 - 1 cm dia.
Cap golden-yellow, covered with rust, flattened, gelatinous scales; convex, becoming expanded, very viscid. Flesh pallid yellow and firm.
Gills yellow, becoming rust at maturity, adnate, crowded.
Spores rust or reddish-brown, smooth, ellipsoid, 5-6.5 x 3-4 μm. Basidia 4-spored. Gill face cystidia fusiform with acute apex.
Stem at first concolorous with cap, becoming more rust, smooth above ring, with bands of rust scales below, more or less equal, typically curved, very viscid. Ring fragile and ephemeral, sub-apical. Flesh yellow, full and tough.
Odour slightly aromatic. **Taste** not distinctive.
Chemical tests none.
Occurrence late summer to autumn; infrequent or rare. (Mendip Hills, near Pilton, Somerset)
■ Inedible.

Pholiota alnicola (Fr.) Sing.   **Strophariaceae**

Medium-sized agaric, with bright yellow greasy cap, lemon-yellow gills, becoming cinnamon and ring zone on stem; solitary or more typically clustered on stumps and other dead wood of broad-leaf trees, favouring birch, alder and willow in damp locations.

Dimensions cap 2 - 8 cm dia; stem 2 - 8 cm tall x 0.5 - 1 cm dia.
Cap bright yellow, sometimes with slight olivaceous tinge at the margin; convex, becoming flattened, smooth but sometimes with pallid velar remnants adhering at the margin, greasy when damp, otherwise dry. Flesh yellow and firm.
Gills pallid yellow, becoming cinnamon at maturity, adnate, crowded. **Spores** brown, smooth, ellipsoid, 8.5-11.5 x 5-5.5 μm. Basidia 4-spored. Gill edge cystidia filiform or narrowly clavate.
Stem pallid lemon-yellow, becoming tinged rust towards base, more or less equal, smooth. Ring zone-like with pallid remnants of veil, very superior. Flesh yellow, fibrous and full.
Odour not distinctive. **Taste** slight, bitter.
Chemical tests none.
Occurrence late summer to winter; infrequent.
■ Inedible.

Pholiota aurivella (Batsch: Fr.) Kummer   **Strophariaceae**
= *Pholiota cerifera* (Karst.) Karst.

Medium to large agaric, with orange-yellow slimy cap covered in darker scales, yellow gills becoming rust and flaky stem; caespitose or in dense clusters on stumps and cut ends of felled trunks, favouring beech.

Dimensions cap 5 - 12 cm dia; stem 3 - 10 cm tall x 0.5 - 0.8 cm dia.
Cap orange-yellow, covered with rust, adpressed, gelatinous scales becoming obscured except at margin; convex, becoming flattened, viscid. Flesh pallid yellow, darker with age and firm.
Gills pallid yellow, becoming rust at maturity, adnate, crowded.
Spores rust or reddish-brown, smooth, ellipsoid, 8-9 x 5-6 µm. Basidia 4-spored. Gill edge cystidia clavate.
Stem concolorous with cap, or more pallid, smooth above ring, with bands of reflexed fibrous scales below, more or less equal, typically curved, viscid. Ring fragile and ephemeral, sub-apical. Flesh yellow, full and tough.
Odour not distinctive. **Taste** not distinctive.
Chemical tests none.
Occurrence late summer to autumn; frequent.
■ Inedible.

Pholiota flammans (Fr.) Kummer   **Strophariaceae**

Medium-sized, bright yellow, scaly agaric, with cottony ring on stem; solitary or caespitose on stumps and logs of coniferous trees.

Dimensions cap 4 - 8 cm dia; stem 5 - 12 cm tall x 0.4 - 1 cm dia.
Cap bright tawny-yellow, covered with more pallid yellow upturned scales with slight lemon tinge; convex, becoming expanded, the margin tending to remain incurved. Flesh pallid yellow, firm.
Gills pallid yellow, becoming more rust-yellow at maturity, adnate or emarginate, crowded. **Spores** rust, smooth, ellipsoid, 4-4.5 x 2-2.5 µm. Basidia 4-spored. Gill face cystidia lanceolate, with thickish walls.
Stem concolorous and scaly as cap, more or less equal or tapering slightly upwards. Ring concolorous, ragged, cottony, sub-apical. Flesh pallid yellow, more or less full.
Odour not distinctive. **Taste** not distinctive.
Chemical tests gill cystidia staining deeply with cotton blue in lactic acid.
Occurrence late summer to autumn; rare. (Culbin Forest, Morayshire)
■ Inedible.

Pholiota gummosa (Lasch) Sing.   **Strophariaceae**

Small or medium-sized agaric with pale greenish-yellow, barely slimy cap, straw gills and scaly stem with ring zone; solitary or caespitose on wood or woody debris, often buried.

Dimensions cap 3 - 8 cm dia; stem 3 - 7 cm tall x 0.4 - 0.9 cm dia.
Cap pallid ochraceous-green with scales, at first whitish then tinged brown; convex, becoming flattened, smooth, slightly viscid when damp, otherwise matt. Flesh pallid lemon-yellow, becoming tinged rust with age, firm.
Gills pallid ochraceous or beige, becoming rust at maturity, adnate, crowded. **Spores** rust, smooth, ellipsoid, 5-7.5 x 3-4 µm. Basidia 4-spored. Gill edge cystidia narrowly clavate.
Stem concolorous with cap, becoming tinged rust towards base, more or less equal, finely scaly and fibrous below ring zone. Ring zone-like with brownish remnants of veil, superior. Flesh pallid lemon-yellow, more rust in stem base, full and fibrous.
Odour not distinctive. **Taste** not distinctive.
Chemical tests none.
Occurrence late summer to autumn; frequent.
■ Inedible.

Pholiota highlandensis (Peck) A H Smith & Hes.
= *Pholiota carbonaria* (Fr.: Fr.) Sing.
= *Flammula carbonaria* (Fr.: Fr.) Kummer
= *Dryophila carbonaria* (Fr.: Fr.) Quél.   **Strophariaceae**

Small or medium-sized agaric with tan slimy cap, and clay-brown or cinnamon gills, ring zone on stem; solitary, in tightly scattered groups, or clustered on burnt ground or, less frequently, woody debris.

Dimensions cap 2 - 7 cm dia; stem 3 - 7 cm tall x 0.4 - 0.9 cm dia.
Cap ochraceous or brownish-tan, sometimes more pallid at the margin; convex becoming flattened, smooth, viscid when damp, otherwise shiny. Flesh pallid yellow and firm.
Gills pallid clay-brown, becoming cinnamon with olivaceous tinge at maturity, adnate, crowded. **Spores** rust, smooth, ellipsoid, 6.5-8 x 3.5-4.5 μm. Basidia 4-spored. Gill edge and gill face cystidia fusiform, comparatively elongated.
Stem pallid yellow, becoming tinged rust towards the base, more or less equal, finely scaly-fibrous below ring zone. Ring zone-like with brownish remnants of veil, superior. Flesh pallid yellow, more rust in stem base, full, fibrous.
Odour not distinctive. **Taste** not distinctive.
Chemical tests none.
Occurrence late summer to winter; frequent.
■ Inedible.

Pholiota squarrosa (Müll.: Fr.) Kummer   **Strophariaceae**
= *Dryophila squarrosa* (Müll.: Fr.) Quél.

Large agaric, straw-yellow covered with coarse rust scales and with ring zone on stem; in dense, often striking, caespitose tufts at the base of living broad-leaf and, very occasionally, coniferous trees.

Dimensions cap 3 - 12 cm dia; stem 5 - 12 cm tall x 1 - 1.5 cm dia.
Cap straw-yellow covered in coarse, upturned, rust scales; convex, becoming flattened but with persistently inrolled margin, not greasy or viscid. Flesh pallid yellow and firm.
Gills pallid yellow, becoming cinnamon at maturity, adnate, crowded.
Spores rust, smooth, ellipsoid, 5.5-9 x 3.5-5 μm. Basidia 4-spored. Gill face cystidia clavate with pointed apex.
Stem concolorous with cap above becoming tinged rust towards the base, more or less equal or tapering downwards, scaly as cap. Ring ragged, almost zone-like, sub-apical. Flesh yellow, more rust in the stem base, almost full with narrow canal, tough.
Odour of radish. **Taste** of radish.
Chemical tests none.
Occurrence late summer to autumn; infrequent.
■ Inedible.

Psilocybe bullacea (Bull.: Fr.) Kummer.   **Strophariaceae**
= *Deconica bullacea* (Bull.: Fr.) Karst.

Small dark brown agaric with greasy flattened cap bearing slightly ragged margin, clay or dark brown gills and slender stem; densely caespitose, typically on manure and rotting straw, also in manured grass.

Dimensions cap 1.5 - 2.5 cm dia; stem 2 - 5 cm tall x 0.1 - 0.4 cm dia.
Cap dark brown, drying more hazel-brown; broadly conical soon becoming flattened, greasy when damp, smooth or striate when damp, margin appendiculate and dentate. Flesh brown and thin.
Gills pallid clay, becoming purplish-brown at maturity, adnate or emarginate, close.
Spores purplish-brown, smooth, ellipsoid, distinct germ pore, 6-7.5 x 4-5 μm. Basidia 4-spored. Gill edge cystidia flask-shaped; gill face cystidia absent.
Stem umber-brown, slightly pruinose at apex, otherwise smooth, slender, more or less equal. Ring absent. Flesh brown, fibrous and full.
Odour not distinctive. **Taste** not distinctive.
Chemical tests none.
Occurrence summer and early autumn; infrequent.
■ Inedible.

Psilocybe coprophila (Bull.: Fr.) Kummer Strophariaceae
= *Deconica coprophila* (Bull.: Fr.) Karst.

Small tawny-brown agaric with greasy domed cap, clay or dark brown gills and short stem; in densely trooping groups, typically on manure of cow, horse and sheep, in grass.

Dimensions cap 0.5 - 1.5 cm dia; stem 1 - 3 cm tall x 0.1 - 0.3 cm dia.
Cap tawny-brown, becoming more buff with age; hemispherical or convex-campanulate, viscid when damp with separable skin, smooth or decorated with tiny whitish velar fragments. Flesh pallid and thin.
Gills pallid clay, becoming vinaceous-brown with pallid margin at maturity, adnate or emarginate, close. **Spores** purplish-brown, smooth, ellipsoid, distinct germ pore, 12-15 x 7-9 µm. Basidia 4-spored. Gill edge cystidia fusiform or flask-shaped; gill face cystidia absent.
Stem whitish, becoming tinged brown, pruinose at apex otherwise silky and decorated with velar fragments, more or less equal or slightly clavate. Ring absent. Flesh pallid brown, fibrous and full.
Odour not distinctive. **Taste** not distinctive.
Chemical tests none.
Occurrence summer and early autumn; infrequent.
■ Inedible.

Psilocybe fimetaria (P Orton) Watl.   Strophariaceae

Small dark brown agaric with greasy convex cap, clay or dark brown gills and slender stem; in small groups on horse dung.

Dimensions cap 0.5 - 3.5 cm dia; stem 2 - 9 cm tall x 0.05 - 0.4 cm dia.
Cap date or sepia-brown with olivaceous tinge, drying more buff, at first decorated with whitish silky scales then smooth with striate margin; conico-convex and more or less umbonate, greasy or viscid when damp with tough separable skin. Flesh pallid buff and thin.
Gills pallid clay, becoming purplish-brown with white floccose edges at maturity, adnate or emarginate, close. **Spores** purplish-brown, smooth, ellipsoid, distinct germ pore, 11-14 x 6.5-7.5 µm. Basidia 4-spored. Gill edge cystidia flask-shaped with pointed tips; gill face cystidia absent.
Stem at first pallid, becoming more brown with age, slightly pruinose at apex, otherwise flaky-fibrillose below then smooth, slender, more or less equal or with somewhat swollen base. Ring ephemeral whitish zone. Flesh pallid but dark brown towards base, fibrous, full.
Odour of meal. **Taste** of meal.
Chemical tests none.
Occurrence summer; localised mainly to Scotland but also in New Forest and elsewhere.
■ Inedible.

Psilocybe semilanceata (Fr.) Kummer   Strophariaceae
= *Agaricus semilanceatus* Fr.   Liberty Cap

Small delicate agaric with distinctive yellowish pointed cap, clay or dark brown gills and long slender stem; in scattered trooping groups, on soil amongst grass in lawns and pastures favouring hilly sites.

Dimensions cap 0.5 - 1.5 cm dia; stem 2.5 -7.5 cm tall x 0.1 - 0.2 cm dia.
Cap pallid ochraceous-brown, hygrophanous, drying buff; conical with distinctive sharply pointed umbo, sticky when damp, smooth. Flesh pallid ochraceous and thin.
Gills pallid clay, becoming purplish-brown at maturity, adnate or emarginate, fairly distant. **Spores** purplish-brown, smooth, ellipsoid, germ pore, 11.5-14.5 x 7-9 µm. Basidia 4-spored. Gill edge cystidia flask-shaped with very elongated and sharp apex; gill face cystidia absent.
Stem pallid cream, occasionally tinged bluish at the base, slender, more or less equal, smooth. Ring absent. Flesh pallid cream, fibrous and full.
Odour not distinctive. **Taste** not distinctive.
Chemical tests none.
Occurrence late summer to autumn; infrequent but locally common.
✚ Poisonous - hallucinogenic. Symptoms can be persistent and dis-tressing.

Psilocybe subcoprophila (Britz.) Sacc.   Strophariaceae
= *Agaricus subcoprophilus* Britz.

Small greyish agaric with greasy conical cap, clay or dark brown gills and slender stem; in dense clusters on manure, often of horses.

Dimensions cap 2 - 4 cm dia; stem 2 - 5 cm tall x 0.2 - 0.4 cm dia.
Cap at first grey then more ochraceous, drying more pallid; broadly conical, soon becoming flattened, somewhat viscid when damp, sulcate. Flesh pallid grey-brown and thin.
Gills pallid clay, becoming darker brown at maturity with white margin, adnate or emarginate, close. **Spores** purplish-brown, smooth, ellipsoid, large germ pore, 14.5-25 x 10.5-13 μm. Basidia 4-spored. Gill edge cystidia flask-shaped with elongated apex; gill face cystidia absent.
Stem concolorous with cap, slightly farinose and striate at apex, otherwise smooth, slender, more or less equal. Ring ephemeral. Flesh brown, fibrous and stuffed.
Odour not distinctive. **Taste** not distinctive.
Chemical tests none.
Occurrence spring; infrequent.
■ Inedible.

Stropharia aeruginosa (Curt.: Fr.) Quél.   Strophariaceae
= *Agaricus aeruginosus* Fr.

Medium-sized agaric with slimy blue-green cap, scaly stem with ring and pale or purple-brown gills; in small trooping groups on soil, often amongst grass, in woods, pastures and heaths.

Dimensions cap 2 - 8 cm dia; stem 4 - 10 cm tall x 0.4 - 1.2 cm dia.
Cap blue-green, becoming more pallid yellow with age; convex or campanulate, glutinous and at first with white cottony scales. Flesh white tinged with blue, firm and moderate.
Gills at first white, becoming clay and dark brown at maturity, adnate, moderately distant. **Spores** purple-brown, smooth, ellipsoid, germ pore, 7-10 x 4-5 μm. Basidia 4-spored. Gill edge cystidia obtusely clavate, or capitate; gill face cystidia lanceolate.
Stem bluish, covered with white woolly scales below ring, fairly slender, more or less equal. Ring white, later tinged brown with spores, membraneous, superior. Flesh white, tinged blue, stuffed or full.
Odour not distinctive. **Taste** not distinctive.
Chemical tests none.
Occurrence summer to autumn; frequent.
✛ Poisonous (probably).

Stropharia aurantiaca (Cke.) P Orton   Strophariaceae

Small to medium agaric with slimy orange-red cap, pale stem with ring and clay or darker brown gills; in small trooping groups, sometimes caespitose, on wood chips, and sawdust, typically in parks and gardens.

Dimensions cap 2 - 6 cm dia; stem 2 - 10 cm tall x 0.2 - 1 cm dia.
Cap orange-red; convex becoming expanded, viscid when damp, otherwise smooth, slightly shiny, sometimes with white velar remnants at the margin. Flesh pallid concolorous or buff, firm and moderate.
Gills at first white, becoming clay with olivaceous tinge at maturity, adnate, close. **Spores** purple-brown, smooth, ellipsoid, germ pore, 11-13 x 6-7.5 μm. Basidia 4-spored. Gill edge cystidia mostly flask-shaped with narrow elongation, sometimes more clavate, thin-walled; gill face cystidia clavate, thin-walled.
Stem pallid ochraceous, tinged cap colour towards base, streaked, slightly thickened at base. Ring yellowish, fragile, superior, ephemeral. Flesh concolorous, stuffed or full.
Odour not distinctive. **Taste** not distinctive.
Chemical tests none.
Occurrence late summer to autumn; rare. (Lyme Park, Cheshire)
■ Inedible.

Stropharia inuncta (Fr.) Quél.   Strophariaceae
= *Agaricus inunctus* Fr.

Smallish agaric with greyish-yellow cap, pale stem and clay or purplish gills; in small trooping groups, in short grass often by sides of paths or at roadsides.

Dimensions cap 3 - 7 cm dia; stem 3 - 7 cm tall x 0.3 - 0.5 cm dia.
Cap at first greyish-violet becoming pallid ochraceous-grey; at first bluntly campanulate, becoming expanded-convex or slightly depressed, at first viscid then dry, smooth. Flesh whitish, firm and thin.
Gills at first pallid, becoming purplish-brown at maturity, adnate, close. **Spores** purple-brown, smooth, ellipsoid, germ pore, 7-8.5 x 4.5-6 µm. Basidia 4-spored. Gill edge cystidia mostly irregular or flask-shaped with short narrow elongation, thin-walled; gill face cystidia clavate but embedded.
Stem pallid, white downy at base, more or less equal. Ring subapical, ephemeral. Flesh white darkening with age, stuffed or full.
Odour not distinctive. **Taste** not distinctive.
Chemical tests gill face cystidia deep blue with cotton blue in lactophenol.
Occurrence summer to autumn; rare. (Forest of Dean, Gloucestershire)
■ Inedible.
Note: may be mistaken for a viscid *Psathyrella*.

Stropharia pseudocyanea (Desm.) Morg. Strophariaceae
= *Agaricus pseudocyaneus* Desm.
= *Agaricus albocyaneus* Fr.

Smallish agaric with slimy pale blue-green or yellowish cap, fragile scurfy stem with woolly ring and clay or purplish gills; in small trooping groups, sometimes clustered, on soil in damp grassy places.

Dimensions cap 2 - 5 cm dia; stem 5 - 7.5 cm tall x 0.4 - 0.6 cm dia.
Cap pallid bluish-green, soon becoming buff with olivaceous tinge; at first campanulate, becoming expanded-convex and bluntly umbonate, very viscid when damp, with whitish velar remnants at margin. Flesh white or pallid bluish-green, firm and thin.
Gills at first buff, becoming purplish-brown at maturity, adnate or notched, close. **Spores** purple-brown, smooth, ellipsoid, minute germ pore, 7-9 x 4-5 µm. Basidia 4-spored. Gill edge cystidia mostly clavate or capitate, thin-walled; gill face cystidia fusiform or flask-shaped.
Stem concolorous with cap, pruinose at apex, more woolly-fibrous below ring, more or less equal, often curved. Ring white, woolly-fibrillose, sub-apical. Flesh white darkening on ageing, hollow or stuffed, very fragile and soft.
Odour not distinctive. **Taste** not distinctive.
Chemical tests none.
Occurrence late summer to autumn; infrequent.
■ Inedible.

Stropharia rugosoannulata Farlow   Strophariaceae
=*Stropharia ferrii* Brés.
=*Stropharia imainana* Benedix

Medium or large fleshy agaric with greasy brownish-yellow cap, robust whitish stem with ring and clay or violet-brown gills; in small trooping groups, sometimes caespitose, on rotting straw at woodland margins, probably introduced to Britain as a commercial crop.

Dimensions cap 4 - 12 cm dia; stem 6 - 12 cm tall x 1 - 3 cm dia.
Cap ivory, becoming brownish with red and grey tinges against a more yellow background; roughly hemispherical, becoming expanded-convex, margin incurved for some time and bearing appendicular velar remnants, greasy when damp, otherwise smooth but rugose towards centre. Flesh pallid and medium.
Gills at first grey, becoming purple-brown at maturity, adnate, crowded. **Spores** purple-brown, smooth, ellipsoid, germ pore, 11-18 x 7.5-10 µm. Basidia 4-spored. Gill edge and gill face cystidia vesicular or flask-shaped with narrower elongated apex.
Stem whitish, fibrillose, stout, widening at base, with white mycelial 'rootlets'. Ring white, pendulous, upper surface fluted, sub-apical. Flesh concolorous, stuffed or full.
Odour not distinctive. **Taste** not distinctive.
Chemical tests contents of gill face cystidia turn yellow with KOH.
Occurrence summer to autumn; rare outside of cultivation. (Westbury-on-Severn, Gloucestershire)
□ Edible.

Stropharia semiglobata (Batsch: Fr.) Quél. **Strophariaceae**
= *Stropharia stercoraria* (Schum.: Fr.) Quél.

Small or variable agaric with domed slimy yellow cap, long slender stem with ring zone and clay or purplish-brown gills; in small trooping groups, sometimes caespitose, on dung.

Dimensions cap 1 - 3 cm dia; stem 6 - 10 cm tall x 0.2 - 0.3 cm dia.
Cap pallid yellow; hemispherical, sometimes with barely defined blunt umbo, viscid when damp, otherwise smooth, slightly shiny. Flesh pallid, delicate and thin.
Gills at first pallid clay, becoming purple-brown at maturity, adnate, close. **Spores** purple-brown, smooth, ellipsoid, germ pore,16-20 x 8-10 µm. Basidia 4-spored. Gill edge cystidia bluntly clavate; gill face cystidia fusiform or flask-shaped with pointed apex.
Stem pallid ochraceous, almost white at apex, very slender, smooth. Ring zone-like, whitish, superior, ephemeral. Flesh concolorous, stuffed or full.
Odour not distinctive. **Taste** not distinctive.
Chemical tests none.
Occurrence summer to autumn; common.
■ Inedible.
Note: the size of fruiting bodies varies according to the substrate on which they grow.

Tubaria conspersa (Pers.: Fr.) Fayod   **Strophariaceae**

Small cinnamon-brown agaric, cap covered with greyish velar fragments; in small trooping groups on soil or wood chips favouring damp situations.

Dimensions cap 0.8 - 2.5 cm dia; stem 2 - 4 cm tall x 0.1 - 0.4 cm dia.
Cap cinnamon or chocolate-brown, covered with greyish silky or scaly velar remnants, slightly ragged at margin; at first convex becoming expanded and slightly depressed. Flesh brown, thin and fragile.
Gills cinnamon with pallid edges, adnate, close. **Spores** pallid yellow, smooth, ellipsoid or lemon-shaped, 7-10 x 4-6 µm. Cystidia clavate, cylindrical or flask-shaped.
Stem concolorous with cap, slender, more or less equal. Ring absent. Flesh brown and fragile.
Odour not distinctive. **Taste** not distinctive.
Chemical tests none.
Occurrence throughout the year but mainly autumn; infrequent.
■ Inedible.

Tubaria furfuracea (Pers.: Fr.) Gill.   **Strophariaceae**

Small tan-brown agaric; in small trooping groups on wood chips and other woody debris.

Dimensions cap 1 - 4 cm dia; stem 2 - 5 cm tall x 0.2 - 0.4 cm dia.
Cap tan or cinnamon-brown, hygrophanous drying more pallid; at first convex, becoming expanded and slightly depressed, margin striate. Flesh brown, thin and fragile.
Gills cinnamon, adnate or slightly decurrent, close. **Spores** pallid ochre-brown, smooth, ellipsoid, 6-8.5 x 4-6 µm. Cystidia clavate.
Stem concolorous with cap, slender, more or less equal. Ring absent. Flesh brown and fragile.
Odour not distinctive. **Taste** not distinctive.
Chemical tests none.
Occurrence throughout the year but mainly autumn; common.
■ Inedible.
Note: the shape of the cystidia is important in distinguishing between this species and *T. hiemalis*.

Tubaria hiemalis Romagn.: Bon   Strophariaceae

Small rust-brown agaric; in small trooping groups on wood chips.

Dimensions cap 1 - 3 cm dia; stem 2 - 4 cm tall x 0.2 - 0.4 cm dia.
Cap rust or deep hazel-brown, hygrophanous drying more pallid; at first hemispherical, becoming expanded and slightly depressed, margin striate or finely sulcate. Flesh brown, thin and fragile.
Gills cinnamon with pallid edges, adnate, close. **Spores** ochre-brown, smooth, ellipsoid, 8-10 x 4-5 μm. Cystidia distinctly capitate, not clavate.
Stem concolorous with cap, slender, more or less equal. Ring absent. Flesh brown, fragile.
Odour not distinctive. **Taste** not distinctive.
Chemical tests none.
Occurrence mainly autumn to spring; frequent.
■ Inedible.
Note: shape of the cystidia is important in distinguishing between this species and *T. furfuracea*.

Tubaria pallidospora J Lange   Strophariaceae

Small pale brown agaric; in small trooping groups in grassy places in or near broad-leaf woods.

Dimensions cap 1 - 1.5 cm dia; stem 2 - 3.5 cm tall x 0.05 - 0.1 cm dia.
Cap pallid brown with ochraceous tinge, hygrophanous drying more pallid; at first convex, becoming expanded and slightly depressed, somewhat pruinose, margin striate. Flesh pallid brown, thin and fragile.
Gills pallid yellowish, adnate, close. **Spores** pallid yellow, smooth, lemon-shaped, 8-10 x 5-6 μm. Gill edge cystidia clavate.
Stem concolorous with cap, slender, more or less equal. Ring absent. Flesh pallid brown and fragile.
Odour not distinctive. **Taste** not distinctive.
Chemical tests none.
Occurrence throughout the year but mainly summer and autumn; infrequent.
■ Inedible.

Crepidotus applanatus (Pers.: Pers.) Kummer
Crepidotaceae

Small white tongue-shaped cap with pale brown gills and rudimentary lateral stalk; in small overlapping groups on decaying wood.

Dimensions cap 1 - 4 cm dia.
Cap white to cream; ligulate, smooth, sometimes faintly striate towards margin. Flesh white, thin, soft and brittle.
Gills at first whitish, becoming buff, decurrent, crowded.
Spores brown, minutely spiny, spherical, 5-6 μm. Basidia 4-spored. Gill edge cystidia clavate; gill face cystidia absent.
Stem white, lateral, rudimentary.
Odour not distinctive. **Taste** not distinctive.
Chemical tests none.
Occurrence late summer to autumn; infrequent.
■ Inedible.

Crepidotus herbarum (Peck) Sacc.   Crepidotaceae
= *Pleurotellus hypnophilus* (Pers.) Sacc.
= *Pleurotellus herbarum* (Peck) Sing.

Very small, whitish kidney-shaped cap with buff gills, more or less sessile; on twigs and other plant debris.

Dimensions cap 0.5 - 1 cm dia.
Cap white or cream; reniform, smooth or slightly downy. Flesh white, thin, watery and brittle.
Gills at first white becoming pallid buff, decurrent, fairly crowded.
Spores yellowish-buff, smooth, elongated-ellipsoid or fusiform, 7-9 x 2.5-3.5 μm. Basidia 4-spored. Gill cystidia absent.
Stem lateral, rudimentary or absent.
Odour not distinctive. **Taste** not distinctive.
Chemical tests none.
Occurrence early summer to autumn; infrequent.
■ Inedible.

Crepidotus lundelli Pil.   Crepidotaceae

Small whitish bracket-like cap with pale brown gills, more or less sessile; on wood and bark mainly of broad-leaf trees.

Dimensions cap 0.5 - 2 cm dia.
Cap white to creamy-ochre; more or less reniform, finely downy or smooth. Flesh white, thin, soft and brittle.
Gills at first whitish, becoming clay-brown, decurrent, fairly distant.
Spores brown, smooth, broadly ellipsoid, 7-8.5 x 4.5-5.5 μm. Basidia 4-spored. Gill edge cystidia clavate; gill face cystidia absent.
Stem lateral, rudimentary or absent.
Odour not distinctive. **Taste** not distinctive.
Chemical tests none.
Occurrence late summer to autumn; infrequent.
■ Inedible.

Crepidotus mollis (Schaeff.:Fr.) Kummer   Crepidotaceae

Small, very soft, whitish bracket-like cap with pale brown gills, more or less sessile; on decaying trunks of broad-leaf trees.

Dimensions cap 1.5 - 6 cm dia.
Cap pallid ochre-brown; more or less fan-shaped, smooth, surface slightly gelatinous, sometimes faintly striate towards margin and with scattered small scales. Flesh white, watery, soft and brittle.
Gills at first pallid greyish-brown becoming more rust, decurrent, fairly crowded. **Spores** pallid snuff-brown, smooth, broadly ellipsoid, 7-10 x 5-7 μm. Basidia 4-spored. Gill edge cystidia clavate; gill face cystidia absent.
Stem lateral, rudimentary or absent.
Odour not distinctive. **Taste** not distinctive.
Chemical tests none.
Occurrence late summer to autumn; infrequent.
■ Inedible.

Crepidotus pubescens Bres.   Crepidotaceae

Small, whitish kidney-shaped cap with pale buff-brown gills, more or less sessile; on twigs and other plant debris.

Dimensions cap 0.5 - 2.5 cm dia.
Cap white to creamy-ochre; reniform, smooth, very finely downy, sometimes faintly striate towards margin. Flesh white, thin, watery and brittle.
Gills at first whitish becoming pallid buff, more or less decurrent, fairly crowded. **Spores** brown, smooth, broadly ellipsoid, 8-12 x 5-6 µm. Basidia 4-spored. Gill edge cystidia clavate or flask-shaped; gill face cystidia absent.
Stem absent.
Odour not distinctive. **Taste** not distinctive.
Chemical tests none.
Occurrence early summer to autumn; infrequent.
■ Inedible.

Crepidotus variabilis (Pers.: Fr.) Kummer   Crepidotaceae
= *Claudopus variabilis* (Pers.: Fr.) Gill.

Small, whitish, irregularly kidney-shaped cap with pale buff gills, more or less sessile; on twigs and other plant debris.

Dimensions cap 0.5 - 2 cm dia.
Cap white; reniform or irregularly fan-shaped, felty or finely hairy, sometimes lobed. Flesh white, thin, soft and brittle.
Gills white at first, becoming pallid buff, decurrent, fairly crowded. **Spores** pinkish-buff, minutely warty, elongated-ellipsoid, 5-7 x 2.5-3.5 µm. Basidia 4-spored. Gill edge cystidia irregularly fusiform or clavate; gill face cystidia absent.
Stem lateral, rudimentary or absent.
Odour not distinctive. **Taste** not distinctive.
Chemical tests none.
Occurrence early summer to autumn; common.
■ Inedible.

Simocybe centunculus (Fr.) Sing.   Crepidotaceae
= *Naucoria centunculus* (Fr.) Kummer

Small brown agaric; trooping, on stumps and logs of broad-leaf trees.

Dimensions cap 0.5 - 2.5 cm dia; stem 1 - 3 cm tall x 0.1 - 0.3 cm dia.
Cap tan-brown sometimes with olivaceous tinge; convex, becoming more flattened, margin not striate, minutely downy or pruinose with filiform cap cystidia. Flesh brown, thin and fragile.
Gills more or less concolorous with cap, edges more pallid and slightly floccose, adnate, close. **Spores** brown, smooth, ellipsoid or bean-shaped, 6-10 x 4-5.5 µm. Gill edge cystidia filiform with swollen apex; gill face cystidia absent.
Stem pallid brown, more or less equal, slender, pruinose. Ring absent. Flesh brown, thin and fragile.
Odour acidic. **Taste** not distinctive.
Chemical tests none.
Occurrence late summer to autumn; infrequent.
■ Inedible.

Cortinarius (Myxacium) delibutus Fr. Cortinariaceae

Medium-sized slimy agaric with yellow cap, cinnamon or rust gills, pale stem with ring zone and slightly swollen base; solitary or in scattered trooping groups on soil, typically with birch or beech.

Dimensions cap 3 - 9 cm dia; stem 5 - 10 cm tall x 0.7 - 1.5 cm dia.
Cap yellow with tawny tinge; at first convex, becoming expanded and sometimes shallowly umbonate, viscid, smooth. Flesh yellowish, with violaceous tinge when young, moderate.
Gills at first violaceous, becoming pallid yellow then cinnamon or rust at maturity, adnate or emarginate, edges irregular, fairly distant.
Spores rust, rough, sub-spherical or broadly ellipsoid, 7-9 x 6-8 μm. Basidia 4-spored.
Stem pallid, with lilaceous tinge above brown cortinal zone, tinged yellow with cortinal remnants below, white downy at the base, more or less equal or slightly swollen at base. Ring absent. Flesh whitish, firm and full.
Odour not distinctive. **Taste** not distinctive.
Chemical tests none.
Occurrence late summer to autumn; infrequent.
■ Inedible.

Cortinarius (Myxacium) mucosus (Bull.: Fr.) Fr.
Cortinariaceae

Medium or large agaric, very slimy, with yellowish-brown cap, cinnamon or rust gills, stoutish white stem with ring zone and slightly swollen base; solitary or in scattered trooping groups on soil, typically with pine or birch.

Dimensions cap 6 - 10 cm dia; stem 6 - 10 cm tall x 1 - 2.5 cm dia.
Cap ochraceous or honey with reddish tinge; at first convex, becoming expanded and sometimes shallowly depressed, very viscid, smooth. Flesh whitish and moderate.
Gills at first pallid whitish then clay or rust at maturity, adnate, close.
Spores rust, rough, ellipsoid, 12-14 x 5.5-6.5 μm. Basidia 4-spored. Gill edge cystidia absent.
Stem white, with rust sub-apical cortinal zone, smooth or faintly scaly below cortina, more or less equal or slightly swollen at base, very viscid. Ring absent. Flesh white, firm and full.
Odour not distinctive. **Taste** not distinctive.
Chemical tests none.
Occurrence late summer to autumn; infrequent.
■ Inedible.

Cortinarius (Myxacium) trivialis J Lange
= *Myxacium collinitum var repandum* Ricken   **Cortinariaceae**

Medium to large agaric with yellowish slimy cap, pale cinnamon or rust gills, yellowish-brown stem with distinctive whitish scales, not bulbous at base; solitary or in scattered trooping groups on soil, typically with alder or willow.

Dimensions cap 4 - 11 cm dia; stem 5 - 12 cm tall x 1 - 2 cm dia.
Cap ochraceous-tan, very viscid when damp; at first conico-convex, becoming expanded, sometimes bluntly umbonate, smooth. Flesh pallid ochraceous and moderate.
Gills at first pallid clay, becoming more ochraceous and then rust at maturity, adnate or emarginate, fairly crowded. **Spores** rust, rough, almond-shaped, 10-13 x 6-7 μm. Basidia 4-spored.
Stem above ring concolorous with cap or darker brown, smooth; below, covered with belts of whitish scaly cortinal remnants, viscid, fairly stout, more or less equal, not bulbous. Ring whitish cortinal zone. Flesh pallid ochraceous, darker brown towards base, fibrous, full or with narrow cavity.
Odour not distinctive. **Taste** not distinctive.
Chemical tests none.
Occurrence late summer to autumn; rare. (Whitley Common, near Guildford, Surrey)
■ Inedible.

Cortinarius (Phlegmacium) caesiocortinatus
Schaeff. apud Moser   **Cortinariaceae**

Medium-sized agaric with brown or yellowish-brown cap, pale clay-brown gills, stem tinged lilac at apex; solitary or scattered on soil, with broad-leaf and coniferous trees.

Dimensions cap 4 - 9 cm dia; stem 4 - 9 cm tall x 1 - 2 cm dia.
Cap chestnut-brown or more ochraceous-brown; at first convex becoming expanded or umbonate. Flesh pallid brown and moderate.
Gills at first pallid lilaceous, becoming clay, adnate, close.
Spores brown, finely warty, sub-spherical or broadly ellipsoid, 8-10 x 7-8.5 µm. Basidia 4-spored.
Stem lilaceous at apex otherwise covered with pallid whitish velar remnants, more or less equal, base somewhat bulbous. Ring absent. Flesh whitish, fibrous and full.
Odour not distinctive. **Taste** not distinctive.
Chemical tests cap skin red with KOH or NH₃.
Occurrence late summer to autumn; infrequent.
◼ Inedible.

Cortinarius (Phlegmacium) crocolitus Quél.
                                       Cortinariaceae

Medium-sized or large agaric with brown or yellowish-brown cap, pale clay-brown gills, stem decorated with yellowish scaly bands; solitary or scattered on soil in broad-leaf woods, often with birch.

Dimensions cap 5 - 12 cm dia; stem 3 - 8 cm tall x 0.6 - 2.5 cm dia.
Cap yellowish-brown, more tawny at centre; at first convex, becoming expanded or umbonate. Flesh ochraceous-buff and moderate.
Gills at first pallid clay or with lilaceous tinge, becoming cinnamon-brown or rust, adnate, close. **Spores** brown, warty, almond-shaped, 10-12.5 x 5.5-7 µm. Basidia 4-spored.
Stem pallid yellowish or darker with age, decorated with ochraceous scaly velar bands, cortina white, more or less equal or slightly bulbous at base. Ring absent. Flesh yellowish, fibrous and full.
Odour not distinctive. **Taste** not distinctive.
Chemical tests flesh yellow with KOH or NH₃.
Occurrence late summer to autumn; infrequent.
◼ Inedible.

Cortinarius (Phlegmacium) nemorensis
(Fr.) J Lange   **Cortinariaceae**

Medium-sized agaric with slimy violet-brown cap, pale stem and violet-tinged flesh; solitary or in trooping groups on soil, with broad-leaf and coniferous trees, favouring beech.

Dimensions cap 4 - 10 cm dia; stem 4 - 8 cm tall x 1 - 3 cm dia.
Cap violaceous-date-brown; at first convex becoming expanded, viscid but often dry and cracking at centre. Flesh pallid, tinged violaceous and moderate.
Gills at first violaceous, becoming clay and then cinnamon, adnate, close. **Spores** rust, finely warty, broadly ellipsoid or almond-shaped, 9-12 x 5-7 µm. Basidia 4-spored.
Stem lilaceous at apex, otherwise covered with pallid whitish velar remnants, cortina pallid violaceous, more or less equal or slightly swollen at base. Ring absent. Flesh pallid. tinged violaceous, fibrous and full.
Odour earthy. **Taste** not distinctive.
Chemical tests flesh chrome yellow with KOH or NH₃.
Occurrence late summer to autumn; infrequent.
◼ Inedible.

Cortinarius (Phlegmacium) varius (Schaeff.) Fr.
Cortinariaceae

Medium to large agaric with yellowish-brown or fawn cap, pallid lilac or clay-brown gills, whitish club-shaped stem with rust-brown thin ring zone; solitary or scattered on calcareous soil with conifers.

Dimensions cap 4 - 10 cm dia; stem 4 - 10 cm tall x 1 - 2 cm dia.
Cap ochraceous with minute brown scaly fibres; at first convex, becoming expanded or umbonate. Flesh whitish tinged ochraceous or buff, moderate.
Gills at first violaceous, becoming pallid buff or clay, adnate, close.
Spores brown, finely warty, broadly ellipsoid or almond-shaped, 10-12 x 5.5-6 µm. Basidia 4-spored.
Stem white, smooth, clavate, with thin brown ring zone and faint brown banding of velar remnants below. Ring absent. Flesh whitish, fibrous and full.
Odour not distinctive. **Taste** not distinctive.
Chemical tests flesh chrome yellow with KOH or NH_3.
Occurrence late summer to autumn; infrequent.
■ Inedible.
Note: the background in the illustration is deceptive - there was Scot's Pine in the immediate vicinity.

Cortinarius (Sericeocybe) anomalus
(Fr. Fr.)Fr.   **Cortinariaceae**

Medium-sized agaric with clay-brown cap, grey-violet gills and somewhat whitish stem with pale yellow belts below; solitary or in scattered trooping groups on soil in broad-leaf and coniferous woods.

Dimensions cap 5 - 8 cm dia; stem 6 - 10 cm tall x 0.8 - 1.5 cm dia.
Cap brownish-clay, sometimes with a violaceous tinge near margin, more date-brown with age; at first convex, becoming expanded, silky smooth or very finely micaceous. Flesh whitish with grey or violaceous tinges, moderate.
Gills at first pallid violaceous, then buff, becoming more rust at maturity, adnate or emarginate, close. **Spores** rust, warty, broadly ellipsoid or sub-spherical, 8-9 x 6-7.5 µm. Basidia 4-spored.
Stem whitish or pallid clay, with violaceous tinge towards apex, more or less equal. Ring absent but with ochraceous velar zones, inferior. Flesh whitish with ochraceous tinges at base, fibrous, full.
Odour not distinctive. **Taste** not distinctive.
Chemical tests none.
Occurrence late summer to autumn; frequent.
■ Inedible.

Cortinarius (Sericeocybe) anomalus
forma **lepidopus** (Cke.) Kühn. & Romagn.   **Cortinariaceae**

Medium-sized agaric with clay-brown cap, grey-violet gills and whitish stem with orange-yellow belts; solitary or in scattered trooping groups on soil, with birch and pine.

Dimensions cap 2.5 - 7 cm dia; stem 5 - 8 cm tall x 0.6 - 1.2 cm dia.
Cap brownish-clay, sometimes with violaceous tinge near margin, more date-brown with age; at first convex, becoming expanded, silky smooth or very finely micaceous. Flesh whitish with grey or violaceous tinges, moderate.
Gills at first pallid violaceous, then buff, becoming more rust at maturity, adnate or emarginate, close. **Spores** rust, warty, broadly ellipsoid or sub-spherical, 6.5-8.5 x 5.5-7µm. Basidia 4-spored.
Stem whitish or pallid clay, with violaceous tinge towards apex, more or less equal. Ring absent but with ochraceous-orange velar zones, inferior. Flesh whitish with ochraceous tinge at base, fibrous, full.
Odour not distinctive. **Taste** not distinctive.
Chemical tests none.
Occurrence late summer to autumn; frequent.
■ Inedible.
Considered by some authorities to be synonymous with *C. anomalus* but apparently specific to birch and pine, somewhat smaller and with more orange cortinal belts.

Cortinarius (Sericeocybe) malachioides P D Orton
Cortinariaceae

Medium to large, pale greyish-lilac agaric with dry cap, club-shaped or bulbous stem with white patches and lilac or brown gills; solitary or in scattered trooping groups on soil, with conifers.

Dimensions cap 3 - 10 cm dia; stem 6 - 7 cm tall x 1.5 - 2.5 cm dia.
Cap pallid lilaceous, fading with age and sometimes with ochraceous tinge, at first somewhat scaly pruinose, finely fibrillose; at first almost sub-spherical becoming expanded with broad umbo. Flesh pallid concolorous and moderate.
Gills at first pallid lilaceous, becoming cinnamon-brown at maturity, adnate or emarginate, close. **Spores** rust, rough, broadly ellipsoid, 9.5-12 x 5.5-6.5 μm. Basidia 4-spored.
Stem concolorous with cap, sheathed with white velar remnants, stoutly clavate or bulbous. Ring absent but sometimes with membraneous ring-like zone. Flesh concolorous with cap, fibrous and full.
Odour not distinctive. **Taste** not distinctive.
Chemical tests none.
Occurrence late summer to autumn; infrequent.
■ Inedible.

Cortinarius (Sericeocybe) ochrophyllus Fr.
Cortinariaceae

Medium-sized agaric with yellowish-brown shiny cap, yellowish-brown gills and pale scaly stem; solitary or in scattered trooping groups on soil, in coniferous woods but with birch.

Dimensions cap 3 - 6 cm dia; stem 5 - 10 cm tall x 0.8 - 1.2 cm dia.
Cap greyish-brown becoming more ochraceous when dry, smooth and somewhat shiny; at first convex becoming expanded. Flesh pallid ochraceous or buff, moderate.
Gills at first yellowish-brown becoming more pallid coffee-brown at maturity, adnate or emarginate, close. **Spores** rust, rough, broadly ellipsoid, 7-8.5 x 5.5-6.6 μm. Basidia 4-spored.
Stem pallid brown, decorated with more ochraceous velar flakes below ring zone, more or less equal or somewhat clavate. Ring absent but with distinct velar zone, superior. Flesh concolorous with cap but darker towards the base, fibrous, full.
Odour not distinctive. **Taste** not distinctive.
Chemical tests none.
Occurrence autumn; infrequent.
■ Inedible.

Cortinarius (Sericeocybe) pholideus (Fr.: Fr.) Fr.
Cortinariaceae

Medium to large agaric with yellowish cap covered in small brown scales, buff or brown gills, brownish scaly stem with ring zone and slightly swollen base; solitary or in scattered trooping groups on soil, in damp broad-leaf woods, typically with birch.

Dimensions cap 3 - 10 cm dia; stem 5 - 12 cm tall x 0.8 - 1.5 cm dia
Cap ochraceous-tan, covered with small darker brown fibrillose scales; at first convex, becoming expanded, slightly greasy. Flesh pallid ochraceous or buff, moderate.
Gills at first pallid violaceous, becoming more rust at maturity, adnate or emarginate, close. **Spores** rust, rough, broadly ellipsoid, 6.5-8.5 x 5-6 μm. Basidia 4-spored.
Stem ochraceous-brown, at first with violaceous tinge towards apex, then darker towards base, decorated with distinctive fibrillose scales below ring zone, more or less equal or somewhat clavate. Ring absent but with distinct brown velar zone, superior. Flesh concolorous with cap but darker towards the base, fibrous and full.
Odour not distinctive. **Taste** not distinctive.
Chemical tests none.
Occurrence late summer to autumn; infrequent.
■ Inedible.

Cortinarius (Telamonia) acutus Fr.   Cortinariaceae

Small sharply umbonate agaric with pale yellow cap, rust gills and slender stem; solitary or in scattered trooping groups on soil in damp situations, with conifers.

Dimensions cap 1 - 3 cm dia; stem 4 - 6 cm tall x 0.1 - 0.2 cm dia.
Cap pallid ochraceous; at first campanulate, becoming expanded with sharply pointed umbo, smooth when dry, striate 1/2 to centre when damp. Flesh pallid and thin.
Gills at first ochraceous, becoming more cinnamon at maturity, adnate or emarginate, edges somewhat flaky, fairly crowded.
Spores rust, finely warty, broadly ellipsoid, 7.5-11.5 x 4-6 µm. Basidia 4-spored. Gill edge cystidia present.
Stem pallid, smooth, more or less equal. Ring of velar remnants forming faint superior ring zone. Flesh pallid but darker brown towards base, fragile and full.
Odour not distinctive. **Taste** not distinctive.
Chemical tests none.
Occurrence late summer to autumn; frequent.
■ Inedible.

Cortinarius (Telamonia) armillatus (Fr.) Fr.
Cortinariaceae

Medium or large agaric with rust-brown bell-shaped fibrous cap, rust gills and reddish belts on stem with swollen base; solitary or in scattered trooping groups on acid soil in heathy situations, typically with birch.

Dimensions cap 4 - 12 cm dia; stem 6 - 15 cm tall x 1 - 3 cm dia.
Cap rust, darkening with age; at first hemispherical, becoming campanulate-expanded, at first smooth becoming radially fibrous-scaly, often with reddish cortinal remnants forming a belt at margin. Flesh pallid and moderate.
Gills at first pallid cinnamon, becoming rust at maturity, adnate or emarginate, fairly crowded. **Spores** rust, finely warty, broadly ellipsoid or almond-shaped, 7-12 x 5-7 µm. Basidia 4-spored. Gill edge cystidia absent.
Stem pallid cap colour streaked with fibrils, more or less equal but markedly swollen at base. Ring of velar remnants forming one or more orange-red median or inferior ring zones. Flesh pallid but darker brown towards base, fibrous and full.
Odour faint, of radish. **Taste** slight, bitter.
Chemical tests none.
Occurrence late summer to autumn; infrequent.
■ Inedible.

Cortinarius (Telamonia) bivelus Fr.   Cortinariaceae

Small or medium-sized agaric with pale or rust-brown cap and gills and whitish bulbous stem with ring zone; solitary or in scattered trooping groups on soil, with birch and conifers.

Dimensions cap 3 - 7 cm dia; stem 5 - 8 cm tall x 1 - 3 cm dia.
Cap pallid reddish-brown, more rust with age, covered with fine brown fibrils; at first hemispherical, becoming expanded-campanulate. Flesh pallid, tinged cap colour, thin.
Gills at first pallid ochraceous, then rust at maturity, adnate or emarginate, crowded. **Spores** rust, warty, broadly ellipsoid, 7-9 x 4-5 µm. Basidia 4-spored. Gill edge cystidia absent.
Stem whitish or pallid clay, stout and strongly bulbous, not rooting. Ring of velar remnants forming single median or inferior zone. Flesh whitish, fibrous and full.
Odour not distinctive. **Taste** not distinctive.
Chemical tests none.
Occurrence late summer to autumn; infrequent.
■ Inedible.

Cortinarius (Telamonia) bulbosus Fr. Cortinariaceae

Small or medium-sized agaric with pale or rust-brown cap and gills, brownish bulbous stem with white sheath; solitary or in scattered trooping groups on soil in broad-leaf and coniferous woods.

Dimensions cap 3 - 7 cm dia; stem 3 - 7 cm tall x 1 - 1.5 cm dia.
Cap reddish-brown, hygrophanous, drying brick, becoming more dull brown with age, finely fibrillose; at first campanulate becoming expanded. Flesh reddish-brown and thin.
Gills at first pallid brown, then cinnamon and rust at maturity, adnate or emarginate, close. **Spores** rust, warty, broadly ellipsoid, 7-9 x 4-5 µm. Basidia 4-spored. Gill edge cystidia absent.
Stem pallid, concolorous with cap, sheathed in white velar remants up to distinct ring zone, bulbous, not rooting. Ring white, median, zone-like or more distinct. Flesh pallid brown, fibrous, full or stuffed.
Odour not distinctive. **Taste** not distinctive.
Chemical tests none.
Occurrence late summer to autumn; infrequent.
■ Inedible.

Cortinarius (Telamonia) cohabitans Karst.
Cortinariaceae

Small or medium-sized chestnut-brown agaric; solitary or in scattered trooping groups on sandy soil favouring dune slacks, with dwarf willow.

Dimensions cap 2 - 4 cm dia; stem 2 - 6 cm tall x 1 - 2.5 cm dia.
Cap dark chestnut-brown, more pallid and with slight lilaceous tinge at margin, strongly hygrophanous; at first convex with inrolled margin, becoming expanded-umbonate, silky-fibrillose when dry. Flesh concolorous and thin.
Gills at first pallid cinnamon, becoming more rust at maturity, adnate or emarginate, close. **Spores** rust, finely warty, broadly ellipsoid, 8-9.5 x 5-5.5 µm. Basidia 4-spored. Gill edge cystidia absent.
Stem violaceous then brown, smooth, more or less equal or slightly bulbous at base. Ring of velar remnants forming distinct superior zone. Flesh concolorous, fibrous and full.
Odour not distinctive. **Taste** not distinctive.
Chemical tests none.
Occurrence late summer to autumn; rare. (Gower, South Wales)
■ Inedible.

Cortinarius (Telamonia) evernius Fr. Cortinariaceae

Small or medium-sized agaric with hygrophanous, pale tawny or reddish-brown cap, amethyst or brown gills and slender whitish scaly stem; solitary or in scattered trooping groups on soil, often amongst moss, with conifers.

Dimensions cap 3 - 9 cm dia; stem 3 - 10 cm tall x 0.8 - 1.3 cm dia.
Cap tawny when dry with violaceous tinge at margin, hygrophanous, becoming reddish-brown or dark umber, silky then more or less smooth; at first convex becoming expanded. Flesh concolorous and thin.
Gills at first amethyst, then clay and finally cinnamon, adnate or emarginate, close. **Spores** rust, almost smooth, ellipsoid, 9 -10.5 x 5-6 µm. Basidia 4-spored. Gill edge cystidia absent.
Stem whitish, with several whitish cottony-scaly belts of velar remnants, more or less equal, then tapering somewhat, fairly slender. Ring white, zone-like. Flesh whitish, fibrous, stuffed or full.
Odour not distinctive. **Taste** not distinctive.
Chemical tests none.
Occurrence late summer to autumn; rare but more common in Scotland. (Culbin Forest, Morayshire)
■ Inedible.

Cortinarius (Telamonia) flexipes Fr.   Cortinariaceae

Small agaric with dark-brown umbonate cap, cinnamon-brown gills and slender brown stem with white belts; solitary or in scattered trooping groups on boggy soil, usually amongst moss, with birch.

Dimensions cap 1 - 3 cm dia; stem 3 - 7 cm tall x 0.3 - 0.5 cm dia.
Cap dull brown, more blackish-brown towards centre and with somewhat pallid margin, at first covered by veil, then only with marginal remnants or more or less smooth; at first bluntly conical, becoming expanded and bluntly umbonate. Flesh pallid concolorous and thin.
Gills at first cinnamon-brown, then darker, adnate or emarginate, somewhat distant. **Spores** rust, very finely warty, ellipsoid, 7-10 x 4.5-5.5 µm. Basidia 4-spored. Gill edge cystidia absent.
Stem brownish, tinged faintly violaceous at apex, smooth above, whitish belted scaly velar remnants below, more or less equal, slender. Ring absent. Flesh brown, fibrous, stuffed or full.
Odour not distinctive. **Taste** not distinctive.
Chemical tests none.
Occurrence late summer to autumn; infrequent.
■ Inedible.

Cortinarius (Telamonia) hemitrichus (Pers.)Fr.
Cortinariaceae

Smallish agaric with greyish-brown, white-flecked umbonate cap, grey-brown gills and slender whitish scaly stem; solitary or in scattered trooping groups on soil, with birch.

Dimensions cap 2 - 4 cm dia; stem 3 - 7 cm tall x 0.3 - 0.5 cm dia.
Cap greyish-brown, more dark brown towards centre and with somewhat pallid margin, at first decorated with fine white flaky scales, then more or less smooth; at first conical, becoming expanded and sharply umbonate. Flesh pallid concolorous and thin.
Gills at first greyish-brown, then cinnamon, adnate or emarginate, close. **Spores** rust, very finely warty, ellipsoid, 7.5-11 x 4-5 µm. Basidia 4-spored. Gill edge cystidia absent.
Stem whitish and smooth above ring, whitish cottony-scaly velar remnants below, more or less equal, slender. Ring thin, cottony-scaly, superior or median. Flesh whitish, fibrous, stuffed or full.
Odour not distinctive. **Taste** not distinctive.
Chemical tests none.
Occurrence late summer to autumn; infrequent.
■ Inedible.

Cortinarius (Telamonia) paleaceus (Weinm.) Fr.
Cortinariaceae

Smallish agaric with grey-brown, white-flecked umbonate cap, grey-brown gills and slender brownish stem with white scales; solitary or in scattered trooping groups on soil in coniferous woods and with birch.

Dimensions cap 2 - 4 cm dia; stem 3 - 7 cm tall x 0.3 - 0.5 cm dia.
Cap brownish-grey, sometimes with ochraceous tinge towards centre, drying pallid fawn, at first decorated with fine white flaky scales then more or less smooth; at first conical, becoming expanded and sharply umbonate. Flesh pallid concolorous and thin.
Gills at first greyish-brown, then cinnamon, adnate or emarginate, close. **Spores** rust, very finely warty, ellipsoid, 6.5-9 x 4-6 µm. Basidia 4-spored. Gill edge cystidia absent.
Stem brownish and fibrous smooth above ring, whitish cottony sheathed with velar remnants below, more or less equal, slender. Ring thin, cottony-scaly, superior or median. Flesh whitish, fibrous, stuffed or full.
Odour of pelargonium. **Taste** not distinctive.
Chemical tests none.
Occurrence late summer to autumn; infrequent.
■ Inedible.

Cortinarius (Telamonia) subbalaustinus R Henry
= *Cortinarius balaustinus* ss. J Lange   **Cortinariaceae**

Medium-sized agaric with tan or yellowish cap, cinnamon-brown gills, brownish streaked stem with slightly swollen base; solitary or in scattered trooping groups on soil, typically with birch.

Dimensions cap 4 - 8 cm dia; stem 4 - 7 cm tall x 0.5 - 1 cm dia.
Cap tan, strongly hygrophanous, drying yellowish; at first convex, becoming expanded, smooth. Flesh ochraceous-brown and fairly thin.
Gills at first ochraceous becoming more cinnamon at maturity, adnate or emarginate, fairly crowded. **Spores** rust, warty, ellipsoid or pip-shaped, 7.5-9 x 4.5-5.5 µm. Basidia 4-spored. Gill edge cystidia absent.
Stem brown, streaked with whitish fibrils, more or less equal or slightly swollen at base. Ring whitish zone, ephemeral. Flesh ochraceous-brown, fibrous and full.
Odour not distinctive. **Taste** not distinctive.
Chemical tests none.
Occurrence late summer to autumn; infrequent.
■ Inedible.

Cortinarius (Telamonia) subviolascens R Henry
Cortinariaceae

Medium-sized agaric with greyish-brown cap, violet-tinged or cinnamon-brown gills and stout, whitish, belted stem; solitary or in scattered trooping groups on soil in confierous woods, typically with spruce.

Dimensions cap 1 - 6 cm dia; stem 3 - 7 cm tall x 1 - 2.5 cm dia.
Cap greyish-brown and at first covered with whitish veil, becoming dirty-brown or fawn at maturity; at first convex becoming more or less flattened, smooth. Flesh clay or fawn and fairly thick.
Gills at first violaceous-blue becoming clay and finally cinnamon, adnate or emarginate, thick, distant. **Spores** rust, finely warty, ellipsoid, 7-11.5 x 5-6 µm. Basidia 4-spored. Gill edge cystidia absent.
Stem white with lilaceous tinge above ring zone when young, otherwise downy-fibrillose. Ring whitish, somewhat booted, zone-like with belts below. Flesh clay or fawn, fibrous, full.
Odour of radish or camphor. **Taste** not distinctive.
Chemical tests none.
Occurrence autumn, infrequent.
■ Inedible

Cortinarius (Telamonia) torvus (Bull.: Fr.)Fr.
Cortinariaceae

Medium to large agaric with pale clay cap, clay or brown gills, and slightly bulbous stem, whitish below ring zone; solitary or in scattered trooping groups on soil, typically with beech or birch.

Dimensions cap 3 - 10 cm dia; stem 4 - 7 cm tall x 1 - 1.5 cm dia.
Cap clay-brown, more pallid when dry, with darker faint radiating fibrils; at first convex, becoming flattened and sometimes slightly umbonate, smooth. Flesh buff and fairly thick.
Gills at first buff with lilaceous tinge, becoming more rust-brown at maturity, adnate or emarginate, close. **Spores** rust, finely warty, ellipsoid, 8-10 x 5-6.5 µm. Basidia 4-spored. Gill edge cystidia absent.
Stem above ring tinged violaceous, whitish-clay below streaked with fibrils and with velar remnants forming irregular bands or patches, more or less equal or slightly bulbous towards base. Ring white, membraneous, superior. Flesh buff with violaceous tinges, fibrous and full.
Odour sweetish, thick. **Taste** slight, bitter.
Chemical tests none.
Occurrence late summer to autumn; infrequent.
■ Inedible.

Cortinarius (Leprocybe) limonius (Fr.: Fr.) Fr.
Cortinariaceae

Largish agaric with yellow cap, rust-brown gills and yellowish stem with brown fibrils; solitary or in scattered trooping groups on soil, with conifers and birch.

Dimensions cap 5 - 9 cm dia; stem 5 - 9 cm tall x 1 - 2 cm dia.
Cap lemon-yellow towards centre, tinged orange-yellow towards margin, slightly hygrophanous, more or less smooth; at first convex, becoming expanded and somewhat shallowly umbonate. Flesh yellowish and moderate.
Gills at first yellowish-brown, then more rust, adnate or emarginate, close. **Spores** rust, very finely warty, broadly ellipsoid or pip-shaped, 7.5-8 x 5.5-6.5 µm. Basidia 4-spored.
Stem yellowish and smooth at apex, otherwise decorated with fine cottony fibrillose cortinal remnants, brownish towards base, more or less equal, tapering at the base, stoutish. Ring absent. Flesh yellowish, fibrous, stuffed or full.
Odour faint, of fruit. **Taste** not distinctive.
Chemical tests none.
Occurrence late summer to autumn; rare, more or less restricted to Scottish Highlands and other montane regions of Europe. (Linn of Dee, Scottish Highlands)
■ Inedible.

Cortinarius (Leprocybe) speciosissimus
Kühn. & Romagn.   **Cortinariaceae**

Medium-sized agaric with tawny-brown, bluntly umbonate cap, gills and stem; solitary or in scattered trooping groups on acid moorland soil, with conifers.

Dimensions cap 3 - 8 cm dia; stem 5 - 12 cm tall x 0.5 - 1.5 cm dia.
Cap tawny or orange-brown, decorated with fine adpressed scales; at first conical-convex, becoming expanded and bluntly umbonate. Flesh concolorous and moderate.
Gills at first pallid ochraceous, becoming rust brown, adnate or emarginate, thick, broad, somewhat distant. **Spores** rust, finely warty, broadly ellipsoid or sub-spherical, 9 -12 x 6.5-8.5 µm. Basidia 4-spored.
Stem concolorous with cap, sometimes decorated below with more ochraceous cortinal remnants, more or less equal or slightly clavate at base. Ring absent. Flesh concolorous, fibrous, stuffed or full.
Odour faint, of radish. **Taste ON NO ACCOUNT SHOULD ANY PART OF THE SPECIMEN BE CONSUMED.**
Chemical tests none.
Occurrence late summer to autumn; rare.
✚ Lethally poisonous.

Cortinarius (Dermocybe) cinnamomeobadius
R Henry
= *Dermocybe cinnamomeobadia* (R Henry) Moser   **Cortinariaceae**

Smallish agaric with chestnut cap, ochre-brown gills and slender yellowish stem; solitary or in scattered trooping groups on soil in coniferous and broad-leaf woods and on heaths.

Dimensions cap 2 - 6 cm dia; stem 3 - 7 cm tall x 0.3 - 0.7 cm dia.
Cap at first ochraceous-rust, becoming chestnut-brown but often with a more ochraceous margin; at first convex becoming expanded-umbonate, finely fibrillose. Flesh ochraceous and fairly thin.
Gills at first orange-yellow or orange-fawn then amber and rust, adnate or emarginate, irregular, fairly crowded. **Spores** rust, very finely warty, ellipsoid, 6.5-8.5 x 4-5 µm. Basidia 4-spored.
Stem ochraceous with olivaceous tinge at apex, streaked with fine brownish fibrils below, more or less equal, base covered in down and sometimes slightly swollen. Ring absent. Flesh ochraceous, fibrous, narrowly hollow or full.
Odour faint, of radish. **Taste** mild.
Chemical tests none.
Occurrence late summer to autumn; infrequent.
■ Inedible.

Cortinarius (Dermocybe) cinnamomeus (L.: Fr.) Fr.
= *Dermocybe cinnamomea* L.: Fr. **Cortinariaceae**

Smallish agaric with olive-brown cap, orange-cinnamon gills and slender yellowish stem with ring zone; solitary or in scattered trooping groups on soil in coniferous woods and with birch on heaths.

Dimensions cap 1.5 - 7 cm dia; stem 2.5 - 9 cm tall x 0.3 - 0.8 cm dia.
Cap hazel or olive-brown, more rust-brown towards centre; at first convex, becoming expanded-umbonate, finely fibrillose. Flesh dirty lemon-yellow or with olivaceous tinge, fairly thin.
Gills at first yellow or ochraceous-orange, then cinnamon, adnate or emarginate, fairly crowded. **Spores** rust, very finely warty, ellipsoid, 6.5-8 x 4-5 μm. Basidia 4-spored.
Stem ochraceous with olivaceous tinge, streaked with fine fibrils, more or less equal, slender. Ring tawny-brown, finely scaly zone. Flesh ochraceous, fibrous, narrowly hollow or full.
Odour faint, of radish. **Taste** bitter.
Chemical tests none.
Occurrence late summer to autumn; infrequent.
■ Inedible.

Cortinarius (Dermocybe) croceifolius Peck
= *Dermocybe croceifolia* (Peck) Moser **Cortinariaceae**

Small agaric with tawny-brown cap, yellowish-tawny gills and slender yellowish stem with faint ring zone; solitary or in scattered trooping groups on soil in coniferous woods.

Dimensions cap 1.5 - 3 cm dia; stem 2.5 - 8.5 cm tall x 0.3 - 0.5 cm dia.
Cap at first ochraceous-brown, becoming more rust; at first convex becoming expanded-umbonate, finely fibrillose. Flesh chrome-yellow and fairly thin.
Gills at first yellow, then tawny and rust at maturity, adnexed, fairly crowded. **Spores** rust, very finely warty, ellipsoid, 6-7.5 x 4-5 μm. Basidia 4-spored.
Stem at first chrome-yellow, becoming discoloured tawny from the base upwards, more or less equal, slender. Ring absent but with faint superior cortinal zone. Flesh chrome-yellow, fibrous, narrowly hollow, stuffed or full.
Odour not distinctive. **Taste** bitter.
Chemical tests none.
Occurrence late summer to autumn; infrequent.
■ Inedible.

Cortinarius (Dermocybe) puniceus P Orton
= *Dermocybe punicea* (P Orton) Moser **Cortinariaceae**

Smallish agaric with wholly blood-red colour except for paler stem base; solitary or in scattered trooping groups on soil with broad-leaf trees.

Dimensions cap 1.5 - 4 cm dia; stem 3 - 6 cm tall x 0.3 - 0.8 cm dia.
Cap purplish-blood-red, sometimes with chestnut tinges towards centre, more or less smooth or finely felty; at first convex becoming expanded and slightly umbonate. Flesh concolorous and thin.
Gills concolorous with cap, adnate or emarginate, close.
Spores rust, very finely warty, ellipsoid, 6.5-8.5 x 4-5 μm. Basidia 4-spored.
Stem concolorous with cap but pallid ochraceous-pink, woolly-downy at base, more or less equal, slender. Ring absent. Flesh concolorous, fibrous, stuffed or full.
Odour not distinctive. **Taste** not distinctive.
Chemical tests none.
Occurrence late summer to autumn; infrequent or rare.
■ Inedible.

Cortinarius (Dermocybe) sanguineus (Wulf.: Fr.) Fr.
= *Dermocybe sanguinea* (Wulf.: Fr.) Moser   **Cortinariaceae**

Smallish agaric, dark blood-red throughout, not swollen at base; solitary or in scattered trooping groups on soil, with conifers.

Dimensions cap 2 - 5 cm dia; stem 3 - 6 cm tall x 0.3 - 0.8 cm dia.
Cap dark blood-red; at first convex, becoming expanded-umbonate, silky fibrillose. Flesh concolorous and fairly thin.
Gills blood-red, then tinged rust, adnate or emarginate, irregular, fairly crowded. **Spores** rust, minutely warty, ellipsoid, 6-9 x 4-6 µm. Basidia 4-spored.
Stem concolorous with cap but more pallid pinkish-downy at base, more or less equal or slightly swollen at base. Ring absent. Flesh concolorous, fibrous, narrowly hollow or full.
Odour not distinctive. **Taste** not distinctive.
Chemical tests none.
Occurrence late summer to autumn; infrequent.
■ Inedible.

Cortinarius (Dermocybe) semisanguineus (Fr.) Gill.
= *Dermocybe semisanguinea* (Fr.) Moser   **Cortinariaceae**

Smallish agaric with yellowish-tan cap, blood-red gills, yellowish stem not swollen at base; solitary or in scattered trooping groups on soil, typically with conifers and birch.

Dimensions cap 2 - 6 cm dia; stem 2 - 10 cm tall x 0.4 - 1.2 cm dia.
Cap olivaceous-tan, sometimes with reddish tinge at centre; at first convex, becoming expanded-umbonate, smooth. Flesh olivaceous and fairly thin.
Gills at first reddish-purple, becoming blood-red and finally more reddish-rust, adnate or emarginate, irregular, fairly crowded.
Spores rust, warty, ellipsoid, 6-8 x 4-5 µm. Basidia 4-spored.
Stem pallid cap colour streaked with olivaceous fibrils below, more or less equal, downy at base and sometimes slightly swollen. Ring absent. Flesh ochraceous, fibrous and full.
Odour faint, of radish. **Taste** slightly bitter.
Chemical tests none.
Occurrence late summer to autumn; frequent.
■ Inedible.

Galerina autumnalis (Peck) Sing. & A H Smith
= *Agaricus antumnalis* Peck   **Cortinariaceae**

Small agaric, with yellowish-tan gelatinous cap and concolorous gills, ring on stem; clustered on stumps and logs of broad-leaf trees.

Dimensions cap 2.5 - 6.5 cm dia; stem 3 - 9 cm tall x 0.3 - 0.8 cm dia.
Cap ochraceous-brown, drying more yellow; convex, becoming flattened with a blunt umbo, smooth, gelatinous. Flesh pallid cap colour and thin.
Gills concolorous with cap, adnate, crowded. **Spores** reddish-ochre, smooth, ellipsoid or almond-shaped, germ pore, 8.5-10.5 x 5-6.5 µm. Basidia 4-spored. Gill edge cystidia fusiform-ventricose or flask-shaped with elongated apex; gill face cystidia similar but larger.
Stem concolorous with cap but darker tan below ring, slender, more or less equal, fibrous. Ring brown, fibrous, superior. Flesh concolorous with cap above, becoming darker tan towards base, stuffed.
Odour faint, of meal. **Taste** not distinctive.
Chemical tests none.
Occurrence late summer to autumn; infrequent.
■ Inedible.

Galerina cephalotricha Kühn.   Cortinariaceae

Very small yellowish-brown agaric; solitary or in small groups amongst moss under conifers or alders.

Dimensions cap 0.5 - 1.5 cm dia; stem 2 - 7 cm tall x 0.05 - 0.2 cm dia.
Cap ochraceous-brown, hygrophanous, drying more pallid; at first narrowly conical, then convex, becoming expanded, striate, not greasy or viscid. Flesh pallid cap colour and thin.
Gills ochraceous-brown, adnate, crowded. **Spores** reddish-ochre, smooth, fusiform or almond-shaped, indistinct germ pore, 9-12 x 5-7µm. Basidia 4-spored. Gill edge cystidia swollen with somewhat clavate extension; gill face cystidia absent.
Stem pallid yellowish, scattered fibrillose towards base, slender, more or less equal. Ring absent. Flesh concolorous with cap, fragile and stuffed.
Odour not distinctive. **Taste** not distinctive.
Chemical tests none.
Occurrence summer to autumn; infrequent.
■ Inedible.

Galerina clavata (Vel.) Kühn.   Cortinariaceae
= *Galera clavata* (Vel.) J Lange

Small ochre-yellow agaric; solitary or in small groups amongst moss in damp grassy places and on lawns after heavy rain.

Dimensions cap 0.5 - 3 cm dia; stem 2 - 7 cm tall x 0.05 - 0.2 cm dia.
Cap bright ochraceous or slightly tawny, hygrophanous, drying more pallid; at first hemispherical or obtusely conical, becoming broadly conical, or slightly umbonate, striate to the disc, smooth. Flesh pallid cap colour, very thin and translucent.
Gills pallid ochraceous with brown tinge, adnate, close or somewhat distant. **Spores** ochraceous-brown, smooth, fusiform or elongated-ellipsoid, no germ pore, 11-15 x 6-8 µm. Basidia 2- or 4-spored. Gill edge cystidia skittle-shaped, sometimes branched at the apex; gill face cystidia absent.
Stem pallid yellowish, whitish pruinose at apex, scattered fibrillose towards base, slender and flexuose, more or less equal. Ring absent. Flesh concolorous with cap, fragile and stuffed.
Odour not distinctive. **Taste** not distinctive.
Chemical tests none.
Occurrence summer to autumn; frequent.
■ Inedible.

Galerina laevis (Pers.) Sing.   Cortinariaceae
= *Galera graminea* Vel.

Very small yellowish-brown agaric; solitary or in small groups amongst grass and moss in lawns and parks.

Dimensions cap 0.5 - 1.2 cm dia; stem 2 - 3.5 cm tall x 0.1 - 0.15 cm dia.
Cap pallid tawny or ochraceous-brown, hygrophanous, drying more pallid; at first campanulate, then convex, becoming expanded, striate when damp. Flesh pallid cap colour and thin.
Gills ochraceous-brown, adnate, somewhat distant.
Spores ochraceous-brown, smooth, fusiform or almond-shaped, germ pore absent, 7-11 x 4-5.5 µm. Basidia 2- or 4- spored. Gill edge cystidia swollen with elongated apex; gill face cystidia absent.
Stem pallid yellowish, slender, more or less equal, silky. Ring absent. Flesh concolorous with cap, fragile and stuffed.
Odour not distinctive. **Taste** not distinctive.
Chemical tests none.
Occurrence summer to autumn; frequent.
■ Inedible.

Galerina pumila (Pers.: Fr.) J Lange   **Cortinariaceae**
= *Galerina mycenopsis* (Fr. Fr.) Kühn.
= *Galera mycenopsis* (Fr.: Fr.) Quél.

Very small tawny-yellow agaric; solitary or in small groups amongst moss.

Dimensions cap 1 - 1.5 cm dia; stem 3 - 5 cm tall x 0.15 - 0.25 cm dia.
Cap ochraceous-tawny, hygrophanous, drying more pallid; at first convex, becoming expanded, striate almost to centre, not greasy or viscid. Flesh pallid cap colour and thin.
Gills ochraceous-brown, adnate, close or somewhat distant.
Spores ochraceous-brown, smooth, fusiform or elongated-ellipsoid, germ pore absent, 9-14 x 6-8 µm. Basidia 4-spored. Gill edge cystidia cylindrical or clavate; gill face cystidia absent.
Stem pallid yellowish, slightly pruinose at apex, scattered fibrillose towards base, slender, more or less equal. Ring absent. Flesh concolorous with cap, fragile and stuffed.
Odour not distinctive. **Taste** not distinctive.
Chemical tests none.
Occurrence summer to autumn; frequent.
■ Inedible.

Gymnopilus hybridus (Fr.: Fr.) Sing.   **Cortinariaceae**
= *Flammula hybrida* (Fr.: Fr.) Gill.

Small to medium yellowish-brown agaric without brown spots on gills and with ring zone on stem; solitary, clustered or densely caespitose on, and close by, stumps and other wood of coniferous and broad-leaf trees.

Dimensions cap 2 - 8.5 cm dia; stem 3 - 8 cm tall x 0.3 - 1 cm dia.
Cap at first saffron, soon apricot and then tawny-rust from centre, somewhat silky-fibrillose; at first convex, becoming expanded and wavy, sometimes with whitish appendiculate velar remnants at margin when young. Flesh concolorous, firm and medium.
Gills cream or ochraceous, becoming more cinnamon at maturity, never spotted, adnate with tooth or slightly emarginate, crowded.
Spores sienna or cinnamon, finely warty, ellipsoid or almond-shaped, 7-9 x 4-5.5 µm. Basidia 4-spored. Gill edge cystidia flask-shaped, capitate, thin-walled; gill face cystidia rare but similar.
Stem concolorous with cap, more or less equal, silky-striate below ring. Ring yellowish then brown, zone-like, sub-apical. Flesh yellowish, becoming rust towards base, firm, stuffed.
Odour not distinctive. **Taste** bitter.
Chemical tests none.
Occurrence late summer to autumn; frequent.
■ Inedible.

Gymnopilus penetrans (Fr.: Fr.) Murr.   **Cortinariaceae**
= *Flammulina penetrans* (Fr.: Fr.) Quél.

Small or medium yellowish-brown agaric without ring on stem; solitary, clustered or densely caespitose on, and close by, stumps and other wood, including sawdust, of coniferous and broad-leaf trees.

Dimensions cap 2.5 - 6 cm dia; stem 2.5 - 6 cm tall x 0.4 - 0.7 cm dia
Cap ochraceous with apricot tinge becoming more cinnamon with age, silky-smooth; at first convex, becoming expanded and wavy. Flesh concolorous, firm and thin.
Gills straw or ochraceous, becoming more cinnamon at maturity but spotted orange-brown, adnate with tooth or slightly emarginate, crowded. **Spores** sienna or cinnamon, finely warty, ellipsoid or almond-shaped, 6.5-9 x 4-5 µm. Basidia 4-spored. Gill edge cystidia flask-shaped, capitate, thin-walled; gill face cystidia rare but similar.
Stem concolorous with cap, slender, more or less equal, silky-striate or fibrillose. Ring absent or very slight zone when young. Flesh yellowish, orange-brown where damaged towards the base, firm and stuffed.
Odour not distinctive. **Taste** bitter.
Chemical tests none.
Occurrence late summer to autumn; common.
■ Inedible.

Gymnopilus spectabilis (Fr.) Gill.　　**Cortinariaceae**

= *Gymnopilus junonius* (Fr.) P Orton
= *Pholiota spectabilis* (Fr.) Kummer

Large, robust, yellowish-brown agaric with ring on stem; solitary, clustered or densely caespitose on, and close by, stumps and other wood of broad-leaf and less frequently, coniferous trees.

Dimensions cap 5 - 20 cm dia; stem 5 - 20 cm tall x 0.5 - 3 cm dia.
Cap sienna with ochraceous or apricot tinges, with cinnamon fibrous scales; at first convex, becoming expanded or broadly umbonate, with yellowish appendiculate velar remnants when young. Flesh straw or cream, firm and thick.
Gills straw or ochraceous, becoming more cinnamon, adnate with tooth or slightly emarginate, crowded. **Spores** sienna or cinnamon, finely warty, lemon- or almond-shaped, 7.5-10 x 5-6 µm. Basidia 4-spored. Gill edge cystidia flask-shaped, capitate, thin-walled; gill face cystidia infrequent, inconspicuous.
Stem concolorous with cap, robust, bulbous-clavate or somewhat ventricose, silky-fibrillose below ring. Ring yellowish, then brown, sheathing, sub-apical. Flesh yellowish, firm and stuffed.
Odour faint or sometimes stronger when cut, fruity. **Taste** bitter.
Chemical tests none.
Occurrence late summer to autumn; frequent.
■ Inedible.

Hebeloma circinans Quél.　　**Cortinariaceae**

Small agaric with tawny-brown greasy cap and buff gills; solitary or in small groups, typically clustered, on calcareous soil with conifers, favouring spruce.

Dimensions cap 2 - 4 cm dia; stem 2 - 4 cm tall x 0.3 - 0.4 cm dia.
Cap pallid tawny-brown, sometimes with pinkish tinge, at first convex, becoming campanulate, then expanded and broadly umbonate, smooth, viscid when damp. Flesh pallid and thin.
Gills pallid buff, becoming more reddish-buff at maturity, adnate, crowded. **Spores** ochraceous-brown, finely warty, almond-shaped, 8.5-12 x 5-6.5 µm. Basidia 4-spored. Gill edge cystidia cylindrical, thin-walled.
Stem whitish, slender, more or less equal, silky, not rooting. Ring absent. Flesh white, firm and stuffed.
Odour not distinctive. **Taste** bitter.
Chemical tests none.
Occurrence late summer to autumn; infrequent.
■ Inedible.

Hebeloma crustuliniforme (Bull.) Quél.　　**Cortinariaceae**

Medium to large agaric with buff or tan greasy cap, clay gills and stoutish pale stem; solitary or in small groups on soil in open mixed woodland.

Dimensions cap 4 - 10 cm dia; stem 4 - 7 cm tall x 1 - 2 cm dia.
Cap buff or ochraceous-tan, darker at centre; at first convex, becoming expanded with a broad low umbo, smooth, greasy when damp. Flesh white and thick.
Gills pallid clay-brown, spotted with age, exuding watery droplets in damp conditions, adnate or adnexed, crowded. **Spores** rust, warty, almond-shaped, 10-12 x 5.5-6.5 µm. Basidia 4-spored. Gill edge cystidia cylindrical or clavate, thin-walled.
Stem whitish, fairly stout, more or less equal, granular towards apex. Ring absent. Flesh white, firm and stuffed.
Odour strong, of radish. **Taste** bitter.
Chemical tests none.
Occurrence late summer to autumn; common.
✧ Poisonous.

Hebeloma funariophilum Moser   Cortinariaceae

Small agaric with pale or yellowish-brown cap, clay gills and pale slender stem; solitary or in small groups on soil of fire sites, often with *Funaria* moss.

Dimensions cap 1 - 3 cm dia; stem 2 - 8 cm tall x 0.2 - 0.4 cm dia.
Cap pallid ochraceous-brown, hygrophanous drying pallid clay; at first convex, becoming expanded, smooth, margin striate when damp. Flesh pallid and thin.
Gills pallid clay-brown with grey tinge, adnate or adnexed, crowded.
Spores rust, warty, almond-shaped, 9-11 x 4.5-5.5 µm. Basidia 4-spored. Gill edge cystidia flask-shaped, thin-walled.
Stem pallid grey-brown, slender, more or less equal, granular towards apex, otherwise silky-fibrous. Ring absent. Flesh pallid, firm and stuffed.
Odour strong, of radish. **Taste** bitter.
Chemical tests none.
Occurrence late summer to autumn; infrequent.
✛ Poisonous.

Hebeloma fusipes Bres.   Cortinariaceae

Small agaric with leather-coloured greasy cap, clay-brown gills and pale slender rooting stem; solitary or in small groups on soil with broad-leaf trees.

Dimensions cap 2 - 4 cm dia; stem 5 - 8 cm tall x 0.4 - 0.6 cm dia.
Cap pallid leather with slightly darker centre; at first convex, becoming expanded or slightly umbonate, smooth, margin minutely denticulate. Flesh pallid and thin.
Gills pallid clay-brown or cinnamon, adnate or adnexed, edges minutely flaky, crowded. **Spores** rust, warty, almond-shaped, 10-14 x 7-9.5 µm. Basidia 4-spored. Gill edge cystidia narrowly flask-shaped, thin-walled.
Stem whitish, slender, more or less equal, slightly pruinose towards apex, otherwise fibrous, rooting spindle-shaped. Ring absent but membraneous-fibrous zone present. Flesh pallid, firm and stuffed.
Odour not distinctive. **Taste** bitter.
Chemical tests none.
Occurrence late summer to autumn; infrequent.
✛ Poisonous.

Hebeloma longicaudum (Pers.: Fr.) Kummer
= *Hebeloma nudipes* (Fr.) Gill.   Cortinariaceae

Medium-sized agaric with pale creamy-brown cap, clay gills and long stoutish pale stem; solitary or in small groups on damp acid soil, with broad-leaf and coniferous trees, often in dense moorland grass.

Dimensions cap 3 - 6 cm dia; stem 7 - 12 cm tall x 0.4 - 0.8 cm dia.
Cap pallid ochraceous with brown tinge; at first convex, becoming expanded with a broad low umbo, smooth, greasy when damp. Flesh white and thick.
Gills pallid clay-brown, spotted with age, exuding watery droplets in damp conditions, adnate or adnexed, crowded. **Spores** rust, finely warty, almond-shaped, 11-13 x 5-7 µm. Basidia 4-spored. Gill edge cystidia filiform or clavate, thin-walled.
Stem concolorous with cap, elongated, more or less equal but slightly bulbous at base, granular towards apex, otherwise finely fibrous. Ring absent. Flesh white, firm and stuffed.
Odour not distinctive or faint, of radish. **Taste** bitter.
Chemical tests none.
Occurrence late summer to autumn; infrequent.
✛ Poisonous.

Hebeloma mesophaeum (Pers.: Fr.) Quél. Cortinariaceae

Smallish agaric, cap with greasy dark brown centre and paler margin, clay gills and pale stem with ring; solitary or in small groups on soil, with conifers, less frequently with broad-leaf trees.

Dimensions cap 2.5 - 4.5 cm dia; stem 4 - 7 cm tall x 0.3 - 0.4 cm dia.
Cap darkish date- or chocolate-brown at centre with pallid whitish margin decorated with fibrous velar remnants when young; at first convex, becoming expanded with a broad low umbo, greasy or viscid when damp. Flesh white and thin.
Gills pallid clay-brown, adnate or adnexed, crowded. **Spores** rust, finely warty, almond-shaped, 8-10 x 5-6 µm. Basidia 4-spored. Gill edge cystidia cylindrical, thin-walled.
Stem pallid buff becoming tinged brown with age, more or less equal. Ring zone-like, fibrillose. Flesh brownish, firm and stuffed.
Odour strong, of radish. **Taste** bitter.
Chemical tests none.
Occurrence late summer to autumn; infrequent.
✣ Poisonous.

Hebeloma pumilum J Lange   Cortinariaceae

Smallish agaric with greasy ochre-brown cap, clay gills and pale stem; solitary or in small groups on soil, with conifers, less frequently with broad-leaf trees.

Dimensions cap 1 - 3 cm dia; stem 2 - 4 cm tall x 0.2 - 0.3 cm dia.
Cap ochraceous-brown with flesh tinge, slightly more pallid towards margin; at first convex, becoming expanded with a broad low umbo, greasy or viscid when damp. Flesh pallid and thin.
Gills pallid clay-brown then slightly darker, adnate or adnexed, crowded. **Spores** rust, finely warty, almond-shaped, 8-9.5 x 4.5-5.5 µm. Basidia 4-spored. Gill edge cystidia clavate.
Stem pallid buff, becoming tinged brown with age, more or less equal. Ring zone-like, ephemeral. Flesh pallid, firm and stuffed.
Odour not distinctive. **Taste** bitter.
Chemical tests none.
Occurrence late summer to autumn; infrequent.
■ Inedible.

Hebeloma radicosum (Bull.: Fr) Ricken   Cortinariaceae
= *Pholiota radicosa* (Bull.: Fr.) Kummer

Medium to large agaric with creamy-brown cap, greasy when damp, clay gills and stoutish pale rooting stem; solitary or in small groups on soil in broad-leaf woods, with oak and beech, probably also associated with underground latrines of rodents e.g. moles.

Dimensions cap 6 - 9 cm dia; stem 5 - 8 cm tall x 1 - 1.5 cm dia
Cap cream with brown tinge; at first convex, becoming expanded, smooth, greasy or viscid when damp. Flesh white and thick.
Gills pallid clay-brown, slightly darker with age, adnate or adnexed, crowded. **Spores** rust, minutely warty, almond-shaped, 9-10 x 5-6 µm. Basidia 4-spored. Gill edge cystidia cylindrical or slightly clavate, thin-walled.
Stem whitish, fairly stout, more or less equal but tapering into a long, tough 'tap root', granular above ring, decorated with brownish fibrous scales below. Ring zone-like, fibrous, superior, ephemeral. Flesh white, firm and stuffed.
Odour of almonds or marzipan. **Taste** mild.
Chemical tests none.
Occurrence late summer to autumn; rare. (Near Street, Somerset)
✣ Poisonous.

Hebeloma sacchariolens Quél.   **Cortinariaceae**

Small to medium-sized agaric with reddish or pale yellow cap and clay gills; solitary or in small groups on soil in damp woodlands.

Dimensions cap 2 - 7 cm dia; stem 4 - 8 cm tall x 0.5 - 1.2 cm dia.
Cap ochraceous-buff, more pallid at margin; at first convex, becoming expanded and barely umbonate, smooth, greasy when damp. Flesh white, fairly thin.
Gills clay-brown, becoming more rust at maturity, adnate, broad, crowded. **Spores** deep rust-brown, warty, almond-shaped, 12-17 x 7-9 µm. Basidia 4-spored. Gill edge cystidia cylindrical, thin-walled.
Stem whitish, more or less equal, granular at apex, otherwise silky-fibrous. Ring absent. Flesh white, firm and stuffed.
Odour sweet scented. **Taste** bitter.
Chemical tests none.
Occurrence late summer to autumn; infrequent.
■ ✣ Inedible, probably poisonous.

Hebeloma sinapizans (Paulet : Fr.) Gill. **Cortinariaceae**

Largish agaric with creamy-buff cap and gills and stout whitish stem; solitary or in small groups on soil in broad-leaf and mixed woods.

Dimensions cap 4 - 12 cm dia; stem 5 - 12 cm tall x 1 - 2 cm dia.
Cap pallid ochraceous-brown, more cream or buff at margin; at first convex, becoming expanded and flattened, smooth, greasy when young. Flesh whitish, thick, extending like a tongue into the stem cavity.
Gills pallid buff becoming more cinnamon at maturity, emarginate, crowded. **Spores** brown, finely warty, almond-shaped, 10-14.5 x 6-8 µm. Basidia 4-spored. Gill edge cystidia broadly clavate, thin-walled.
Stem pallid, faintly banded with brownish scales, stout, more or less equal, with bulbous base. Ring absent. Flesh white, firm, stuffed becoming hollow.
Odour strong, of radish or raw potato. **Taste** not distinctive.
Chemical tests none.
Occurrence late summer to autumn; infrequent.
✣ Poisonous.

Inocybe acuta Boud.   **Cortinariaceae**

Small, fibrous, brown agaric with acute umbo, brownish gills and slightly bulbous stem; solitary or in trooping groups on soil in mixed woods.

Dimensions cap 1 - 3.5 cm dia; stem 2 - 4 cm tall x 0.2 - 0.5 cm dia.
Cap umber-brown; at first campanulate, becoming flattened with a pronounced nipple-like umbo, finely radially fibrillose and cracking slightly with age. Flesh whitish and unchanging.
Gills pallid clay, becoming brown, emarginate, crowded.
Spores snuff-brown, smooth, angular-lumpy, 9-11 x 6-6.5 µm. Basidia 4-spored. Gill edge cystidia cylindrical or flask-shaped, some with encrusted apices.
Stem concolorous with cap, pruinose at apex, more or less equal, base slightly bulbous. Ring absent. Flesh whitish, discolouring brown, fibrous.
Odour not distinctive. **Taste** not distinctive.
Chemical tests none.
Occurrence late summer to autumn; infrequent.
■ Inedible.

Inocybe agardhii (Lund.) P Orton   **Cortinariaceae**

Medium-sized, yellowish-brown agaric with slightly umbonate cap, olive-brown gills and ring zone on stem; solitary or in trooping groups on sandy soil, with willow.

Dimensions cap 2 - 7 cm dia; stem 5 - 8 cm tall x 0.5 - 1 cm dia.
Cap ochraceous-brown, darkening to cinnamon, with coarse radial fibres; at first convex becoming flattened with a broad umbo, cracking radially with age. Flesh ochraceous-buff and unchanging.
Gills pallid clay, becoming brown with olivaceous tinge, adnate with decurrent tooth, close or crowded. **Spores** snuff-brown, smooth, ellipsoid or bean-shaped, 8-10 x 4.5-6 μm. Basidia 4-spored. Gill edge cystidia pear-shaped, thick walled with encrusted apices.
Stem concolorous with cap, fibrous beneath ring zone, stoutish, more or less equal, base slightly bulbous. Ring zone-like, indistinct. Flesh ochraceous-buff and unchanging.
Odour not distinctive. **Taste** not distinctive.
Chemical tests none.
Occurrence summer to autumn; infrequent.
■ Inedible.

Inocybe assimilata (Britz.) Sacc.   **Cortinariaceae**

= *Inocybe umbrina* Bres.
= *Astrosporina umbrina* (Bres.) Rea

Small dark brown agaric with slightly umbonate cap; solitary or in trooping groups on soil in broad-leaf and coniferous woods.

Dimensions cap 1.5 - 3 cm dia; stem 3 - 5 cm tall x 0.3 - 0.5 cm dia.
Cap umber-brown or purplish-brown with radial fibres, later cracked; at first convex, becoming flattened with a broad umbo, cracking radially with age; with cortina when young. Flesh whitish.
Gills at first clay or grey-brown, becoming darker brown, adnate with decurrent tooth, close or crowded. **Spores** snuff-brown, smooth, not star-shaped but bluntly nodulose, 6-9 x 5-6 μm. Basidia 4-spored. Gill edge cystidia narrowly ellipsoid, thick-walled with encrusted apices.
Stem pallid concolorous with cap, fibrous, barely pruinose at apex, more or less equal, with white bulbous base, almost marginate. Ring absent. Flesh whitish but more brown in base, unchanging.
Odour not distinctive. **Taste** not distinctive.
Chemical tests none.
Occurrence summer to autumn; infrequent.
■ Inedible.

Inocybe asterospora Quél.   **Cortinariaceae**

Medium-sized agaric with dull brown, radially streaked, umbonate cap, pale grey-brown gills, stem with a whitish bulbous base; solitary or in trooping groups on soil, often calcareous, with broad-leaf trees, favouring oak, also with conifers.

Dimensions cap 3 - 6 cm dia; stem 4 - 6 cm tall x 0.3 - 1.2 cm dia.
Cap dull brown, with darker radial fibres; conical or campanulate with a low umbo, cracking radially. Flesh white and unchanging.
Gills pallid greyish-white, becoming brown with more pallid edge, adnexed, irregular, crowded. **Spores** snuff-brown, star-shaped with 5-8 irregular rays, 9-12 x 8-10 μm. Basidia 4-spored. Gill edge cystidia fusiform, thick-walled with encrusted apices.
Stem more or less concolorous, with longitudinal fibres, strongly pruinose throughout, slender, more or less equal, with whitish marginate basal bulb. Ring absent. Flesh white and unchanging.
Odour of meal. **Taste** not distinctive.
Chemical tests none.
Occurrence late summer to autumn; common.
■ Inedible.
Possibly to be confused with *I.splendens* but distinguished by having stellate spores.

Inocybe bongardii (Weinm.) Quél.   **Cortinariaceae**

Medium-sized agaric with pinkish-brown scaly cap and pale brown gills, fruity odour; solitary or in trooping groups on soil, often calcareous, with broad-leaf and coniferous trees.

Dimensions cap 3 - 6 cm dia; stem 5 - 10 cm tall x 0.5 - 1 cm dia.
Cap pallid cinnamon-brown with pink tinge, covered with darker flattened fibrous scales; at first conical-convex, becoming campanulate with a low umbo. Flesh white or buff with pink tinge, firm and moderate.
Gills pallid olive-grey, becoming brown with pallid edges, adnexed, irregular, crowded. **Spores** snuff-brown, smooth, bean-shaped; 10-12 x 6-7 µm. Basidia 4-spored. Gill edge cystidia cylindrical or clavate, thin-wallled, not encrusted.
Stem more or less concolorous with cap, bruising reddish, slender, more or less equal without basal bulb. Ring absent. Flesh white or buff with pink tinge, full.
Odour strong, fruity, reminiscent of ripe pears. **Taste** not distinctive.
Chemical tests none.
Occurrence late summer to autumn; infrequent.
✠ Poisonous.

Inocybe brunneorufa Stangl & Veselsky   **Cortinariaceae**

Small brown agaric with whitish umbo, clay or cinnamon gills and stem with pronounced white basal bulb; solitary or in trooping groups on calcareous soil, with broad-leaf trees, favouring beech.

Dimensions cap 1.5 - 2 cm dia; stem 2 - 4 cm tall x 0.2 - 0.5 cm dia.
Cap hazel-brown, with somewhat reddish-brown centre covered with white bloom; campanulate, becoming expanded umbonate. Flesh whitish and unchanging.
Gills clay, becoming olivaceous-brown, adnexed or adnate, close.
Spores snuff-brown, broadly ellipsoid, 8-11 x 6.5-8 µm. Basidia 4-spored. Gill edge cystidia flask-shaped, thick-walled, not encrusted.
Stem hazel-brown, pruinose, more or less equal, with white, abrupt basal bulb. Ring absent. Flesh whitish and unchanging.
Odour not distinctive. **Taste** not distinctive.
Chemical tests none.
Occurrence late summer to autumn; infrequent.
■ Inedible.

Inocybe cookei Bres.   **Cortinariaceae**

Smallish straw-yellow umbonate agaric with pallid gills and stem with bulbous base; solitary or in trooping groups on soil, often calcareus, in mixed woods.

Dimensions cap 2 - 6 cm dia; stem 4 - 6 cm tall x 0.4 - 0.8 cm dia.
Cap pallid ochraceous or straw-yellow, covered densely with long silky radiating fibres; conical or campanulate, becoming expanded with prominent umbo. Flesh whitish, becoming concolorous with age.
Gills whitish, then pallid ochraceous with grey tinge and finally cinnamon, adnexed, close. **Spores** snuff-brown, smooth, ellipsoid, 7-8 x4-5 µm. Basidia 4-spored. Gill edge cystidia pear-shaped, thin-walled, not encrusted.
Stem whitish with ochraceous tinge, more or less equal, with whitish marginate basal bulb. Ring absent. Flesh white or straw yellow with age.
Odour not distinctive or faint, of honey. **Taste** not distinctive.
Chemical tests none.
Occurrence late summer to autumn; infrequent.
✠ Poisonous.

Inocybe corydalina Quél.   **Cortinariaceae**

Smallish brown or green umbonate agaric with grey-brown gills and whitish stem; solitary or in trooping groups on calcareous soil, with beech.

Dimensions cap 3 - 7 cm dia; stem 3 - 8 cm tall x 0.5- 1 cm dia.
Cap pallid ochraceous, covered densely with darker brown radiating fibres, typically greenish at centre and sometimes with greenish tinge throughout, campanulate, becoming expanded-umbonate. Flesh white or dirty-yellow with age.
Gills clay, becoming dirty grey-brown with more pallid margin, adnexed, close. **Spores** snuff-brown, smooth, ellipsoid or lemon-shaped, 8-11 x 5-7 µm. Basidia 4-spored. Gill edge cystidia fusiform, thick-walled with encrusted apices.
Stem whitish with brown tinge and sometimes green at base, more or less equal, with slightly swollen base. Ring absent. Flesh white or dirty-yellow with age.
Odour faint, suggestive of pear blossom. **Taste** not distinctive.
Chemical tests none.
Occurrence late summer to autumn; infrequent.
■ Inedible.

Inocybe dulcamara (A. & S.: Pers.) Kummer
Cortinariaceae

Smallish, brown, umbonate agaric with olive-brown gills and smooth spores; solitary or in trooping groups on soil, often sandy, with coniferous trees.

Dimensions cap 2 - 4 cm dia; stem 3 - 4 cm tall x 0.5 - 0.8 cm dia.
Cap ochraceous-brown with darker reddish-brown radiating fibres; conical or campanulate with a slight umbo, sometimes splitting radially. Flesh pallid and unchanging.
Gills clay, becoming olivaceous and finally cinnamon-brown with more pallid margin, adnexed or adnate, broad, close. **Spores** snuff-brown, ellipsoid or reniform, 9-10.5 x 5.5-6 µm. Basidia 4-spored. Gill edge cystidia pear-shaped, thin-walled, without encrustations.
Stem whitish with brown tinge, finely woolly, more or less equal. Ring absent. Flesh pallid and unchanging.
Odour not distinctive. **Taste** not distinctive.
Chemical tests none.
Occurrence late summer to autumn; infrequent.
✛ Poisonous.

Inocybe fastigiata (Schaeff.: Fr.) Quél.   **Cortinariaceae**
= *Inocybe pseudofastigiata* Rea

Smallish, yellowish-brown, conical agaric with olive-brown gills and smooth spores; solitary or in trooping groups on soil, often calcareous, in broad-leaf woods.

Dimensions cap 2 - 6 cm dia; stem 3 - 8 cm tall x 0.3 - 1 cm dia.
Cap straw with darker brown silky radiating fibres; conical or campanulate with a distinct sharp umbo, splitting radially. Flesh white and unchanging.
Gills clay, becoming olivaceous-brown with more pallid margin, adnexed or adnate, close. **Spores** snuff-brown, broadly ellipsoid, 8-15 x 4.5-7.5 µm. Basidia 4-spored. Gill edge cystidia cylindrical, thin-walled, without encrustations.
Stem whitish with ochraceous-brown tinge, more or less equal. Ring absent. Flesh white and unchanging.
Odour faint, of meal. **Taste** not distinctive or slightly bitter.
Chemical tests none.
Occurrence late summer to autumn; frequent.
✛ Poisonous.

Inocybe fibrosa (Sow.) Gill.   Cortinariaceae

Medium-sized, whitish agaric with grey-brown gills and angular spores; solitary or in trooping groups on soil, often calcareous, in coniferous or mixed woods.

Dimensions cap 4 - 8 cm dia; stem 6 - 10 cm tall x 1 - 2 cm dia.
Cap ivory-white with ochraceous tinge; conical or campanulate with a low umbo and irregular margin, silky, cracking radially. Flesh white and unchanging.
Gills pallid yellowish-white becoming grey-brown, adnexed, irregular, close. Cortina absent. **Spores** snuff-brown, more or less 8-cornered, 8-12 x 6-7 µm. Basidia 4-spored. Cystidia flask-shaped with apical projections, thick-walled with apical encrustations.
Stem more or less concolorous with cap, pruinose at apex, more or less equal. Ring absent. Flesh white and unchanging.
Odour acidic. **Taste** not distinctive.
Chemical tests none.
Occurrence late summer to autumn; infrequent.
■ Inedible.

Inocybe fibrosoides Kühn.   Cortinariaceae

Smallish, pale brown, conical agaric with grey-brown gills, pale stem with bulbous base and angular spores; solitary or in trooping groups on soil in coniferous woods.

Dimensions cap 4 - 8 cm dia; stem 4 - 9 cm tall x 0.5 - 1 cm dia.
Cap pallid brownish or creamy-brown; conical or campanulate with a low umbo, radially silky-fibrous. Flesh white and unchanging.
Gills white, becoming grey-brown, adnexed, close. Cortina absent. **Spores** snuff-brown, angular, 8.5-12 x 5.5-9 µm. Basidia 4-spored. Cystidia flask-shaped with apical projections, same septate, thick-walled with apical encrustations.
Stem whitish, pruinose at apex, more or less equal, base bulbous, abrupt. Ring absent. Flesh white and unchanging.
Odour not distinctive. **Taste** not distinctive.
Chemical tests none.
Occurrence late summer to autumn; infrequent.
■ Inedible.

Inocybe flocculosa (Berk.) Sacc.   Cortinariaceae
= *Inocybe gausapata* Kühn.

Small·agaric with brown, softly scaly, umbonate cap, pale yellowish-brown gills and stem and smooth spores; solitary or in trooping groups on soil under broad-leaf trees, favouring oak.

Dimensions cap 2 - 4 cm dia; stem 3 - 5 cm tall x 0.2 - 0.5 cm dia.
Cap dark brown covered with more pallid, soft, scaly fibres; conical or campanulate with a low umbo, cracking towards centre. Flesh white and unchanging.
Gills pallid brown, becoming more ochraceous rust-brown, slightly emarginate, crowded. **Spores** snuff-brown, almond-shaped, smooth, 8-11.5 x 4.5-6 µm. Basidia 4-spored. Cystidia fusiform, thick-walled, with apical encrustations.
Stem more or less concolorous with cap, fibrous, pruinose at apex, downy at base, more or less equal. Ring absent. Flesh white, turning slightly reddish.
Odour of meal. **Taste** not distinctive.
Chemical tests none.
Occurrence late summer to autumn; infrequent.
■ Inedible.

Inocybe furfurea Kühn.   Cortinariaceae

Small agaric with brown, fibrous, umbonate cap, pale greyish gills, powdery pinkish stem and smooth spores; solitary or in trooping groups on soil with broad-leaf trees.

Dimensions cap 1 - 2.5 cm dia; stem 1 - 3 cm tall x 0.1 - 0.4 cm dia.
Cap reddish-brown, darker, almost black, towards centre, covered with soft fibres when young, then granular; flattish-convex with a low umbo, tending to fine radial and concentric cracking. Flesh white and unchanging.
Gills pallid grey, adnate, deeply ventricose, fairly distant.
Spores snuff-brown, almond-shaped, smooth, 8 -10 x 5-6 µm. Basidia 4-spored. Cystidia thin-walled, more or less cylindrical, with apical encrustations.
Stem pinkish-brown, whitish towards base, fibrillose, pruinose throughout, more or less equal or slightly bulbous at base. Ring absent. Flesh pallid and fibrous.
Odour not distinctive. **Taste** not distinctive.
Chemical tests none.
Occurrence late summer to autumn; infrequent.
■ Inedible.

Inocybe geophylla (Fr.: Fr.) Kummer   Cortinariaceae

Small, sharply umbonate agaric with whitish cap and stem, pale clay gills and smooth spores; solitary or in trooping groups on soil in broad-leaf and coniferous woods, often by pathsides.

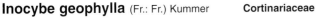

Dimensions cap 1.5 - 3.5 cm dia; stem 1 - 6 cm tall x 0.3 - 0.6 cm dia.
Cap white with yellowish tinge; conical then expanded with prominent, more or less acute umbo, smooth, silky. Flesh white, unchanging and thin.
Gills cream, becoming buff, adnexed, crowded. **Spores** snuff-brown, smooth, almond-shaped, 8-10.5 x 5-6 µm. Basidia 4-spored. Cystidia broadly fusiform, thick-walled with apical encrustations.
Stem concolorous with cap, slender, more or less equal, silky-fibrous. Ring absent. Flesh white, unchanging and stuffed.
Odour of meal or earthy. **Taste** not distinctive.
Chemical tests none.
Occurrence summer to autumn; common.
✤ Poisonous.

Inocybe geophylla var lilacina (Peck) Gill.   Cortinariaceae

Small, sharply umbonate agaric with lilac cap and stem, pale clay gills and smooth spores; solitary or in trooping groups on soil in broad-leaf and coniferous woods.

Dimensions cap 1.5 - 3.5 cm dia; stem 1 - 6 cm tall x 0.3 - 0.6 cm dia.
Cap lilaceous, occasionally with ochraceous tinge at umbo; conical then expanding with a prominent, more or less acute umbo, smooth, silky. Flesh violaceous, unchanging and thin.
Gills cream, becoming buff, adnexed, crowded. **Spores** snuff-brown, smooth, slightly almond-shaped, 7-10 x 5.5-6 µm. Basidia 4-spored. Cystidia broadly fusiform, thick-walled with apical encrustations.
Stem concolorous with cap, slender, more or less equal, silky-fibrous . Ring absent. Flesh violaceous and unchanging.
Odour of meal or earthy. **Taste** not distinctive.
Chemical tests none.
Occurrence summer to autumn; frequent.
✤ Poisonous.

Inocybe griseolilacina J Lange   **Cortinariaceae**

Small, grey or lilac-brown, umbonate agaric with brown gills and smooth spores; solitary or in trooping groups on soil with broad-leaf trees.

Dimensions cap 1.5 - 2.5 cm dia; stem 3 - 6 cm tall x 0.3 - 0.4 cm dia.

Cap grey-brown with lilaceous tinge, particularly when young, at first fibrous, then with small shaggy scales; conical or campanulate becoming expanded-umbonate. Flesh lilaceous or grey-brown and unchanging.

Gills clay becoming snuff-brown, adnate, close. **Spores** snuff-brown, broadly ellipsoid, 8.5-11 x 5-6 μm. Basidia 4-spored. Cystidia fusiform, thick-walled with apical encrustations.

Stem at first lilaceous, becoming more brown with age, covered in white cottony fibres, more or less equal. Ring absent. Flesh concolorous with that of cap, unchanging.

Odour strong, of meal. **Taste** not distinctive.

Chemical tests none.

Occurrence late summer to autumn; infrequent.

■ Inedible.

Inocybe hystrix (Fr.) Karst.   **Cortinariaceae**

Small, brown, pointed-scaly umbonate agaric with clay gills and smooth spores; solitary or in small groups on soil, often calcareous, in broad-leaf woods, favouring beech.

Dimensions cap 2 - 4 cm dia; stem 3 - 8 cm tall x 0.4 - 0.8 cm dia.

Cap pallid, covered densely with dark brown, pointed, erect scales; convex or campanulate with a blunt umbo. Flesh pallid and unchanging.

Gills whitish, then clay with a more pallid margin, adnexed or adnate, close. **Spores** snuff-brown, broadly ellipsoid, 9-11.5 x 4.5-6.5 μm. Basidia 4-spored. Cystidia fusiform, thick-walled with apical encrustations.

Stem concolorous with cap, covered with dark brown reflexed scales, more or less equal. Ring absent. Flesh pallid and unchanging.

Odour not distinctive. **Taste** not distinctive.

Chemical tests none.

Occurrence late summer to autumn; infrequent or rare.

✢ Poisonous.

Inocybe jurana (Pat.) Sacc.   **Cortinariaceae**

Smallish agaric with brown, fibrous, splitting, conical cap, pale brownish gills and stem and smooth spores; solitary or in trooping groups on calcareous soil in mixed woods but favouring beech.

Dimensions cap 3 - 7 cm dia; stem 4 - 8 cm tall x 0.5 - 1 cm dia.

Cap reddish- or vinaceous-brown; conical-campanulate, fibrous, radially splitting. Flesh whitish, unchanging or tinged pinkish-brown.

Gills cream, becoming grey-brown with whitish, slightly fimbriate edges, adnate or free, crowded. **Spores** snuff-brown, ellipsoid, 10-15 x 5.5-7.5 μm. Basidia 4-spored. Cystidia thin-walled, clavate, without apical encrustations.

Stem whitish above, tinged reddish- or vinaceous-brown below, more or less equal or weakly bulbous at base. Ring absent. Flesh whitish, tinged vinaceous in base.

Odour of slight, meal. **Taste** not distinctive.

Chemical tests none.

Occurrence late summer to autumn; infrequent.

■ Inedible.

Inocybe lacera (Fr.: Fr.) Kummer   Cortinariaceae

Small umbonate agaric with yellowish-brown, scaly cap, pale brownish gills and stem and smooth spores; solitary or in trooping groups, often on sandy soil with coniferous trees, and on old moss-covered fire sites.

Dimensions cap 1 - 4 cm dia; stem 2 - 4 cm tall x 0.2 - 0.5 cm dia.
Cap variable, ochraceous or dark brown, scaly-fibrous, radially splitting; convex then expanded with a slight umbo. Flesh whitish and unchanging.
Gills cream, becoming grey-brown with whitish edges, adnexed or emarginate, crowded. **Spores** snuff-brown, ellipsoid, 11-17 x 4.5-7 µm. Basidia 4-spored. Cystidia fusiform, thick-walled with apical encrustations.
Stem whitish or pallid brown, slender, more or less equal. Ring absent. Flesh whitish and unchanging.
Odour of meal. **Taste** not distinctive.
Chemical tests none.
Occurrence late summer to autumn; infrequent.
■ Inedible.
Distinguished by the unusually large ellipsoid spores. Several forms of the species are recognised.

Inocybe lanuginella Schroet. apud Cohn   Cortinariaceae
= *Inocybe decipientoides* Peck

Small umbonate agaric with dull brown, slightly scaly cap, pale gills and stem and angular spores; solitary or in trooping groups on soil, often amongst grass, in broad-leaf and coniferous woods.

Dimensions cap 1 - 3 cm dia; stem 2 - 4 cm tall x 0.3 - 0.6 cm dia.
Cap dull brown, with darker radial scaly fibres; conical or campanulate, then expanded with a sharp umbo. Flesh white and unchanging.
Gills cream, becoming ochraceous with more pallid edge, adnexed or emarginate, irregular, crowded. **Spores** snuff-brown, elongated-ellipsoid, angular, 8-9 x 5-6 µm. Basidia 4-spored. Cystidia spherical or pear-shaped with apical encrustations.
Stem pallid, concolorous with cap, with longitudinal fibres, slender, more or less equal . Ring absent. Flesh pallid, flushed brown and unchanging.
Odour of meal. **Taste** not distinctive.
Chemical tests none.
Occurrence summer to autumn; infrequent.
■ Inedible.

Inocybe lanuginosa (Bull.: Fr.) Kummer   Cortinariaceae

Smallish umbonate agaric with dark-brown shaggy cap, pale gills, dark scaly stem and angular spores; solitary or in trooping groups on soil and rotting debris in coniferous or mixed woods.

Dimensions cap 3 - 5 cm dia; stem 3 - 6 cm tall x 0.4 - 0.8 cm dia.
Cap dark brown, decorated with small erect scales; convex then expanded-flattened. Flesh pallid and unchanging.
Gills pallid, becoming dirty beige, adnexed or emarginate, crowded. **Spores** snuff-brown, oblong, with 8-10 nodules, 8-10 x 5-6 µm. Basidia 4-spored. Cystidia vesicular or pear-shaped, with or without apical encrustations.
Stem brown, more pallid at apex, floccose-scaly, more or less equal. Ring absent. Flesh pallid, flushed brown, unchanging.
Odour not distinctive. **Taste** not distinctive.
Chemical tests none.
Occurrence late summer to autumn; infrequent.
■ Inedible.

Inocybe leptocystis Atk.   Cortinariaceae

Smallish umbonate agaric with brown cap, pale gills and stem and smooth spores; solitary or in trooping groups on soil with coniferous trees.

Dimensions cap 1.5 - 5.5 cm dia; stem 2 - 5 cm tall x 0.2 - 0.6 cm dia.
Cap tawny-brown, with darker radiating fibrils; conical or campanulate, then expanded, with a blunt umbo, smooth or slightly downy-granular. Flesh white and unchanging.
Gills pallid cinnamon, adnexed or emarginate, crowded.
Spores snuff-brown, smooth, ellipsoid, 7.5-11 x 4.5-6 µm. Basidia 4-spored. Cystidia cylindrical with long necks, fairly thin-walled, yellowish, without apical encrustations.
Stem cream, becoming tinged brown, not or barely pruinose, more or less equal but somewhat bulbous at base. Ring absent. Flesh pallid, unchanging.
Odour not distinctive. **Taste** not distinctive.
Chemical tests none.
Occurrence late summer to autumn; infrequent.
■ Inedible.
Distinguished by the long-throated cystidia and pallid gills.

Inocybe longicystis Atk.   Cortinariaceae

Smallish umbonate agaric with brown shaggy cap, brown gills with fringed edges, pale stem and angular spores; solitary or in trooping groups on soil, typically amongst mosses, with or close by coniferous trees.

Dimensions cap 1 - 5 cm dia; stem 4 - 6 cm tall x 0.4 - 0.7 cm dia.
Cap cinnamon or ochraceous-brown, more pallid at margin, with darker radiating scaly fibres; conical or campanulate, then expanded with a blunt umbo. Flesh white and unchanging.
Gills cinnamon, with a more pallid, distinctly fimbriate margin, adnexed or emarginate, crowded. **Spores** snuff-brown, oblong, nodulose, 8-10 x 5-6 µm. Basidia 4-spored. Cystidia cylindrical, distinctively long and abundant, thin-walled, without apical encrustations.
Stem pallid brown, with whitish, floccose, longitudinal fibres, stoutish, more or less equal . Ring absent. Flesh pallid, flushed brown and unchanging.
Odour not distinctive. **Taste** not distinctive.
Chemical tests none.
Occurrence late summer to autumn; infrequent.
■ Inedible.
Distinguished by the long cystidia which give the gill edge its fringed appearance.

Inocybe lucifuga Heim   Cortinariaceae

Small, straw-coloured, scaly, umbonate agaric with clay or olive-brown gills and smooth spores; solitary or in trooping groups on acid soil with conifers in moorland areas.

Dimensions cap 2 - 4 cm dia; stem 3 - 6 cm tall x 0.3 - 0.8 cm dia.
Cap pallid straw or beige, fibrous scaly; convex then expanded with a slight umbo. Flesh white and unchanging.
Gills clay, becoming olivaceous-brown with irregular pallid edge, adnexed, crowded.
Spores snuff-brown, smooth, ellipsoid, 7.5-13 x 4.5-7 µm. Basidia 4-spored. Cystidia fusiform, thick-walled with apical encrustations.
Stem concolorous with cap, pruinose at apex, slightly fibrillose, more or less equal. Ring absent. Flesh white and unchanging.
Odour strong, of radish. **Taste** not distinctive.
Chemical tests none.
Occurrence late summer to autumn; infrequent.
■ Inedible.

Inocybe maculata Boud.   **Cortinariaceae**

Smallish, brown, fibrous, umbonate agaric with clay or olive-brown gills and smooth spores; solitary or in trooping groups on calcareous soil with beech.

Dimensions cap 2 - 8 cm dia; stem 3 - 8 cm tall x 0.5 - 1.2 cm dia.
Cap chestnut-brown, with long radial fibres and, for a long time, with white downy 'bloom'; conical then campanulate with distinct umbo. Flesh white and unchanging.
Gills clay, becoming olivaceous-brown, adnate, crowded.
Spores snuff-brown, smooth, ellipsoid, 9-12 x 4.5-6.5 µm. Basidia 4-spored. Cystidia clavate, thin-walled, without apical encrustations.
Stem concolorous with cap, slightly fibrillose, more or less equal, with or without basal bulb. Ring absent. Flesh white and unchanging.
Odour allegedly of truffles. **Taste** not distinctive.
Chemical tests none.
Occurrence late summer to autumn; infrequent.
✣ Poisonous.

Inocybe napipes J Lange   **Cortinariaceae**

Smallish, brown umbonate agaric with grey-brown gills, pale brown stem with bulbous base and angular spores; solitary or in trooping groups on soil in mixed woods.

Dimensions cap 2 - 5 cm dia; stem 3 - 7 cm tall x 0.5 - 0.8 cm dia.
Cap chestnut or umber-brown covered densely with radiating fibres; campanulate, becoming expanded with somewhat acute umbo. Flesh white or tinged buff with age.
Gills clay, becoming dirty grey-brown, adnexed, close. **Spores** snuff-brown, smooth, with prominent nodules, 8-10 x 5.5-7.5 µm. Basidia 4-spored. Cystidia flask-shaped, thick-walled with apical encrustations.
Stem whitish and pruinose at apex, tinged cap colour below, more or less equal, with white bulbous base. Ring absent. Flesh white, tinged brown with age.
Odour not distinctive. **Taste** not distinctive.
Chemical tests none.
Occurrence late summer to autumn; frequent.
✣ Poisonous.

Inocybe patouillardi Bres.   **Cortinariaceae**

Medium-sized agaric with ivory, umbonate cap streaked red, pinkish-cream gills, pale stem and smooth spores; solitary or in trooping groups on soil, often calcareous, under broad-leaf trees, favouring beech.

Dimensions cap 3 - 8 cm dia; stem 3 - 10 cm tall x 1 - 2 cm dia.
Cap ivory, covered with russet or reddish radial fibres; conical or campanulate with a low umbo. Flesh white and unchanging.
Gills rose-pink, becoming cream and darkening to olivaceous-brown in older specimens, sometimes bruising red. **Spores** cigar brown, reniform, smooth, 10-13 x 5.5-7 µm. Basidia 4-spored. Cystidia sub-cylindrical, thin-walled without apical encrustations.
Stem white, sometimes staining red, more or less smooth and typically with a slight marginate basal bulb. Ring absent. Flesh white and unchanging.
Odour not distinctive when young but more or less rank in older specimens. **Taste ON NO ACCOUNT SHOULD ANY PART OF THIS SPECIES BE EATEN.**
Chemical tests none.
Occurrence summer to autumn; infrequent.
✚ Lethally poisonous even in very small quantities.

Inocybe pelargonium Kühn.   **Cortinariaceae**

Small, pale, straw-coloured umbonate agaric with grey-brown gills, whitish stem and smooth spores; solitary or in trooping groups on soil, typically with spruce.

Dimensions cap 2 - 4 cm dia; stem 2 - 6 cm tall x 0.3 - 0.7 cm dia.
Cap pallid ochraceous covered with fine radiating fibres; campanulate, becoming expanded-umbonate. Flesh whitish, unchanging.
Gills whitish, becoming dirty grey-brown, adnexed, close.
Spores snuff-brown, smooth, ellipsoid, 7-11 x 4-6 µm. Basidia 4-spored. Cystidia fusiform, thick-walled with apical encrustations.
Stem whitish, pruinose throughout, more or less equal, with slightly swollen base, sometimes more abrupt. Ring absent. Flesh whitish.
Odour of *Pelargonium*. **Taste** not distinctive.
Chemical tests none.
Occurrence late summer to autumn; infrequent.
■ Inedible.

Inocybe phaeoleuca Kühn.   **Cortinariaceae**

Smallish, brown umbonate agaric with pale gills, whitish stem and smooth spores; solitary or in trooping groups on soil with broad-leaf trees, favouring beech.

Dimensions cap 3 - 5.5 cm dia; stem 2 - 5 cm tall x 0.5 - 0.8 cm dia
Cap chestnut or umber-brown, darker at centre, covered densely with radiating fibres, breaking up somewhat into fine adpressed scales; hemispherical or convex, becoming expanded with a low umbo. Flesh whitish or tinged brown with age.
Gills white, becoming pallid ochraceous-brown, adnexed, close.
Spores snuff-brown, smooth, ellipsoid, 8.5-12 x 5.5-7 µm. Basidia 4-spored. Cystidia fusiform, thick-walled with apical encrustations.
Stem whitish, tinged ochraceous where bruised, pruinose throughout, more or less equal, with slightly swollen base. Ring absent. Flesh white or tinged brown with age.
Odour not distinctive. **Taste** not distinctive.
Chemical tests none.
Occurrence late summer to autumn; infrequent.
■ Inedible.

Inocybe praetervisa Quél.   **Cortinariaceae**

Small umbonate agaric with yellowish-brown fibrous cap, pale olive gills, pale brown stem and angular spores; solitary or in trooping groups on soil in broad-leaf and coniferous woods.

Dimensions cap 2.5 - 5.5 cm dia; stem 5 - 6 cm tall x 0.4 - 0.6 cm dia.
Cap ochraceous-brown, radially fibrillose, splitting; conical or campanulate then expanded with umbo. Flesh white or yellowish, unchanging.
Gills whitish, becoming pallid olivaceous-brown, adnexed or emarginate, irregular, crowded. **Spores** snuff-brown, oblong, with rounded knobbly projections, 8.5-13 x 6.5-9 µm. Basidia 4-spored. Cystidia tapering or fusiform with yellowish contents, thin-walled, without apical encrustations.
Stem pallid straw, then tinged brown, slightly pruinose throughout, more or less equal, with non-marginate bulbous base. Ring absent. Flesh yellowish and unchanging.
Odour faint, of meal. **Taste** not distinctive.
Chemical tests none.
Occurrence summer to autumn; infrequent.
■ ✛ Inedible, possibly poisonous.

Inocybe pusio Karst.   Cortinariaceae

Small agaric with brown fibrous cap, pale violet-tinged gills and stem apex and smooth spores; solitary or in trooping groups on soil in broad-leaf woods.

Dimensions cap 1 - 3.5 cm dia; stem 2 - 4.5 cm tall x 0.3 - 0.6 cm dia.
Cap ochraceous or pallid with dark brown radial fibres and smooth brown centre; convex then expanded, slightly umbonate, margin incurved when young. Flesh pallid and unchanging.
Gills at first with lilaceous tinge, becoming greyish-brown, with fimbriate margin, adnate or more or less free, crowded.
Spores snuff-brown, ellipsoid, 7.5-11 x 4.5-6 µm. Basidia 4-spored. Cystidia fusiform, thin-walled, without apical encrustations.
Stem greyish-brown, tinged violaceous at apex, more or less equal or slightly bulbous at base. Ring absent. Flesh pallid, flushed brown and unchanging.
Odour of meal. **Taste** not distinctive.
Chemical tests none.
Occurrence summer to autumn; infrequent.
■ Inedible.

Inocybe serotina Peck   Cortinariaceae

Smallish agaric with clay-coloured flattish cap, pale greyish-brown gills, whitish stout stem and smooth spores; solitary or in trooping groups favouring sand dunes.

Dimensions cap 2 - 5 cm dia; stem 4 - 6 cm tall x 0.5 - 1.2 cm dia.
Cap whitish-clay or pallid ochraceous, smooth; flattened-convex with a shallow blunt umbo. Flesh pallid, unchanging and firm.
Gills grey-brown, becoming brown, adnexed, close. **Spores** snuff-brown, more or less boat-shaped, smooth, 13-17 x 5.5-7 µm. Basidia 4-spored. Cystidia broad, thin-walled, without apical encrustations.
Stem whitish, with fine, light brown fibrils, stout, more or less equal, typically deeply buried in sand. Ring absent. Flesh pallid, unchanging fibrous and full.
Odour not distinctive. **Taste** not distinctive.
Chemical tests none.
Occurrence late summer to autumn; localised.
■ Inedible.

Inocybe sindonia (Fr.) Karst.   Cortinariaceae
= *Inocybe eutheles* (Berk. & Br.) Quél.
= *Inocybe kuehneri* Stangl & Veselsky

Pale cream umbonate agaric with clay gills and smooth spores; solitary or in trooping groups on soil in coniferous woods.

Dimensions cap 2 - 6 cm dia; stem 3 - 8 cm tall x 0.3 - 1 cm dia.
Cap pallid cream, straw or buff with silky radiating fibres, slightly scaly at centre; conical or campanulate, then expanded with a distinct umbo. Flesh white and unchanging.
Gills clay with white edge, adnexed or adnate, close. **Spores** pallid brown, broadly ellipsoid, 8.5 -10.5 x 4 -5.5 µm. Basidia 4-spored. Cystidia fusiform, thick-walled with sparse apical encrustations.
Stem concolorous with cap, pruinose throughout, more or less equal or slightly swollen at base. Ring absent. Flesh white and unchanging.
Odour not distinctive. **Taste** not distinctive.
Chemical tests none.
Occurrence late summer to autumn; frequent.
✛ Poisonous.

Inocybe soluta Vel.   Cortinariaceae
= *Inocybe brevispora* Huijsm.

Medium-sized brown agaric with slightly umbonate cap, cinnamon gills, pale brown stem and angular spores; solitary or in trooping groups on soil with conifers, favouring pine and spruce.

Dimensions cap 2 - 4 cm dia; stem 2 - 3 cm tall x 0.3 - 0.6 cm dia.
Cap chestnut or darker brown, radially fibrillose; at first convex, becoming flattened with a broad umbo; with cortina when young. Flesh pallid brown and unchanging.
Gills pallid clay, becoming cinnamon or rust-brown, adnate with decurrent tooth, crowded. **Spores** snuff-brown, angular but with few corners, some more or less square, 6-9 x 4.5-7 µm. Basidia 4-spored. Cystidia flask-shaped, thin-walled, without apical encrustations.
Stem pallid, concolorous with cap, smooth, stoutish, more or less equal, base not bulbous. Ring absent. Flesh pallid brown and unchanging.
Odour not distinctive. **Taste** not distinctive.
Chemical tests none.
Occurrence late summer to autumn; infrequent.
■ Inedible.

Inocybe splendens Heim   Cortinariaceae

Smallish agaric with dark brown umbonate cap, pale greyish-brown gills; brown stem with a white bulbous base and smooth spores; solitary or in trooping groups on soil in open mixed woodland.

Dimensions cap 3 - 5 cm dia; stem 3 - 4 cm tall x 0.7 - 1.2 cm dia.
Cap dark brown, sometimes tinged purplish, with long, fibrous scales; conical or convex, with a blunt umbo, cracking radially. Flesh pallid and unchanging.
Gills whitish, becoming pallid greyish-brown, adnexed, close.
Spores snuff-brown, more or less boat-shaped, smooth, 9-12 x 5.5-6 µm. Basidia 4-spored. Cystidia fusiform, thick-walled with apical encrustations.
Stem more or less concolorous with cap, with longitudinal fibres, pruinose at apex, slender, more or less equal, with whitish marginate basal bulb. Ring absent. Flesh white and unchanging.
Odour not distinctive. **Taste** not distinctive.
Chemical tests none.
Occurrence late summer to autumn; rare or overlooked.
■ Inedible.
Possibly to be confused with *I. asterospora* but distinguished by having smooth spores.

Inocybe tigrina Heim   Cortinariaceae

Smallish agaric, cap with dark brown scales on a pale background, buff-brown gills; pale stem and smooth spores; solitary or in trooping groups on soil in broad-leaf or mixed woods.

Dimensions cap 2 - 5 cm dia; stem 3 - 6 cm tall x 0.2 - 0.5 cm dia.
Cap pallid, with darker brown fibrous scales; conical or campanulate, with a low umbo, cracking radially. Flesh white and unchanging.
Gills pallid, becoming buff-brown, adnexed, crowded. **Spores** snuff-brown, almond-shaped, smooth, 8-11 x 4.5-5.5 µm. Basidia 4-spored. Cystidia fusiform, thick-walled with apical encrustations.
Stem pallid, tinged brown with age, pruinose at apex, slender, more or less equal. Ring absent. Flesh white and unchanging.
Odour not distinctive. **Taste** not distinctive.
Chemical tests none.
Occurrence late summer to autumn; infrequent.
■ Inedible.

Inocybe vulpinella Bruylants   **Cortinariaceae**

Small argaric with rust-brown, umbonate cap, whitish stem and smooth spores; solitary or in trooping groups on sand, often with willows, in coastal dunes.

Dimensions cap 2 - 3.5 cm dia; stem 2.5 - 3.5 cm tall x 0.3 - 0.6 cm dia.
Cap rust-brown, darker at centre, fibrous, slightly scaly at centre; conical or campanulate, then expanded with a blunt umbo. Flesh brown and unchanging.
Gills brown, adnexed or adnate, close. **Spores** snuff-brown, narrowly ellipsoid, 12-16 x 6-8 µm. Basidia 4-spored. Cystidia fusiform, thick-walled with apical encrustations.
Stem whitish, then browning from base up, base remaining white, more or less equal, but slightly swollen at base. Ring absent. Flesh white, then brown, otherwise unchanging.
Odour not distinctive. **Taste** not distinctive.
Chemical tests none.
Occurrence late summer to autumn; localised.
■ Inedible.

Naucoria escharoides (Fr.: Fr.) Kummer   **Cortinariaceae**
= *Agaricus escharoides* Fr.: Fr.
= *Hylophila escharoides* (Fr.: Fr.) Quél.
= *Alnicola escharoides* (Bull.: Fr.) Kühn.

Small, yellowish-brown agaric; trooping on damp soil with alder, less commonly with willow and birch.

Dimensions cap 1 - 3 cm dia; stem 1 - 5 cm tall x 0.1 - 0.3 cm dia.
Cap ochraceous-brown, more bright yellow towards centre; convex, fibrous, margin not striate, at first slightly fibrillose then smooth. Flesh yellowish, thin and fragile.
Gills at first ochraceous, becoming more cinnamon at maturity, adnate, close. **Spores** brown, minutely roughened, ellipsoid, 8-11 x 4.5-6 µm. Basidia 4-spored. Gill edge cystidia filiform with swollen base, long projection and pointed apex; gill face cystidia absent.
Stem tawny, more pallid at apex, darker brown towards base, slender, more or less equal, faintly striate, minutely pruinose at apex. Ring absent. Flesh dirty brown, thin and fragile.
Odour not distinctive. **Taste** not distinctive.
Occurrence late summer to autumn; frequent.
■ Inedible.

Naucoria luteofibrillosa (Kühn.) Kühn. & Romagn.
                            Cortinariaceae

Small agaric with pale, whitish coffee scurfy cap and pale gills; trooping on damp soil with alder.

Dimensions cap 1 - 2 cm dia; stem 2 - 4 cm tall x 0.2 - 0.25 cm dia
Cap pallid coffee or somewhat darker, minutely fibrillose scurfy, more so near margin with velar fragments; convex becoming more flattened and slightly umbonate. Flesh pallid, thin and fragile.
Gills pallid cinnamon, adnate, close. **Spores** brown, minutely roughened, ellipsoid, 12-13 x 9-10 µm. Basidia 4-spored. Gill edge cystidia with swollen base, short projection and blunt apex; gill face cystidia absent.
Stem dark brown, more pallid at apex, more or less equal, slender, decorated with whitish cortinal remnants. Ring absent. Flesh brown, thin and fragile.
Odour not distinctive. **Taste** not distinctive.
Occurrence late summer to autumn; frequent.
■ Inedible.

Naucoria pseudoscolecina Reid   Cortinariaceae

Small, golden brown agaric; trooping on damp soil with alder and possibly with other broad-leaf trees inhabiting swampy ground.

Dimensions cap 1 - 1.5 cm dia; stem 1.5 - 3 cm tall x 0.1 - 0.15 cm dia.
Cap golden-brown, more reddish-brown at centre; convex, becoming more flattened and slightly umbonate, margin striate, smooth, slightly viscid when damp. Flesh brown, thin and fragile.
Gills cinnamon, adnate, close. **Spores** brown, clearly roughened, ellipsoid, 8-11 x 5-6.5 μm. Gill edge cystidia with swollen base, short projection and blunt apex; gill face cystidia absent.
Stem dark brown, more or less equal, slender, smooth. Ring absent. Flesh dirty brown, thin and fragile.
Odour not distinctive. **Taste** not distinctive.
Occurrence late summer to autumn; rare. (Forest of Dean, Gloucestershire)
■ Inedible.

Naucoria scolecina (Fr.) Quél.   Cortinariaceae
= *Agaricus scolecinus* Fr.
= *Hylophila scolecina* (Fr.) Quél.
= *Alnicola badia* Kühn.

Small, reddish-brown agaric; trooping on damp soil with alder.

Dimensions cap 1 - 2 cm dia; stem 1.5 - 5 cm tall x 0.1 - 0.2 cm dia.
Cap reddish-brown; convex becoming more flattened, margin striate, at first minutely downy, then smooth. Flesh brown, thin and fragile.
Gills at first pallid brown, becoming rust-brown at maturity, adnate, close. **Spores** brown, minutely roughened, ellipsoid, 8-15 x 5-7 μm. Gill edge cystidia filiform with swollen base and blunt apex; gill face cystidia absent.
Stem brown, more or less equal, slender, sometimes markedly white cottony at base. Ring absent. Flesh dirty brown, thin and fragile.
Odour not distinctive. **Taste** not distinctive.
Occurrence late summer to autumn; rare or locally frequent. (Forest of Dean, Gloucestershire)
■ Inedible.

Naucoria striatula P Orton   Cortinariaceae

Small, dull brown agaric; trooping on damp soil with alder.

Dimensions cap 1 - 2 cm dia; stem 1.5 - 5 cm tall x 0.1 - 0.2 cm dia.
Cap cinnamon-brown, hygrophanous, drying pallid ochraceous; convex becoming more flattened with slight umbo, striate almost to centre, at first with minute velar, fragments then smooth. Flesh brown, thin and fragile.
Gills at first pallid brown, becoming cinnamon at maturity, adnate, close. **Spores** brown, warty, ellipsoid, 8.5-13 x 4.5-6 μm. Basidia 4-spored. Gill edge cystidia filiform with swollen base and blunt apex; gill face cystidia absent.
Stem pallid ochraceous above, darker brown below, more or less equal, slender, with fine ochraceous fibrils. Ring absent. Flesh pallid brown, thin and fragile.
Odour not distinctive. **Taste** not distinctive.
Occurrence late summer to autumn; frequent.
■ Inedible.

Naucoria subconspersa Kühn.   **Cortinariaceae**

Small, dark brown agaric; trooping on damp soil with alder and willow.

Dimensions cap 1 - 2 cm dia; stem 2.5 - 5 cm tall x 0.1 - 0.2 cm dia.
Cap dark umber-brown with more pallid yellowish-brown margin; convex, becoming more flattened and slightly umbonate, margin striate, finely granular, finely felty when dry. Flesh brown, thin and fragile.
Gills pallid cinnamon with pinkish tinge, adnate, close.
Spores brown, minutely roughened, ellipsoid, 8-12 x 4.5-6 µm. Basidia 4-spored. Gill edge cystidia filiform with swollen base, long projection and slightly swollen apex; gill face cystidia absent.
Stem dark brown, more or less equal, slender, decorated with pallid fibres. Ring absent. Flesh dirty brown, thin and fragile.
Odour not distinctive. **Taste** not distinctive.
Occurrence late summer to autumn; frequent.
■ Inedible.

Rozites caperata (Pers.: Fr.) Karst.   **Cortinariaceae**
= *Cortinarius caperatus* (Pers.: Fr.) Fr.
= *Pholiota caperata* (Pers.: Fr.) Kummer

Medium to large agaric with yellowish cap covered in silky fibres, pale brown gills, whitish stem with ring, and slightly swollen base; solitary or scattered on soil amongst heather with conifers.

Dimensions cap 5 - 12 cm dia; stem 4 - 10 cm tall x 1.0 - 1.5 cm dia.
Cap ochraceous, tinged brown, covered with fine silky-fibrous velar remnants; at first convex, becoming expanded or umbonate. Flesh whitish, tinged ochraceous or buff, moderate.
Gills pallid buff or clay, adnate, crowded. **Spores** ochraceous-brown, finely warty, broadly ellipsoid, 11-14 x 7-9 µm. Basidia 4-spored. Cystidia absent.
Stem whitish, fibrous, more or less equal or slightly swollen at base. Ring whitish, narrow and superior. Flesh whitish-buff, fibrous and full.
Odour not distinctive. **Taste** not distinctive.
Chemical tests none.
Occurrence late summer to autumn; rare. Locally frequent in Scottish Highlands and other montane regions of Europe, very occasional elsewhere.
□ Edible.

HOMOBASIDIOMYCETES
RUSSULALES AND LAMELLATE PORIALES

The Russulales [280] comprise fleshy, often robust, mushrooms. The Russulaceae by and large possess brightly-coloured, convex or flattened caps with whitish gills and stems. The gills often have a brittle, waxy feel and the flesh also breaks easily with a crumbly texture due to the presence of rounded cellular elements known as *sphaerocysts*. The majority of species fall into the sub-family Genuinae in which, typically, the caps are coloured with shades of red, purple, yellow or green, *lamellules* or intermediate gills are absent and gill colour ranges from white to ochre, sometimes greying or blackening. Stems are typically white but may possess a rosy flush or turn grey on bruising. Members of the smaller sub-family, the Compactae [10], possess abundant lamellules and the caps are usually white, brownish or blackening. The flesh may have a reddish coloration when cut. Amongst the Russulaceae, taste (hot, acrid or bland) and the application of various chemicals provide significant determining clues.

Spores are patterned with an assortment of warts, ridges and connections to produce a complete or partial network. The colour of the spores *en masse* is important and the accepted colour coding, based on the work of Romagnesi (*Les Russules d'Europe*), ranging from A to H is given in square brackets. The colour chart against which to compare the spore print may be found on page 16.

The Lactariaceae are characterised by the production of a milky latex which is particularly apparent if the gills are damaged. The taste of the latex and any colour changes which take place when it comes into contact with the air are amongst the determining characters. As in the Russulaceae, the caps and stems possess abundant sphaerocysts which give a brittle, sometimes crumbly texture; the gills, however, do not generally possess sphaerocysts in the same quantity and therefore may lack some of the brittleness. There is often no strong colour difference between cap, gills and stem and the predominating colours are creams, yellows and reddish-browns. Cap texture is a further significant factor when determining *Lactarius* species.

Spores are ornamented similarly to those amongst the Russulaceae and the same basis of colour coding applies.

Included at the end of this section is the small group of lamellate Poriales [30] including, most notably, the *Pleurotus* and *Lentinus* genera.

Lactarius acerrimus Britz.   Russulaceae

Large agaric with tawny-buff, slightly zoned cap, pale gills and exuding very hot milk; solitary or in scattered groups, on soil under broad-leaf trees, favouring oak.

Dimensions cap 8 - 15 cm dia; stem 2 - 6 cm tall x 2 - 4 cm dia.
Cap ochraceous-buff, with vague, darker, tawny, concentric bands; at first convex with inrolled margin, later irregularly flattened or slightly depressed, at first finely downy then smooth. Flesh whitish, thick and firm.
Gills at first buff, becoming tawny or cinnamon, slightly decurrent, narrow, distant but strongly wrinkled. **Spores** hyaline (ochraceous-cream [D-E] in the mass), ornamented with warts interconnected by a network of thin and thick ridges, broadly ellipsoid or sub-spherical, 10.5-13.5 x 8.5-10 µm. Basidia 2-spored. Cystidia cylindrical or filiform.
Stem pallid buff, becoming more ochraceous with age, more or less equal, sometimes eccentric, smooth. Ring absent. Flesh whitish, firm, full but with cavities.
Odour faint, of fruit. **Taste** very hot.
Chemical tests none. Milk white, unchanging.
Occurrence summer to autumn; rare. (Westonbirt Arboretum, Gloucestershire)
■ Inedible.

Lactarius azonites (Bull.) Fr.   Russulaceae

Medium-sized agaric with greyish, slightly downy cap, pale gills and exuding milk; solitary or in scattered groups, on soil under broad-leaf trees, favouring oak.

Dimensions cap 3 - 8 cm. dia; stem 3 - 7 cm tall x 0.5 - 1.5 cm dia.
Cap greyish, with buff or sepia tinges; at first convex with slightly inrolled margin, later irregularly flattened or slightly depressed with faint umbo, finely downy when dry. Flesh whitish, becoming rose when cut (up to 5 minutes delay), later more orange or coral, thick and fairly brittle.
Gills at first white, becoming pallid yellowish or buff, slightly decurrent, narrow, close or crowded. **Spores** hyaline (ochraceous-cream [E] in the mass), ornamented with warts interconnected by a network of thick ridges, sub-spherical, 8-9 x 7.5-8.5 µm. Basidia 4-spored. Cystidia cylindrical or filiform.
Stem whitish, becoming greyish-buff, bruising faintly rose, more or less equal, finely downy. Ring absent. Flesh whitish, otherwise reacting as in cap, stuffed.
Odour not distinctive. **Taste** mild then hot or acrid.
Chemical tests none. Milk white, slowly rose on skin (no change in air).
Occurrence late summer to autumn; infrequent.
■ Inedible.
Similar to *L. fuliginosus* but more pallid greyish in the cap.

Lactarius blennius (Fr.: Fr.) Fr.   Russulaceae

Medium to large agaric with greyish-green, spotted, slimy cap, pale gills and exuding milk; solitary or in scattered groups on soil under broad-leaf trees, favouring beech.

Dimensions cap 4 - 10 cm dia; stem 4 - 5 cm tall x 1 - 1.7 cm dia.
Cap greenish-olive or greenish-grey with sepia tinge and darker drop-like blotches in concentric bands; at first flattened-convex with slightly incurved margin, later irregularly flattened and depressed, very viscid when damp, sticky when dry. Flesh whitish and fairly thick.
Gills at first whitish, becoming pallid cream or buff, bruising more brownish, slightly decurrent, narrow, crowded. **Spores** hyaline (cream [C] in the mass), ornamented with low warts interconnectecd by ridges with limited cross connections, ellipsoid, 7-8 x 5.5-6.5 µm. Basidia 4-spored. Cystidia cylindrical or filiform.
Stem whitish, becoming creamy-buff, bruising more brownish, more or less equal, fairly stout, smooth, viscid. Ring absent. Flesh whitish and stuffed.
Odour not distinctive. **Taste** very hot or acrid.
Chemical tests none. Milk white, drying pallid grey.
Occurrence summer to autumn; very common.
■ Inedible.

Lactarius brittanicus Reid　　　**Russulaceae**

Medium-sized agaric with apricot or brownish cap, cinnamon gills and exuding milk; solitary or in scattered groups, on soil under broad-leaf trees, favouring beech.

Dimensions cap 3 - 8 cm dia; stem 3 - 8 cm tall x 0.8 - 1.8 cm dia.
Cap apricot or orange with brown tinge; at first flattened-convex with slightly inrolled margin, later slightly depressed with faint umbo, smooth and typically puckered. Flesh whitish-buff and fairly thin.
Gills ochraceous or pallid cinnamon, ageing or bruising darker brown, decurrent, narrow, close. **Spores** hyaline (whitish with salmon tinge [B] in the mass), ornamented with isolated warts, a few interconnectecd by ridges, sub-spherical, 6-8 µm. Basidia 4-spored. Cystidia cylindrical or filiform.
Stem reddish-brown, darker below, more or less equal or fusiform, smooth. Ring absent. Flesh whitish-buff, stuffed or hollow.
Odour not distinctive. **Taste** mild, becoming faintly hot or acrid.
Chemical tests none. Milk white, slowly yellow on handkerchief.
Occurrence late summer to autumn; infrequent.
■ Inedible.

Lactarius camphoratus (Bull.: Fr.) Fr.　　**Russulaceae**

Smallish agaric with reddish-brown cap, pale gills and exuding milk, curry-scented; solitary or in extensive scattered groups on soil under conifers, less typically under broad-leaf trees.

Dimensions cap 2.5 - 5 cm dia; stem 3 - 5 cm tall x 0.4 - 0.7 cm dia.
Cap reddish or bay-brown; at first convex with slightly incurved margin, later flattened or slightly depressed with small umbo, smooth, matt. Flesh pallid, concolorous with cap, thin, granular and brittle.
Gills pallid, concolorous with cap, decurrent, narrow, crowded.
Spores hyaline (cream [C-D] in the mass), ornamented with warts, a few interconnected by ridges, sub-spherical, 7.5-8.5 x 6.5-7.5 µm. Basidia 4-spored. Cystidia cylindrical or filiform.
Stem pallid, concolorous with cap, more or less equal, finely downy. Ring absent. Flesh concolorous, stuffed, becoming hollow.
Odour at first of bugs, later and when dry, strongly of curry. **Taste** mild.
Chemical tests none. Milk watery, clouded, white.
Occurrence late summer to autumn; frequent.
□ Edible particularly as flavouring.

Lactarius cilicioides (Fr.) Fr.　　　**Russulaceae**

Large creamy-buff agaric with shaggy fringe to cap and exuding milk; solitary or in scattered groups on soil under birch.

Dimensions cap 10 - 20 cm dia; stem 4 - 8 cm tall x 2 - 4 cm dia.
Cap pallid yellowish-buff, with more ochraceous fibrils and with faint banding towards margin; at first convex with incurved margin, later flattened and depressed, slightly sticky, the fibrils forming a prominent matted fringe to the cap margin. Flesh pallid yellow, thick, hard and brittle.
Gills ochraceous or buff, decurrent, narrow, close. **Spores** hyaline (pallid ochraceous [F] in the mass), ornamented with a network of both thick and thin ridges and occasional warts, broadly ellipsoid, 7.5-8.5 x 5.5-6.5 µm. Basidia 4-spored. Cystidia cylindrical or filiform.
Stem concolorous with cap, more or less equal or tapering slightly downwards, stout, smooth. Ring absent. Flesh pallid yellow, stuffed, soon becoming hollow, hard.
Odour not distinctive. **Taste** mild then hot, acrid.
Chemical tests none. Milk white, then rapidly sulphur-yellow.
Occurrence late summer to autumn; rare. (Thetford Warren, Norfolk)
✚ Probably poisonous.

Lactarius cimicarius (Batsch) Gill.   **Russulaceae**

Smallish agaric with dark brown cap, orange-brown gills and exuding watery milk; solitary or in scattered groups on soil with oak and beech.

Dimensions cap 3 - 7 cm dia; stem 2 - 6 cm tall x 0.6 - 1 cm dia.
Cap dark umber-brown with chestnut tinge; at first flattened-convex, later irregularly expanded or slightly depressed. Flesh buff, thin and fairly brittle.
Gills at first saffron-orange, then brick or cinnamon, somewhat decurrent, narrow, close. **Spores** hyaline (cream [C] in the mass), ornamented sparsely with warts, interconnected by prominent ridges, broadly ellipsoid, 7-9 x 6-8 µm. Basidia 4-spored. Cystidia cylindrical or filiform.
Stem pallid, concolorous with cap, more or less equal, smooth. Ring absent. Flesh buff, stuffed becoming hollow.
Odour oily, of bugs. **Taste** mild.
Chemical tests none. Milk white, watery, unchanging.
Occurrence late summer to autumn; locally frequent.
■ Inedible.

Lactarius circellatus Fr.   **Russulaceae**

Medium to large agaric with greyish-brown cap, yellowish gills and exuding milk; solitary or in scattered groups on soil specifically under hornbeam.

Dimensions cap 5 - 10 cm dia; stem 3 - 6 cm tall x 0.8 - 2 cm dia.
Cap pallid greyish-brown with umber or violaceous tinges and a silvery sheen; at first flattened-convex, later irregularly expanded or slightly depressed. Flesh whitish, thick and fairly brittle.
Gills at first whitish, then pallid yellowish-cream, finally ochraceous, adnexed or slightly decurrent, narrow, close. **Spores** hyaline (cream with salmon tinge [D] in the mass), ornamented sparsely with warts interconnected by thick ridges running more or less across the spore, broadly ellipsoid, 6-8 x 5.5-6.5 µm. Basidia 4-spored. Cystidia cylindrical or filiform.
Stem whitish or pallid, concolorous with cap, more or less equal or fusiform, smooth. Ring absent. Flesh whitish, stuffed, becoming hollow.
Odour not distinctive. **Taste** mild then hot or acrid.
Chemical tests none. Milk white, unchanging.
Occurrence late summer to autumn; locally frequent.
■ Inedible.

Lactarius controversus (Pers.) Fr.   **Russulaceae**

Large creamy agaric with pinkish spots, inrolled cap margin and exuding milk; solitary or in scattered groups on soil with poplar.

Dimensions cap 8 - 15 cm dia; stem 3 - 7 cm tall x 2 - 4 cm dia.
Cap pallid buff, tinged with vinaceous blotches arrranged concentrically; at first convex with inrolled margin becoming depressed or infundibuliform, at first downy, becoming smooth. Flesh white, thick and firm.
Gills pallid buff, decurrent, narrow, close. **Spores** hyaline (pallid pinkish-cream [B] in the mass), ornamented with a network of both thick ridges and prominent warts, broadly ellipsoid, 6-7.5 x 4.5-6 µm. Basidia 4-spored. Cystidia cylindrical or filiform.
Stem concolorous with cap, more or less equal or tapering slightly downwards, stout, smooth. Ring absent. Flesh white and full.
Odour not distinctive. **Taste** mild then hot.
Chemical tests none. Milk white, unchanging.
Occurrence late summer to autumn; infrequent.
□ Edible.

Lactarius deliciosus (L.: Fr.) S F Gray   Russulaceae

Fairly large, buff agaric with distinctive salmon-pink blotches, pale carrot colour gills and exuding carroty milk; solitary or in scattered groups on soil under conifers favouring pine and spruce.

Dimensions cap 3 - 10 cm dia; stem 3 - 6 cm tall x 1.5 - 2 cm dia.
Cap pallid buff, sometimes with faint greenish tinge and with concentric salmon-pink blotches; at first convex with inrolled margin, becoming depressed, slightly sticky. Flesh pallid yellowish, sometimes stained carrot, ageing dull greyish-green, thick, firm and brittle.
Gills at first apricot or saffron, becoming carrot, dull green on bruising, slightly decurrent, narrow, close. **Spores** hyaline (pallid ochraceous [F] in the mass), ornamented with a network of thick and thin ridges, warts absent, ellipsoid, 7-9 x 6-7 µm. Basidia 4-spored. Cystidia cylindrical or filiform.
Stem concolorous with cap, more or less equal, smooth, stout. Ring absent. Flesh pallid yellowish, reacting as in cap, stuffed, becoming hollow.
Odour faint, of fruit. **Taste** mild, then slightly bitter.
Chemical tests none. Milk carrot colour.
Occurrence summer to autumn; infrequent. (Frequent in Scotland)
☐ Edible and good.

Lactarius deterrimus Gröger   Russulaceae

Largish buff agaric with distinctive salmon-pink blotches, pale carrot gills and exuding carroty milk; solitary or in scattered groups on soil under conifers, favouring pine and spruce.

Dimensions cap 3 - 10 cm dia; stem 3 - 6 cm tall x 1.5 - 2 cm dia.
Cap pallid buff, with greenish tinge and concentric salmon-pink blotches; at first convex with inrolled margin, becoming depressed, slightly sticky. Flesh pallid yellowish, sometimes stained carrot, ageing dull greyish-green, thick, firm and brittle.
Gills at first apricot or saffron, becoming carrot, dull green on bruising, slightly decurrent, narrow, close. **Spores** hyaline (pallid ochraceous [F] in the mass), ornamented with a network of thick and thin ridges, warts absent, ellipsoid, 7-10 x 6-7.5 µm. Basidia 4-spored. Cystidia cylindrical or filiform.
Stem concolorous with cap, more or less equal, smooth, stout. Ring absent. Flesh pallid yellowish, reacting as in cap, stuffed, never hollow.
Odour faint, of fruit. **Taste** mild then slightly bitter.
Chemical tests none. Milk carrot colour, discolouring vinaceous in 30 minutes.
Occurrence summer to autumn; frequent.
☐ Edible but bitter.

Lactarius flavidus Boud.   Russulaceae
= *Lactarius aspideus var flavidus* (Boud.) Neuhoff.

Smallish straw-yellow agaric, bruising reddish-purple and exuding milk; solitary or in scattered groups on soil favouring limestone and under beech or oak.

Dimensions cap 3 - 5 cm dia; stem 3 - 6 cm tall x 0.6 - 2 cm dia.
Cap dull ochraceous or straw; at first flattened-convex with incurved margin, becoming depressed, smooth, slightly sticky when damp. Flesh very pallid straw, discolouring purplish, thick, granular and brittle.
Gills pallid straw, tinged reddish-purple where bruised, slightly decurrent, sometimes forked near stem, narrow, crowded.
Spores hyaline (pallid cream [B] in the mass), ornamented with low warts connected by ridges forming an incomplete network, ellipsoid, 8.5-10 x 7.5-9 µm. Basidia 4-spored. Cystidia cylindrical or filiform.
Stem pallid, concolorous with cap, more or less equal, sticky when damp. Ring absent. Flesh very pallid straw and reacting as in cap, stuffed, becoming hollow.
Odour faint, oily or of bugs. **Taste** slight, bitter.
Chemical tests none. Milk white or slightly cream.
Occurrence late summer to autumn; rare. (Priddy Forest, Somerset)
■ Inedible.

Lactarius fluens Boud.   **Russulaceae**

Medium-sized agaric with pale olive-brown greasy cap, pallid gills and exuding copious amounts of milk; solitary or in scattered groups on soil under beech.

Dimensions cap 5 - 10 cm dia; stem 3 - 7 cm tall x 1.5 - 2 cm dia.
Cap brightish olive-brown or olive-green, zoned; at first convex, later flattened or depressed with somewhat irregular margin, at first finely downy then smooth when dry, greasy or viscid when damp. Flesh whitish with brown tinge, firm and brittle.
Gills cream, discolouring reddish or olivaceous-brown, adnate or barely decurrent, close. **Spores** hyaline (cream [C] in the mass), ornamented with a network of thick and thin ridges, warts absent, broadly ellipsoid, 7-9 x 6-6.5 μm. Basidia 4-spored. Cystidia cylindrical or filiform.
Stem pallid, concolorous with cap, tapering downwards, smooth. Ring absent. Flesh whitish, stuffed, becoming hollow.
Odour not distinctive. **Taste** acrid.
Chemical tests none. Milk white, unchanging, copious.
Occurrence late summer to autumn; infrequent.
■ Inedible.

Lactarius fuliginosus (Fr.) Fr.   **Russulaceae**

Small or medium-sized agaric with dull brown velvety cap, paler brown stem, salmon gills and exuding milk; solitary or in scattered groups, on soil in woods with beech.

Dimensions cap 4 - 8 cm dia; stem 3 - 8 cm tall x 0.6 - 1.5 cm dia.
Cap dull umber-brown with sepia tinge; at first convex, later flattened or slightly depressed, smooth but with sulcate margin, thin, dry, suede-like. Flesh whitish and thin.
Gills at first ochraceous-buff then more salmon-brown, decurrent, narrow, close. **Spores** hyaline (pallid ochraceous [F] in the mass), ornamented with a network of coarse and thin ridges, sub-spherical, 8-9 x 7-8 μm. Basidia 4-spored. Cystidia cylindrical or filiform.
Stem pallid, concolorous with cap, more or less equal, smooth. Ring absent. Flesh whitish, stuffed becoming hollow.
Odour not distinctive. **Taste** mild or slightly bitter.
Chemical tests none. Milk white, then pinkish on contact with flesh.
Occurrence late summer to autumn; infrequent.
■ Inedible.

Lactarius fulvissimus Romagn.   **Russulaceae**

Medium-sized agaric with pale tawny-brown cap and gills and exuding milk; solitary or in scattered groups on soil under broad-leaf trees, less commonly in mixed woods.

Dimensions cap 5 - 7 cm dia; stem 2 - 5 cm tall x 0.8 - 1 cm dia.
Cap fulvous or tawny-brown, often with more pallid margin; at first convex, later flattened or slightly depressed, smooth but with sulcate margin, slightly sticky. Flesh white with fulvous tinge, thin, granular and brittle.
Gills fulvous or buff, decurrent, narrow, close. **Spores** hyaline (yellowish cream [D-E] in the mass), ornamented with delicate warts, some interconnected by a network of thin ridges, sub-spherical, 7-7.5 x 6-6.5 μm. Basidia 4-spored. Cystidia cylindrical or filiform.
Stem concolorous with cap, more or less equal, stout, smooth. Ring absent. Flesh white with fulvous tinge, stuffed becoming hollow.
Odour not distinctive. **Taste** mild then slightly acrid.
Chemical tests none. Milk white, yellow on handkerchief in 2 minutes.
Occurrence late summer to autumn; infrequent.
■ Inedible.

Lactarius glyciosmus (Fr.: Fr.) Fr.　　Russulaceae

Small to medium-sized agaric with greyish-lilac cap, buff gills and exuding milk, with distinctive aroma of coconut; solitary or in scattered groups on soil under birch.

Dimensions cap 2 - 5 cm dia; stem 2 - 6 cm tall x 0.4 - 1.2 cm dia
Cap greyish-lilac or more buff; at first convex and with somewhat incurved margin, later flattened or slightly depressed and sometimes with a small central umbo, matt or finely downy when dry. Flesh whitish-buff, thin, granular and brittle.
Gills at first pallid yellowish or buff then concolorous with cap, decurrent, narrow, crowded. **Spores** hyaline (cream [C-D] in the mass), ornamented with small warts, some interconnected by a network of thin ridges, broadly ellipsoid, 7-8.5 x 5.5-6.5 µm. Basidia 4-spored. Cystidia cylindrical or filiform.
Stem pallid buff, more or less equal or tapering slightly upwards, smooth. Ring absent. Flesh whitish with buff tinge, soft, stuffed becoming hollow, very brittle.
Odour strong, of coconuts. **Taste** mild, then slightly hot and acrid.
Chemical tests none. Milk white, unchanging.
Occurrence late summer to autumn; common.
☐ Edible.

Lactarius helvus (Fr.) Fr.　　Russulaceae

Medium-sized or large agaric with cinnamon-brown cap, buff gills and exuding milk; solitary or in scattered groups on soil with conifers, often at moorland altitudes.

Dimensions cap 6 - 12 cm dia; stem 6 - 12 cm tall x 0.5 - 3 cm dia.
Cap cinnamon-brown with reddish tinge; at first convex, later flattened or slightly depressed, sometimes with a small umbo, covered with fine adpressed scales giving a slightly crusty appearance. Flesh whitish with buff tinge, moderate, at first firm then soft.
Gills pallid buff with ochraceous tinge, becoming more ochraceous-brown with age, slightly decurrent, narrow, close. **Spores** hyaline (whitish [A-B] in the mass), ornamented with small warts, some interconnected by a network of thin ridges, broadly ellipsoid, 7-9 x 5.5-6.5 µm. Basidia 4-spored. Cystidia cylindrical or filiform.
Stem concolorous with cap or more reddish-brown, more or less equal or slightly clavate. Ring absent. Flesh whitish with buff tinge, stuffed becoming hollow.
Odour not distinctive in fresh specimens, of stock cubes when drying. **Taste** mild.
Chemical tests none. Milk watery, sparse.
Occurrence late summer to autumn; locally common. More prevalent in Scotland and northern Europe.
■ ✤ Inedible, possibly poisonous.

Lactarius hepaticus Plowright　　Russulaceae

Small to medium-sized agaric with dull reddish-brown cap, buff gills and exuding milk; solitary or in scattered groups on soil under pines.

Dimensions cap 2 - 6 cm dia; stem 3 - 6 cm tall x 0.4 - 0.8 cm dia.
Cap liver or dull chestnut-brown; at first convex, later flattened or slightly depressed, sometimes with a small central umbo and puckering, matt when dry. Flesh whitish with buff tinge, thin, granular and brittle.
Gills at first pallid yellowish or buff, becoming deeper coloured with age, slightly decurrent, narrow, close. **Spores** hyaline (cream [B-C] in the mass), ornamented with prominent warts, some interconnected by a network of thick or thin ridges, broadly ellipsoid, 8-9 x 6-7 µm. Basidia 4-spored. Cystidia cylindrical or filiform.
Stem reddish-brown, more pallid above, more or less equal. Ring absent. Flesh whitish with buff tinge, stuffed becoming hollow.
Odour not distinctive. **Taste** bitter and acrid.
Chemical tests none. Milk white, yellow on handkerchief in 2 minutes.
Occurrence late summer to autumn; locally common. More prevalent in southern Britain.
■ Inedible.

Lactarius mitissimus (Fr.) Fr.   **Russulaceae**

Smallish agaric with orange-brown cap, buff gills and exuding milk; solitary or in scattered groups on soil under coniferous and broad-leaf trees.

Dimensions cap 3 - 6 cm dia; stem 3 - 7 cm tall x 0.6 - 1.2 cm dia.
Cap bright orange or apricot-brown; at first convex with incurved margin, later flattened or slightly depressed and sometimes with a small central umbo, slightly greasy. Flesh whitish, thin and firm.
Gills pallid ochraceous or buff, adnate or slightly decurrent, narrow, close. **Spores** hyaline (pale cream [C] in the mass), ornamented with warts mostly interconnected by a network of thin ridges, sub-spherical, 8-9.5 x 6.5-7.5 μm. Basidia 4-spored. Cystidia cylindrical or filiform.
Stem pallid, concolorous with cap, more or less equal, or tapering slightly downwards, smooth. Ring absent. Flesh whitish, stuffed becoming hollow.
Odour not distinctive. **Taste** mild.
Chemical tests none. Milk white, unchanging, copious.
Occurrence late summer to autumn; frequent.
■ Inedible.

Lactarius piperatus (Scop.: Fr.) S F Gray   **Russulaceae**

Large creamy-white agaric, exuding milk; solitary or in scattered groups on soil under broad-leaf trees.

Dimensions cap 6 - 16 cm dia; stem 3 - 7 cm tall x 2 - 3 cm dia.
Cap whitish-cream; at first convex and markedly incurved, later depressed or infundibuliform, smooth and matt when dry. Flesh whitish, thick, granular and brittle.
Gills at first white, becoming cream with age, decurrent, narrow, crowded. **Spores** hyaline (white [A] in the mass), ornamented with small warts mostly finely interconnected, ellipsoid, 7.5-9.5 x 6.5-8.5 μm. Basidia 4-spored. Cystidia cylindrical or filiform.
Stem concolorous with cap, more or less equal or tapering slightly downwards, more or less smooth. Ring absent. Flesh white, stuffed becoming hollow.
Odour not distinctive. **Taste** very hot and acrid.
Chemical tests none. Milk white, unchanging, copious.
Occurrence late summer to autumn; common.
■ Inedible.

Lactarius pubescens (Fr. ex Krombh.) Fr.   **Russulaceae**

Medium to large agaric with creamy-buff, hairy cap, whitish gills, and short stout stem and exuding milk; solitary or in scattered groups on sandy soil under or near birch.

Dimensions cap 4 - 10 cm dia; stem 3 - 6 cm tall x 1 - 2.3 cm dia.
Cap cream with slightly darker buff or rosy tinges; at first convex, later flattened and depressed, the margin remaining inrolled, dry and coarsely shaggy-hairy at the margin. Flesh whitish, thick, granular and brittle.
Gills white, tinged salmon-pink, slightly decurrent, narrow, close.
Spores hyaline (pallid cream with salmon tinge [C-D] in the mass), ornamented with warts, some interconnected by an irregular network of ridges, ellipsoid, 6.5-8.5 x 5.5-6.5 μm. Basidia 4-spored. Cystidia cylindrical or filiform.
Stem concolorous with cap, sometimes with buff banding towards apex, stout, more or less equal or tapering slightly downwards, smooth. Ring absent. Flesh whitish, stuffed becoming hollow.
Odour faint, of *Pelargonium*. **Taste** slight, acrid.
Chemical tests none. Milk white, sparse, unchanging.
Occurrence late summer to autumn; infrequent or rare. (Whitley Common, near Guildford, Surrey)
✢ Poisonous.

Lactarius pyrogalus (Bull.: Fr.) Fr.   **Russulaceae**

Small or medium-sized agaric with dull fawn cap, yellowish gills and exuding milk; solitary or in scattered groups on soil under hazel.

Dimensions cap 5 - 10 cm dia; stem 4 - 6 cm tall x 0.7 - 1.5 cm dia
Cap pallid dull fawn, sometimes with greyish or yellowish tinges and with faint concentric zoning; at first convex, later flattened or slightly depressed, matt when dry, slightly sticky when damp. Flesh whitish, fairly thick, granular and brittle.
Gills at first pallid, yellowish or buff, more ochraceous with age, slightly decurrent, narrow, close. **Spores** hyaline (pallid ochraceous [F] in the mass), ornamented with warts most of which are interconnected by a network of thick ridges, sub-spherical or broadly ellipsoid, 7-8 x 5.5-6.5 µm. Basidia 4-spored. Cystidia cylindrical or filiform.
Stem pallid, concolorous with cap, more or less equal or slightly fusiform, slightly sticky when damp. Ring absent. Flesh whitish, stuffed sometimes becoming hollow.
Odour not distinctive. **Taste** hot and acrid.
Chemical tests milk yellowish-orange on slide with KOH. Milk white, unchanging.
Occurrence late summer to autumn; frequent.
■ ✛ Inedible, possibly poisonous.

Lactarius quietus (Fr.) Fr.   **Russulaceae**

Medium-sized agaric with dull reddish-brown, concentrically-banded cap, pale gills and exuding milk; solitary or in scattered groups on soil specifically under oak.

Dimensions cap 3 - 8 cm dia; stem 4 - 9 cm tall x 1 - 1.5 cm dia.
Cap dull cinnamon or reddish-brown, with limited concentric spotty bands of darker colour; at first convex, later flattened or slightly depressed, matt when dry. Flesh whitish with buff tinge, thick, firm.
Gills at first buff or pallid brown, becoming darker reddish-brown sometimes with mauve tinge, slightly decurrent, fairly narrow, crowded. **Spores** hyaline (cream with salmon tinge [C-D] in the mass), ornamented with warts mostly interconnected by a network of ridges, broadly ellipsoid, 7.5-9 x 6.5-7.5 µm. Basidia 4-spored. Cystidia cylindrical or filiform.
Stem more or less concolorous with cap, more or less equal, sometimes slightly furrowed . Ring absent. Flesh whitish, stuffed becoming hollow.
Odour faint, of oil. **Taste** slight, bitter.
Chemical tests none. Milk white or slightly cream.
Occurrence late summer to autumn; very common.
■ Inedible.

Lactarius resimus Fr.   **Russulaceae**

Medium-sized or large agaric with whitish cap and gills and exuding milk; solitary or in scattered groups on acid soil under conifers or birches.

Dimensions cap 7 - 16 cm dia; stem 4 - 9 cm tall x 0.5 - 1.5 cm dia.
Cap whitish, without concentric zones, sometimes bruising yellowish, margin finely woolly only when young; at first convex with inrolled margin, later flattened or slightly depressed. Flesh whitish, fairly thick, granular and brittle.
Gills concolorous with cap, more or less decurrent, narrow, fairly crowded. **Spores** hyaline (white [A-B] in the mass), ornamented with warts mostly interconnected by a network of ridges, ellipsoid, 7-9 x 5.5-7 µm. Basidia 4-spored. Cystidia cylindrical or filiform.
Stem concolorous with cap, more or less equal, smooth. Ring absent. Flesh white, stuffed becoming hollow.
Odour faint, of fruit. **Taste** acrid, bitter.
Chemical tests none. Milk white, discolouring yellowish after several minutes.
Occurrence late summer to autumn; infrequent.
■ Inedible.

Lactarius rufus (Scop.: Fr.) Fr.   **Russulaceae**

Small to medium-sized agaric with dull reddish-brown cap, pale flesh-coloured gills and exuding milk; solitary or in scattered groups on soil under conifers, favouring pines occasionally with birches.

Dimensions cap 3 - 10 cm dia; stem 4 - 8 cm tall x 0.5 - 2 cm dia.
Cap reddish or bay-brown, at first convex with inrolled margin, later flattened or slightly depressed and typically with a small central umbo, matt when dry. Flesh whitish, fairly thick, granular and brittle.
Gills at first pallid yellowish, then more concolorous with cap, more or less decurrent, narrow, fairly crowded. **Spores** hyaline (whitish-cream with slight salmon tinge [B] in the mass), ornamented with warts mostly interconnected by a network of thin ridges, ellipsoid, 8-9.5 x 6.5-7.5 µm. Basidia 4-spored. Cystidia cylindrical or filiform.
Stem concolorous with cap, or more pallid, more or less equal, smooth. Ring absent. Flesh white, stuffed becoming hollow.
Odour not distinctive. **Taste** at first mild then, after a delay, very hot and acrid.
Chemical tests none. Milk white, unchanging.
Occurrence early summer to autumn; common.
■ Inedible.

Lactarius ruginosus Romagn.   **Russulaceae**

Medium to large agaric with dull umber-brown cap, buff gills and exuding milk; solitary or in scattered groups on soil under beech.

Dimensions cap 4 - 9 cm dia; stem 4 - 10 cm tall x 0.7 - 2 cm dia.
Cap cigar, sepia or umber-brown; at first convex with incurved margin, later flattened or slightly depressed, finely downy. Flesh whitish, salmon-pink after 2 to 3 minutes when cut, thick, granular and brittle.
Gills ochraceous or buff with salmon-pink tinge especially where bruised, slightly decurrent, narrow, fairly crowded. **Spores** hyaline (pallid ochraceous [E] in the mass), ornamented with prominent warts, some interconnected by a network of deep ridges interspersed with finer ones, broadly ellipsoid, 7-9 x 7-8 µm. Basidia 4-spored. Cystidia cylindrical or filiform.
Stem concolorous with cap but more pallid, more or less equal, finely downy. Ring absent. Flesh whitish as in cap, stuffed becoming hollow.
Odour not distinctive. **Taste** at first mild, becoming slightly bitter.
Chemical tests none. Milk white, pinkish on skin contact (no change in air).
Occurrence late summer to autumn; infrequent.
☐ Edible but poor.

Lactarius salmonicolor Heim & Lecl.   **Russulaceae**

Medium-sized, distinctively orange agaric, exuding milk; solitary or in scattered groups on soil under fir.

Dimensions cap 3 - 7 cm dia; stem 3 - 7 cm tall x 0.6 - 1.3 cm dia.
Cap wholly ochraceous-orange; at first convex with incurved margin, later slightly depressed, smooth, matt. Flesh concolorous, moderate, granular and brittle.
Gills pallid orange, adnate or slightly decurrent, narrow, fairly crowded. **Spores** hyaline (cream with salmon tinge [D] in the mass), ornamented with large prominent warts interconnected by a network of thinnish ridges, ellipsoid, 9-12 x 6.5-7.5 µm. Basidia 4-spored. Cystidia cylindrical or filiform.
Stem concolorous with cap but more pallid, particularly above, more or less equal or slightly fusiform, smooth. Ring absent. Flesh concolorous and stuffed.
Odour not distinctive. **Taste** not distinctive.
Chemical tests none. Milk orange, wine-red on gills only when exposed to air.
Occurrence late summer to autumn; rare. (Westonbirt Arboretum, Gloucestershire)
■ Inedible.

Lactarius sphagneti (Fr.) Neuh.   **Russulaceae**

Small agaric with reddish-brown cap, yellowish gills and exuding watery milk; solitary or in scattered groups in *Sphagnum* moss with conifers, favouring pines.

Dimensions cap 3 - 5 cm dia; stem 3 - 5 cm tall x 0.5 - 1 cm dia.
Cap reddish-brown with more pallid margin, tinged pinkish; at first convex with inrolled margin, later flattened or slightly depressed and sometimes with a small central umbo, matt when dry. Flesh pallid, fairly thin, granular and brittle.
Gills at first pallid pinkish then more concolorous with cap, more or less decurrent, narrow, fairly crowded. **Spores** hyaline (whitish-cream with slight salmon tinge [B] in the mass), ornamented with warts mostly interconnected by a network of thin ridges, ellipsoid, 8-9.5 x 6.5-7.5 µm. Basidia 4-spored. Cystidia cylindrical or filiform.
Stem concolorous with cap or more pallid at the apex, more or less equal, smooth. Ring absent. Flesh pallid, stuffed becoming hollow.
Odour not distinctive. **Taste** mild.
Chemical tests none. Milk watery, then yellow after several minutes.
Occurrence late summer to autumn; infrequent.
■ Inedible.

Lactarius subdulcis (Pers.: Fr.) S F Gray   **Russulaceae**

Smallish agaric with pale reddish-brown cap, buff gills and exuding milk; solitary or in scattered groups on soil under broad-leaf trees, favouring beech.

Dimensions cap 3 - 7 cm dia; stem 3 - 7 cm tall x 0.6 - 1.3 cm dia.
Cap at first dull russet or cinnamon-brown becoming more buff with age; at first convex with incurved margin, later slightly depressed sometimes with small umbo, smooth or slightly sulcate at margin, matt. Flesh whitish or buff, thin, granular and brittle.
Gills at first whitish, then buff with rose tinge, adnate or slightly decurrent, narrow, fairly crowded. **Spores** hyaline (cream with salmon tinge [C-D] in the mass), ornamented with large prominent warts interconnected by a network of thinnish ridges, ellipsoid, 7.5-9.5 x 6.5-8 µm. Basidia 4-spored. Cystidia cylindrical or filiform.
Stem concolorous with cap but more pallid, particularly above, more or less equal or slightly fusiform, sometimes finely grooved. Ring absent. Flesh whitish or buff, stuffed becoming hollow.
Odour faint, oily. **Taste** at first mild, becoming slightly bitter.
Chemical tests none. Milk white, unchanging, copious.
Occurrence late summer to autumn; common.
□ Edible.

Lactarius tabidus Fr.   **Russulaceae**
= *Lactarius thejogalus* (Bull.) Fr.

Small agaric with orange-brown cap, cinnamon gills and exuding milk; solitary or in scattered groups on soil under broad-leaf trees, favouring birch.

Dimensions cap 2 - 4 cm dia; stem 3 - 5 cm tall x 0.4 - 1 cm dia.
Cap rust-brown with ochraceous or orange tinges, becoming more pallid with age; flatly convex, typically with a small central papilla, striate or sulcate at margin, matt. Flesh whitish or buff, thin, granular and brittle.
Gills at first ochraceous-buff, becoming cinnamon, adnate or slightly decurrent, narrow, fairly crowded. **Spores** hyaline (pallid cream [B-C] in the mass), ornamented with warts interconnected by occasional fine ridges, ellipsoid, 7-9.5 x 6-7 µm. Basidia 4-spored. Cystidia cylindrical or filiform.
Stem concolorous with cap but more brick-red towards base, more or less equal or tapering slightly upwards. Ring absent. Flesh whitish or buff, stuffed becoming hollow.
Odour not distinctive. **Taste** at first mild, becoming slightly bitter.
Chemical tests none. Milk white, becoming slightly yellow on hand-kerchief.
Occurrence late summer to autumn; common.
□ Edible.

Lactarius torminosus (Schaeff.: Fr.) S F Gray   Russulaceae

Largish salmon-pink, distinctively woolly agaric, exuding milk; solitary or in scattered groups on soil under broad-leaf trees, favouring birch.

Dimensions cap 4 - 12 cm dia; stem 4 - 8 cm tall x 1 - 2 cm dia.
Cap buff with pinkish tinge or salmon with darker concentric zones; convex, later depressed or infundibuliform, margin inrolled, woolly or hairy, matt. Flesh white, thick, granular and brittle.
Gills pallid buff or salmon, decurrent, narrow, crowded. **Spores** hyaline (pallid yellowish-cream with salmon tinge [C-D] in the mass), ornamented with low warts some of which are interconnected by fine ridges to form an incomplete network, ellipsoid, 7.5-10 x 6-7.5 μm. Basidia 4-spored. Cystidia cylindrical or filiform.
Stem concolorous with cap but more pallid, more or less equal or fusiform, at first finely downy, then smooth or slightly pitted. Ring absent. Flesh white, stuffed becoming hollow.
Odour faint, of turpentine. **Taste** hot and acrid.
Chemical tests none. Milk white, unchanging.
Occurrence late summer to autumn; frequent.
✚ Poisonous.

Lactarius trivialis (Fr.: Fr.) Fr.   Russulaceae

Largish agaric with violet cap and pale gills and stem, exuding milk; solitary or in scattered groups on boggy soil under coniferous trees and birches.

Dimensions cap 6 - 15 cm dia; stem 6 - 12 cm tall x 1 - 2 cm dia.
Cap at first purplish-violet with darker zones, becoming brownish-grey or pinkish-beige; convex later, barely depressed, greasy or viscid when damp. Flesh whitish, thick, brittle.
Gills pallid, greyish-green where damaged, adnate or barely decurrent, narrow, crowded. **Spores** hyaline (pallid yellowish-cream [E] in the mass), ornamented with low warts, some interconnected by ridges forming an incomplete network, ellipsoid, 8.5-10.5 x 7-8.5 μm. Basidia 4-spored. Cystidia cylindrical or filiform.
Stem pallid with violaceous or hazel tinges, more or less equal or fusiform, greasy when damp. Ring absent. Flesh white, stuffed becoming hollow.
Odour not distinctive. **Taste** hot and acrid.
Chemical tests none. Milk white, becoming greyish-green on exposure to air.
Occurrence late summer to autumn; infrequent.
■ Inedible.

Lactarius turpis (Weinm.) Fr.   Russulaceae
= *Lactarius necator* (Bull.: Fr.) Karst.
= *Lactarius plumbeus* (Bull.: Fr.) S F Gray

Large, dull olive-brown agaric, typically slimy or sticky and exuding milk; solitary or in scattered groups on soil in damp places under birch.

Dimensions cap 5 - 20 cm dia; stem 4 - 8 cm tall x 1 - 2.5 cm dia.
Cap dull dark-olive or umber-brown with slightly more pallid margin; at first convex with inrolled margin, later flattened and then depressed, at first felty or woolly, becoming viscid. Flesh whitish, discolouring brown, thick, granular, brittle.
Gills buff, becoming sepia-tinged especially where bruised, decurrent, fairly narrow, crowded. **Spores** hyaline (cream with salmon tinge [D] in the mass), ornamented with ridges mainly across the spore but in a well developed network, ellipsoid, 7.5-8.5 x 6-7 μm. Basidia 4-spored. Cystidia cylindrical or filiform.
Stem concolorous with cap but more pallid, more or less equal or tapering towards the base, slightly pitted, viscid. Ring absent. Flesh whitish as in cap, stuffed becoming hollow.
Odour not distinctive. **Taste** hot, acrid.
Chemical tests flesh purple or violet with KOH or NH_3. Milk white, unchanging in air, copious.
Occurrence late summer to autumn; common.
■ Inedible.

Lactarius uvidus (Fr.) Fr.   **Russulaceae**

Small or medium-sized agaric with greyish-lilac cap, whitish gills bruising lilac and exuding milk; solitary or in scattered groups on acid soil with birch or willow and with conifers.

Dimensions cap 3 - 9 cm dia; stem 3 - 6 cm tall x 0.5 - 2.5 cm dia.
Cap buff, or lilaceous with darker marking; at first convex with incurved margin, later slightly depressed, smooth, viscid when damp. Flesh whitish or buff, becoming lilaceous after a delay where cut, moderate, granular, firm.
Gills at first whitish, then buff bruising lilaceous, adnate or slightly decurrent, narrow, fairly crowded. **Spores** hyaline (cream [D] in the mass), ornamented with large prominent warts interconnected by a network of thick and thin ridges, broadly ellipsoid, 8.5-10.5 x 7.5-8.5 μm. Basidia 4-spored. Cystidia cylindrical or filiform.
Stem concolorous with cap but more pallid, more or less equal, sometimes finely veined. Ring absent. Flesh whitish or buff, reacting as in cap, stuffed becoming hollow.
Odour not distinctive. **Taste** slight, bitter.
Chemical tests none. Milk white, slowly turning lilaceous.
Occurrence late summer to autumn; infrequent, mainly in Scotland.
■ Inedible.

Lactarius vellereus (Fr.) Fr.   **Russulaceae**

Large creamy-white agaric, finely woolly and exuding milk; solitary or in scattered groups on soil in broad-leaf woods.

Dimensions cap 10 - 25 cm dia; stem 4 - 7 cm tall x 2 - 4 cm dia.
Cap creamy-white becoming tinged buff with age; at first convex with inrolled margin, later depressed or infundibuliform, dry, slightly woolly or downy. Flesh white, becoming tinged buff or yellowish, very thick and hard.
Gills white becoming cream, decurrent, narrow, fairly distant.
Spores hyaline (white [A] in the mass), ornamented with small warts connected by fine ridges forming an incomplete network, subspherical, 7.5-9.5 x 6.5-8.5 μm. Basidia 4-spored. Cystidia cylindrical or filiform.
Stem concolorous with cap, more or less equal or tapering towards base, finely woolly or downy. Ring absent. Flesh white, hard more or less throughout.
Odour not distinctive. **Taste** mild.
Chemical tests none. Milk white, unchanging in air, copious.
Occurrence late summer to autumn; common.
■ Inedible.
Note: the variety *L. vellereus var bertillonii* produces a more acrid taste but is otherwise identical.

Lactarius vietus (Fr.) Fr.   **Russulaceae**

Small to medium-sized agaric, with greyish cap, buff gills and exuding milk; solitary or in scattered groups on soil in damp places under birch.

Dimensions cap 2 - 7 cm dia; stem 3 - 8 cm tall x 0.5 - 1.3 cm dia.
Cap greyish with buff or yellowish-brown tinges; at first flattened-convex with incurved margin and then depressed, sometimes with a blunt umbo, slightly viscid when damp. Flesh whitish-buff, granular and brittle.
Gills buff becoming straw or dirty brown where bruised, slightly decurrent, fairly narrow, crowded. **Spores** hyaline (pallid cream with salmon tinge [A-B] in the mass), ornamented with a network of ridges, ellipsoid, 8-9.5 x 6.5-7.5 μm. Basidia 4-spored. Cystidia cylindrical or filiform.
Stem dirty white with greyish tinge, more or less equal or tapering towards base which may be clavate, slender, viscid. Ring absent. Flesh whitish-buff, stuffed, very brittle.
Odour not distinctive. **Taste** hot, acrid.
Chemical tests none. Milk white, drying greenish or brownish on the gills.
Occurrence late summer to autumn; frequent.
■ Inedible.

Lactarius violascens Fr.   **Russulaceae**

Medium or large agaric with zoned violet-brown cap, yellowish gills and exuding milk; solitary or in scattered groups on soil under broadleaf trees and shrubs.

Dimensions cap 5 - 8 cm dia; stem 4 - 8 cm tall x 0.75 - 1 cm dia.
Cap violaceous-grey-brown, somewhat zoned; at first convex with incurved margin, later depressed, smooth, matt, greasy when damp. Flesh pallid concolorous, firm and brittle.
Gills ochraceous, darker and brown-spotted when old, adnate or slightly decurrent, narrow, fairly crowded. **Spores** hyaline (yellowish cream [D-E] in the mass), ornamented with a network of thick ridges, ellipsoid, 7-11 x 6-8 µm. Basidia 4-spored. Cystidia cylindrical or filiform.
Stem concolorous with cap, more or less equal or slightly clavate, smooth. Ring absent. Flesh pallid concolorous, stuffed becoming hollow.
Odour not distinctive. **Taste** mild or slightly bitter.
Chemical tests none. Milk white, slowly turning lilac.
Occurrence late summer to autumn; infrequent.
■ Inedible.

Lactarius volemus Fr.   **Russulaceae**

Medium or large agaric with tawny-brown cap, golden-yellow gills and exuding milk; solitary or in scattered groups on soil under broad-leaf and coniferous trees.

Dimensions cap 5 - 11 cm dia; stem 4 - 12 cm tall x 1 - 3 cm dia.
Cap tawny-brown with apricot tinge; at first convex with incurved margin, later depressed, smooth or slightly downy. Flesh whitish, firm and brittle.
Gills concolorous with cap but more pallid, adnate or slightly decurrent, narrow, fairly crowded. **Spores** hyaline (whitish [A-B] in the mass), ornamented with a network of thick ridges, spherical, 8-10 µm. Basidia 4-spored. Cystidia cylindrical or filiform.
Stem concolorous with cap but more pallid, particularly above, more or less equal, smooth or faintly downy. Ring absent. Flesh whitish or buff, stuffed becoming hollow.
Odour faint, of fish. **Taste** mild.
Chemical tests none. Milk white, unchanging, copious.
Occurrence late summer to autumn; infrequent.
☐ Edible.

Russula aeruginea Lindblad.: Fr.   **Russulaceae**

Largish agaric with pale green cap, buff, brittle gills and white stem; solitary or in scattered groups on soil under birch.

Dimensions cap 4 - 9 cm dia; stem 4 - 8 cm tall x 0.7 - 2 cm dia.
Cap pallid grass-green with yellowish or brownish tinges or more rusty spotting, darker at centre; at first convex, later flattened and slightly depressed, smooth or with slight radial veining, margin sulcate, cuticle peeling 1/2 to centre. Cystidia cylindrical or fusiform. Flesh white, moderately thick, granular and brittle.
Gills buff, becoming yellowish tinged, adnexed, forked, fairly narrow, crowded. **Spores** hyaline (yellowish cream in the mass [D-E]), ornamented with blunt warts (up to 0.6 µm), partly connected by fine lines in an incomplete network, ellipsoid, 6-10 x 5-7 µm. Basidia 4-spored. Cystidia cylindrical or fusiform.
Stem white, sometimes with rust spotting, more or less equal, smooth. Ring absent. Flesh white, fairly firm but brittle and full.
Odour not distinctive. **Taste** mild or slightly hot.
Chemical tests flesh slightly pink with $FeSO_4$; cap cystidia purple with sulpho-vanillin.
Occurrence summer to autumn; frequent.
☐ Edible.

Russula albonigra (Krombh.) Fr.   Russulaceae
= *Russula anthracina* Romagn.

Large agaric, dirty-white, blackening, with brittle crowded gills; solitary or in scattered groups on soil under coniferous and broad-leaf trees.

Dimensions cap 7 - 12 cm dia; stem 3 - 6 cm tall x 1.5 - 3 cm dia.
Cap at first dirty-white, becoming brownish-black; at first flattened; convex with incurved margin, later somewhat depressed, smooth or slightly sticky, cuticle peeling 3/4 to centre. Cystidia absent. Flesh white, blackening when damaged or cut, moderately thin, granular, brittle.
Gills whitish-buff but darkening in cap, slightly decurrent, narrow, crowded. **Spores** hyaline (white in the mass [A]), ornamented with small warts (up to 0.4 µm), largely connected by fine lines forming a more or less complete network, sub-spherical or broadly ellipsoid, 7-9 x 7-8 µm. Basidia 4-spored. Cystidia cylindrical or filiform.
Stem concolorous with cap, more or less equal, stout, smooth. Ring absent. Flesh concolorous, fairly firm, brittle and full.
Odour not distinctive. **Taste** mild or slightly hot and bitter.
Chemical tests flesh more or less salmon pink with $FeSO_4$; no reaction with sulpho-vanillin.
Occurrence summer to autumn; infrequent.
☐ Edible.

Russula alutacea (Pers.: Fr.) Fr.   Russulaceae

Large agaric with pale yellowish-wine cap, buff brittle gills and whitish stem; solitary or in scattered groups on soil under broad-leaf trees.

Dimensions cap 7 - 13 cm dia; stem 3 - 10 cm tall x 1.5 - 4 cm dia.
Cap straw with vinaceous, olive or brownish tinges, pallid and more greenish at centre or more concolorous throughout; at first sub-spherical, becoming convex, later flattened and slightly depressed, rather irregular, smooth, cuticle peeling 1/3 to centre. Cystidia absent. Flesh white, thick, granular and brittle.
Gills buff or straw, adnexed, forked, interveined, moderately crowded. **Spores** hyaline (ochraceous in the mass [G-H]), ornamented with low warts (up to 0.8 µm), partly connected by ridges forming an incomplete network, ellipsoid, 8-11 x 7-9 µm. Basidia 4-spored. Cystidia cylindrical or filiform.
Stem white, sometimes tinged rose at base, more or less equal or tapering slightly upwards, smooth. Ring absent. Flesh white, fairly hard, brittle and full.
Odour not distinctive. **Taste** mild.
Chemical tests flesh salmon-pink with $FeSO_4$, no reaction with sulpho-vanillin.
Occurrence summer; infrequent.
☐ Edible.
Note: spore ornamentation differs from *R. olivacea*.

Russula aquosa Lecl.   Russulaceae

Small or medium agaric with a somewhat washed out cherry-red cap, fragile white gills and stem; solitary or in scattered groups on marshy ground with mosses.

Dimensions cap 3 - 9 cm dia; stem 4 - 9 cm tall x 1 - 2 cm dia.
Cap cherry or lilaceous-red but typically washed out; at first convex, becoming flattened or irregular, smooth or slightly sticky, cuticle peeling 2/3 to centre. Cystidia clavate, non-septate. Flesh watery white, moderately thick, very fragile.
Gills white, almost free, narrow, somewhat distant. **Spores** hyaline (white in the mass [A]), ornamented with warts (up to 0.7 µm), connected by fine lines forming a partial network, sub-spherical or broadly ellipsoid, 7-9 x 6-7 µm. Basidia 4-spored. Cystidia cylindrical or filiform.
Stem white or tinged brown, clavate, smooth, very fragile. Ring absent. Flesh concolorous, watery, fragile and full.
Odour not distinctive. **Taste** mild or slightly hot.
Chemical tests flesh slowly violet with phenol; cap cystidia moderately purple with sulpho-vanillin.
Occurrence summer to autumn; infrequent.
■ Inedible.

Russula atropurpurea (Krombh.) Britz.   **Russulaceae**

Largish agaric with dark purplish cap, pale cream brittle gills and white stem; solitary or in scattered groups on soil under broad-leaf trees, favouring beech or oak, occasionally with conifers.

Dimensions cap 4 - 10 cm dia; stem 3 - 6 cm tall x 1 - 2 cm dia.
Cap purplish-red, almost black at centre, sometimes mottled ochraceous with age; at first convex, later flattened and slightly depressed, smooth, slightly sticky when damp, cuticle not peeling readily. Cystidia cylindrical or somewhat clavate, abundant. Flesh white, moderately thick, granular and brittle.
Gills pallid cream, adnexed, forked, fairly broad, crowded.
Spores hyaline (whitish in the mass [A-B]), ornamented with warts, connected by fine ridges into almost complete network, broadly ellipsoid, 7-9 x 6-7 µm. Basidia 4-spored. Cystidia cylindrical or filiform.
Stem white, sometimes greying slightly with age, more or less equal, stout, smooth. Ring absent. Flesh white, at first fairly firm but brittle, becoming softer, full.
Odour faint, of apples. **Taste** mild to fairly hot.
Chemical tests flesh salmon-pink with $FeSO_4$; cap cystidia purple with sulpho-vanillin.
Occurrence summer to autumn; common.
☐ Edible only after cooking.

Russula aurata (With.) Fr.   **Russulaceae**

Largish agaric with reddish-orange cap, pale yellow, brittle gills and golden-yellow stem; solitary or in scattered groups on soil under broad-leaf trees.

Dimensions cap 4 - 9 cm dia; stem 4 - 8 cm tall x 1 - 2.5 cm dia.
Cap scarlet, blood-red or reddish-orange with golden yellow tinge; at first convex, later flattened and slightly depressed, smooth, margin faintly sulcate when mature, cuticle peeling 1/2 to centre. Cystidia absent. Flesh whitish, thick, granular and brittle.
Gills pallid ochraceous, adnexed or free, broad, fairly distant.
Spores hyaline (ochraceous in the mass [H]), ornamented with conical warts (up to 1.5 µm), partly connected by fine or thicker lines in an incomplete network, ellipsoid, 7-10 x 6-8 µm. Basidia 4-spored. Cystidia fusiform, clavate or cylindrical.
Stem white, becoming golden-yellow, more or less equal, smooth. Ring absent. Flesh whitish, fairly firm but brittle and full.
Odour not distinctive. **Taste** mild.
Chemical tests no reaction with sulpho-vanillin.
Occurrence summer to autumn; infrequent.
☐ Edible.

Russula badia Quél.   **Russulaceae**

Medium-sized or large agaric with deep purplish-brown cap, brittle cream gills and whitish stem; solitary or in scattered groups on soil with conifers.

Dimensions cap 7 - 11 cm dia; stem 5 - 11 cm tall x 1 - 2.5 cm dia.
Cap deep livid purple with bay or brownish tinges; at first flattened-convex, later slightly depressed, smooth, greasy when damp, cuticle scarcely peeling. Cystidia cylindrical or clavate. Flesh white, moderate, granular and firm.
Gills at first white, then pallid ochraceous, adnexed, greasy and flexible, close. **Spores** hyaline (pallid ochraceous in the mass [F-G]), ornamented with conical warts (up to 1 µm), some joined by crests but with virtually no network, ellipsoid, 8-11 x 6.5-8 µm. Basidia 4-spored. Cystidia cylindrical or filiform.
Stem white or with pink tinge, more or less equal or slightly clavate towards base, pruinose above, somewhat veined. Ring absent. Flesh white, hard, stuffed or full.
Odour not distinctive, possibly of cedarwood. **Taste** hot after some time.
Chemical tests cap cystidia purple with sulpho-vanillin.
Occurrence summer; rare. More or less limited to Scotland. (Culbin Forest, Morayshire)
■ Inedible.

Russula betularum Hora   Russulaceae

Small to medium-sized agaric with pale delicate pinkish-yellow cap, brittle white gills and stem; solitary or in scattered groups on soil often in damp places, under birch.

Dimensions cap 2 - 5 cm dia; stem 2.5 - 6.5 cm tall x 0.5 - 1 cm dia.
Cap delicate pallid rose or pink with yellowish or buff tinges, sometimes wholly deeper pink; at first flattened-convex, later slightly depressed, smooth or with slight radial veining, margin typically sulcate and with small warty bumps, sticky when damp, cuticle peeling fully. Cystidia cylindrical or clavate. Flesh white, thin, granular and fragile.
Gills white, more or less free, fairly distant. **Spores** hyaline (white in the mass [A]), ornamented with blunt warts (up to 0.7 μm), mostly connected by fine lines in a near complete network, ellipsoid, 8-10 x 7.5-8 μm. Basidia 4-spored. Cystidia cylindrical or filiform.
Stem white, more or less equal or slightly clavate towards base, longer than cap diameter, smooth. Ring absent. Flesh white, brittle, fragile, stuffed.
Odour not distinctive. **Taste** hot.
Chemical tests cap cystidia purple with sulpho-vanillin.
Occurrence summer; frequent.
✣ Poisonous.

Russula brunneoviolacea Crawshay   Russulaceae

Small to medium-sized agaric with violet and brown-tinged cap, cream brittle gills and whitish stem; solitary or in scattered groups on soil under broad-leaf trees, favouring oak.

Dimensions cap 3 - 7 cm dia; stem 3 - 6 cm tall x 0.75 - 1.5 cm dia.
Cap livid violaceous or purple with brown tinge; at first flattened-convex with incurved margin, later somewhat depressed, smooth with striate margin, cuticle peeling 3/4 to centre. Cystidia cylindrical or narrowly clavate. Flesh white, moderately thin and soft.
Gills cream, discolouring brown, almost free, narrow, crowded.
Spores hyaline (cream in the mass [C-E]), ornamented with warts of variable size, occasionally connected by fine lines, broadly ellipsoid, 7-9 x 6-7.5 μm. Basidia 4-spored. Cystidia cylindrical or filiform.
Stem white, discolouring brown, more or less equal or narrowly clavate, smooth. Ring absent. Flesh concolorous, soft, brittle and full.
Odour not distinctive. **Taste** mild.
Chemical tests cap cystidia moderately purple with sulpho-vanillin.
Occurrence summer; infrequent.
☐ Edible.

Russula claroflava Grove   Russulaceae

Largish agaric with bright yellow cap, brittle yellow gills and white stem; solitary or in scattered groups on soil, often in damp places, under birch.

Dimensions cap 4 - 10 cm dia; stem 4 - 10 cm tall x 1 - 2 cm dia.
Cap bright or ochraceous-yellow, discolouring grey with age; at first convex, later flattened and slightly depressed, smooth, slightly sticky when damp, margin finely sulcate in older specimens, cuticle peeling 1/2 to centre. Cystidia absent. Flesh white, moderately thick, granular, soft, slowly greying where damaged.
Gills pallid ochraceous, turning grey where damaged, more or less free or adnexed, forked, fairly crowded. **Spores** hyaline (pallid ochraceous in the mass [F]), ornamented with warts (up to 1 μm), connected by fine lines in an extensive network, ellipsoid, 9-10 x 7.5-8 μm. Basidia 4-spored. Cystidia cylindrical or filiform.
Stem white, reacting as in cap, more or less equal, fairly stout, smooth. Ring absent. Flesh white, grey with age, fairly soft, stuffed.
Odour not distinctive. **Taste** mild or slightly hot.
Chemical tests flesh reddish, then grey with $FeSO_4$; no reaction with sulpho-vanillin.
Occurrence; frequent.
☐ Edible.
Note: *R. claroflava* sometimes appears in spring.

Russula cyanoxantha (Schaeff.) Fr.   Russulaceae

Large agaric, cap colour green to purple, distinctively flexible white gills; solitary or in scattered groups on soil under broad-leaf trees.

Dimensions cap 5 - 15 cm dia; stem 5 - 10 cm tall x 1.5 - 3 cm dia.
Cap variable through purple, wine, olive, green and brown; at first sub-spherical, becoming convex, later flattened and slightly depressed, smooth with faint radiating veins, slightly greasy when damp, cuticle peeling 1/2 to centre. Cystidia small, short, narrow, scattered. Flesh white, thick, granular, fairly hard and brittle.
Gills white or pallid cream, adnexed or slightly decurrent, occasionally forked, flexible (not brittle), narrow, crowded.
Spores hyaline (white in the mass [A]), ornamented with small warts (up to 0.5 μm), occasionally connected, broadly ellipsoid, 6-8 x 5-6 μm. Basidia 4-spored. Cystidia cylindrical or filiform.
Stem white, or occasionally tinged purple, more or less equal or slightly clavate at the base, stout, smooth. Ring absent. Flesh white, fairly firm and stuffed.
Odour not distinctive. **Taste** mild.
Chemical tests flesh greenish with $FeSO_4$ (or no reaction); cap cystidia hardly reacting with sulpho-vanillin.
Occurrence summer to autumn; common.
☐ Edible.

Russula cyanoxantha Var **peltereaui** Sing.   Russulaceae

Large agaric with wholly green cap colour, a variety of *Russula cyanoxantha* and otherwise similar in all respects.
Occurrence summer to autumn; frequent.
☐ Edible.

Russula decolorans (Fr.) Fr.   Russulaceae

Medium or large agaric with distinctively coloured orange-brown cap, pale yellow gills and white stem, all parts greying or blackening; solitary or in scattered groups on soil under coniferous trees.

Dimensions cap 4.5 - 11 cm dia; stem 4.5 - 10 cm tall x 1 - 2.5 cm dia.
Cap reddish, orange-brown, or fulvous, blackening; at first sub-spherical, then flattened-convex with incurved margin, later somewhat depressed, smooth or slightly sticky, cuticle peeling only at margin. Cystidia clavate or cylindrical, septate. Flesh white, greying rapidly when damaged or cut, moderately thick, fleshy and firm.
Gills pallid ochraceous, blackening in cap, adnexed, strongly interveined, narrow, close. **Spores** hyaline (pallid ochraceous in the mass [E-F]), ornamented with warts of varying height, occasionally connected by fine lines, ellipsoid, 9-14 x 7-12 μm. Basidia 4-spored. Cystidia cylindrical or filiform.
Stem white, greying readily when bruised or with age, more or less equal or somewhat clavate, stout, smooth. Ring absent. Flesh concolorous, firm, brittle and full.
Odour not distinctive. **Taste** mild.
Chemical tests cap cystidia purple with sulpho-vanillin.
Occurrence summer to autumn; very localised in Scottish Highlands and other montane regions of Europe.
☐ Edible.

Russula delica Fr.   Russulaceae

Medium-sized agaric with whitish cap discolouring yellowish-brown, brittle whitish gills and white stem also discolouring; solitary or in scattered groups on soil under broad-leaf and coniferous trees.

Dimensions cap 5 - 17 cm dia; stem 2 - 6 cm tall x 2 - 4 cm dia.
Cap whitish or tinged yellowish-brown; at first convex, soon depressed or infundibuliform with inrolled margin, matt, dry. Cystidia cylindrical or filiform. Flesh white, unchanging, thick, granular and brittle.
Gills whitish, sometimes with a slight blue-green tinge towards the stem, slightly decurrent, brittle, somewhat narrow, crowded.
Spores hyaline (white in the mass [A-B]), ornamented with warts (up to 1.5 μm), connected occasionally by fine lines (sometimes more abundant, forming a network), broadly ellipsoid, 8-12 x 7-9 μm. Basidia 4-spored. Cystidia cylindrical or filiform.
Stem white, tinged as cap, stout, more or less equal, smooth. Ring absent. Flesh white, robust and full.
Odour slight, of oil, or fish. **Taste** hot, acrid, bitter.
Chemical tests cap cystidia barely reacting with sulpho-vanillin.
Occurrence late summer to autumn; frequent.
■ Inedible.

Russula densifolia Gill.   Russulaceae

Medium-sized agaric with whitish cap discolouring dirty-brown, brittle creamy-white gills and white stem also discolouring; solitary or in scattered groups on soil under broad-leaf and coniferous trees.

Dimensions cap 3 - 10 cm dia; stem 4 - 9 cm tall x 0.7 - 3 cm dia.
Cap at first whitish, becoming discoloured dirty brown and finally blackish though pallid for a long time towards margin; at first convex, soon depressed and later flattened-depressed, matt, dry. Cystidia narrow, few scattered. Flesh white, slowly discolouring reddish then black, thick, granular and brittle.
Gills whitish becoming pallid cream, adnate or slightly decurrent, brittle, narrow, crowded. **Spores** hyaline (white in the mass [A]), ornamented with small warts (up to 0.5 μm), connected by fine lines in a more or less complete network, sub-spherical or ellipsoid, 7-9 x 6-7 μm. Basidia 4-spored. Cystidia cylindrical or filiform.
Stem white, discolouring as in cap, stout, more or less equal, smooth. Ring absent. Flesh white, reacting as in cap, robust, full.
Odour not distinctive. **Taste** very hot.
Chemical tests flesh pink, then greenish with $FeSO_4$; no reaction with sulpho-vanillin.
Occurrence summer to autumn; frequent.
□ Edible.

Russula emetica (Schaeff.: Fr.) S F Gray   Russulaceae
The Sickener

Medium-sized agaric with distinctive scarlet cap, brittle creamy-white gills and white stem; solitary or in scattered groups on soil specifically under pines.

Dimensions cap 3 - 10 cm dia; stem 4 - 9 cm tall x 0.7 - 2 cm dia.
Cap scarlet or cherry-red, sometimes with more pallid areas; at first convex, later flattened and slightly depressed, slightly sticky when damp, sometimes sulcate at margin on ageing, cuticle peeling fully. Cystidia narrow, clavate. Flesh white, cap colour beneath cuticle, thin, granular and brittle.
Gills whitish, becoming pallid straw, more or less free, brittle, narrow. **Spores** hyaline (white in the mass [A]), ornamented with large warts (up to 1.2 μm), connected by fine lines forming a more or less complete network, ellipsoid, 9-11 x 7.5-8.5 μm. Basidia 4-spored. Cystidia cylindrical or filiform.
Stem white, more or less equal or slightly clavate at base, smooth. Ring absent. Flesh white, fragile and stuffed.
Odour slight, of fruit. **Taste** very hot.
Chemical tests flesh salmon-pink with $FeSO_4$; cap cystidia strongly purple with sulpho-vanillin.
Occurrence summer to autumn; infrequent but locally common.
✣ Poisonous.

Russula emeticella (Sing.) Hora

Smallish agaric with pinkish cap and white gills and stem; solitary or in scattered groups on soil under broad-leaf and, rarely, coniferous trees.

Dimensions cap 2.5 - 6 cm dia; stem 2.5 - 7 cm tall x 1 - 2 cm dia.
Cap pallid pink or darker cherry-red, sometimes more cream at centre; at first convex, later flattened and slightly depressed, slightly shiny or matt, more or less wholly smooth, cuticle peeling fully. Cystidia conical. Flesh white or pinkish under cuticle, thin, granular and brittle.
Gills whitish, adnexed, narrow, interveined, brittle, fairly crowded.
Spores hyaline (white in the mass [A]), ornamented with large warts (up to 1 μm), more or less isolated, broadly ellipsoid, 8.5-10 x 6.5-8 μm. Basidia 4-spored. Cystidia cylindrical or filiform.
Stem white, more or less equal or slightly clavate at base, smooth. Ring absent. Flesh white, soft, very fragile, stuffed.
Odour slight, of coconuts. **Taste** very hot.
Chemical tests flesh salmon with $FeSO_4$; cap cystidia strongly purple with sulpho-vanillin.
Occurrence summer to autumn; frequent.
□ Edible.

Russula erythropus Peltereau   Russulaceae

Medium or large agaric with wine-coloured cap, buff gills and pink-flushed stem; solitary or in scattered groups on soil under coniferous trees.

Dimensions cap 6 - 11 cm dia; stem 4 - 8 cm tall x 1.5 - 3 cm dia.
Cap vinaceous, livid or blood-red, or with olivaceous tinge; at first convex, becoming expanded and later somewhat depressed, smooth, cuticle peeling only at margin. Cystidia sparse, fusiform or cylindrical. Flesh white, moderately thin, granular, at first firm but soon soft.
Gills buff, adnexed, narrow, interveined, close. **Spores** hyaline (pallid ochraceous in the mass [E-F]), ornamented with warts (up to 0.75 μm), occasionally connected by lines, ellipsoid, 8-10 x 7-8 μm. Basidia 4-spored. Cystidia cylindrical or filiform.
Stem white, typically with rosaceous flush, tinged brown with age, more or less equal or slightly clavate, stout, decorated with elongated reticulation. Ring absent. Flesh concolorous, fairly firm, brittle and full.
Odour distinct, of crab. **Taste** mild.
Chemical tests flesh dull green with $FeSO_4$; no reaction with sulpho-vanillin.
Occurrence autumn; infrequent, more likely in northern regions.
□ Edible.

Russula farinipes Romell   Russulaceae

Medium-sized agaric with pale yellowish cap and gills, stem slender and tinged cap colour; solitary or in scattered groups on soil under or close by broad-leaf trees.

Dimensions cap 3 - 6 cm dia; stem 3 - 6 cm tall x 1 - 1.5 cm dia.
Cap pallid ochraceous or ivory, sometimes with brown spotting; at first convex, later flattened and slightly depressed, slightly greasy when damp, margin markedly striate, otherwise smooth, cuticle peeling only at the margin. Cystidia large, fusiform. Flesh white, becoming pallid straw, granular and brittle.
Gills pallid yellowish-buff, adnexed or slightly decurrent, brittle, narrow. **Spores** hyaline (white in the mass [A]), ornamented with small isolated warts (up to 0.5 μm), broadly ellipsoid, 6-8 x 5-7 μm. Basidia 4-spored. Cystidia cylindrical or filiform.
Stem white, flushed straw, fairly slender, smooth or slightly pruinose at apex. Ring absent. Flesh white, becoming pallid straw, brittle and full.
Odour slight, of fruit. **Taste** very hot.
Chemical tests flesh salmon-pink with $FeSO_4$; cap cystidia moderately purple with sulpho-vanillin.
Occurrence late summer to autumn; frequent.
■ Inedible.

Russula fellea (Fr.) Fr.   **Russulaceae**

Medium-sized agaric, straw with paler gills and stem, smelling of geraniums; solitary or in scattered groups on soil specifically under beech.

Dimensions cap 4 - 9 cm dia; stem 2 - 6 cm tall x 1 - 2 cm dia.
Cap straw with buff or ochraceous tinges; at first convex, later flattened and sometimes with broad umbo, slightly sticky when damp, sometimes striate at margin on ageing, cuticle peeling at the margin only. Cystidia abundant, narrow, clavate. Flesh white, fairly thick, granular and brittle.
Gills pallid straw, adnexed, brittle, narrow. **Spores** hyaline (pallid cream in the mass [A-C]), ornamented with warts (up to 0.7 µm), connected by fine lines forming a more or less complete network, ellipsoid, 7.5-9 x 6-7 µm. Basidia 4-spored. Cystidia cylindrical or filiform.
Stem pallid concolorous with cap, more or less equal or tapering slightly upwards, smooth. Ring absent. Flesh white, fairly full.
Odour strong, of geranium. **Taste** very hot.
Chemical tests flesh cream or buff with $FeSO_4$; cap cystidia strongly purple with sulpho-vanillin.
Occurrence late summer to autumn; common.
■ Inedible.

Russula foetens (Pers.: Fr.) Fr.   **Russulaceae**

Medium or large agaric with honey cap, brittle creamy-white gills and white stem; solitary or in scattered groups on soil under broad-leaf or coniferous trees.

Dimensions cap 5 - 12 cm dia; stem 5 - 12 cm tall x 1.5 - 4 cm dia.
Cap honey or ochraceous-brown; at first sub-spherical, later convex and then flattened and sometimes slightly depressed, sticky or viscid when damp, margin striate and with warty bumps, cuticle peeling 1/4 or 1/2 to centre. Cystidia cylindrical, fusiform or tapered. Flesh white with russet tinge where cut, thick, brittle and firm.
Gills cream with brown spotting, adnexed, brittle, distant.
Spores hyaline (cream in the mass [B-D]), ornamented with large warts (up to 1.5 µm), more or less isolated, sub-spherical, 8-10 x 7-9 µm. Basidia 4-spored. Cystidia cylindrical or filiform.
Stem whitish or buff, more or less equal or slightly fusiform, stout, furrowed or wrinkled. Ring absent. Flesh white reacting as in cap, brittle, full but with irregular cavities.
Odour strong, oily or rancid. **Taste** gills hot; stem mild or bitter.
Chemical tests flesh salmon-pink with $FeSO_4$; no reaction with sulpho-vanillin.
Occurrence summer to autumn; common.
■ Inedible.

Russula fragilis (Pers: Fr.) Fr.   **Russulaceae**

Small to medium-sized agaric with pale purplish cap, brittle white gills and white stem; solitary or in scattered groups on soil with broad-leaf and coniferous trees.

Dimensions cap 2 - 6 cm dia; stem 2 - 6 cm tall x 0.5 - 1.5 cm dia.
Cap violaceous, purple or reddish-purple with olivaceous or yellowish tinges at centre; at first convex, later flattened or slightly depressed, sometimes finely striate at margin on ageing, cuticle peeling 1/2 or 3/4 to centre. Cystidia cylindrical or clavate. Flesh white, thin, granular and fragile.
Gills white, adnexed, slightly crenulated edges, brittle, narrow.
Spores hyaline (white in the mass [A-B]), ornamented with warts (up to 0.5 µm), connected by fine lines, sub-spherical, 6-8 µm. Basidia 4-spored. Cystidia cylindrical or filiform.
Stem white, more or less equal or somewhat clavate, smooth. Ring absent. Flesh white, fragile and stuffed.
Odour slight, of fruit. **Taste** very hot.
Chemical tests cap cystidia strongly purple with sulpho-vanillin.
Occurrence autumn; frequent.
■ Inedible.

Russula gracillima Schaeff.   **Russulaceae**

Small to medium-sized agaric with pale pinkish-green cap, brittle creamy gills and whitish stem; solitary or in scattered groups on soil specifically under birch.

Dimensions cap 2 - 6 cm dia; stem 3 - 7 cm tall x 0.5 - 1 cm dia.
Cap pallid pink with greenish or olivaceous tinges at centre; at first convex, later flattened, sometimes with a slight umbo, slightly sticky when damp, sometimes striate at margin on ageing, cuticle peeling 1/3 or 1/2 to centre. Cystidia cylindrical or clavate. Flesh white, thin, granular and brittle.
Gills cream, adnexed or slightly decurrent, brittle, narrow.
Spores hyaline (pale cream in the mass [C-D]), ornamented with large warts (up to 1 μm), very occasionally connected by fine lines, ellipsoid, 7-9 x 5-7 μm. Basidia 4-spored. Cystidia cylindrical or filiform.
Stem white, sometimes flushed greyish-rose, more or less equal, slender, smooth. Ring absent. Flesh white, fragile and stuffed.
Odour slight, of fruit. **Taste** very hot.
Chemical tests flesh salmon-pink with FeSO$_4$; cap cystidia moderately purple with sulpho-vanillin.
Occurrence summer to autumn; infrequent.
■ Inedible.

Russula heterophylla (Fr.) Fr.   **Russulaceae**

Medium or largish agaric with greenish or brownish cap, pale cream gills and white stem; solitary or in scattered groups on soil under broad-leaf trees.

Dimensions cap 5 - 10 cm dia; stem 3 - 6 cm tall x 1 - 3 cm dia.
Cap variable through tan, brown and green with ochraceous tinge; at first sub-spherical, later convex or slightly depressed, smooth with radiating, anastomosing veins, cuticle barely peeling. Cystidia cylindrical, fusiform or clavate. Flesh white, granular and brittle.
Gills white or pallid cream, adnexed, brittle, with slight oily feel, crowded. **Spores** hyaline (white in the mass [A]), ornamented with small warts (up to 0.5 μm), not connected into a network, subspherical or broadly ellipsoid, 5 -7 x 4-6 μm. Basidia 4-spored. Cystidia cylindrical or filiform.
Stem white, tinged brown with age, more or less equal, stout, smooth. Ring absent. Flesh white, brittle and full.
Odour not distinctive. **Taste** mild.
Chemical tests flesh salmon-pink with FeSO$_4$; cap cystidia faintly purple with sulpho-vanillin.
Occurrence late summer to autumn; infrequent.
☐ Edible.
Similar to *R. cyanoxantha v. peltereaui* but FeSO$_4$ reaction differs. Note: possessing the smallest spores in the genus.

Russula laurocerasii Melzer   **Russulaceae**

Largish agaric with ochre-brown cap, brittle creamy-white gills and white stem, smelling of almonds; solitary or in scattered groups on soil under broad-leaf or coniferous trees, often favouring oaks.

Dimensions cap 4 - 8 cm dia; stem 4 - 8 cm tall x 1.5 - 3.5 cm dia.
Cap ochraceous-brown or honey; at first sub-spherical, becoming convex, later flattened and slightly depressed, rather glutinous when damp, markedly striate at margin with small warty bumps, cuticle peeling 1/4 or 1/2 to centre. Cystidia cylindrical, fusiform or tapered. Flesh white, becoming russet where cut, thick, granular, brittle and firm.
Gills cream with brown spotting, adnexed, brittle, fairly narrow, distant. **Spores** hyaline (cream in the mass [B-D]), ornamented with warts (up to 1.2 μm.), connected by fine lines in a more or less complete network and by wings (up to 2 μm), sub-spherical, 8-9.5 x 8-8.5 μm. Basidia 4-spored. Cystidia cylindrical or filiform.
Stem whitish-buff, more or less equal or slightly fusiform, stout, wrinkled. Ring absent. Flesh white, reacting as in cap, brittle, full but with irregular cavities.
Odour strong, of bitter almonds. **Taste** either mild or hot.
Chemical tests flesh salmon-pink with FeSO$_4$; no reaction with sulpho-vanillin.
Occurrence summer to autumn; frequent.
■ Inedible.

Russula lepida Fr. **Russulaceae**

Medium to large agaric with red cap, brittle pale cream gills and white or pinkish stem; solitary or in scattered groups on soil under broad-leaf trees, favouring beech.

Dimensions cap 4 - 10 cm dia; stem 3 - 7 cm tall x 1.5 - 3.5 cm dia.
Cap red, more pallid yellowish in patches; at first convex, later flattened and slightly depressed, matt or slightly pruinose, cuticle barely peeling. Cystidia cylindrical, fusiform or sub-clavate. Flesh white, fairly thick, granular and brittle.
Gills pallid cream, occasionally with reddish tinge at margin, adnexed, brittle, fairly crowded. **Spores** hyaline (pallid cream in the mass [B-C]), ornamented with abundant warts (up to 0.5 μm), connected by ridges and lines into a network, sub-spherical, 8-9 x 7-8 μm. Basidia 4-spored. Cystidia cylindrical or filiform.
Stem white or flushed red, more or less equal or slightly clavate at base, pruinose. Ring absent. Flesh white, hard, fragile, full or stuffed.
Odour slight, of fruit. **Taste** mild, of cedarwood pencils, sometimes slightly bitter.
Chemical tests no reaction with sulpho-vanillin.
Occurrence summer to autumn; frequent.
■ Inedible.

Russula lutea (Huds.: Fr.) S F Gray **Russulaceae**

Medium-sized agaric with distinctive golden-yellow cap, brittle saffron gills and white stem; solitary or in scattered groups on soil under broad-leaf trees.

Dimensions cap 2 - 7 cm dia; stem 2 - 6 cm tall x 0.5 - 1.5 cm dia.
Cap egg- or golden-yellow with occasional apricot tinges, at first convex, later flattened and slightly depressed, slightly shiny or matt, more or less wholly smooth, cuticle peeling 3/4 or fully to centre. Cystidia absent. Flesh white, thin, granular and brittle.
Gills deep saffron-yellow, adnexed, strongly interveined, brittle, fairly crowded. **Spores** hyaline (ochraceous in the mass [H]), ornamented with large warts (up to 1 μm), more or less isolated apart from the occasional fine line, broadly ellipsoid, 7.5-9 x 6-8 μm. Basidia 4-spored. Cystidia cylindrical or filiform.
Stem white, more or less equal or slightly clavate at base, smooth. Ring absent. Flesh white, soft, fragile and stuffed.
Odour slight with age, of fruit, eg. apricots. **Taste** mild.
Chemical tests flesh brownish-pink with $FeSO_4$; no reaction with sulpho-vanillin.
Occurrence summer to autumn; frequent.
□ Edible.

Russula mairei Sing. **Russulaceae**

Medium-sized agaric with distinctive scarlet cap, brittle white gills and stem; solitary or in scattered groups on soil, specifically under beech.

Dimensions cap 3 - 9 cm dia; stem 2.5 - 4.5 cm tall x 1 - 1.5 cm dia.
Cap scarlet or cherry-red, sometimes pallid pink wholly or in part; at first convex, later flattened and slightly depressed, slightly sticky when damp, otherwise matt and smooth, (not striate at margin), cuticle peeling 1/3 to centre. Cystidia clavate. Flesh white, cap coloured beneath cuticle, thick, granular and firm.
Gills whitish with green tinge, becoming cream, adnexed, brittle, rounded in profile, crowded. **Spores** hyaline (whitish in the mass [A]), ornamented with warts (up to 0.5 μm), connected by fine lines forming a more or less complete small-mesh network, ellipsoid, 9-11 x 7.5-8.5 μm. Basidia 4-spored. Cystidia cylindrical or filiform.
Stem white, more or less equal, occasionally slightly clavate at base, smooth. Ring absent. Flesh white and full.
Odour slight when young, of coconut. **Taste** very hot.
Chemical tests flesh salmon-pink with $FeSO_4$; cap cystidia strongly purple with sulpho-vanillin.
Occurrence summer to autumn; common.
✚ Poisonous.

Russula mustelina Fr.   **Russulaceae**

Medium or large agaric with brown cap, cream gills, soon discolouring brown; solitary or in scattered groups on soil under conifers.

Dimensions cap 5 - 10 cm dia; stem 3 - 8 cm tall x 1.5 - 4 cm dia.
Cap ochraceous-brown, darker brown towards centre; convex with strongly incurved or inrolled margin, later somewhat expanded, smooth or slightly sticky, cuticle peeling 1/2 to centre. Cystidia sparse, cylindrical or narrowly clavate. Flesh white, brown when damaged or cut, moderately thick, hard and brittle.
Gills cream or straw, becoming spotted or discoloured brown, adnate or emarginate, forking, soft and flexible, narrow, crowded. **Spores** hyaline (pallid cream in the mass [B-D]), ornamented with small warts (up to 0.4 µm), somewhat connected by fine lines forming a partial network, sub-spherical or broadly ellipsoid, 7-10 x 6-8 µm. Basidia 4-spored. Cystidia cylindrical or filiform.
Stem whitish, staining brown, more or less equal, stout, pruinose near the apex, otherwise smooth. Ring absent. Flesh concolorous, hard, brittle, with cavities.
Odour not distinctive. **Taste** mild.
Chemical tests flesh brownish-pink with $FeSO_4$; no reaction with sulpho-vanillin.
Occurrence summer to autumn; infrequent and localised to highland regions.
■ Inedible.

Russula nigricans (Bull.) Fr.   **Russulaceae**

Large agaric with distinctive blackening cap and thick, distant gills; solitary or in scattered groups on soil under broad-leaf and coniferous trees.

Dimensions cap 5 - 20 cm dia; stem 3 - 8 cm tall x 1 - 4 cm dia.
Cap at first dirty-white, discolouring brown and then black; at first convex with incurved margin, later flattened and depressed, dry, smooth, cuticle peeling 3/4 to centre. Cystidia absent. Flesh white, becoming greyish when exposed to air and finally blackening, thick.
Gills straw becoming grey with pinkish tinge and finally black, adnate, thick, brittle, distant. **Spores** hyaline (white in the mass [A]), ornamented with small warts (up to 0.5 µm), connected by fine lines in a partly complete network, sub-spherical or broadly ellipsoid, 7-8 x 6-7 µm. Basidia 4-spored. Cystidia cylindrical or filiform.
Stem white, changing as in cap, more or less equal or tapering slightly downwards, smooth. Ring absent. Flesh white changing as in cap, brittle and full.
Odour slight, of fruit. **Taste** slowly hot.
Chemical tests flesh pink, then greenish, with $FeSO_4$; no reaction with sulpho-vanillin.
Occurrence summer to autumn; common.
☐ Edible but poor.

Russula nitida (Pers.: Fr.) Fr.   **Russulaceae**

Smallish agaric with variable purplish cap, brittle cream gills and white or faintly purplish stem; solitary or in scattered groups on soil in damp areas under birch.

Dimensions cap 2 - 6 cm dia; stem 2 - 9 cm tall x 0.5 - 2 cm dia.
Cap very variable, but essentially purplish with red, grey, buff or green tinges; at first convex, later flattened and slightly depressed, striate at margin, cuticle peeling 1/2 or 2/3 to centre. Cystidia abundant, cylindrical or clavate. Flesh white, thin, granular and brittle.
Gills cream or straw, sometimes with red tinge on edges, adnexed, brittle, somewhat distant, interveined. **Spores** hyaline (pallid ochraceous in the mass [E-G]), ornamented with spines (up to 1.5 µm), with very limited connections, ellipsoid, 8 -11 x 6-9 µm. Basidia 4-spored. Cystidia cylindrical or filiform.
Stem white, or tinged cap colour, more or less equal or clavate, smooth. Ring absent. Flesh white, fragile and stuffed.
Odour not distinctive. **Taste** mild or slightly hot.
Chemical tests cap cystidia strongly purple with sulpho-vanillin.
Occurrence late summer to autumn; frequent.
☐ Edible.

Russula ochroleuca (Pers.: Sécr.) Fr.   Russulaceae

Medium-sized agaric with ochre-yellow cap, brittle creamy-white gills and white stem; solitary or in scattered groups on soil under broad-leaf and coniferous trees.

Dimensions cap 4 - 10 cm dia; stem 4 - 7 cm tall x 1.5 - 2.5 cm dia.
Cap ochraceous, occasionally with greenish tinge; at first convex, later flattened and slightly depressed, slightly sticky when damp, striate at margin on ageing, cuticle peeling 2/3 to centre. Cystidia absent. Flesh white, thin, granular and brittle.
Gills cream, adnexed, brittle, narrow. **Spores** hyaline (pallid cream in the mass [A]), ornamented with large warts (up to 1.2 μm.), connected by fine lines in a complete network, ellipsoid, 8 -10 x 7-8 μm. Basidia 4-spored. Cystidia cylindrical or filiform.
Stem white, greying slightly with age, more or less equal or tapering slightly upwards, smooth. Ring absent. Flesh white, fragile and stuffed.
Odour not distinctive. **Taste** mild or fairly hot.
Chemical tests flesh salmon-pink with FeSO$_4$; no reaction with sulpho-vanillin.
Occurrence late summer to autumn; common.
☐ Edible.

Russula paludosa Britz.   Russulaceae

Medium-sized or large agaric with red cap, brittle cream-yellow gills and whitish stem; solitary or in scattered groups on acid or moorland soil with coniferous trees.

Dimensions cap 5 - 14 cm dia; stem 5 - 14 cm tall x 1 - 3 cm dia.
Cap variable, reddish with apricot, bay, blood or scarlet tinges; at first convex, later flattened and depressed, shiny or somewhat sticky when damp, cuticle peeling 1/2 or 3/4 to centre. Cystidia sparse, cylindrical, non-septate. Flesh white, moderate, granular, and brittle.
Gills cream or pallid ochraceous, adnexed, brittle, narrow, strongly interveined. **Spores** hyaline (ochraceous-cream in the mass [E-F]), ornamented with large warts (up to 1.25 μm), connected by fine lines in an incomplete network, ellipsoid, 8-10 x 7-8 μm. Basidia 4-spored. Cystidia cylindrical or filiform.
Stem white or tinged pink, more or less equal or tapering slightly upwards, smooth. Ring absent. Flesh white, fragile and stuffed.
Odour not distinctive. **Taste** mild.
Chemical tests cap cystidia moderately purple with sulpho-vanillin.
Occurrence late summer to autumn; probably restricted to Scottish Highlands in Britain, montane or moorland regions elsewhere in Europe.
☐ Edible.

Russula parazurea Schaeff.   Russulaceae

Medium-sized agaric with grey-blue cap, pale gills and stem; solitary or in scattered groups on soil under broad-leaf trees.

Dimensions cap 3 - 8 cm dia; stem 3 - 7 cm tall x 1 - 2 cm dia.
Cap greyish blue; at first convex, later flattened and slightly depressed, typically matt and finely pruinose, cuticle peeling 1/2 or 3/4 to centre. Cystidia cylindrical or tapered. Flesh white, moderate, granular and brittle.
Gills at first whitish, becoming pallid buff, occasionally forked, adnexed, brittle, narrow. **Spores** hyaline (cream in the mass [C-D]), ornamented with small warts (up to 0.5 μm), mostly connected by lines forming a network, ellipsoid, 6-8.5 x 5-6.5 μm. Basidia 4-spored. Cystidia cylindrical or filiform.
Stem white, more or less equal, stout, smooth. Ring absent. Flesh white, brittle and stuffed.
Odour not distinctive. **Taste** mild.
Chemical tests no reaction with FeSO$_4$; cap cystidia moderately purple with sulpho-vanillin.
Occurrence late summer to autumn; infrequent.
☐ Edible.

Russula pectinata (Bull.) Fr.   **Russulaceae**

Medium-sized agaric with yellowish cap, cream gills and pale stem; solitary or in scattered groups on soil under or close to broad-leaf trees, favouring oak but also under conifers.

Dimensions cap 3 - 6 cm dia; stem 3 - 5 cm tall x 1 - 2 cm dia.
Cap ochraceous or straw, often with rust spotting; at first convex, later flattened and slightly depressed, slightly sticky when damp, markedly striate at margin with small denticular warts, cuticle peeling 2/3 to centre. Cystidia cylindrical or tapered, occasionally with apical appendage. Flesh white, thin, granular and brittle.
Gills at first whitish, becoming cream, adnexed, brittle, narrow.
Spores hyaline (yellowish in the mass [D-F]), ornamented with small warts (up to 0.5 µm), connected occasionally in pairs by fine lines, broadly ellipsoid, 8-9 x 5.5-6 µm. Basidia 4-spored. Cystidia cylindrical or filiform.
Stem whitish or pallid cap colour, more or less equal, stout, smooth. Ring absent. Flesh white, brittle and stuffed.
Odour slight, of oil or fish. **Taste** of oil.
Chemical tests flesh salmon-pink with $FeSO_4$; cap cystidia faintly purple with sulpho-vanillin.
Occurrence late summer to autumn; infrequent.
■ Inedible.

Russula pectinatoides Peck ss. Sing.   **Russulaceae**

Medium-sized agaric with yellowish cap, cream gills and pale stoutish stem; solitary or in scattered groups on soil under or close to broad-leaf trees.

Dimensions cap 3 - 8 cm dia; stem 3 - 5 cm tall x 1 - 2 cm dia.
Cap ochraceous or straw, often with rust spotting; at first convex, later flattened and slightly depressed, slightly greasy (not sticky) when damp, markedly striate at margin with small denticular warts, cuticle peeling 2/3 to centre. Cystidia cylindrical or tapered, occasionally with apical appendage. Flesh white, thin, granular and brittle.
Gills at first whitish, becoming cream, adnexed, brittle, narrow.
Spores hyaline (yellowish in the mass [D-F]), ornamented with small warts (up to 0.5 µm), connected occasionally in pairs by fine lines, broadly ellipsoid, 6.5-8 x 5-6 µm. Basidia 4-spored. Cystidia cylindrical or filiform.
Stem whitish or pallid cap colour but with rust spotting and more general rust at base, more or less equal, stout, smooth. Ring absent. Flesh white, brittle and stuffed.
Odour slight, of fruit. **Taste** not distinctive.
Chemical tests flesh salmon pink with $FeSO_4$; cap cystidia faintly purple with sulpho-vanillin.
Occurrence late summer to autumn; infrequent.
■ Inedible. Note: similar to *R. pectinata* but taste differs.

Russula puellaris Fr.   **Russulaceae**

Smallish agaric with pale purple or wine-red cap, cream gills and pale stem; solitary or in scattered groups on soil, specifically under broad-leaf or coniferous trees.

Dimensions cap 3 - 5 cm dia; stem 2 - 6 cm tall x 0.5 - 1.5 cm dia.
Cap variable, ranging from vinaceous to brick but also washed out, typically darker at centre; at first convex, later flattened and slightly depressed, margin sulcate or pectinate, cuticle peeling 1/2 or 2/3 to centre. Cystidia mainly clavate, also cylindrical. Flesh white, thin, granular and brittle bruising pallid ochraceous.
Gills at first whitish, becoming pallid ochraceous, adnexed, brittle, narrow. **Spores** hyaline (cream in the mass [D-E]), ornamented with blunt warts (up to 1.2 µm), more or less isolated, broadly ellipsoid, 6.5-9 x 5.5-7 µm. Basidia 4-spored. Cystidia cylindrical or filiform.
Stem whitish or with ochraceous tinge in older specimens, with elongate reticulations, more or less equal. Ring absent. Flesh white, brittle and stuffed, reacting as in cap.
Odour not distinctive. **Taste** mild.
Chemical tests flesh salmon-pink with $FeSO_4$; cap cystidia strongly purple with sulpho-vanillin.
Occurrence late summer to autumn; frequent.
☐ Edible.

Russula puellula Ebb., Möll & Schaeff.   **Russulaceae**

Smallish agaric with red cap, cream gills and white stem; solitary or in scattered groups on soil, specifically with beech, often on dryish soil.

Dimensions cap 3 - 7 cm dia; stem 3 - 5 cm tall x 1 - 1.5 cm dia.
Cap typically red with brick, pink or purplish tinges; at first convex, later flattened and slightly depressed, smooth, cuticle peeling 1/2 to centre. Cystidia cylindrical or narrowly clavate, septate. Flesh white, medium and granular.
Gills at first whitish, becoming cream, adnexed, brittle, narrow.
Spores hyaline (cream in the mass [C-D]), ornamented with blunt warts (up to 0.7 μm), occasionally connected by lines or ridges, broadly ellipsoid, 6-9.5 x 5-7.5 μm. Basidia 4-spored. Cystidia cylindrical or filiform.
Stem whitish, more or less equal, smooth. Ring absent. Flesh white, brittle and stuffed.
Odour not distinctive. **Taste** mild.
Chemical tests flesh salmon pink with $FeSO_4$; cap cystidia weakly purple with sulpho-vanillin.
Occurrence late summer to autumn; rare. (Stourhead, Wiltshire)
■ Inedible.

Russula pulchella Borszcow   **Russulaceae**
= *Russula depallens* (Pers.) Fr.

Medium-sized agaric with pinkish-green cap, cream gills and pale stem; solitary or in scattered groups on soil, specifically under or close to birch.

Dimensions cap 4 - 9 cm dia; stem 3 - 5 cm tall x 1 - 2 cm dia.
Cap typically from rose to vinaceous but also buff and with greenish tinge at centre, paling almost wholly after rain; at first convex, later flattened and slightly depressed, slightly sticky when damp, cuticle peeling 1/2 to centre. Cystidia cylindrical, fusiform or narrowly clavate. Flesh white, thickish at centre otherwise thin, granular and brittle.
Gills at first whitish, becoming cream, adnexed, brittle, narrow.
Spores hyaline (deep cream in the mass [E-F]), ornamented with blunt warts (up to 0.7 μm), partly connected by lines or ridges in an incomplete network, broadly ellipsoid, 8-10 x 6-7 μm. Basidia 4-spored. Cystidia cylindrical or filiform.
Stem whitish or with grey tinge in older specimens, more or less equal, stout, smooth. Ring absent. Flesh white, brittle and stuffed.
Odour not distinctive. **Taste** fairly hot.
Chemical tests flesh salmon-pink with $FeSO_4$; cap cystidia moderately purple with sulpho-vanillin.
Occurrence late summer to autumn; infrequent.
■ Inedible.

Russula queletii Fr.   **Russulaceae**

Medium to large agaric with purplish-red cap and stem, pale cream gills; solitary or in scattered groups on soil (favouring calcareous), generally under pine or spruce.

Dimensions cap 4 - 10 cm dia; stem 3 - 7 cm tall x 1 - 1.5 cm dia.
Cap purplish or brownish-red, discolouring readily after rain or with age; at first convex, later flattened and slightly depressed, margin sulcate otherwise smooth, cuticle peeling 1/2 or 2/3 to centre. Cystidia cylindrical often with processes. Flesh white, becoming pallid straw, granular and brittle.
Gills pallid cream, spotting greenish where damaged, adnexed, brittle, narrow. **Spores** hyaline (cream in the mass [C-E]), ornamented with small isolated warts (up to 0.5 μm), broadly ellipsoid, 8-10 x 7-9 μm. Basidia 4-spored. Cystidia cylindrical or filiform.
Stem carmine red, more or less clavate, smooth or with fine reticulations. Ring absent. Flesh white, becoming pallid straw, brittle and full.
Odour slight, of fruit (apples). **Taste** very hot.
Chemical tests flesh salmon pink with $FeSO_4$; cap cystidia strongly purple with sulpho-vanillin.
Occurrence summer to autumn; infrequent.
■ Inedible.

Russula rosea Quél.   Russulaceae

= *Russula aurora* Krombh. ss. Melz. & Zv.

Small to medium-sized agaric with delicate flesh-pink cap, white gills and stem; solitary or in scattered groups on soil, specifically under broad-leaf trees.

Dimensions cap 4 - 9 cm dia; stem 4 - 7 cm tall x 1 - 2 cm dia.
Cap flesh pink with more cream centre, less frequently reddish; at first convex, later flattened, margin tending to remain incurved, smooth but margin faintly striate in older specimens, cuticle peeling 1/2 to centre. Cystidia absent. Flesh white, granular and brittle.
Gills at first white, becoming very pallid cream, adnexed, brittle, forked, narrow. **Spores** hyaline (very pallid cream in the mass [B]), ornamented with small warts (up to 0.5 μm), partly connected by fine lines and ridges forming an incomplete network, broadly ellipsoid, 6-8 x 5-6.5 μm. Basidia 4-spored. Cystidia cylindrical or filiform.
Stem white, smooth but faintly pruinose, particularly near the apex. Ring absent. Flesh white, solid and brittle.
Odour slight, of fruit. **Taste** mild.
Chemical tests flesh salmon-pink with $FeSO_4$; stem carmine with sulpho-vanillin on dried material; no reaction with sulpho-vanillin on cap.
Occurrence late summer to autumn; locally frequent.
■ Inedible.
Very similar to *R. minutula* but larger.

Russula sanguinea (Bull.) Fr.   Russulaceae

Medium to large agaric with purplish-red cap, cream gills and stem flushed cap colour; solitary or in scattered groups on soil, specifically under coniferous trees favouring pine.

Dimensions cap 4 - 10 cm dia; stem 4 - 10 cm tall x 1 - 3 cm dia.
Cap purplish-red, sometimes with whitish areas; at first convex, later flattened and slightly depressed, dry, smooth or finely sulcate, cuticle peeling marginally. Cystidia cylindrical or subclavate, sometimes capitate. Flesh white, granular and brittle.
Gills at first cream, then pallid ochraceous, adnexed or more or less decurrent, brittle, forked, narrow. **Spores** hyaline (cream in the mass [C-F]), ornamented with small warts (up to1 μm), partly connected by fine lines and ridges forming an incomplete network, broadly ellipsoid, 7-9 x 6-8 μm. Basidia 4-spored. Cystidia cylindrical or filiform.
Stem flushed cap colour, more or less equal or fusiform, reddish, pruinose and longitudinally veined. Ring absent. Flesh white, brittle and full.
Odour faint, of fruit. **Taste** somewhat hot.
Chemical tests no reaction with $FeSO_4$; cap cystidia barely purple with sulpho-vanillin.
Occurrence summer to autumn; infrequent.
■ Inedible.

Russula sardonia Fr.   Russulaceae

Medium to large agaric with purplish-red cap, primrose gills and stem tinged cap colour; solitary or in scattered groups on soil, specifically under pines.

Dimensions cap 4 - 10 cm dia; stem 3 - 8 cm tall x 1 - 1.5 cm dia.
Cap purplish or brownish-red with greenish or ochraceous tinges; at first convex, later flattened and slightly depressed, slightly sticky when damp, smooth, cuticle peeling marginally. Cystidia narrowly cylindrical or fusiform. Flesh white, becoming pallid straw, granular and brittle.
Gills at first primrose, becoming pallid golden yellow, adnexed or more or less decurrent, brittle, forked, narrow sometimes weeping.
Spores hyaline (cream in the mass [C-F]), ornamented with small warts (up to 0.5 μm), partly connected by fine lines and ridges forming an incomplete network, broadly ellipsoid, 7-9 x 6-8 μm. Basidia 4-spored. Cystidia cylindrical or filiform.
Stem white, flushed greyish-rose or lilac, more or less clavate, smooth. Ring absent. Flesh white, becoming pallid straw, brittle and full.
Odour slight, of fruit. **Taste** very hot.
Chemical tests flesh salmon-pink with $FeSO_4$; cap cystidia strongly purple with sulpho-vanillin; flesh and gills rose pink with NH_3.
Occurrence summer to autumn; locally frequent.
■ Inedible.

Russula solaris Ferd. & Winge   **Russulaceae**

Medium-sized agaric with pale yellow cap and gills and white stem; solitary or in scattered groups on soil, specifically under beech.

Dimensions cap 2 - 7 cm dia; stem 3 - 6 cm tall x 0.8 - 1.5 cm dia.
Cap pallid yellow or straw, more pallid towards margin; at first convex, later flattened and slightly depressed, slightly sticky when damp, finely sulcate at the margin on ageing, cuticle peeling 2/3 to centre. Cystidia cylindrical or clavate. Flesh white, thin, granular and brittle.
Gills pallid straw, adnexed, brittle, narrow. **Spores** hyaline (yellowish-cream in the mass [D-E]), ornamented with small warts (up to 0.5 µm.), connected occasionally in pairs by fine lines, broadly ellipsoid, 7-8 x 5-7 µm. Basidia 4-spored. Cystidia cylindrical or filiform.
Stem whitish or pallid cap colour, more or less equal, smooth. Ring absent. Flesh white, stuffed and fragile.
Odour slight, of fruit. **Taste** hot.
Chemical tests flesh salmon-pink with $FeSO_4$; cap cystidia purple with sulpho-vanillin.
Occurrence late summer to autumn; rare. (Bridgham near Thetford, Norfolk)
✛ Probably poisonous.

Russula sororia (Fr.) Romell   **Russulaceae**

Small to medium-sized agaric with pale brown cap and whitish gills and stem; solitary or in scattered groups on soil, specifically under oak.

Dimensions cap 3 - 6 cm dia; stem 3 - 6 cm tall x 1 - 2 cm dia.
Cap pallid sepia-brown sometimes with greyish tinge and typically more pallid towards margin; at first convex, later flattened and slightly depressed, slightly sticky when damp, sulcate at margin with warty bumps, cuticle peeling 1/2 to centre. Cystidia tapered. Flesh white, thin, granular and brittle.
Gills dirty white or cream, adnexed, brittle, narrow. **Spores** hyaline (cream in the mass [B-D]), ornamented with warts (up to 0.7 µm), connected occasionally by fine lines, sub-spherical 7-9 x 5-7 µm. Basidia 4-spored. Cystidia cylindrical or filiform.
Stem whitish, more or less equal, stout, smooth. Ring absent. Flesh white, stuffed and fragile.
Odour rancid or of cheese. **Taste** oily becoming slowly hot.
Chemical tests flesh salmon-pink with $FeSO_4$; cap cystidia faintly purple with sulpho-vanillin; stem pale rose with formalin.
Occurrence late summer to autumn; infrequent.
■ Inedible.

Russula turci Bres.   **Russulaceae**

Variable-sized agaric with dark plum-coloured cap, brittle yellow gills and white stem; solitary or in scattered groups on soil with conifers.

Dimensions cap 3 - 10 cm dia; stem 3 - 7 cm tall x 1 - 2.5 cm dia.
Cap livid purple or dark vinaceous, typically darker at centre, sometimes with brown tinge, or more pallid; at first convex then more flattened, later somewhat depressed, viscid when moist, otherwise matt or slightly pruinose, cuticle peeling 1/3 to centre. Cystidia absent. Flesh white, moderately thick, granular and brittle.
Gills saffron, adnexed, strongly interveined, close. **Spores** hyaline (pallid ochraceous in the mass [G]), ornamented with small warts (up to 0.4 µm), largely connected by lines forming a more or less complete network, sub-spherical or broadly ellipsoid, 7-9 x 6-8 µm. Basidia 4-spored. Cystidia cylindrical or filiform.
Stem white, occasionally with rosaceous tinge, more or less equal or slightly clavate, stout, faintly pruinose. Ring absent. Flesh concolorous, fairly firm, brittle and full.
Odour of iodoform at stem base (not always obvious). **Taste** mild.
Chemical tests no reaction with $FeSO_4$ or sulpho-vanillin.
Occurrence summer to autumn; common in Scottish Highlands and other montane regions of Europe, rare elsewhere.
☐ Edible.

Russula versicolor J Schaff.   **Russulaceae**

Smallish agaric typically pale with purplish cap, cream gills and whitish stem; solitary or in scattered groups on soil specifically under or close to birch.

Dimensions cap 2 - 5 cm dia; stem 2 - 5 cm tall x 1 - 1.5 cm dia
Cap purplish-violet with variable reddish, yellowish, olivaceous or brown tinges; first convex, becoming flattened-convex and later slightly depressed, at first smooth then with sulcate margin, slightly sticky when damp, cuticle peeling to centre. Cystidia absent. Flesh white, thin and brittle.
Gills pallid cream, adnexed, brittle, narrow, crowded.
Spores hyaline (ochraceous-cream in the mass [E-F]), with variable ornamentation of warts and ridges, broadly ellipsoid, 6-9 x 4-7 µm. Basidia 4-spored. Cystidia cylindrical or filiform.
Stem whitish, discolouring yellow, more or less equal, stout, smooth. Ring absent. Flesh white, stuffed and brittle.
Odour not distinctive. **Taste** mild or fairly hot.
Chemical tests flesh salmon-pink with $FeSO_4$; cap cystidia strongly purple with sulpho-vanillin.
Occurrence summer to autumn; frequent.
■ Inedible.

Russula vesca Fr.   **Russulaceae**

Medium or large agaric with variable violet to buff cap, cuticle shrinking from the margin, pale cream gills and whitish stem; solitary or in scattered groups on soil under broad-leaf trees.

Dimensions cap 5 - 10 cm dia; stem 3 - 10 cm tall x 1.5 - 2.5 cm dia.
Cap variable from violaceous to buff, generally pallid, sometimes with greenish tinge; first convex, becoming flattened-convex and later slightly depressed, smooth, cuticle peeling 1/2 to centre and retracting from margin. Cystidia cylindrical or fusiform. Flesh white, medium and firm.
Gills pallid cream, adnexed, narrow, forked near the stem and slightly interveined, crowded. **Spores** hyaline (white in the mass [A]), with ornamentation of small warts occasionally joined by ridges, broadly ellipsoid, 6-8 x 5-6 µm. Basidia 4-spored. Cystidia cylindrical or filiform.
Stem whitish sometimes with violaceous tinge, more or less equal, smooth. Ring absent. Flesh white, stuffed, firm and brittle.
Odour not distinctive. **Taste** mild, nutty.
Chemical tests flesh rapidly deep salmon-pink with $FeSO_4$; cap cystidia barely purple with sulpho-vanillin.
Occurrence summer to autumn; frequent.
☐ Edible.

Russula veternosa Fr.   **Russulaceae**

Medium or large agaric with pinkish-red cap, egg-yellow gills and whitish stem; solitary or in scattered groups on soil under beech.

Dimensions cap 3 - 10 cm dia; stem 2 - 7 cm tall x 1 - 2 cm dia.
Cap pink or coral-red, tinged ochraceous or buff, particularly at centre; convex, becoming flattened and later slightly depressed, smooth, cuticle peeling1/2 or 3/4 to centre. Cystidia cylindrical or clavate. Flesh white, medium and firm.
Gills ochraceous, adnexed, narrow, forked, interveined, brittle, crowded. **Spores** hyaline (ochraceous in the mass [G-H]), with spines up to 1.5 µm, virtually no connections, broadly ellipsoid, 7-9 x 6-8 µm. Basidia 4-spored. Cystidia cylindrical or filiform.
Stem whitish, more or less equal, smooth. Ring absent. Flesh white, stuffed and brittle.
Odour faint, of honey. **Taste** mild or fairly hot.
Chemical tests cap cystidia strongly purple with sulpho-vanillin.
Occurrence autumn; infrequent.
■ Inedible.

Russula violacea Quél. **Russulaceae**

Smallish agaric with violaceous or greenish cap, pale cream gills and whitish stem, smelling of *Pelargonium*; solitary or in scattered groups on soil under broad-leaf trees.

Dimensions cap 2 - 5 cm dia; stem 2 - 5 cm tall x 1 - 1.5 cm dia.
Cap violaceous with green or brown tinges; first convex, becoming flattened-convex and later slightly depressed, at first smooth then with striate margin, slightly sticky when damp, cuticle peeling 1/4 to centre. Cystidia cylindrical or clavate. Flesh white, thin and brittle.
Gills pallid cream, adnexed, brittle, narrow, crowded.
Spores hyaline (cream in the mass [B-D]), ornamented with spines which may be joined by ridges, broadly ellipsoid, 7.5-8.5 x 6-7 μm. Basidia 4-spored. Cystidia cylindrical or filiform.
Stem whitish, more or less equal, smooth. Ring absent. Flesh white, stuffed and brittle.
Odour of pelargonium. **Taste** hot.
Chemical tests flesh salmon-pink with $FeSO_4$; cap cystidia purple with sulpho-vanillin; no reaction with NH_3.
Occurrence summer to early autumn; infrequent.
■ Inedible.
Note several varieties are recognised.

Russula violeipes Quél. **Russulaceae**

Medium-sized agaric with yellowish-green and purplish cap, yellow gills and stem flushed purple; solitary or in scattered groups on soil under broad-leaf trees.

Dimensions cap 4 - 8 cm dia; stem 4 - 7 cm tall x 1 - 3 cm dia.
Cap straw or yellowish with green, olive or purple tinges at the margin; at first sub-spherical, becoming flattened-convex and later slightly depressed, slightly pruinose, otherwise smooth, cuticle peeling minimally. Cystidia absent. Flesh white, thick, hard and brittle.
Gills pallid straw or buff, slightly decurrent, brittle, narrow, greasy, typically forked near the stem, crowded. **Spores** hyaline (cream in the mass [C-D]), ornamented with warts (up to 1 μm), connected by lines or ridges forming a well-developed network, sub-spherical, 6.5-9 x 6-8 μm. Basidia 4-spored. Cystidia cylindrical or filiform.
Stem pallid cap colour flushed violet or purple, farinose near the apex, more or less equal but tapering at base, stout, smooth. Ring absent. Flesh white, stuffed and brittle.
Odour slight, of shrimps. **Taste** mild.
Chemical tests flesh salmon-pink with $FeSO_4$, brownish with phenol; no reaction with sulpho-vanillin on cap.
Occurrence summer to early autumn; infrequent.
☐ Edible.

Russula virescens (Schaeff.: Zant.) Fr. **Russulaceae**

Medium or large agaric with pale green cap covered with darker patches, cream gills and stem; solitary or in scattered groups on soil under broad-leaf trees favouring beech.

Dimensions cap 5 - 12 cm dia; stem 4 - 9 cm tall x 2 - 4 cm dia.
Cap dull or verdigris-green to almost cream, decorated with darker verdigris green floccose patches; at first convex, becoming flattened-convex and later slightly lobed, cracking into darker patches, cuticle peeling 1/2 to centre. Cystidia absent. Flesh white, thick and brittle.
Gills cream, adnate, brittle, interveined, close. **Spores** hyaline (white in the mass [A-B]), ornamented with warts (up to 0.5 μm), either unconnected or connected by lines forming a network, sub-spherical, 7-9 x 6-7 μm. Basidia 4-spored. Cystidia cylindrical or filiform.
Stem creamy-white, browning slightly, farinose near apex, otherwise smooth, stout, more or less equal. Ring absent. Flesh white, stuffed and brittle.
Odour not distinctive. **Taste** mild.
Chemical tests almost no reaction with sulpho-vanillin on cap.
Occurrence summer to early autumn; infrequent.
☐ Edible.

Russula xerampelina (Schaeff.) Fr.   **Russulaceae**

Large agaric with reddish-purple cap, ochre gills and whitish stem; solitary or in scattered groups on soil under broad-leaf trees, favouring beech and oak.

Dimensions cap 5 - 14 cm dia; stem 3 - 11 cm tall x 1 - 3 cm dia.
Cap reddish but frequently tinged with purple, cinnamon or straw; at first convex, later flattened and slightly depressed, slightly sticky when damp otherwise dry and matt, sulcate at margin on ageing, cuticle peeling barely 1/4 to centre. Cystidia mainly clavate, also cylindrical, tapering or fusiform. Flesh white, fairly firm, thickish and brittle.
Gills pallid ochraceous, adnexed, thick, brittle, broad. **Spores** hyaline (deep yellowish-cream in the mass [E-F]), ornamented with largish warts (up to 1.2 μm) connected by occasional fine lines, ellipsoid, 8-11 x 6.5-9 μm. Basidia 4-spored. Cystidia cylindrical or filiform.
Stem whitish, sometimes tinged cap colour and bruising brownish, more or less equal or tapering upwards, smooth or slightly veined. Ring absent. Flesh white, partly stuffed and hard.
Odour of fish or crabs. **Taste** mild.
Chemical tests flesh dull green with $FeSO_4$; no reaction with sulpho-vanillin on cap.
Occurrence late summer to early autumn; common.
☐ Edible.

Russula xerampelina var **olivascens** (Schaeff.) Fr.
Russulaceae

Similar to *R. xerampelina* in habitat, general characteristics, spore size and chemical reactions but with predominantly greyish-green cap colour; solitary or in scattered groups on soil under broad-leaf trees, favouring beech and oak.

Lentinus lepideus (Fr.: Fr.) Fr.   **Lentinaceae**
= *Panus lepideus* (Fr.: Fr.) Corn.

Medium or large whitish scaly cap on slender stem; solitary or clustered on dead, often treated, conifer wood.

Dimensions cap 5 - 12 cm dia; stem 3 - 8 cm tall x 1 - 2 cm dia.
Cap whitish-cream decorated with brown scales concentrically arranged; at first convex, becoming flattened or depressed, margin regular, remaining incurved. Flesh white, thin, at first soft then tough.
Gills at first white, becoming ochraceous, finely toothed, adnate-decurrent down to a ring-like zone, broad, close. **Spores** hyaline, smooth, elongated-ellipsoid or cylindrical, non-amyloid, occasionally with droplets, 7.5-12 x 3-4.5 μm. Basidia 2- or 4-spored. Cystidia-like gill edge cells cylindrical.
Stem concolorous cap with white membraneous velar remnants towards apex when young but increasingly brown towards base, decorated with fine brown scales, typically in belt-like rings, slender, more or less equal. Ring white, fibrous, sub-apical. Flesh white, elastic or firm and full.
Odour faint, of aniseed. **Taste** not distinctive.
Chemical tests none.
Occurrence late summer to early autumn; rare in Britain, more wide-spread in Europe. (New Forest, Hampshire)
■ Inedible.
Specimen shown is atypical in that the cap is disproportionately small.

Lentinus tigrinus (Bull.: Fr.) Fr.   **Lentinaceae**

Medium-sized, whitish, scaly, shell-shaped cap on slender often eccentric stem; in small clusters on rotting wood of willow and, less commonly, poplar.

Dimensions cap 3 - 7 cm dia; stem 3 - 5 cm tall x 0.4 - 0.6 cm dia.
Cap white, covered with fine brownish scales concentrically arranged; at first convex, becoming flattened and usually depressed at the point of attachment, margin regular. Flesh white, thin, at first weakly elastic then more leathery.
Gills at first white, becoming ochraceous, finely toothed, decurrent, narrow, close or somewhat distant. **Spores** hyaline, smooth, elongated-ellipsoid or cylindrical, non-amyloid, with droplets, 6-8 x 3-3.5 μm. Basidia 4-spored. Cystidia-like gill edge cells cylindrical.
Stem concolorous with cap, with fine, blackish, cobwebby scales, typically lateral or eccentric, slender, more or less equal. Sometimes with spherical ring-zone when young. Ring absent. Flesh white, firm, tough and full.
Odour not distinctive or slightly acidic. **Taste** not distinctive.
Chemical tests none.
Occurrence summer; rare. (Welney near Wisbech, Cambs)
■ Inedible.

Lentinus torulosus (Pers.: Fr.) Lloyd   **Lentinaceae**
= *Panus torulosus* (Pers.: Fr.) Fr.
= *Pleurotus conchatus* (Bull.: Fr.) Pil.

Medium-sized, lilac or yellowish-brown, funnel-shaped cap with pale yellow gills and short thick eccentric stem; in small, dense clusters on stumps and other wood of broad-leaf trees.

Dimensions cap 4 - 8 cm dia; stem 1 - 2 cm tall x 1 - 2 cm dia.
Cap ochraceous-brown with violaceous tinge when young; at first convex, soon becoming concave and infundibuliform with somewhat irregular margin, at first smooth, becoming cracked with age. Flesh pallid and thin.
Gills at first lilaceous or buff, becoming ochraceous, deeply decurrent, narrow, close. **Spores** hyaline, smooth, ellipsoid, non-amyloid, occasionally with droplets, 6-7 x 3-3.5 μm. Basidia 4-spored. Gill edge cystidia thick-walled, clavate; gill face cystidia similar but more slender.
Stem at first with lilaceous bloom, then concolorous with cap, lateral or eccentric, stout, tapering downwards. Ring absent. Flesh pallid, firm and full.
Odour not distinctive. **Taste** not distinctive.
Chemical tests none.
Occurrence summer to early autumn; infrequent.
■ Inedible.

Pleurotus cornucopiae (Paulet : Pers.) Rolland
= *Pleurotus sapidus* (Schulz.) Sacc.   **Lentinaceae**

Medium to large, cream, funnel-shaped cap, fruiting body with longish eccentric stem, in caespitose tufts on stumps and other wood of broad-leaf trees favouring oak and elm.

Dimensions cap 5 - 12 cm dia; stem 2 - 5 cm tall x 1 - 2.5 cm dia.
Cap wholly cream when young, later with ochraceous tinge; at first convex, soon becoming depressed and infundibuliform with somewhat irregular, split or lobed margin, at first whitish, pruinose then smooth. Flesh white and thin.
Gills concolorous with cap or pallid buff, deeply decurrent, narrow, close. **Spores** hyaline, smooth, elongated-ellipsoid, non-amyloid, with droplets, 8-11 x 3-3.5 μm. Basidia 4-spored. Cystidia absent.
Stem concolorous with cap, lateral or eccentric, stout, tapering downwards, curved, typically fused at base. Ring absent. Flesh white, firm and full.
Odour slight, of aniseed. **Taste** not distinctive.
Chemical tests none.
Occurrence summer to early autumn; infrequent.
□ Edible.

Pleurotus dryinus (Pers.: Fr.) Kummer   Lentinaceae
= *Pleurotus corticatus* (Fr.) Quél.

Medium to large, cream, scaly cap with cream gills and eccentric stem; in caespitose tufts, parasitic on stumps and other wood of broad-leaf and coniferous trees.

Dimensions cap 5 - 15 cm dia; stem 2 - 6 cm tall x 1 - 4 cm dia.
Cap wholly cream when young, later with greyish tinge; at first convex, becoming more or less flattened, fibrillose-scaly, becoming more so with age, margin incurved and decorated with velar remnants when young. Flesh whitish, tough and thick.
Gills concolorous with cap, deeply decurrent and running down the stem as ridges, narrow, close, sometimes forked. **Spores** hyaline, smooth, cylindrical-ellipsoid, non-amyloid, with droplets, 9.5-14 x 3.5-4.5 µm. Basidia 4-spored. Cystidia absent.
Stem concolorous with cap, lateral or eccentric, stout, tapering downwards, curved, typically fused at base. Ring absent. Flesh white, firm and full.
Odour slight, of aniseed. **Taste** not distinctive.
Chemical tests none.
Occurrence autumn; infrequent.
■ Inedible

Pleurotus ostreatus (Jacq.: Fr.) Kummer   Lentinaceae
= *Pleurotus columbinus* Quél.   Oyster Mushroom

Medium to large, distinctive, bluish-grey or brown oyster-shaped cap, pale gills and short eccentric stem; in caespitose tufts on stumps, trunks and felled timber of broad-leaf trees, favouring beech.

Dimensions cap 6 - 14 cm dia; stem 2 - 3 cm tall x 1 - 2 cm dia.
Cap colour variable through bluish-grey, brownish-grey and wholly brown; at first convex, becoming slightly depressed, oyster-shaped with somewhat wavy, lobed or split margin, smooth, shiny. Flesh white and moderate.
Gills at first white, becoming tinged pallid ochraceous, decurrent, narrow, close.
Spores hyaline, smooth, ellipsoid or sub-cylindrical, non-amyloid, with droplets, 7.5-11 x 3-4 µm. Basidia 4-spored. Cystidia absent.
Stem white, woolly at base, lateral or eccentric, stout, tapering downwards. Ring absent. Flesh white, firm and full.
Odour not distinctive. **Taste** not distinctive.
Chemical tests none.
Occurrence throughout the year but fruiting mainly summer and autumn; common.
☐ Edible and excellent.

Phyllotus porrigens (Pers.: Fr.) Karst.   Lentinaceae
= *Pleurotus porrigens* (Pers.: Fr.) Kühn. & Romagn.
= *Pleurocybella porrigens* (Pers.: Fr.) Sing.

Small, white, shell-shaped cap and gills, sessile; in caespitose tufts on stumps and rotting fallen wood of coniferous trees, favouring pine.

Dimensions cap 2 - 5 cm dia x 2 - 7 cm tall.
Cap white; at first shallowly convex, then flattened or depressed, shell-shaped, narrowing at point of attachment to the substrate, smooth. Flesh white, elastic and moderate.
Gills white, becoming cream with age, decurrent, narrow, close.
Spores hyaline, smooth, sub-spherical, non-amyloid, with droplets, 5-7.5 x 4.5-6.5 µm. Basidia 4-spored. Cystidia absent.
Stem absent. Ring absent.
Odour not distinctive. **Taste** not distinctive.
Chemical tests none.
Occurrence late summer to early autumn; in Britain limited to Scottish Highlands, generally montane elsewhere in Europe.
■ Inedible.

Pleurotus pulmonarius (Fr.) Quél.   **Lentinaceae**

Medium-sized, whitish, shell-shaped cap, and gills with very short eccentric stem; in caespitose tufts on stumps, trunks and felled timber of broad-leaf trees.

Dimensions cap 2 - 10 cm dia; stem 1 - 1.5 cm tall x 0.75 - 1.25 cm dia.

Cap whitish-cream, becoming pallid greyish-brown with age; at first shallowly convex, then flattened, undulating with somewhat wavy, lobed or split margin, narrowing at point of attachment with stem, smooth. Flesh white, elastic and moderate.

Gills at first white, becoming tinged pallid ochraceous, decurrent, narrow, close.

Spores hyaline, smooth, ellipsoid or sub-cylindrical, non-amyloid, with droplets, 7.5-11 x 3-4 µm. Basidia 4-spored. Cystidia absent.

Stem white, lateral or eccentric, very short, woolly, stout. Ring absent. Flesh white, firm and full.

Odour not distinctive. **Taste** not distinctive.

Chemical tests none.

Occurrence late summer to early autumn; infrequent.

☐ Edible.

HOMOBASIDIOMYCETES BOLETALES

Fungi in which the gills are largely replaced by tubes opening through pores, the Boletales [150] include under the Kew classification several families which may appear unrelated since they are morphologically dissimilar.

Amongst the pore-bearing boletes are some of the largest fruiting bodies found within the mushroom-type fungi and the order is classified into families according to spore characteristics, overall profile, texture of caps and stems and colour changes in the cut or bruised tissues (the latter provide significant determining features). Members with tubes which remain whitish or at most greyish-pink fall into the *Leccinum* and *Tylopilus* genera. *Leccinum* members possess non-bulbous stems with dark scales whilst *Tylopilus*, limited to one European species, has a netted stem and intensely bitter taste. Boletes with coloured tubes fall mainly into the *Boletus*, *Xerocomus* and *Suillus* genera, the latter being distinguished by largely slimy caps whilst *Boletus* and *Xerocomus* tend to be dry. Many amongst the *Boletus* genus develop massive caps with swollen stems.

Amongst the less typical members several families bear true gills,including the Gomphidiaceae, Paxillaceae and Hygrophoropsidaceae but all have spore and fruiting body characteristics more closely allied with the boletes than the agarics. The Hygrophoropsidaceae may, at first glance, be confused with the *Cantharellus* but are closely allied to the Paxillaceae which includes the common Roll Rim (*Paxillus involutus*).

Phylloporus rhodoxanthus, a member of the Xerocomaceae, is an oddity that has developed bright yellow pseudo-gills.

Rhizopogon was formerly included amongst the Gastromycetes and in many guide books it is still to be found in the Hymenogastraceae family. It is however, a highly modified development of the *Suillus* genus in which the 'mushroom' arrangement has been replaced by a *peridium* and *gleba*.

The Coniophoraceae, located on pages 128-129, look like encrusting brackets but possess a distinctly bolete spore structure. When determining any bolete, colour changes in the cut flesh should be noted. These may take up to ten minutes to become fully apparent but are important determining features.

The majority of boletes are edible and include some exceptionally palatable members, most notably the Cep (*Boletus edulis*). As a rule-of-thumb those with reddish tubes and pores are generally to be avoided and some, including *Boletus satanas,* are dangerously toxic.

Hygrophoropsis aurantiaca (Wulf: Fr.) Maire

= *Cantharellus aurantiacus* Wulf: Fr.   **Hygrophoropsidaceae**
= *Clitocybe aurantiaca* (Wulf.: Fr.) Studer   False Chanterelle

Smallish orange-yellow agaric with shallowly funnel-shaped cap, reminiscent of a chanterelle but with true gills; in small troops on soil, typically with conifers or on heaths.

Dimensions cap 2 - 8 cmdia; stem 3 - 5 cm tall x 0.5 - 1 cm dia.
Cap orange-yellow; at first convex, becoming shallowly infundibuliform, finely downy, typically remaining more or less incurved at margin. Flesh concolorous, thin and tough.
Gills orange, decurrent, narrow, forked, crowded. **Spores** hyaline, smooth, broadly ellipsoid, non-amyloid, with droplets, 5-7 x 3.5-5 μm. Basidia 4-spored. Cystidia absent.
Stem concolorous with cap, stout, more or less equal, smooth, typically curved. Ring absent. Flesh concolorous, cartilaginous and tough.
Odour not distinctive. **Taste** not distinctive.
Chemical tests none.
Occurrence late summer to autumn; common.
■ Inedible. Can cause distressing (but non-fatal) symptoms if eaten by mistake for *Cantharellus cibarius*.

Hygrophoropsis pallida (Peck.) Kreisel

Hygrophoropsidaceae

Very similar in appearance and habit to *H. aurantiaca* but generally paler and more funnel-shaped.

Dimensions cap 2 - 7 cm dia; stem 2 - 5 cm tall x 0.5 - 0.8 cm dia.
Cap pallid orange-yellow; at first convex, becoming infundibuliform, finely downy, typically remaining more or less incurved at margin. Flesh concolorous, thin and tough.
Gills cream, decurrent, narrow, forked, crowded. **Spores** hyaline, smooth, broadly ellipsoid, non-amyloid, with droplets, 5-7 x 3.5-5 μm. Basidia 4-spored. Cystidia absent.
Stem concolorous, with cap, more or less equal, smooth, typically curved. Ring absent. Flesh concolorous, cartilaginous and tough.
Odour not distinctive. **Taste** not distinctive.
Chemical tests none.
Occurrence late summer to autumn; rare. (Westonbirt Arboretum, Gloucestershire)
■ Inedible.

Omphalotus olearius (DC.: Fr.) Sing.   **Paxillaceae**

= *Clitocybe olearia* (DC.: Fr.) Maire
= *Pleurotus olearius* (DC.: Fr.) Gill.

Large orange-yellow, luminous agaric with shallowly funnel-shaped cap; tufted on roots and at base of trunks of various broad-leaf trees, including olive, oak, chestnut and beech.

Dimensions cap 8 - 12 cm dia; stem 4 - 14 cm tall x 1 - 2.5 cm dia
Cap intense orange-yellow; at first convex, becoming shallowly infundibuliform, smooth, typically remaining more or less incurved at margin. Flesh concolorous, moderate, tough, luminous due to atromentic acid.
Gills deeper orange, decurrent, narrow, forked, crowded.
Spores hyaline, smooth, sub-spherical, non-amyloid, 5-7 x 5-6 μm. Basidia 4-spored.
Stem pallid, concolorous with cap, stout, tapering downwards, smooth, typically curved and eccentric. Ring absent. Flesh concolorous and tough.
Odour not distinctive. **Taste** not distinctive.
Chemical tests none.
Occurrence late summer to autumn; frequent in Mediterranean region on olive wood, very rare elsewhere.
✛ Poisonous.

Paxillus atrotomentosus (Batsch : Fr.) Fr.   **Paxillaceae**

Large or massive agaric with brown cap, buff gills and dark brown velvety, sometimes eccentric stem; solitary or in small tufts on rotten, sometimes buried, coniferous wood.

Dimensions cap 12 - 28 cm dia; stem 3 - 9 cm tall x 2 - 5 cm dia.
Cap snuff-brown with irregular reddish-ochre patches; at first convex becoming expanded and depressed but with persistently inrolled margin, finely downy, becoming smooth with age. Flesh cream, thick and firm.
Gills at first buff, becoming ochraceous and spotted rust with age, decurrent, branched close to the stem giving netted appearance, crowded. **Spores** sienna-brown, smooth, ellipsoid, non-amyloid, some with droplets, 5-6.5 x 3-4.5 µm. Basidia 4-spored. Cystidia absent.
Stem at first olivaceous and finely downy, becoming dark brown and more coarsely downy, equal, very stout, eccentric. Ring absent. Flesh ochraceous or buff, solid and firm.
Odour not distinctive. **Taste** not distinctive.
Chemical tests none.
Occurrence late summer to autumn; infrequent.
■ Inedible.

Paxillus involutus (Batsch : Fr.) Fr.   **Paxillaceae**
Roll Rim

Large agaric with brown, sometimes greasy cap and stem, and ochre-brown gills; solitary or in trooping groups on soil in broad-leaf woods, favouring birch and on acid heath.

Dimensions cap 5 - 12 cm dia; stem 3 - 7 cm tall x 0.8 - 1.2 cm dia.
Cap at first fulvous with olivaceous tinge, becoming more hazel-brown; at first convex , becoming expanded and depressed but with persistently inrolled margin, downy especially at margin, becoming smooth with age, also greasy at centre when damp. Flesh pallid ochraceous, thick, firm.
Gills at first pallid ochraceous, becoming sienna-brown and spotted rust with age or on bruising, decurrent, crowded. **Spores** sienna-brown, smooth, ellipsoid, non-amyloid, without droplets, 8-10 x 5-6 µm. Basidia 4-spored. Cystidia fusiform with brown contents.
Stem concolorous with cap, bruising darker brown, smooth, equal or tapering downwards, sometimes slightly eccentric. Ring absent. Flesh ochraceous, becoming more fulvous in stem base, solid and firm.
Odour not distinctive. **Taste** with caution, acidic.
Chemical tests none.
Occurrence late summer to autumn; common.
✚ Dangerously poisonous.

Boletus aereus Bull.: Fr.   **Boletaceae**

Medium to large bolete with dull brown cap, dirty white pores and stout swollen stem; solitary or in small groups on soil under broad-leaf trees, favouring beech and oak.

Dimensions cap 7 - 16 cm dia; stem 6 - 8 cm tall x 2 - 4 cm dia.
Cap dull cigar, or sepia-brown; convex or bun-shaped, at first minutely downy, becoming smooth but finely cracked creating a rough texture. Flesh white, unchanging or dirty olivaceous or slightly vinaceous where cut, thick and firm.
Pores white or cream, discolouring dirty olivaceous or vinaceous, circular. Tubes concolorous with pores but becoming more sulphur-yellow, depressed. **Spores** olive-green or brown, smooth, sub-fusiform, with droplets, 13.5-15.5 x 4-5.5 µm. Basidia 4-spored. Cystidia fusiform.
Stem pallid background covered with a network which is brownish at apex, buff around the middle and rust below, stout and bulbous. Ring absent. Flesh white and firm.
Odour strong, earthy. **Taste** not distinctive.
Chemical tests flesh pallid rust with NH_3.
Occurrence summer to early autumn; rare. (Overton, Norfolk)
☐ Edible.

Boletus albidus Rocques   Boletaceae

Medium to large bolete with clay cap, yellow or olive-green pores and stout swollen stem; solitary or in small groups on calcareous soil under broad-leaf trees, favouring beech.

Dimensions cap 8 - 16 cm dia; stem 5 - 8 cm tall x 3 - 4 cm dia.
Cap pallid clay with olivaceous tinge; convex or bun-shaped, at first minutely downy, then cracked. Flesh white or pallid yellow, slowly blue then fading where cut, thick and firm.
Pores lemon-yellow, blue on bruising and discolouring olivaceous with age, circular, small. Tubes concolorous with pores, depressed.
Spores olive-green or brown, smooth, sub-fusiform, with droplets, 12-16 x 4.5-6 μm. Basidia 4-spored. Cystidia fusiform.
Stem background at first yellow, becoming clay, covered with a whitish network sometimes reduced to the upper third, sometimes with reddish flush in lower third, stout and bulbous, slightly 'rooting'. Ring absent. Flesh whitish or pallid vinaceous, firm.
Odour not distinctive. **Taste** bitter, then foul.
Chemical tests no reaction with NH_3.
Occurrence summer to early autumn; rare. (Blenheim, Oxon)
■ Inedible.

Boletus appendiculatus Schaeff.   Boletaceae

Medium to large bolete with reddish-buff cap, chrome-yellow pores and stout swollen netted stem; solitary or in small groups on calcareous soil under broad-leaf trees, favouring beech and oak.

Dimensions cap 8 - 13 cm dia; stem 8 - 12 cm tall x 3 - 4 cm dia
Cap bay-brown with ochraceous or sienna tinges; convex or bun-shaped, at first downy, then cracked. Flesh pallid yellow, unchanging and firm.
Pores chrome-yellow, darker with age, slightly blue on bruising, circular, small. Tubes concolorous with pores, depressed.
Spores olive-green or brown, smooth, sub-fusiform, with droplets, 12-15 x 3.5-4 μm. Basidia 4-spored. Cystidia fusiform.
Stem background chrome or lemon-yellow with faint reddish line mid-way, flushed reddish below, covered throughout with pallid network, stout and bulbous. Ring absent. Flesh pallid yellow with slight reddish tinge at base, firm and full.
Odour not distinctive. **Taste** not distinctive.
Chemical tests clay-pink or fulvous with NH_3.
Occurrence summer to early autumn; infrequent in south, rare elsewhere. (Stourhead, Gloucestershire)
□ Edible.

Boletus calopus Fr.   Boleteceae
= *Boletus pachypus* Fr.

Medium to large bolete with dull greyish cap, dirty cream pores and stout swollen netted stem; solitary or in small groups on soil under broad-leaf trees, favouring beech.

Dimensions cap 6 - 12 cm dia; stem 7 - 9 cm tall x 3.5 - 4.5 cm dia.
Cap smoke-grey with olivaceous tinge; at first hemispherical, then convex, at first suede-like then smooth. Flesh pallid yellow, pallid blue where cut, thick and firm.
Pores dirty cream, or yellow, blue on bruising, circular, small. Tubes concolorous with pores, depressed. **Spores** olive-green or brown, smooth, sub-fusiform, with droplets, 12-16 x 4.5-5.5 μm. Basidia 4-spored. Cystidia fusiform.
Stem lemon-yellow at apex, otherwise reddish, decorated with a white or vinaceous network, stout and bulbous. Ring absent. Flesh pallid yellow, pallid blue where cut and sometimes slightly red in base, firm.
Odour not distinctive. **Taste** bitter.
Chemical tests no reaction with NH_3.
Occurrence summer to early autumn; infrequent.
■ Inedible.

Boletus edulis Bull.: Fr.

Boletaceae
Cep. King Bolete

Large or massive bolete with dull brown cap, dirty white pores and stout swollen stem; solitary or in small groups on soil under broadleaf or coniferous trees.

Dimensions cap 8 - 20 cm dia; stem 6 - 8 cm tall x 2 - 4 cm dia.
Cap dull cigar- or bay-brown; convex or bun-shaped, at first with a whitish bloom, becoming smooth and dry or greasy, slightly viscid when damp. Flesh white, unchanging or dirty straw or slightly vinaceous where cut, thick and firm.
Pores white or cream, discolouring dirty yellowish-grey, circular. Tubes concolorous with pores but becoming more greyish-yellow, depressed. **Spores** olive-green or brown, smooth, ellipsoid-fusiform, with droplets, 14-17 x 4.5-5.5 µm. Basidia 4-spored. Cystidia fusiform.
Stem pallid cap colour background, covered with a white network, stout and bulbous. Ring absent. Flesh white, firm and unchanging.
Odour not distinctive. **Taste** not distinctive.
Chemical tests no reaction with NH₃.
Occurrence summer to autumn; common.
❑ Edible and good.

Boletus erythropus (Fr.) Krombh.

Boletaceae

Large or massive bolete with dull, dark brown cap and orange-red pores and stem; solitary or in small groups on acid soil under or near coniferous or, less frequently, broad-leaf trees.

Dimensions cap 8 - 20 cm dia; stem 5 - 14 cm tall x 2 - 5 cm dia.
Cap dull, dark, snuff-brown, sometimes with olivaceous tinge, more ochraceous at margin; convex or bun-shaped, becoming expanded with age, at first slightly downy becoming smooth and dry or greasy, slightly viscid when damp. Flesh yellow, rapidly dark blue where cut, thick, fairly firm.
Pores at first orange-red, becoming more rust with age, bruising rapidly dark blue or black, circular, small. Tubes lemon-yellow, rapidly turning dark blue where cut, free. **Spores** olive-green or brown, smooth, ellipsoid, with droplets, 12-15 x 4-6 µm. Basidia 4-spored. Cystidia clavate, tinged brown.
Stem pallid cap colour or yellowish background covered densely with orange-red dots, stout and more or less bulbous. Ring absent. Flesh yellow, reacting as in cap, fairly firm.
Odour not distinctive. **Taste** not distinctive.
Chemical tests flesh brownish with NH₃.
Occurrence late summer to early autumn; frequent.
❑ Edible only after cooking.

Boletus impolitus Fr.

Boletaceae
= *Boletus obsonium* Fr.

Large bolete with yellowish-brown cap, yellow pores and robust yellowish stem, flesh not changing colour; solitary or in small groups on soil under broad-leaf trees, favouring oak, often on mown grass.

Dimensions cap 5 - 16 cm dia; stem 5 - 10 cm tall x 3 - 5 cm dia.
Cap at first olivaceous-brown, becoming more yellowish; convex or bun-shaped, at first slightly downy, becoming smooth and dry, often cracking. Flesh pallid yellow, deeper above tubes, unchanging, thick and firm.
Pores at first lemon-yellow becoming chrome-yellow, tinged brown when old, unchanging, small. Tubes concolorous with pores, unchanging, adnexed. **Spores** olive-green or brown, smooth, subfusiform, with droplets, 10-16 x 4.5-5.5 µm. Basidia 4-spored. Cystidia irregularly fusiform or ventricose.
Stem pallid, cap colour background, darker towards base, covered with fine reddish punctate dots, more or less equal or stoutly bulbous. Ring absent. Flesh concolorous and reacting as in cap.
Odour sour, reminiscent of phenol. **Taste** not distinctive.
Chemical tests flesh ochraceous with NH₃.
Occurrence summer to early autumn; infrequent or rare. (Chard, Somerset)
❑ Edible.

Boletus luridus Schaeff.   Boletaceae

Large bolete with snuff-brown cap, orange-red pores and netted red stem, all parts bruising dark bluish-black; solitary or in small groups on soil under broad-leaf trees, favouring beech and oak, also in open grassland.

Dimensions cap 6 - 14 cm dia; stem 8 - 14 cm tall x 1 - 3 cm dia.
Cap dull brown, with snuff or olivaceous tinges; convex or bun-shaped, at first slightly downy, becoming smooth and dry. Flesh lemon-yellow turning rapidly bluish-black where cut, thick and firm.
Pores orange-red, more yellow at margin, turning rapidly bluish-black where bruised, circular, small. Tubes yellowish-green, turning bluish-black where cut, free. **Spores** olive-green or brown, smooth, ellipsoid, with droplets, 11-15 x 4.5-6.5 μm. Basidia 4-spored. Cystidia fusiform or narrowly ventricose.
Stem pallid, cap colour background covered with a reddish-orange or brownish network, more or less equal. Ring absent. Flesh concolorous and reacting as in cap.
Odour not distinctive. **Taste** not distinctive.
Chemical tests flesh ochraceous with NH_3.
Occurrence summer to autumn; infrequent.
☐ Edible when cooked.

Boletus pinophilus Pil. & Dermek   Boletaceae
= *Boletus pinicola* (Vitt.) Venturi

Large bolete with dark brown cap, cream pores, reddish-brown netted bulbous stem; solitary or scattered on soil specifically with Scots Pine.

Dimensions cap 6 - 15 cm dia; stem 5 - 15 cm tall x 4 - 7 cm dia.
Cap dark brown with reddish tinge; convex or bun-shaped, becoming expanded with age, dull, sometimes slightly wrinkled. Flesh white, tinged cap colour beneath the cuticle, unchanging, thick, fairly firm.
Pores white then cream, becoming olivaceous-brown with age, circular, small. Tubes concolorous, adnate, 10-25 mm long.
Spores pallid olivaceous-yellow, smooth, cylindrical-ellipsoid, with droplets, 15.5-20 x 4.5-5.5 μm. Basidia 4-spored. Gill edge cystidia clavate or fusiform; gill face cystidia fusiform, sparse.
Stem pallid, with fine reddish-brown dots arranged into a network which is whitish at apex, stout, bulbous. Ring absent. Flesh white, unchanging, fairly firm.
Odour faint, of spice. **Taste** not distinctive.
Chemical tests flesh greenish with $FeSO_4$.
Occurrence late summer to early autumn; infrequent in Scottish Highlands, rare elsewhere.
☐ Edible.

Boletus purpureus Fr.   Boletaceae
= *Boletus rhodoxanthus* (Krombh.) Kallenb.

Large or massive bolete with pinkish-beige cap, blood-red pores and red and yellow stem, all parts bruising dark blue; solitary or in small groups often fused on soil under oak or beech.

Dimensions cap 6 - 25 cm dia; stem 6 - 15 cm tall x 3 - 6 cm dia.
Cap cream, becoming pinkish-beige; convex or bun-shaped, becoming expanded with age, dull, slightly viscid when moist, finely downy, bruising blackish. Flesh lemon-yellow, blue where cut, thick and fairly firm.
Pores yellow, very soon blood-red, bruising dark blue or black, circular, small. Tubes yellow, dark blue where cut, adnate, 10-25 mm long. **Spores** pallid olivaceous-yellow, smooth, cylindrical-ellipsoid, with droplets, 10.5-15.5 x 4-5 μm. Basidia 4-spored. Cystidia mainly fusiform, occasionally vesicular.
Stem golden-yellow with dark purplish-red network of dots, stout and more or less bulbous. Ring absent. Flesh yellow, reacting as in cap, fairly firm.
Odour faint, sour. **Taste** not distinctive.
Chemical tests flesh greenish with $FeSO_4$.
Occurrence late summer to early autumn; rare. (New Forest, Hampshire)
■ Inedible.

Boletus queletii Schulz.   Boletaceae

Large or massive bolete with dull brown cap, orange-red pores and stem, all parts bruising dark blue; solitary or in small groups on calcareous soil under broad-leaf trees.

Dimensions cap 6 - 15 cm dia; stem 6 - 13 cm tall x 2 - 3.5 cm dia.
Cap snuff-brown, sometimes with reddish tinge; convex or bun-shaped, becoming expanded with age, dull, slightly polished when dry, margin narrowly overhanging. Flesh pallid yellow, blue where cut, greenish after 10 minutes, thick, fairly firm.
Pores yellow, becoming orange-rust, bruising dark blue or black, circular, small. Tubes lemon-yellow, dark blue where cut, adnate,10-20 mm long. **Spores** pallid olivaceous-yellow, smooth, cylindrical-ellipsoid, with droplets, 9-15 x 4.5-6 µm. Basidia 2- or 4-spored. Cystidia fusiform, clavate or vesicular.
Stem yellowish with fine orange-red floccose dots, stout and more or less bulbous. Ring absent. Flesh yellow, reacting as in cap but with dark purplish or yellowish tinges in base, fairly firm.
Odour faint, sour. **Taste** slight, sour.
Chemical tests flesh greenish with NH_3.
Occurrence late summer to early autumn; rare. (Stourhead, Wilts.)
☐ Inedible.
Note: avoid confusion with *B. erythropus*.

Boletus reticulatus Schaeff.   Boletaceae
= *Boletus aestivalis* Fr.   Summer King Bolete

Large or massive bolete with dull brown cap, dirty white pores and stout swollen stem; solitary or in small groups on soil under broad-leaf or coniferous trees.

Dimensions cap 8 - 30 cm dia; stem 6 - 12 cm tall x 3 - 5 cm dia.
Cap pale cigar-brown or cinnamon; convex or bun-shaped, at first suede-like, with a whitish bloom, becoming smooth and dry. Flesh white, unchanging or dirty straw, darker brown beneath cuticle, thick and firm.
Pores white or cream, discolouring dirty yellowish-grey, circular. Tubes concolorous with pores but becoming more greyish-yellow, depressed. **Spores** olive-green or brown, smooth, sub-fusiform, with droplets, 14-17 x 4.5-5.5 µm. Basidia 3- or 4-spored. Cystidia fusiform, clavate, capitate or vesicular.
Stem pallid, cap colour background covered wholly with a white network, becoming more brown with age, stout and bulbous. Ring absent. Flesh white, firm and unchanging.
Odour not distinctive. **Taste** not distinctive.
Chemical tests no reaction with NH_3.
Occurrence summer to autumn; infrequent.
☐ Edible and good.
Note: the cap is a paler brown than in *B. edulis*.

Boletus satanas Lenz.   Boletaceae
Satan's Bolete

Large or massive bolete with chalk-white cap, orange-red pores and bulbous red and yellow netted stem, all parts bruising blue; solitary or in small groups on calcareous soil with oak or beech.

Dimensions cap 8 - 25 cm dia; stem 6 - 9 cm tall x 5 - 10 cm dia.
Cap chalk-white; convex or bun-shaped, becoming expanded with age, at first finely downy, then smooth. Flesh chalk-white or pallid cream, sometimes flushed rose where eaten, slowly pallid blue where cut, thick and firm.
Pores red, soon blood-red but orange at margin, bruising greenish, circular, small. Tubes yellow then olivaceous, blue where cut, adnate,10-25 mm long. **Spores** olivaceous-brown, smooth, fusiform-ellipsoid, with droplets, 11-14 x 4.5-6 µm. Basidia 4-spored. Gill edge cystidia ventricose; gill face cystidia fusiform.
Stem yellow above with red net, wholly red from halfway, stout and very bulbous. Ring absent. Flesh pallid lemon above, dirty white below, reacting as in cap but red in base, fairly firm.
Odour of spice. **Taste DO NOT ATTEMPT TO TASTE ANY PART OF THE SPECIMEN.**
Chemical tests flesh straw with NH_3.
Occurrence late summer to early autumn; very rare. (Somerset)
✚ Lethally poisonous.

Boletus satanoides Smotl.   Boletaceae
= *Boletus splendidus* Martin

Large or massive bolete with reddish-grey cap, orange-red pores and red and yellow netted stem, all parts bruising blue; solitary or in small groups on soil under oak.

Dimensions cap 6 - 20 cm dia; stem 6 - 8 cm tall x 2 - 4 cm dia.
Cap at first pallid coffee, becoming more grey and developing reddish flush; convex or bun-shaped, becoming expanded with age, smooth, bruising blackish-blue. Flesh pallid yellow, blue where cut, thick and fairly firm.
Pores yellow, very soon orange-red, bruising blue, circular, small. Tubes yellow, blue where cut, adnate, 10-25 mm long. **Spores** olivaceous-brown, smooth, fusiform-ellipsoid, with droplets, 10.5-12.5 x 4.5-5 µm. Basidia 4-spored. Cystidia fusiform or lageniform.
Stem yellow with orange net above, dark red middle area and coral base, stout and more or less bulbous. Ring absent. Flesh yellow, reacting as in cap but red in base, fairly firm.
Odour of spice, then foul. **Taste** not distinctive.
Chemical tests flesh straw with NH_3.
Occurrence late summer to early autumn; rare. (Windsor Great Park, Berkshire)
✛ Poisonous.

Leccinum aurantiacum S F Gray   Boletaceae
= *Leccinum rufum* (Schaeff.) Kreisel

Medium to large bolete with deep orange cap, whitish pores and scaly stem, darkening throughout where cut or bruised; solitary or scattered on soil with aspen (poplar).

Dimensions cap 5 - 10 cm dia; stem 8 - 14 cm tall x 1.5 - 4.5 cm dia.
Cap dark orange-rust with apricot tinge; convex or bun-shaped, at first downy then smooth, matt, cuticle somewhat overhanging margin. Flesh creamy-white then vinaceous or sepia where cut, thick and firm.
Pores white or cream, darkening vinaceous where bruised, circular, very small. Tubes concolorous with pores, vinaceous where cut, depressed. **Spores** ochraceous-buff, smooth, sub-fusiform, with droplets, 14-16.5 x 4-5 µm. Basidia 2- or 4-spored. Cystidia clavate or fusiform and typically clustered on the gill edge.
Stem dirty white, covered with woolly scales in irregular network, at first white then rust, stoutish, more or less equal or swollen towards base. Ring absent. Flesh creamy-white, becoming vinaceous and livid in base, firm and full.
Odour not distinctive. **Taste** not distinctive.
Chemical tests no reaction with NH_3.
Occurrence late summer to early autumn; rare. (Newbury, Berkshire)
☐ Edible.

Leccinum carpini (Schulz.) Moser.: Reid   Boletaceae
= *Boletus carpini* (Schulz.) Pearson.

Medium to large bolete with dull mid-brown cap, dirty white pores and scaly stem, turning blue-grey where cut; solitary or scattered on soil, usually under hornbeams.

Dimensions cap 3 - 9 cm dia; stem 8 - 9 cm tall x 0.8 - 1.1 cm dia.
Cap dull snuff-brown, sometimes more pallid or with olivaceous tinges; convex or bun-shaped, at first smooth then cracked, matt, cuticle sometimes drawn away from margin in older specimens. Flesh greyish coral-pink, blackening where cut, thick and firm.
Pores at first whitish, becoming buff, darkening where bruised, circular, small. Tubes concolorous with pores, coral-pink to blackish where cut, depressed. **Spores** ochraceous-brown, smooth, sub-fusiform, with droplets, 15-19 x 5-6 µm. Basidia 4-spored. Gill edge cystidia fusiform or clavate; gill face cystidia similar but sparse.
Stem pallid whitish-buff, covered with woolly scales, buff towards apex, darker brown below, slender, more or less equal or slightly swollen two thirds towards base. Ring absent. Flesh concolorous with and reacting as in cap, firm and full.
Odour not distinctive. **Taste** not distinctive.
Chemical tests flesh red with formalin.
Occurrence summer; rare. (Forest of Dean, Gloucestershire)
☐ Edible.

Leccinum crocipodium (Letellier) Watl.   **Boletaceae**

= *Boletus crocipodius* Letellier
= *Leccinum nigrescens* (Richon & Roze) Sing.

Medium to large bolete with pale brown cap, yellowish pores and scaly stem, darkening throughout where cut or bruised; solitary or scattered on soil with oak.

Dimensions cap 4 - 10 cm dia; stem 6 - 12 cm tall x 1.5 - 2.5 cm dia.
Cap hazel-brown with ochraceous tinge; convex or bun-shaped, at first downy then cracked, matt, cuticle slightly overhanging margin. Flesh straw, brick, then blackening where cut, thick and firm.
Pores at first lemon-yellow, darkening brown where bruised, circular, small. Tubes concolorous with pores, blackish where cut, depressed.
Spores ochraceous-brown, smooth, sub-fusiform, with droplets, 12-17.5 x 4.5-7 µm. Basidia 4-spored. Cystidia fusiform.
Stem straw, covered with woolly scales in irregular network, buffor cinnamon, slender, more or less equal or slightly swollen towards base. Ring absent. Flesh grey, becoming vinaceous, firm and full.
Odour not distinctive. **Taste** not distinctive.
Chemical tests olivaceous with NH$_3$.
Occurrence summer; infrequent.
☐ Edible.

Leccinum holopus (Rostk.) Watl.   **Boletaceae**

= *Krombholziella holopoda* (Rostk.) Pil.

Smallish bolete with white cap, pores and scaly stem, not darkening where cut or bruised; solitary or in scattered groups on soil with birch.

Dimensions cap 3 - 6 cm dia; stem 5 - 10 cm tall x 0.7 - 1.2 cm dia.
Cap whitish or very pallid buff with greenish-grey tinge when older; convex or bun-shaped, at first downy then smooth, matt, cuticle slightly overhanging margin. Flesh white, unchanging, thick and firm.
Pores white, becoming dingy with age, circular, small. Tubes concolorous with pores, unchanging, depressed. **Spores** ochraceous, smooth, sub-fusiform, with droplets, 14.5-20.5 x 4.5-6.5 µm. Basidia 4-spored. Cystidia fusiform or ventricose.
Stem concolorous with cap, covered with woolly scales which may become pallid brownish, slender, more or less equal or slightly swollen towards base. Ring absent. Flesh whitish, firm and fibrous.
Odour not distinctive. **Taste** not distinctive.
Chemical tests fulvous with NH$_3$.
Occurrence late summer to early autumn; rare. (Culbin Forest, Morayshire)
☐ Edible.

Leccinum quercinum (Pil.) Green & Watl.   **Boletaceae**

= *Krombholziella quercina* (Pil.) Sutara

Medium to large bolete with brown cap, buff pores and scaly stem; solitary or in small scattered groups on soil with oak.

Dimensions cap 6 - 14 cm dia; stem 8 - 16 cm tall x 2 - 3.5 cm dia.
Cap date - or chestnut-brown;convex or bun-shaped, cuticle slightly overhanging margin, smooth or fibrillose scaly. Flesh white with vinaceous tinge, thick and firm.
Pores at first whitish, becoming buff, vinaceous where bruised and with age, circular, small. Tubes concolorous with pores, depressed.
Spores snuff-brown, smooth, fusiform, without droplets, 12-15 x 3.5-5 µm. Basidia 4-spored. Gill edge cystidia fusiform or ventricose; gill face cystidia absent.
Stem pallid whitish-buff background, covered with pallid woolly scale, becoming rust or dark brown, more or less equal or slightly swollen towards base. Ring absent. Flesh pallid but with grey-green tinge and firm.
Odour not distinctive. **Taste** not distinctive.
Chemical tests none.
Occurrence summer; rare. (Reading, Berkshire)
☐ Edible.

Leccinum rigidipes new species claimed by P Orton (1988)
Boletaceae

Medium to large bolete with brown cap, dirty-white pores and rigid scaly stem; solitary or in small scattered groups on soil, specifically with birch.

Dimensions cap 5 - 12 cm dia; stem 8 - 13 cm tall x 1 - 3 cm dia.
Cap brown with rust or cinnamon tinges; convex or bun-shaped, at first finely downy, becoming smooth, greasy when moist. Flesh dirty-white, becoming slightly buff, thick and firm.
Pores dirty white or greyish-brown, circular, small. Tubes concolorous with pores, adnate. **Spores** ochraceous-brown, smooth, fusiform, with droplets, 14-19 x 4.5-6 µm. Basidia 4-spored. Cystidia narrowly lageniform.
Stem whitish, covered with snuff-brown scales at apex, blackish below,tapering slightly upwards. Ring absent. Flesh concolorous with cap, very firm.
Odour faint, aromatic. **Taste** slight, sour.
Chemical tests no reaction with formalin.
Occurrence late summer to early autumn; frequent.
☐ Edible.
Very similar to *L. scabrum* but with differences in rigidity of stem, greasiness of cap and colour of stem scales.

Leccinum scabrum (Bull.: Fr.) S F Gray   **Boletaceae**

Medium to large bolete with brown cap, dirty-white pores and scaly stem; solitary or in small scattered groups on soil specifically with birch .

Dimensions cap 5 - 15 cm dia; stem 6 - 15 cm tall x 1.5 - 3 cm dia.
Cap brown, with reddish or greyish tinges; convex or bun-shaped, at first finely downy, becoming smooth, somewhat scurfy when older. Flesh dirty-white, unchanging, thick and firm.
Pores dirty white or greyish-brown, circular, small. Tubes concolorous with pores, adnate. **Spores** ochraceous-brown, smooth, fusiform, with droplets,12.5-17.5 x 5-5.5 µm. Basidia 4-spored. Cystidia narrowly lageniform.
Stem whitish, covered with grey-brown scales, more or less equal or tapering slightly upwards. Ring absent. Flesh concolorous with cap, firm.
Odour faint, aromatic. **Taste** slight, sour.
Chemical tests no reaction with formalin.
Occurrence late summer to early autumn; common.
☐ Edible.

Leccinum versipelle (Fr. & Hok.) Snell.   **Boletaceae**
= *Boletus versipellis* Fr. & Hok.
= *Leccinum testaceoscabrum* (Sécr.) Sing.

Large or massive bolete with distinctive orange cap, greyish-yellow pores and scaly stem; solitary or in small scattered groups on soil specifically with birch and on heaths.

Dimensions cap 8 - 20 cm dia; stem 8 - 20 cm tall x 1.5 - 4 cm dia.
Cap tawny-orange, convex; or bun-shaped, at first slightly downy then smooth and sometimes slightly viscid, cuticle overhanging cap margin. Flesh white becoming dark vinaceous where cut, thick and firm.
Pores at first whitish, then buff, darkening rust where bruised, circular, small. Tubes concolorous with pores, vinaceous where cut, depressed. **Spores** ochraceous-brown, smooth, sub-fusiform, with droplets, 12.5-15 x 4-5 µm. Basidia 4-spored. Cystidia fusiform or clavate.
Stem pallid whitish background covered with brown or brownish-black woolly scales, stout, more or less equal or clavate. Ring absent. Flesh white, turning blue-green then black where cut, firm.
Odour not distinctive. **Taste** not distinctive.
Chemical tests flesh brick-red with NH_3.
Occurrence summer to early autumn; common.
☐ Edible.

Pulveroboletus lignicola (Kallenb.) Pil.   Boletaceae
= *Boletus lignicola* Kallenb.

Small bolete with orange-brown cap, yellowish pores and more or less smooth stem; solitary or in small groups in association with buried wood of coniferous trees and apparently with *Phaeolus schweinitzii*.

Dimensions cap 3 - 8 cm dia; stem 3 - 8 cm tall x 0.5 - 1.5 cm dia.
Cap yellowish-brown with orange tinge; convex or bun-shaped, at first minutely downy, becoming smooth, slightly greasy when damp. Flesh lemon-yellow, at first blue above tubes where cut, then fading, thick and soft.
Pores at first yellow, becoming more rust with age, circular or irregular, small. Tubes concolorous with pores, depressed.
Spores pallid yellow, smooth, ellipsoid, with droplets, 6.5-10 x 3-4 µm. Basidia 4-spored. Cystidia fusiform or lageniform.
Stem yellow with rust tinge and with yellow mycelial fragments at base, downy or slightly punctate, more or less equal or tapering downwards. Ring absent. Flesh yellow and firm.
Odour faint, of aromatic resin. **Taste** not distinctive.
Chemical tests tubes dark blue-green with Melzer's reagent.
Occurrence autumn; very rare. (New Forest, Hampshire)
■ Inedible.

Tylopilus felleus (Fr.) Karst.   Boletaceae
= *Boletus felleus* Fr.

Medium to large bolete with brown cap, pale pores, and stem covered with brown network, distinguished by very bitter taste; solitary or in small groups on soil under broad-leaf or coniferous trees.

Dimensions cap 6 - 12 cm dia; stem 7 - 10 cm tall x 2 - 3 cm dia.
Cap snuff- or fulvous-brown; convex or bun-shaped, at first slightly downy then smooth and dry. Flesh whitish, with pinkish tinge beneath cap cuticle, unchanging, thick and firm.
Pores pallid with faint coral-pink tinge when mature, bruising brownish, small. Tubes concolorous with pores, adnate. **Spores** clay-pink or vinaceous, smooth, sub-fusiform, with droplets, 11-15 x 4-5 µm. Basidia 2- or 4-spored. Cystidia fusiform or clavate.
Stem pallid background with brown reticulation, stout and slightly bulbous. Ring absent. Flesh concolorous and reacting as in cap, firm.
Odour not distinctive. **Taste** very bitter.
Chemical tests no reaction with NH_3.
Occurrence summer to autumn; infrequent.
■ Inedible.

Phylloporus rhodoxanthus (Schwartz) Bres.
= *Phylloporus pelletieri* (Lév.) Quél.
= *Paxillus paradoxus* (Kchb.) Cleland   Xerocomaceae

Smallish bolete with dark, reddish-brown cap, distinctive bright yellow gills and stem flushed with cap colour; solitary on soil, often acidic, near coniferous and broad-leaf trees.

Dimensions cap 4 - 6 cm dia; stem 2 - 6 cm tall x 0.8 - 1.5 cm dia.
Cap dark reddish-brown, finely downy; at first hemispherical, becoming flattened with a low blunt umbo and incurved margin. Flesh pallid yellow, cap coloured beneath cuticle, thick and spongy.
Gills at first lemon-yellow, becoming bright chrome-yellow at maturity, regularly anastomosing into pseudo-pores, broad, adnate-decurrent. **Spores** pallid greenish-yellow, smooth, ellipsoid, with droplets, 10-14 x 3.5-5.5 µm. Basidia 4-spored. Cystidia fusiform or clavate, abundant.
Stem yellowish but flushed cap colour in mid-section, smooth, more or less equal or tapering below. Ring absent. Flesh concolorous with cap and quite tough.
Odour not distinctive. **Taste** not distinctive.
Chemical tests flesh purplish-red with NH_3.
Occurrence summer to early autumn; rare. (Stourhead, Wiltshire)
■ Inedible.

Xerocomus armeniacus (Quél.) Quél.　**Xerocomaceae**

Smallish bolete with apricot-brown cap, yellow pores bruising blue; solitary or in small groups on soil under or in the vicinity of pine or oak, often in open grassland.

Dimensions cap 3 - 7 cm dia; stem 3 - 6 cm tall x 0.5 - 1.5 cm dia.
Cap at first apricot, becoming dull reddish-brown with age; convex or bun-shaped, at first downy, becoming matt, soon cracking. Flesh yellowish, becoming blue where cut, thick and firm.
Pores yellow, then olivaceous, blue where bruised, angular, large. Tubes concolorous with pores, adnate or depressed.
Spores olivaceous-brown, smooth, sub-fusiform, without droplets, 9-15 x 4-6 µm. Basidia 4-spored. Cystidia fusiform.
Stem yellow at apex, reddish-brown below, finely striate, more or less equal, tapering at the base. Ring absent. Flesh yellow, becoming blue where cut, vinaceous at base and firm.
Odour not distinctive. **Taste** not distinctive.
Chemical tests none.
Occurrence late summer to early autumn; infrequent.
☐ Edible.

Xerocomus badius (Fr.) Gilb.　**Xerocomaceae**
= *Boletus badius* Fr.

Large bolete with bay-brown cap, lemon-yellow pores bruising distinctively bluish-green; solitary or in small groups on soil under broad-leaf or coniferous trees.

Dimensions cap 4 - 14 cm dia; stem 5 - 12 cm tall x 1 - 4 cm dia.
Cap bay-brown, sometimes with brick-red or ochraceous tinges; convex or bun-shaped, at first downy, becoming smooth and polished, somewhat viscid when damp. Flesh white or lemon-yellow, becoming faintly blue where cut particularly above the pores, thick and firm.
Pores lemon-yellow or cream, bluish-green where bruised, angular, large. Tubes concolorous with pores, adnate or depressed.
Spores olivaceous-brown, smooth, thick-walled, sub-fusiform, without droplets, 13-15 x 4.5-5.5 µm. Basidia 4-spored. Cystidia fusiform.
Stem pallid cap colour with fine cottony fibrils, fairly stout, more or less equal. Ring absent. Flesh white or lemon-yellow, becoming faintly blue where cut particularly, in apex and firm.
Odour not distinctive. **Taste** not distinctive.
Chemical tests none.
Occurrence late summer to early autumn; common.
☐ Edible and good.

Xerocomus chrysenteron (Bull.) Quél.　**Xerocomaceae**
= *Boletus chrysenteron* Bull.

Medium to large bolete, with dull brown cap, cracking to show pinkish-red flesh beneath, greenish-yellow pores and red-streaked stem; solitary or in small groups on soil under broad-leaf trees.

Dimensions cap 4 - 11 cm dia; stem 4 - 8 cm tall x 1 - 1.5 cm dia.
Cap dull brown or pallid sepia, with olivaceous tinge; convex or bun-shaped, at first slightly downy, becoming smooth and dry, but also crazed, revealing coral-red flesh beneath, slightly sticky when damp. Flesh cream, unchanging or slightly blue above tubes where cut, thick and soft.
Pores lemon-yellow, greenish with age and sometimes bruising greenish, angular, large. Tubes concolorous with pores, adnate or depressed. **Spores** olive-green or brown, smooth, thick-walled, sub-fusiform, without droplets, 12-15 x 3.5-5 µm. Basidia 4-spored. Cystidia fusiform.
Stem lemon-yellow or buff background covered with coral-red fibrils in the lower 2/3, more or less equal. Ring absent. Flesh cream, turning slightly blue in base of stem where cut and soft.
Odour not distinctive. **Taste** not distinctive.
Chemical tests no reaction with NH_3.
Occurrence summer to autumn; common.
☐ Edible.

Xerocomus lanatus (Rostk.) Sing.　　　Xerocomaceae
= *Boletus lanatus* Rostk.

Large bolete with light brown cap, lemon-yellow pores, bruising distinctively bluish-green; solitary or in small groups on soil under broad-leaf or mixed trees favouring birch.

Dimensions cap 4 - 9 cm dia; stem 5 - 8 cm tall x 1 - 2 cm dia.
Cap buff or cinnamon-brown, darker where bruised; convex or bun-shaped, at first downy, becoming smooth. Flesh white with a brown line beneath the cuticle, more or less unchanging, thick and somewhat spongy.
Pores lemon-yellow or cream, bluish-green where bruised, angular, large. Tubes concolorous with pores, adnate or depressed.
Spores olivaceous-brown, smooth, thick-walled, sub-fusiform, without droplets, 9-11.5 x 3.5-4.5 μm. Basidia 4-spored. Cystidia fusiform - cylindrical.
Stem cap colour, more pallid at apex and base, decorated with an irregular network of reddish veins, more or less equal or slightly clavate. Ring absent. Flesh whitish, lemon-yellow in the base, firm and full.
Odour not distinctive. **Taste** not distinctive.
Chemical tests quickly greenish on the cap surface with NH_3.
Occurrence late summer to early autumn; frequent.
■ Inedible.

Xerocomus parasiticus (Fr.) Quél.　　　Xerocomaceae
= *Boletus parasiticus* (Bull.) Fr.

Smallish straw yellow bolete; parasitic on *Scleroderma citrinum*, solitary or in clusters of two or three.

Dimensions cap 2 - 4 cm dia; stem 1 - 4 cm tall x 0.5 - 1 cm dia.
Cap pallid sienna or straw with olivaceous tinge; convex or bun-shaped, at first finely downy, becoming smooth and dry, then cracking. Flesh pallid lemon-yellow, unchanging, thick and firm.
Pores lemon-yellow becoming more rust with age, circular, compound. Tubes concolorous with pores, adnate or decurrent. **Spores** olive-green or brown, smooth, thick-walled, elongated sub-fusiform, with droplets,11-21 x 3.5-5 μm. Basidia 4-spored. Gill edge cystidia fusiform or clavate; gill face cystidia fusiform.
Stem concolorous with cap, tapering downwards and typically curved up and around the host. Ring absent. Flesh pallid lemon-yellow with rust tinge at base and firm.
Odour not distinctive. **Taste** not distinctive.
Chemical tests flesh browning with NH_3.
Occurrence autumn; rare. (New Forest, Hampshire)
☐ Edible.

Xerocomus porosporus Imler　　　Xerocomaceae
= *Boletus porosporus* (Imler) Watl.

Medium-sized bolete with olive-brown cap, often cracking, and pale yellow pores bruising bluish; solitary or in small groups on soil under broad-leaf or coniferous trees.

Dimensions cap 4 - 8 cm dia; stem 4 - 8 cm tall x 2 - 3 cm dia.
Cap at first faintly yellowish then dark olive-brown, sepia with age; convex or bun-shaped, downy, becoming smooth and dry. Flesh pallid lemon-yellow or whitish, sometimes dirty blue where cut, thick and soft.
Pores pallid lemon-yellow, becoming more olivaceous with age and discolouring dirty bluish where bruised, angular, compound, large. Tubes concolorous with pores, adnate with tooth. **Spores** olive-green or brown, smooth, thick-walled, sub-fusiform,with distinctive truncated pore, without droplets, 13-15 x 4.5-5.5 μm. Basidia 4-spored. Cystidia fusiform, some more or less capitate, sparse.
Stem pallid cap colour background covered with darker fibrils, more or less equal or tapering downwards. Ring absent. Flesh pallid lemon-yellow or whitish, more chrome at apex, rust at base, soft.
Odour not distinctive. **Taste** not distinctive.
Chemical tests none.
Occurrence autumn; rare. (Stourhead, Wiltshire)
■ Inedible.
Note: the truncation on the spore is unique amongst European boletes.

Xerocomus pruinatus (Fr. & Hok.) Quél.   **Xerocomaceae**
= *Boletus pruinatus* Fr. & Hok.

Medium to large bolete with chestnut-brown cap displaying distinctive whitish 'bloom', yellow pores and reddish-tinged stem; solitary or in small groups on soil under broad-leaf or coniferous trees but favouring beech.

Dimensions cap 4 - 10 cm dia; stem 9 - 10 cm tall x 2 - 3 cm dia.
Cap chestnut or dark russet-brown; convex or bun-shaped, at first with a whitish bloom, otherwise smooth and dry. Flesh lemon-yellow becoming more chrome-yellow with age, turning slowly bluish-green where cut, thick and firm.
Pores at first lemon-yellow becoming chrome-yellow, bruising slowly bluish, angular, small. Tubes concolorous and reacting as in cap, adnate. **Spores** olive-green or brown, smooth, thick-walled, sub-fusiform, without droplets, 11.5-14 x 4.5-5.5 µm. Basidia 4-spored. Gill edge cystidia clavate or fusiform; gill face cystidia similar but sparse.
Stem pallid cap colour background, covered with fine reddish dots, stout and sub-fusiform or tapering slightly downwards. Ring absent. Flesh concolorous and reacting as in cap but with brownish tinge at base, firm.
Odour not distinctive. **Taste** not distinctive.
Chemical tests flesh fulvous-brown with NH_3.
Occurrence summer to autumn; infrequent or rare. (Alfriston Forest, Sussex).
☐ Edible.

Xerocomus pulverulentus (Opat.) Gilb.   **Xerocomaceae**
= *Boletus pulverulentus* Opat.

Medium-sized bolete with dull brown cap, sulphur-yellow pores and reddish-streaked stem, all parts bruising bluish-black; solitary or in small groups on soil under broad-leaf trees, favouring oak and holly.

Dimensions cap 4 - 9 cm dia; stem 5 - 6.5 cm tall x 0.8 - 1.4 cm dia.
Cap dull snuff-brown, sometimes with reddish or olivaceous tinge; convex or bun-shaped, at first slightly downy, becoming smooth and dry. Flesh lemon- or chrome-yellow, immediately dark blue or black-ish where cut, thick and firm.
Pores dirty sulphur-yellow, rapidly turning bluish where bruised, circular, small. Tubes concolorous with pores and turning rapidly blue-black where cut, depressed. **Spores** olive-green or brown, smooth, thick-walled, sub-fusiform, with droplets,10-14 x 3.5-6.5 µm. Basidia 4-spored. Cystidia fusiform or ventricose.
Stem pallid cap colour background, covered with reddish fibrils, tapering downwards. Ring absent. Flesh concolorous with, and reacting as in, cap.
Odour not distinctive. **Taste** not distinctive.
Chemical tests flesh ochraceous with NH_3.
Occurrence late summer to autumn; rare. (Stourhead, Wiltshire)
☐ Edible and good.

Xerocomus rubellus (Krombh.) Moser   **Xerocomaceae**
= *Boletus versicolor* Rostk.
= *Xerocomus chrysenteron var versicolor* (Rostk.) Mass.

Small to medium-sized bolete, with striking red cap, yellowish pores and red-streaked stem; solitary or in small groups on soil under broad-leaf trees.

Dimensions cap 3 - 6 cm dia; stem 4 - 8 cm tall x 1 - 1.5 cm dia.
Cap scarlet, tinged vinaceous with age; convex or bun-shaped, cuticle at first slightly downy, becoming smooth and dry, cracking. Flesh buff, unchanging or slightly blue above tubes where cut, thick and soft.
Pores lemon-yellow, greenish with age and sometimes bruising greenish, angular, medium. Tubes concolorous with pores, adnate or depressed. **Spores** olive-green or brown, smooth, thick-walled, ellipsoid or sub-fusiform, without droplets,11-14 x 4.5-5.5 µm. Basidia 4-spored. Cystidia narrowly lageniform.
Stem pallid yellow or buff background covered with red fibrils in the lower 2/3, yellow in base, fibrillose, more or less equal. Ring absent. Flesh buff, turning slightly blue where cut and soft.
Odour not distinctive. **Taste** slight, sour.
Chemical tests flesh pinkish with NH_3.
Occurrence late summer to early autumn; infrequent.
■ Inedible.

Xerocomus subtomentosus (L.: Fr.) Quél.
= *Boletus subtomentosus* L.: Fr.   **Xerocomaceae**

Medium to large bolete with brown cap, chrome-yellow pores and yellowish stem; solitary or in small groups on soil under broad-leaf trees and in mixed woods.

Dimensions cap 5 - 12 cm dia; stem 3 - 8 cm tall x 1 - 2 cm dia.
Cap fulvous-brown; convex or bun-shaped, at first downy, becoming smooth, sometimes cracking. Flesh white or pallid yellow , with faint brownish zone beneath cap cuticle, unchanging, thick and soft.
Pores chrome-yellow, becoming tinged olivaceous with age, angular, large. Tubes concolorous with pores, adnate. **Spores** olive-green or brown, smooth, thick-walled, sub-fusiform, without droplets,10-13 x 3.5-5 µm. Basidia 4-spored. Cystidia cylindrical or clavate.
Stem pallid cap colour, sometimes with brick-red tinge, slender, slightly bulbous. Ring absent. Flesh concolorous with cap but browner.
Odour not distinctive. **Taste** not distinctive.
Chemical tests flesh brownish with NH_3.
Occurrence autumn; infrequent.
☐ Edible.

Chalciporus piperatus (Bull.:Fr.) Bat. **Strobilomycetaceae**
= *Boletus piperatus* Fr.
= *Suillus piperatus* (Fr.) Kuntze

Medium-sized bolete with greasy sienna cap, angular tawny pores, stem with rust flush and bright yellow base; solitary or in small groups on soil with coniferous and broad-leaf trees.

Dimensions cap 4 - 8 cm dia; stem 4 - 7 cm tall x 0.5 - 2 cm dia.
Cap sienna with ochraceous or cinnamon tinges; convex or bun-shaped, then expanded, smooth, shiny when dry, greasy when damp. Flesh buff or pallid ochraceous, unchanging, thick and soft.
Pores at first cinnamon, becoming rust at maturity, angular, soon widening. Tubes concolorous with pores, slightly decurrent.
Spores ochraceous or cinnamon, smooth, thick-walled, elongated ellipsoid or sub-fusiform, without droplets, 8-11 x 3-4 µm. Basidia 4-spored. Cystidia cylindrical or slightly ventricose.
Stem pallid ochraceous or cinnammon, discolouring rust, base lemon-yellow, taperingslightly upwards. Ring absent. Flesh buff but chrome-yellow in base, unchanging, soft.
Odour not distinctive. **Taste** peppery.
Chemical tests flesh green with $FeSO_4$.
Occurrence late summer to early autumn; infrequent.
☐ Edible.

Porphyrellus porphyrosporus (Fr.) Gilb.
= *Porphyrellus pseudoscaber* (Sécr.) Sing.
= *Boletus pseudoscaber* Sécr.
= *Boletus porphyrosporus* (Fr. & Hok.) Bat.   **Strobilomycetaceae**

Medium to large bolete with dull brown velvety cap and stem, buff or brown pores; solitary or in small groups on soil with broad-leaf and, occasionally, coniferous trees.

Dimensions cap 4 - 15 cm dia.; stem 4 - 14 cm tall x 1 - 3 cm dia.
Cap hazel, sepia or snuff-brown; convex, becoming irregularly expanded, at first downy, becoming smooth with age. Flesh dirty buff, tinged vinaceous, blue-green above the tubes, thick, firm or some-what spongy.
Pores at first buff, soon vinaceous-olive, blue-green where bruised, small, rounded. Tubes concolorous with pores, adnate. **Spores** vinaceous-brown, smooth, thick-walled, ellipsoid or sub-fusiform, with droplets, 12-16 x 5-6.5 µm. Basidia 4-spored. Gill edge cystidia clavate, in groups; gill face cystidia fusiform.
Stem concolorous with cap, more or less equal, at first downy then smooth. Ring absent. Flesh concolorous and reacting as in cap, firm, stuffed.
Odour sour. **Taste** sour.
Chemical tests purplish or chestnut with NH_3.
Occurrence summer to early autumn; rare. (Abergavenny, Gwent)
■ Inedible.

Suillus bovinus (L.: Fr.) Kuntze   **Strobilomycetaceae**
= *Boletus bovinus* L.: Fr.

Medium-sized bolete with slimy yellowish cap, olive-green angular pores and pale stem; solitary or in small groups on soil under coniferous trees, favouring Scots Pine.

Dimensions cap 3 - 10 cm dia; stem 4 - 6 cm tall x 0.5 - 1 cm dia.
Cap ochraceous or pinkish-clay, more pallid at margin; convex or bun-shaped, smooth, sticky or viscid when damp. Flesh white with pinkish-clay tinge, thick and soft.
Pores at first pallid olive or buff, becoming more ochraceous with age, angular, compound, unequal, large. Tubes greyish with vinaceous tinge, more or less decurrent. **Spores** pallid olive-green or brown, smooth, thick-walled, ellipsoid sub-fusiform, without droplets, 8-10 x 3-4 µm. Basidia 4-spored. Cystidia cylindrical - clavate, sparse.
Stem pallid yellowish-sienna, more or less equal or tapering at base. Ring absent. Flesh white with pinkish-clay tinges at base, soft.
Odour of fruit. **Taste** not distinctive.
Chemical tests none.
Occurrence autumn; common.
☐ Edible.
Note: the mycelium of this species is pink.

Suillus collinitus Fr.   **Strobilomycetaceae**
= *Suillus fluryi* Huijsman

Largish bolete with slimy reddish-brown cap, chrome-yellow pores and stem with pink tinge at base; solitary or in small groups on soil under coniferous trees, favouring Scots Pine.

Dimensions cap 7 - 11 cm dia; stem 4 - 7 cm tall x 1 - 2 cm dia.
Cap reddish or chestnut-brown, radially fibrillose; convex or bun-shaped, smooth, sticky or viscid when damp. Flesh pallid yellow, thick and soft.
Pores at first chrome-yellow, becoming more olivaceous with age and bruising brown, angular, small. Tubes concolorous, more or less decurrent. **Spores** pallid yellow, smooth, thick-walled, ellipsoid, with droplets, 7.5-9 x 3.5-4.5 µm. Basidia 4-spored. Gill edge cystidia clavate; gill face cystidia absent.
Stem concolorous with pores but punctate with fine darker dots and tinged pink at base, more or less equal or thickened at base. Ring absent. Flesh pallid yellow and soft.
Odour sour. **Taste** not distinctive.
Chemical tests flesh coral with NH_3.
Occurrence autumn; rare. (Gower Coast, South Wales)
☐ Edible.
Note: the mycelium of this species is pink.

Suillus flavidus (Fr.: Fr.) Sing.   **Strobilomycetaceae**

Smallish bolete with slimy yellow cap, yellowish pores and stem with jelly-like ring; solitary or trooping on soil, specifically under pines.

Dimensions cap 3 - 8 cm dia; stem 3 - 8 cm tall x 0.3 - 0.8 cm dia.
Cap yellow with ochraceous or lemon tinges; at first flatly conical, then convex or flattened, smooth or somewhat sulcate, viscid even in dry conditions. Flesh yellowish, unchanging, spongy, thin other than in centre.
Pores golden-yellow, becoming more dingy with age, angular, large. Tubes concolorous with pores, adnate. **Spores** more or less hyaline, smooth, thick-walled, ellipsoid, with droplets, 7-8.5 x 3-4 µm. Basidia 4-spored. Cystidia cylindrical or clavate, with apical encrustations.
Stem yellowish, more or less equal, with longitudinal fibrils below ring. Ring pallid brown, gelatinous, sub-apical. Flesh yellow, unchanging, somewhat fragile.
Odour not distinctive. **Taste** not distinctive.
Chemical tests none.
Occurrence summer to autumn; in Britain limited to Scottish Highlands, elsewhere in Europe rare. (Culbin Forest, Morayshire)
☐ Edible but poor.

Suillus granulatus (Fr.) Kuntze **Strobilomycetaceae**
= *Boletus granulatus* Fr.

Medium-sized bolete with slimy yellowish cap, small yellow pores exuding distinctive droplets and pale stem; solitary or in small groups on soil under or near coniferous trees.

Dimensions cap 3 - 9 cm dia; stem 4 - 8 cm tall x 0.7 - 1 cm dia.
Cap ochraceous or rust; convex or bun-shaped, smooth, shiny when dry, sticky or viscid when damp. Flesh pallid lemon-yellow, unchanging, moderate and soft.
Pores at first pallid lemon-yellow or buff, exuding droplets which darken on drying, circular, small. Tubes concolorous with pores, adnate or more or less decurrent. **Spores** ochraceous or sienna-brown, smooth, thick-walled, ellipsoid, sub-fusiform, with droplets, 8-10 x 2.5-3.5 µm. Basidia 4-spored. Gill edge cystidia narrowly clavate; gill face cystidia more or less absent.
Stem pallid yellowish, granular towards apex, exuding milky droplets, with pinkish tinge towards base, more or less equal or slightly swollen at base. Ring absent. Flesh straw-yellow, unchanging and soft.
Odour not distinctive. **Taste** not distinctive.
Chemical tests none.
Occurrence autumn; common.
☐ Edible.

Suillus grevillei (Klotzsch) Sing. **Strobilomycetaceae**
= *Boletus elegans* Schum.: Fr. Larch Bolete

Medium to large bolete, slimy, with yellow cap, small pores, whitish ring on stem; solitary or in small groups on soil, with larch.

Dimensions cap 3 - 10 cm dia; stem 5 - 7 cm tall x 1.5 - 2 cm dia.
Cap chrome-yellow, rust with age; convex or bun-shaped, smooth, shiny and sometimes wrinkled when dry, sticky or viscid when damp. Flesh pallid lemon-yellow, unchanging, moderate, soft.
Pores at first pallid lemon-yellow, becoming more ochraceous and tinged rust where bruised, angular, small. Tubes pallid yellow, ageing more ochraceous, slightly decurrent. **Spores** ochraceous or sienna-brown, smooth, thick-walled, ellipsoid, sub-fusiform, with droplets, 8-11 x 3-4 µm. Basidia 2- or 4-spored. Cystidia cylindrical or slightly ventricose.
Stem concolorous with cap, tinged more rust below, slightly woolly or granular, sometimes with a netted appearance, slender, more or less equal or slightly swollen at base. Ring pallid whitish, superior, pointing upwards. Flesh concolorous with cap but darker towards base, unchanging and soft.
Odour not distinctive. **Taste** not distinctive.
Chemical tests none.
Occurrence late summer to early autumn; common.
☐ Edible.

Suillus laricinus (Berk.) Kuntze **Strobilomycetaceae**
= *Suillus aeruginascens* (Sécr.) Snell
= *Boletus viscidus* L.: Fr.

Medium to large bolete with pale slimy flecked cap, dirty yellowish pores and pale stem with ring; solitary or in small groups on soil, with larch.

Dimensions cap 3 - 10 cm dia; stem 5 - 10 cm tall x 1 - 2 cm dia.
Cap pallid buff with olivaceous or hazel-brown blotching; convex or bun-shaped, smooth, margin sometimes with velar remnants, viscid when damp, satiny when dry. Flesh cream, unchanging or slightly blue where cut, moderate and spongy.
Pores at first pallid, buff, tinged sulphur-yellow, then vinaceous, angular, large. Tubes concolorous with pores, more or less decurrent.
Spores snuff-brown tinged vinaceous, smooth, thick-walled, ellipsoid or sub-fusiform, without droplets, 10-12 x 4-5.5 µm. Basidia 4-spored. Cystidia cylindrical or clavate with browish encrustations.
Stem pallid straw above ring, greyish below, more or less equal. Ring white or grey-brown, sub-apical. Flesh cream with olivaceous tinge at base, unchanging or slightly blue where cut, firm.
Odour not distinctive. **Taste** not distinctive.
Chemical tests none.
Occurrence summer to autumn; rare in Britain, more common in Europe. (Forest of Dean, Gloucestershire)
☐ Edible but poor.

Suillus luteus (L.: Fr.) S F Gray   Strobilomycetaceae
= *Boletus luteus* Fr.   Slippery Jack

Medium to large bolete with slimy brown cap, small yellow pores, and darkish zoned ring; solitary or in small groups on soil with coniferous trees, favouring Scots Pine.

Dimensions cap 5 - 10 cm dia; stem 5 - 10 cm tall x 2 - 3 cm dia.
Cap chestnut or sepia-brown; convex or bun-shaped, smooth, shiny and somewhat wrinkled when dry, very glutinous or mucilaginous when damp, margin sometimes retaining velar remnants. Flesh whitish with yellow tinge, unchanging, moderate and soft.
Pores at first pallid lemon-yellow or straw, becoming more dirty yellow or sienna-brown with age, circular, medium. Tubes concolorous with pores, adnate. **Spores** ochraceous or buff, smooth, thin-walled, elongated, ellipsoid, sub-fusiform, with droplets, 7-10 x 3-3.5 µm. Basidia 4-spored. Gill edge cystidia cylindrical-clavate; gill face cystidia absent.
Stem pallid straw soon discolouring with glandular sepia-brown dots, stout, more or less equal or slightly swollen at base. Ring at first whitish, darkening with age, vinaceous-brown below, large, lax. Flesh whitish with vinaceous tinge in base, unchanging and soft.
Odour not distinctive. **Taste** not distinctive.
Chemical tests none.
Occurrence autumn; common.
☐ Edible but poor.

Suillus tridentinus (Bres.) Sing.   Strobilomycetaceae
= *Boletus tridentinus* Bres.
= *Ixocomus tridentinus* (Bres.) Bat.

Medium to large bolete with slimy, orange cap, angular yellowish pores and stem with ring; solitary or in small groups on soil with larch.

Dimensions cap 5 - 12 cm dia; stem 4 - 7.5 cm tall x 1.5 - 2 cm dia.
Cap orange or rust, yellowish towards margin; convex, becoming expanded with small darker brown adpressed scales embedded in gluten and sometimes indistinct. Flesh lemon-yellow with salmon-pink tinge, unchanging or salmon deepening, moderate and soft.
Pores at first yellowish, becoming more orange and finally rust with age, angular, compound, broad except at margin. Tubes concolorous, adnate and more or less decurrent. **Spores** brown, smooth, thin-walled, elongated-ellipsoid or sub-fusiform, with droplets, 10-13 x 4-5 µm. Basidia 4-spored. Cystidia cylindrical or clavate.
Stem yellowish-orange decorated with fine rust netting below ring, more or less equal or tapering slightly upwards. Ring yellowish, sometimes zone-like. Flesh lemon-yellow, unchanging and soft.
Odour not distinctive. **Taste** not distinctive.
Chemical tests flesh vinaceous with NH_3.
Occurrence autumn; infrequent or rare. (Stourhead, Wiltshire)
■ Inedible.

Suillus variegatus (Fr.) Kuntze   Strobilomycetaceae
= *Boletus variegatus* Fr.

Medium to large bolete with slimy, tawny-yellow, scaly cap, small olive-green pores and yellowish stem; solitary or in small groups on soil under or near coniferous trees.

Dimensions cap 6 - 13 cm dia; stem 5 - 9 cm tall x 1.5 - 2 cm dia.
Cap ochraceous or olivaceous-brown; convex or bun-shaped, at first slightly downy, becoming more greasy or sticky with age, with small darker brown adpressed scales. Flesh pallid ochraceous, unchanging or occasionally with blue tinge, moderate and soft.
Pores at first ochraceous with olivaceous tinge, becoming more cinnamon-brown with age, sometimes bluish where bruised, sub-angular, compound, medium. Tubes dark ochraceous or buff, adnate. **Spores** brown, smooth, elongated-ellipsoid, sub-fusiform, with droplets, 9-11 x 3-4 µm. Basidia 4-spored. Cystidia cylindrical or clavate with limited brownish encrustations.
Stem pallid ochraceous, more lemon-yellow towards apex, more rust below, more or less equal or tapering slightly upwards. Ring absent. Flesh pallid ochraceous, unchanging or occasionally with blue tinges, soft.
Odour strong, fungoid. **Taste** not distinctive.
Chemical tests none.
Occurrence summer to early autumn; infrequent.
☐ Edible.

Strobilomyces floccopus (Wahl.: Fr.) Karst.

= *Boletus floccopus* Wahl.: Fr.
= *Boletus strobilaceus* Fr.   **Strobilomycetaceae**

Medium to large bolete with greyish-brown shaggy scaly cap, whitish-grey pores and scaly stem; solitary or in small scattered groups on soil in broad-leaf or coniferous woods.

Dimensions cap 5 - 12 cm dia; stem 8 - 12 cm tall x 1 - 2 cm dia.
Cap pallid greyish-brown, sometimes with olivaceous tinge; convex or bun-shaped, with large, thick, ragged scales, some overhanging margin. Flesh white, turning coral-pink, then vinaceous and finally brown where cut, thick and firm.
Pores at first whitish, becoming pallid grey, reddening where bruised, angular, large. Tubes concolorous with pores and reacting similarly where cut, depressed. **Spores** black with violaceous tinge, reticulate, sub-spherical or broadly ellipsoid, 10-12 x 8.5-11 µm. Basidia 2- or 4-spored. Cystidia fusiform-ventricose.
Stem concolorous with cap but more pallid grey above, covered with shaggy scales, more or less equal. Ring rough, sheathing. Flesh concolorous with and reacting as in cap, firm.
Odour not distinctive. **Taste** not distinctive.
Chemical tests none.
Occurrence summer; rare. (Forest of Dean, Gloucestershire)
☐ Edible.

Chroogomphus rutilus (Schaeff.: Fr.) Miller

= *Gomphidius rutilus* (Schaeff.: Fr.) Lund. & Nanff.
= *Gomphidius viscidus* L.: Fr.   **Gomphidiaceae**

Medium to large bolete with reddish-brown slimy cap, pale or purplish gills and yellowish stem; solitary or in scattered groups in coniferous woods, favouring grassy pathsides adjacent to pines.

Dimensions cap 3 - 12 cm dia; stem 5 - 10 cm tall x 0.4 - 0.8 cm dia.
Cap reddish-brown or brick with vinaceous tinge; convex or bun-shaped, sometimes bluntly umbonate, heavily viscid when damp, becoming shiny when dry. Flesh vinaceous, moderate and firm.
Gills at first buff with olivaceous tinge, becoming purplish when mature, deeply decurrent, close. **Spores** black with fuscous tinge, sub-fusiform, without droplets, 15-24 x 5.5-7 µm. Basidia 4-spored. Cystidia cylindrical, with some encrustations.
Stem yellowish-buff, more pallid with vinaceous tinge above, bright chrome-yellow towards base, covered with cottony velar remnants, slightly viscid, more or less equal but narrowing abruptly at apex. Ring absent. Flesh concolorous with cap but deep chrome-yellow at base, firm and solid.
Odour not distinctive. **Taste** not distinctive.
Chemical tests none.
Occurrence autumn; locally common.
☐ Edible but uninspiring.

Gomphidius roseus (Fr.) Karst.   **Gomphidiaceae**

Small to medium bolete with slimy coral-red cap, greyish-white gills and stem with ring zone; solitary or in scattered groups on soil in coniferous woods, probably only with pines.

Dimensions cap 3 - 5 cm dia; stem 2.5 - 4.5 cm tall x 0.4 - 1 cm dia.
Cap coral-red with vinaceous tinge; convex or bun-shaped, becoming expanded and flattened, heavily viscid when damp, becoming shiny when dry. Flesh dirty white, moderate and firm.
Gills at first whitish, becoming tinged olivaceous-grey when mature, deeply decurrent, thick. **Spores** black with fuscous tinge, sub-fusiform, without droplets, 15.5-20 x 5-5.5 µm. Basidia 4-spored. Cystidia cylindrical with brownish encrustations.
Stem dirty white, tinged cap colour, viscid or dry, more or less equal. Ring white, glutinous, zone-like. Flesh concolorous with cap but sometimes dirty yellow at base, firm, solid.
Odour not distinctive. **Taste** not distinctive.
Chemical tests flesh vinaceous with NH_3; yellowish-green with $FeSO_4$.
Occurrence autumn; infrequent.
■ Inedible.

Gyrodon lividus (Bull.: Fr.) Sacc.   Gyrodontaceae

= *Boletus lividus* Fr.
= *Uloporus lividus* (Fr.) Quél.
= *Gyrodon sistotremoides* (Fr.) Opat.

Medium to large bolete with slimy yellowish cap, sulphur-yellow pores and yellow stem; solitary or in small groups, with alder.

Dimensions cap 4 - 10 cm dia; stem 5 - 8 cm tall x 1 - 2 cm dia.
Cap yellow, straw or pallid cinnamon; convex, becoming irregularly expanded, at first viscid, becoming dry with age. Flesh pallid yellow, tinged blue above tubes, firm or somewhat spongy.
Pores dark sulphur-yellow, greenish where bruised and discolouring brown with age, labyrinthine. Tubes concolorous with pores but becoming more sulphur-yellow, decurrent. **Spores** yellowish, smooth, thin-walled, broadly ellipsoid, with droplets, 4.5-6.5 x 3.5-4.5 µm. Basidia 2- or 4- spored. Gill edge cystidia fusiform or clavate, sparse; gill face cystidia more or less absent.
Stem concolorous with cap, then brownish, more or less equal, slender, fibrillose. Ring absent. Flesh pallid yellow, rust in base.
Odour not distinctive. **Taste** not distinctive.
Chemical tests negative with NH_3.
Occurrence summer and early autumn; rare. (Forest of Dean, Gloucestershire)
☐ Edible.

Gyroporus castaneus (Bull.: Fr.) Quél.   Gyrodontaceae

Bolete of variable size with tawny brown cap and stem, and whitish pores; solitary or in small groups on soil, typically with oak.

Dimensions cap 3 - 10 cm dia; stem 3 - 9 cm tall x 1 - 3 cm dia.
Cap tawny-brown or cinnamon; convex, becoming irregularly expanded, at first downy, becoming smooth with age. Flesh white, unchanging, thin and firm.
Pores whitish, discolouring straw with age, small. Tubes concolorous with pores, more or less free. **Spores** pallid straw, smooth, thick-walled, ellipsoid, with droplets, 8-11 x 4.5-6 µm. Basidia 4-spored. Gill edge cystidia fusiform; gill face cystidia absent.
Stem concolorous with cap or more pallid at apex, more or less equal, smooth. Ring absent. Flesh white, unchanging, stuffed or full.
Odour not distinctive. **Taste** not distinctive.
Chemical tests negative with NH_3.
Occurrence summer and early autumn; rare. (New Forest, Hampshire)
☐ Edible.

Rhizopogon luteolus Fr.   Rhizopogonaceae

Small yellowish-brown tuberous fruiting body; in clusters, half-buried in sandy soil.

Dimensions 2 - 5 cm dia.
Fruiting body peridium ochraceous, roughly sub-spherical, covered with brownish mycelial threads, thick, tough, cracking. Gleba (spore mass) at first pallid, becoming olivaceous when mature.
Spores olivaceous-brown, elongated, ellipsoid, smooth, 6-10 x 2.5-3.5 µm.
Odour not distinctive. **Taste** not distinctive.
Chemical tests none.
Occurrence autumn; in Britain infrequent other than in Scottish Highlands, in Europe more frequent.
■ Inedible

HOMOBASIDIOMYCETES   GASTROMYCETES

This major group is characterised typically, though not always, by the development of the spore-bearing tissue, not in the form of a distinct hymenium but as a somewhat amorphous mass of scattered cells *(gleba)* enclosed wholly within a sac-like wall *(peridium)* which may be sessile or become raised up on a stem. The immature structure is usually rounded, onion-shaped, thimble-shaped or more irregular.

In the Lycoperdales [80], which include the Puffballs and Earthstars, the fruiting body takes the form of a rounded, thin, papery peridium which ruptures, either by an apical pore or by breaking down more irregularly to expose the mass of spore-forming gleba. In the Lycoperdaceae, the fertile tissue is either more or less sessile, attached to the substrate by thin mycelial cords, or is elevated on a sterile pedestal, and some members achieve, or exceed, the size of footballs, the ripe gleba containing tens of millions of spores. In the Geastraceae, a multi-layered peridium develops in which the thick outer layers *(exoperidium)* split into segments, reflex back and down, and carry the inner papery sac *(endoperidium)* enclosing the gleba, upwards on stellate arms.

In the limited order of Tulostomatales [5] the gleba is carried upwards on a tough stem and the spores are dispersed by air currents when the papery peridium ruptures either by an apical pore *(Tulostoma)* or by breaking away wholly *(Battarraea)*, whilst in the Phallales [15], including the Stinkhorns, the spores are embedded in a foul-smelling, slimy gleba, which erupts from a soft, leathery 'egg', either (Phallaceae) as a phallic-shaped 'head' borne on a spongy stem, or (Clathraceae) on the surface of a net-like, more or less sessile sphere, or of squid-like arms.

The Sclerodermatales [15] which include the Earthballs may, at first sight, appear similar to the Lycoperdales but the peridium is much tougher with a thick and leathery feel and it breaks down irregularly. In the very rare species, *Pisolithus*, a number of *periodioles* are borne collectively in a knobbly mass raised up on an irregular and partly subterranean stem. In the Nidulariales [15] or Birdsnest fungi, the small sessile fruiting body is more or less thimble-shaped and contains a number of tiny peridioles, like miniature bird's eggs, protected until they mature by a membrane stretched across the top of the thimble.

Bovista limosa Rostrup   **Lycoperdaceae**

Small white or brownish rounded structure attached to the substrate by mycelial threads; solitary or in scattered troops on dry, sandy or calcareous soil, favouring dunes.

Dimensions 0.5 - 1.5 cm dia x 0.5 - 1.5 cm tall.
Fruiting body exoperidium whitish, smooth, falling away in flakes at maturity to reveal reddish-brown papery endoperidium; sub-spherical, opening by a raised, conical, corrugated apical pore, slightly pointed below and mounted on a short straight sterile 'stem'; becoming detached and blowing about in the wind. Gleba (spore mass) at first white and firm, becoming reddish-brown, powdery.
Spores brown, minutely warty, spherical, 4-6 µm. Basidia 4-spored. Cystidia absent.
Odour not distinctive. **Taste** not distinctive.
Chemical tests none.
Occurrence summer to autumn; very rare - in Britain only reported from dunes in Lancashire and South Wales. (Gower Peninsula, South Wales)
■ Inedible.

Bovista plumbea Pers.: Pers.   **Lycoperdaceae**

Small, white, rounded structure attached to the substrate by several strands; solitary or in scattered troops on lawns, in short grass and pastures.

Dimensions 2 - 3 cm dia x 2 - 3 cm tall.
Fruiting body exoperidium white, smooth, falling away in flakes at maturity to reveal dark grey papery endoperidium; sub-spherical, slightly pointed below, the fertile head merging abruptly into a cluster of mycelial strands which may rupture allowing the fruiting body to roll around in the wind (brown remains often persisting through the winter), opening by an apical pore. Gleba (spore mass) at first white and firm, becoming clay-brown and finally olive-brown, powdery.
Spores brown, smooth, ellipsoid or sub-spherical, 4.5-6 x 4.5-5.5 µm. Basidia 4-spored. Sterigmata up to 15 µm. Cystidia absent.
Odour not distinctive. **Taste** not distinctive.
Chemical tests none.
Occurrence late summer to autumn; infrequent.
☐ Edible. For culinary purposes the specimens must be used when white throughout.

Bovista pusilla (Batsch) Pers.   **Lycoperdaceae**

Small, white or brownish rounded structure attached to the substrate by mycelial threads; solitary or in scattered troops on acid soil in grasslands.

Dimensions 1 - 3 cm dia x 1 - 3 cm tall.
Fruiting body exoperidium whitish, becoming tan or brown, at first smooth, soon coarsely branny-flaky and splitting into distinct patches, falling away in flakes at maturity to reveal greyish-brown papery endoperidium; sub-spherical, opening by a raised, conical, corrugated apical pore, wrinkled below and attached by grey-brown mycelial cords. Gleba (spore mass) at first white and firm, becoming olivaceous-brown, powdery. **Spores** brown, minutely warty, spherical, 3.5-5.5 µm. Basidia 4-spored. Cystidia absent.
Odour not distinctive. **Taste** not distinctive.
Chemical tests none.
Occurrence summer to autumn; infrequent.
■ Inedible.

Calvatia gigantea (Batsch : Pers.) Lloyd   **Lycoperdaceae**
Giant Puff Ball

= *Langermannia gigantea* (Batsch : Pers.) Rostk.
= *Lasiosphaera gigantea* (Batsch : Pers.) Smarda
= *Lycoperdon giganteum* Batsch : Pers.

White, sometimes very large, rounded structure attached by a
mycelial cord; typically in ones and twos or in small troops on soil
adjacent to wooded areas, gardens, hedgerows and parks.

Dimensions 7 - 80 cm dia x 7 - 80 cm tall.
Fruiting body exoperidium white, smooth, flaking away to reveal
smooth whitish endoperidium which slowly turns brown and disintegrates as the spore mass matures; irregularly sub-spherical, without
pedestal but attached to the substrate by mycelial cords which rupture, allowing the mature fruiting body to roll about in the wind. Gleba
(spore mass) at first white and firm, becoming olive-brown, powdery.
Spores olive-brown, minutely warty, spherical, sometimes with
droplets, 3.5-5.5 µm. Basidia 4-spored. Cystidia absent.
Odour not distinctive. **Taste** not distinctive.
Chemical tests none.
Occurrence summer to autumn; uncommon but may be locally more
frequent.
☐ Edible and good. For culinary purposes the specimens must be
used when white throughout.

Handkea excipuliformis (Pers.) Kreisel   **Lycoperdaceae**

= *Calvatia excipuliformis* (Pers.) Perdek
= *Lycoperdon saccatum* Schaeff.: Fr.
= *Lycoperdon excipuliformis* Schaeff.: Pers.

Pale buff or brown, pestle-shaped structure; often solitary, but also in
small troops on soil in pastures, woodlands, heaths and wasteland.

Dimensions 3 - 10 cm dia x 8 - 20 cm tall.
Fruiting body exoperidium pallid buff becoming dull brown, covered
with short spines or warts, falling away to reveal ochraceous,
smooth, papery endoperidium, splitting irregularly from the apex,
leaving a brown cup and stem; the fertile head tapering downwards
into a sterile basal region which is clearly demarcated and which
occurs in varying lengths (remains persisting through the winter).
Gleba (spore mass) at first white and firm, becoming ochraceous-
brown, then purple brown, powdery.
Spores olive-brown, warty, spherical, 5.5 - 3.5 µm. Basidia 4-spored.
Sterigmata up to 12 µm. Cystidia absent.
Odour not distinctive. **Taste** not distinctive.
Chemical tests none.
Occurrence late summer to autumn; common.
☐ Edible. For culinary purposes the specimens must be used when
white throughout.

Handkea utriformis (Bull.: Pers.) Kreisel   **Lycoperdaceae**

= *Calvatia caelata* (Bull.: Pers.) Morg.
= *Calvatia utriformis* (Bull.: Pers.) Jaap
= *Lycoperdon caelatum* Bull.: Pers.

Whitish pear-shaped structure; solitary or in small troops on soil in
dry grassy places, heaths, and edges of woodland.

Dimensions 6 - 12 cm dia.
Fruiting body exoperidium white or pallid grey, coarsely warty, the
warts falling away to reveal smooth papery endoperidium with a reticulate pattern, squatly pear-shaped or sub-spherical, becoming grey-
brown with age, splitting irregularly from the apex, finally leaving a
brownish cup and stem; fertile head tapering towards a short, stout,
sterile basal region (remains may persist through the winter). Gleba
(spore mass) at first white and firm, becoming yellowish-brown and
finally olive-brown, powdery. **Spores** olive-brown, finely warty,
spherical, 4-5 µm. Basidia 1-4-spored. Sterigmata up to 5 µm.
Cystidia absent.
Odour not distinctive. **Taste** not distinctive.
Chemical tests none.
Occurrence summer to autumn; infrequent.
☐ Edible. For culinary purposes the specimens must be used when
white throughout.

Lycoperdon echinatum Pers.   **Lycoperdaceae**

Smallish, whitish or brown, spiny, rounded structure on a short pedestal; typically in troops on soil in broad-leaf woods and on heaths.

Dimensions 2 - 5 cm dia x 3 - 6 cm tall.
Fruiting body exoperidium whitish then brown, covered with comparatively long spines (4-7 mm), convergent in groups at tips, falling away to reveal brown papery endoperidium decorated with a reticulate pattern; sub-spherical opening by an apical pore, the fertile head tapering or pinching down into a distinct, short, sterile, spongy basal region. Gleba (spore mass) at first white and firm, becoming sepia-brown, powdery. **Spores** brown, minutely spiny, spherical, 4-6 μm. Basidia 2-4-spored. Sterigmata up to 5 μm. Cystidia absent.
Odour not distinctive. **Taste** not distinctive.
Chemical tests none.
Occurrence summer to autumn; infrequent.
■ Inedible.
The only member of the family with distinctly long spines.

Lycoperdon foetidum Bon.   **Lycoperdaceae**
= *Lycoperdon perlatum var nigrescens* Pers.
= *Lycoperdon nigrescens* (Pers.) Lloyd

Smallish, dark brown, spiny, rounded structure on a short pedestal; typically in troops on acid soil on heaths, and in coniferous and mixed woods.

Dimensions 1 - 4 cm dia x 1.5 - 3 cm tall.
Fruiting body exoperidium pallid brown covered with dark brown spines, fused in groups at the, falling away to reveal brown papery endoperidium decorated with a faint reticulate pattern; sub-spherical opening by an apical pore, the fertile head tapering or pinching down into a distinct but very short, sterile, spongy basal region. Gleba (spore mass) at first white and firm becoming sepia brown, powdery. **Spores** brown, minutely spiny, spherical, 4-5 μm. Basidia 2-4-spored. Sterigmata up to 5 μm. Cystidia absent.
Odour not distinctive. **Taste** not distinctive.
Chemical tests none.
Occurrence summer to autumn; locally frequent.
■ Inedible.

Lycoperdon lividum Pers.   **Lycoperdaceae**
= *Lycoperdon spadiceum* Pers.

Smallish, ochre-brown, rounded structure on a pedestal; typically in troops on sandy soil on heaths, dunes and pastures.

Dimensions 1 - 2.5 cm dia x 1.5 - 3 cm tall.
Fruiting body exoperidium greyish or ochre-brown, covered with coarse warts, falling away to reveal brown endoperidium decorated with a faint reticulate pattern; sub-spherical opening by an apical pore, the fertile head tapering or pinching down into a distinct, sterile, spongy basal region. Gleba (spore mass) at first white and firm becoming brown, powdery. **Spores** olive-brown, minutely warty or spiny, spherical, 3.5-4.5 μm. Basidia 1-3-spored. Sterigmata up to 20 μm. Cystidia absent.
Odour not distinctive. **Taste** not distinctive.
Chemical tests none.
Occurrence summer to autumn; infrequent.
■ Inedible.

Lycoperdon mammaeforme Pers.: Pers.

Lycoperdaceae

Smallish, white, rounded structure decorated with woolly patches showing pink beneath, on a pedestal; solitary or in troops on calcareous soil under broad-leaf trees, favouring beech.

Dimensions 3 - 5 cm dia x 3 - 5 cm tall.
Fruiting body exoperidium white, splitting into large woolly or floccose patches and revealing endoperidium, pallid pinkish when young, more brown with age, and opening by a distinct pore; pear-shaped or capitate, narrowing into a distinct sterile, spongy, basal region which may be very short or more distinct. Gleba (spore mass) at first white and firm, becoming yellowish-brown, powdery.
Spores brown, minutely spiny, spherical, 4.5-5.5 μm. Basidia 2-4-spored.
Sterigmata generally not seen. Cystidia absent.
Odour not distinctive. **Taste** not distinctive.
Chemical tests none.
Occurrence autumn; rare. (Leigh Woods, Avon)
■ Inedible.

Lycoperdon molle Pers.: Pers.   **Lycoperdaceae**

Smallish, pale, ochre-brown rounded structure on a pedestal; solitary or more typically in troops on soil in broad-leaf or coniferous woods.

Dimensions 2 - 4 cm dia x 2 - 6 cm tall.
Fruiting body exoperidium ochre-brown, covered with short soft spines, falling away to reveal creamy-brown endoperidium; sub-spherical opening by an apical pore, the fertile head tapering down into a distinct, sterile, basal region. Gleba (spore mass) at first white and firm, becoming olive-brown and finally dark brown, powdery.
Spores olive-brown, coarsely warty, spherical, 4-5 μm. Basidia 2-4-spored. Sterigmata up to 20 μm. Cystidia absent.
Odour not distinctive. **Taste** not distinctive.
Chemical tests none.
Occurrence summer to autumn; infrequent and more southern European in distribution.
■ Inedible.

Lycoperdon perlatum Pers.: Pers.   **Lycoperdaceae**
= *Lycoperdon gemmatum* Batsch

Smallish, white or dull brown, rounded structure on a pedestal; typically in troops on soil in mixed woodland.

Dimensions 2.5 - 6 cm dia x 2 - 9 cm tall.
Fruiting body exoperidium white, becoming ochre-brown, covered with short pyramidal warts which fall off to reveal endoperidium decorated with a reticulate pattern; sub-spherical opening by an apical pore, the fertile head tapering down into a distinct, sterile, basal region. Gleba (spore mass) at first white and firm, becoming olive-brown, powdery. **Spores** olive-brown, minutely warty, spherical, 3.5-4.5 μm. Basidia 2-4-spored. Sterigmata up to 7 μm. Cystidia absent.
Odour not distinctive. **Taste** not distinctive.
Chemical tests none.
Occurrence summer to autumn; common.
□ Edible. For culinary purposes the specimens must be used when white throughout.

Lycoperdon pyriforme Schaeff.: Pers.   **Lycoperdaceae**

Smallish, white, or grey-brown, pear-shaped structure on a pedestal; typically in troops on rotten wood or stumps, sometimes submerged so that the fruiting bodies appear to be growing on soil.

Dimensions 1.5 - 4 cm dia x 1 - 5 cm tall.
Fruiting body exoperidium of short spines, granules or warts, white becoming grey-brown which falls away to reveal whitish, smooth, inner peridium opening by an apical pore; the fertile head tapering down into a distinct, spongy, sterile, basal region. Gleba (spore mass) at first white and firm, becoming olive-brown, powdery.
Spores olive-brown, smooth, spherical, 3-4 µm. Basidia 2-4-spored. Sterigmata up to 10 µm. Cystidia absent. The brown capillitium threads are unusual for the genus in that they bear no trace of hyaline pores.
Odour not distinctive. **Taste** not distinctive.
Chemical tests none.
Occurrence summer to autumn; common.
☐ Edible. For culinary purposes the specimens must be used when white throughout.

Lycoperdon umbrinum Pers.: Pers.   **Lycoperdaceae**

Smallish, dark brown, rounded structure on a short pedestal; typically trooping on soil in or near coniferous woods.

Dimensions 2 - 5 cm dia x 2 - 4 cm tall.
Fruiting body exoperidium of short, dark brown spines which falls away to reveal endoperidium, whitish yellow, thin, smooth not reticulated, opening by an apical pore; the fertile head tapering down into a distinct, sterile, basal region. Gleba (spore mass) at first white and firm, becoming olive-yellow or olive-brown, powdery.
Spores yellowish-olive, minutely warty, spherical, 4.5-5.5 µm. Basidia 2-spored. Sterigmata very slender, up to 7 µm. Cystidia absent.
Odour not distinctive. **Taste** not distinctive.
Chemical tests none.
Occurrence summer to autumn; common.
☐ Edible. For culinary purposes the specimens must be used when white throughout.

Vascellum pratense (Pers.) Kreisel.   **Lycoperdaceae**
= *Vascelium depressum* (Bon.) Smarda
= *Lycoperdon depressum* Bon.
= *Lycoperdon hiemale* Vitt.

Smallish white or pallid yellowish-brown, more or less rounded structure on a pedestal; typically in small troops on lawns and in other grassy places with short grass.

Dimensions 2 - 4 cm dia x 2 - 5 cm tall.
Fruiting body exoperidium at first white, becoming ochraceous and finally pallid brown, coarsely scurfy, falling away to reveal pallid ochraceous, smooth, papery endoperidium opening at first by a small apical pore; the fertile head tapering down into a spongy, sterile, basal region clearly demarcated by a membrane (eventually the upper part breaks away wholly leaving a brown bowl and stem). Gleba (spore mass) at first white and firm, then olive-brown, powdery. **Spores** olive-brown, finely warty, sub-spherical, 3.5-4.5 x 3.5-4 µm. Basidia 1-4-spored. Sterigmata slender, up to 5 µm. Cystidia absent.
Odour not distinctive. **Taste** not distinctive.
Chemical tests none.
Occurrence summer to autumn; common.
☐ Edible. For culinary purposes the specimens must be used when white throughout.

Geastrum coronatum Pers.   Geastraceae

= *Geastrum limbatum* Fr.

Greyish stalked 'bulb' surmounting a star-shaped reflexed brownish base which raises the spore sac above the surrounding substrate; sometimes solitary but more often in small trooping groups on soil amongst leaf litter in broad-leaf and coniferous woods.

Dimensions unopened bulb 3 - 5 cm dia, fully expanded 5 - 10 cm dia from the tips of rays; spore sac 1 - 2.5 cm dia.
Fruiting body dirty grey-brown, coarsely scaly, bulb-shaped, sessile, the thick, brittle outer peridium splitting at maturity into 5 - 8 pointed starfish-like rays which reflex back to reveal the grey-brown inner fleshy layer; inner peridium (spore sac) sub-spherical, thin and papery, surmounting a short thick stem, opening by a non-elevated apical pore. Gleba (spore mass) at first pallid and firm, becoming brown, powdery. **Spores** dark brown, warty, sub-spherical, 3.5-4.5 µm. Basidia 2-4-spored. Cystidia absent.
Odour not distinctive. **Taste** not distinctive.
Chemical tests none.
Occurrence late summer to early autumn; rare. (Thetford Forest, Norfolk)
■ Inedible.
Note Breitenbach and Kränzlin make the species synonymous with *G. quadrifidum*.

Geastrum fimbriatum Fr.   Geastraceae

= *Geastrum sessile* (Sow.) Pouz.

Cream 'bulb' surmounting a star-shaped reflexed concolorous base which raises the spore sac above the surrounding substrate; sometimes solitary but more often in small trooping groups on soil amongst leaf litter in broad-leaf woods.

Dimensions unopened bulb 2 - 3 cm dia, fully expanded 2 - 5 cm dia from the tips of rays; spore sac 0.8 - 1.5 cm dia.
Fruiting body creamy-ochraceous, coarsely scaly, bulb-shaped, sessile, the thick, brittle outer peridium splitting at maturity into 5 - 8 pointed starfish-like rays which reflex back to reveal the cream inner fleshy layer; inner peridium (spore sac) sub-spherical, thin and papery, sessile, opening by a slightly elevated apical pore. Gleba (spore mass) at first pallid and firm, becoming brown, powdery.
Spores dark brown, finely spiny, sub-spherical, 2.5-3.5 µm. Basidia 2-4-spored. Cystidia absent.
Odour not distinctive. **Taste** not distinctive.
Chemical tests none.
Occurrence late summer to early autumn; infrequent.
■ Inedible.

Geastrum fornicatum (Huds.) Fr.   Geastraceae

Greyish-brown 'bulb' surmounting a whitish star-shaped base, which raises the spore sac above the surrounding substrate; sometimes solitary but more often in small trooping groups on rich soil amongst organic debris in broad-leaf woods and hedgerows, often favouring hollow rotting stumps.

Dimensions unopened bulb 1.5 - 4 cm dia, fully expanded 3 - 8.5 cm dia from the tips of rays; spore sac 1 - 3 cm dia.
Fruiting body outer peridium brown, coarsely granular, bulb-shaped, sessile, splitting at maturity into 4 tall, narrow, pointed starfish-like rays, resting on a basal membrane which splits into four similar upturned rays with the tips of the two sets meeting; upper rays reflex back to reveal a brownish inner layer; inner peridium (spore sac) brownish-grey corm-shaped surmounting a short stem, up to 3 mm long, with a distinct pallid apical collar, thin, papery, opening by an apical fringed pore. Gleba (spore mass) at first pallid and firm, becoming brown, powdery. **Spores** dark brown, warty, sub-spherical, 3.5-4.5 µm. Basidia 2-4-spored. Cystidia absent.
Odour not distinctive. **Taste** not distinctive.
Chemical tests none.
Occurrence late summer to autumn; rare. (Abergavenny, Gwent)
■ Inedible.

Geastrum pectinatum Pers.   Geastraceae
= *Geastrum plicatum* Berk.
= *Geastrum tenuipes* Berk.

Grey pointed 'bulb' surmounting a whitish star-shaped base, which raises the spore sac above the surrounding substrate; sometimes solitary, more often in small trooping groups, on rich soil in broad-leaf and coniferous woods.

Dimensions unopened bulb 1.5 - 4 cm dia, fully expanded 3 - 8.5 cm dia from the tips of rays; spore sac 1 - 3 cm dia.
Fruiting body outer peridium brown, coarsely granular, bulb-shaped, sessile, splitting at maturity into 6 - 12 pointed starfish-like rays, resting on an ephemeral basal membrane which reflex back to reveal a yellowish-brown inner layer; inner peridium (spore sac) bluish-grey corm-shaped surmounting a short stem, up to 5 mm long, with a distinct pallid apical collar, thin, papery, opening by an apical pore on a conical, furrowed 'beak' without a 'halo'. Gleba (spore mass) at first pallid and firm, becoming brown, powdery.
Spores dark brown, warty or spiny, sub-spherical, 5.5-7.5 µm. Basidia 2-4-spored. Cystidia absent.
Odour not distinctive. **Taste** not distinctive.
Chemical tests none.
Occurrence late summer to autumn; rare. (Ebbor Gorge, Somerset)
■ Inedible.

Geastrum quadrifidum Pers.: Pers.   Geastraceae
= *Lycoperdon coronatum* Scop.

Grey pointed 'bulb' surmounting a whitish star-shaped base, which raises the spore sac above the surrounding substrate; sometimes solitary, more often in small trooping groups, on soil amongst leaf litter in coniferous woods.

Dimensions unopened bulb 1.5 - 3 cm dia, fully expanded 2 - 4 cm dia from the tips of rays; spore sac 0.5 - 2.5 cm dia.
Fruiting body outer peridium brown, coarsely granular, bulb-shaped, sessile, splitting at maturity into 4 - 8 pointed starfish-like fleshy rays, resting on a basal membrane embedded in the substrate which reflex back to reveal a whitish inner layer; inner peridium (spore sac) bluish-grey bulb-shaped surmounting a short stem, up to 2.5 mm long, with a distinct apical collar, thin, papery, opening by an apical pore surrounded by a pallid 'halo'. Gleba (spore mass) at first pallid and firm, becoming brown, powdery. **Spores** dark brown, warty or spiny, sub-spherical, 3.5-5 µm. Basidia 2-4-spored. Cystidia absent.
Odour not distinctive. **Taste** not distinctive.
Chemical tests none.
Occurrence late summer to autumn; rare. (Nettlecombe, West Somerset)
■ Inedible.

Geastrum rufescens Pers.: Pers.   Geastraceae
= *Geastrum vulgatum* Vitt.

Creamy 'bulb' surmounting a star-shaped concolorous base, which raises the spore sac above the surrounding substrate; sometimes solitary, more often in small trooping groups, on soil amongst leaf litter in broad-leaf and coniferous woods.

Dimensions unopened bulb 3 - 4 cm dia, fully expanded 5 - 8 cm dia from the tips of rays; spore sac 2 - 3 cm dia.
Fruiting body at first pale creamy-pink, becoming more uniformly brown, coarsely scaly, bulb-shaped and sessile, the thick, brittle outer peridium splitting at maturity into 4 - 8 pointed starfish-like rays which reflex back to reveal the creamy-pink inner fleshy layer, soon becoming wholly cream coloured; inner peridium (spore sac) sub-spherical, thin and papery, sessile, opening by a fringed apical pore. Gleba (spore mass) at first pallid and firm, becoming brown, powdery.
Spores dark brown, warty, spherical, 3.5-4.5 µm. Basidia 2-4-spored. Cystidia absent.
Odour not distinctive. **Taste** not distinctive.
Chemical tests none.
Occurrence late summer to early autumn; infrequent.
■ Inedible.
The main distinction between this species and *G. triplex* lies in the absence of a saucer sub-tending the spore sac.

Geastrum schmidelii Vitt.   Geastraceae
= *Geastrum nanum* Pers.

Brown pointed 'bulb' surmounting a whitish star-shaped base, which
raises the spore sac above the surrounding substrate; sometimes
solitary, more often in small trooping groups, on light sandy soil
amongst grass and moss, often in sand dunes.

Dimensions unopened bulb 2 - 4 cm dia, fully expanded 2 - 5 cm
dia from the tips of rays; spore sac 1 - 3 cm dia.
Fruiting body outer peridium brown, coarsely granular, bulb-shaped,
sessile, splitting at maturity into 5 - 8 pointed starfish-like rays,
resting on an ephemeral basal membrane which reflex back to reveal
a yellowish-brown inner layer; inner peridium (spore sac) greyish-
brown corm-shaped surmounting a short stem, up to 3 mm long, thin,
papery, opening by an apical pore on a conical, furrowed 'beak' without a 'halo'. Gleba (spore mass) at first pallid and firm, becoming
brown, powdery. **Spores** dark brown, warty or spiny, sub-spherical,
4-5 μm. Basidia 2-4-spored. Cystidia absent.
Odour not distinctive. **Taste** not distinctive.
Chemical tests none.
Occurrence rare. (Aberystwyth, Dyfed, Wales)
■ Inedible.
The fruiting bodies may persist in good condition over winter.

Geastrum striatum DC.   Geastraceae

Dull greyish-brown pointed 'bulb' surmounting a more brownish star-
shaped base, which raises the spore sac above the surrounding substrate; sometimes solitary, more often in small trooping groups, on
rich soil amongst organic debris in broad-leaf and coniferous woods,
also in hedgerows and hollow rotting stumps.

Dimensions unopened bulb 2 - 4 cm dia; when fully expanded
3 - 6 cm dia from the tips of rays; spore sac 2 - 3 cm dia.
Fruiting body outer peridium brown, coarsely fibrous, scaly, bulb-
shaped, sessile, splitting at maturity into 6 - 9 pointed starfish-like
rays, resting on an ephemeral basal membrane which reflex back to
reveal a more pallid brown inner layer; inner peridium (spore sac)
brownish-grey bulb-shaped surmounting a short stem, up to 5 mm
long, with a distinct pallid apical collar, thin, papery, opening by an
apical pore on a conical, furrowed 'beak' surrounded by a 'halo'.
Gleba (spore mass) at first pallid, firm, becoming brown, powdery.
Spores dark brown, finely warty, sub-pherical, 3.5-4.5 μm. Basidia
2-4-spored. Cystidia absent.
Odour not distinctive. **Taste** not distinctive.
Chemical tests none.
Occurrence late summer to autumn; rare. (Pilton, Somerset)
■ Inedible.
The fruiting bodies may persist in good condition over winter.

Geastrum triplex Jungh.   Geastraceae

Creamy white 'bulb' surmounting a star-shaped base which raises
the spore sac above the surrounding substrate; sometimes solitary,
more often in small trooping groups, on soil amongst leaf litter in
broad-leaf woods.

Dimensions unopened bulb 3 - 5 cm dia, fully expanded 5 - 10 cm
dia from the tips of rays; spore sac 2.5 - 4 cm dia.
Fruiting body at first brownish, coarsely scaly, bulb-shaped, sessile,
the thick, brittle outer peridium splitting at maturity into 4 - 8 pointed
starfish-like rays which reflex back to reveal the cream-coloured inner
fleshy layer; inner peridium (spore sac) creamy-greyish, sub-spherical, thin and papery, sessile, resting on concolorous saucer-like
base, opening by an apical pore surrounded by a distinct, more pallid
ring or halo. Gleba (spore mass) at first pallid and firm, becoming
brown, powdery. **Spores** dark brown, warty, spherical, 3.5-4.5 μm.
Basidia 2-4-spored. Sterigmata long, up to 20 μm. Cystidia absent.
Odour not distinctive. **Taste** not distinctive.
Chemical tests none.
Occurrence late summer to autumn; infrequent.
■ Inedible.
The rays typically split transversely leaving the spore sac and bowl
resting on what appears as a layered base.

Myriostoma coliformis (With. :Pers.) Corda. **Geastraceae**
= *Geastrum coliformis* With.: Pers.

Pale brown 'bulb' surmounting a star-shaped reflexed buff or brown base which raises the spore sac above the surrounding substrate; sometimes solitary but more often in small trooping groups on sandy soil on woodland margins.

Dimensions unopened bulb 2 - 5 cm dia, fully expanded 3 - 10 cm dia from the tips of rays; spore sac 1.5 - 5 cm dia.
Fruiting body brown, fleshy, bulb-shaped, sessile, the brittle outer peridium splitting at maturity into 5 - 12 pointed starfish-like rays which reflex back to reveal the pallid buff inner layer, later becoming brown; inner peridium (spore sac) greyish brown, sub-spherical, thin and papery, sessile, opening by several pores reminiscent of a pepper pot. Gleba (spore mass) at first pallid and firm, becoming brown, powdery. **Spores** brown, irregularly warty, sub-spherical, 4-6 µm. Basidia 2-4-spored. Cystidia absent.
Odour strong when drying, of curry or stock cubes. **Taste** not distinctive.
Chemical tests none.
Occurrence late summer to autumn; not in Britain, localised in Europe
■ Inedible.

Tulostoma brumale Pers.: Pers.　　　**Tulostomataceae**
= *Tulostoma mammosum* Fr.

Small, pale, grey-brown, more or less rounded head surmounting a tough brown stem; in small trooping groups on sandy soil and dunes.

Dimensions spore sac 0.8 - 1.2 cm dia; stem 2 - 4 cm tall x 0.2 - 0.3 cm dia.
Fruiting body peridium sub-spherical with small papillate projection perforated by an ostiole, exoperidium soon disappearing, leaving a papery endoperidium; borne on an ochraceous-brown, stiff, smooth stem. The dried fruiting bodies often persist. Gleba (spore mass) at first whitish and firm, becoming brown, powdery. **Spores** yellowish, finely warty, spherical, 3.5-5 µm. Basidia 2-4-spored. Cystidia absent.
Odour not distinctive. **Taste** not distinctive.
Chemical tests none.
Occurrence summer to early autumn; rare. (Gower Coast, South Wales)
■ Inedible.

Battarraea phalloides (Dicks.) Pers.　　　**Battarraeaceae**

Brown, more or less rounded spore mass surmounting a shaggy ochre brown stem; in small trooping groups on sandy soil, probably associated with submerged rotting wood.

Dimensions spore sac 2.5 - 9 cm dia; stem 10 - 25 cm tall x 1 - 2.5 cm dia.
Fruiting body peridium sub-spherical, borne on a shaggy, stiff, ochre-brown stem terminating in a basal, largely submerged volva. The dried fruiting bodies often persist for more than one season. Gleba (spore mass) at first whitish and firm, becoming brown, powdery. **Spores** brown, finely warty, sub-spherical (occasionally ovoid), 5-6.5 µm. Basidia 2-4-spored. Cystidia absent.
Odour not distinctive. **Taste** not distinctive.
Chemical tests none.
Occurrence summer to autumn; very rare in southern Britain, more frequent in central Europe. (Blyford, Suffolk)
■ Inedible.
Note: the size of this species is deceptive compared with the much smaller species in the illustration of *T. brumale*.

Crucibulum laeve (Huds.) Kam.   Nidulariaceae
= *Crucibulum vulgare* Tul.   Common Bird's Nest Fungus

Small brown cup containing whitish 'eggs' in the bottom; in clusters on wood, twigs and other organic debris including overwintering cereal stubble (may be present in large numbers).

Dimensions 0.4 - 0.8 cm dia x 0.3 - 0.7 cm tall.
Fruiting body receptacle at first sub-spherical, becoming deeply cup-shaped, outer surface brownish-yellow, inner surface silvery-white one-layered, covered by an ochraceous membrane (operculum) which ruptures at maturity to reveal several greyish-white lens-shaped 'eggs' (peridioles) at the base of the cup and attached by fine mycelial strands. The receptacle becomes blackish-brown at maturity. Flesh soft but resistant. **Spores** hyaline, smooth, ovoid, non-amyloid, 7-10 x 3-5 μm. Basidia 2-4-spored. Cystidia absent.
Odour not distinctive. **Taste** not distinctive.
Chemical tests none
Occurrence autumn to winter; infrequent.
■ Inedible.

Cyathus olla (Batsch: Pers.) Pers.   Nidulariaceae

Small greyish-brown trumpet containing whitish 'eggs' in the bottom; in dense clusters, on twigs and other organic debris favouring dead stems of *Compositae*; also on soil.

Dimensions 0.6 - 0.8 cm dia x 0.8 - 1.5 cm tall.
Fruiting body receptacle at first cylindrical or top-shaped, outer surface ochraceous-grey covered in finely felty hairs, inner surface greyish-silver, covered by a whitish membrane (operculum) which ruptures at maturity to reveal several greyish-white lens-shaped 'eggs' (peridioles) at the base of a trumpet-shaped cup and attached by fine mycelial strands. Flesh soft but resistant. **Spores** hyaline, smooth, ovoid-ellipsoid, non-amyloid, 10-14 x 6-8 μm. Basidia 2-4-spored. Cystidia absent.
Odour not distinctive. **Taste** not distinctive.
Chemical tests none
Occurrence spring to autumn; infrequent.
■ Inedible.

Cyathus striatus (Huds.) Pers.   Nidulariaceae

Small brown 'teacup' containing whitish 'eggs' in the bottom; in dense clusters, on soil, twigs and other organic debris.

Dimensions 0.8 - 1.2 cm dia x 0.8 - 1.5 cm tall.
Fruiting body receptacle inverted bell- or cup-shaped, outer surface reddish-brown covered in tufts of coarse hairs, inner surface greyish and fluted, covered by a whitish membrane (operculum) which ruptures at maturity to reveal several greyish-white lens-shaped 'eggs' (peridioles) at the base of the cup and attached by fine mycelial strands. Flesh soft but resistant. **Spores** hyaline, smooth, ovoid, non-amyloid, 15-22 x 3.5-12 μm. Basidia 4-spored; cystidia absent.
Odour not distinctive. **Taste** not distinctive.
Chemical tests none
Occurrence spring to autumn; infrequent.
■ Inedible.

Pisolithus arhizus (Pers.) Rausch.   Sclerodermataceae
= *Pisolithus arenarius* A. & S.
= *Pisolithus tinctorius* (Mich.: Pers.) Coker & Couch

Brown, knobbly fruiting body narrowing into a buried, thick, base tinged yellow; solitary in well-drained soil, including sand and gravel pits and coal waste heaps.

Dimensions 5 - 10 cm dia x 8 - 25 cm tall.
Fruiting body brown, irregular knobbly structure, reminiscent of fused horse droppings, the outer wall rupturing to reveal the pebble-shaped spore-bearing peridioles; narrowing into a thick, subterranean, basal region with distinctive partial chrome-yellow coloration. Spore mass brown, powdery. **Spores** brown, finely warty, spherical, 7-11.5 µm. Basidia 2-4-spored. Cystidia absent.
Odour not distinctive. **Taste** not distinctive.
Chemical tests none.
Occurrence autumn; very rare in Britain, more frequent in central Europe. (New Forest, Hampshire)
■ Inedible.

Scleroderma areolatum Ehrenb.   Sclerodermataceae
= *Scleroderma lycoperdoides* Schw.

Yellowish-brown, hard, rounded structure attached by thick rooting strands; typically in small troops on bare soil or amongst sparse moss and other vegetation in damp places.

Dimensions 2 - 4 cm dia.
Fruiting body yellowish-brown peridium covered with smooth darker brown scales leaving a dotted and netted pattern where rubbed away, thin, leathery, tough, opening by an irregular pore or fissure; the sub-spherical to spherical fruit body attached to the sub-strate by a dense mass of thick mycelial cords. Gleba (spore mass) at first whitish and firm, then marbled white and brown, finally deep purple brown, powdery. **Spores** dark brown, decorated with long spines, spherical, 9-14 µm. Basidia 2-4-spored. Cystidia absent.
Odour not distinctive. **Taste** not distinctive.
Chemical tests none.
Occurrence late summer to autumn; infrequent.
■ Inedible.

Scleroderma citrinum Pers.   Sclerodermataceae
= *Scleroderma aurantium* (Vaill.) Pers.   Common Earth Ball
= *Scleroderma vulgare* Horn.

Yellowish, hard, rounded scaly structure attached by thin threads; in small troops on bare, typically sandy, soil or amongst moss on heaths and in mixed woods, often favouring banks.

Dimensions 2 - 10 cm dia.
Fruiting body dirty-yellowish or ochre-brown peridium decorated with coarse scales, thick, leathery, tough, opening by an irregular fissure when mature; the sub-spherical to spherical fruit body attached to the substrate by mycelial threads or cords. Gleba (spore mass) at first whitish and firm, then marbled white and brown, finally deep purple-brown or black, powdery. **Spores** dark brown, with a netted ornamentation, spherical, 9-13 µm. Basidia 2-4-spored. Cystidia absent.
Odour strong, of gas or acetylene. **Taste** not distinctive.
Chemical tests none.
Occurrence late summer to autumn; very common.
■ Inedible. Some Continental authors advocate use of the young powdered fruiting body as a condiment but the practice is not to be recommended.

Scleroderma verrucosum (Bull.) Pers.
Sclerodermataceae

Yellowish, hard, rounded scaly structure tapering into a thick-ribbed basal region; in small troops on bare, typically sandy, soil on heaths and in mixed woods.

Dimensions 2.5 - 6 cm dia.
Fruiting body dirty-yellowish or ochre-brown peridium, decorated with small brownish scales, thin, leathery, tough, opening by an irregular fissure when mature; the sub-spherical to spherical fruit body attached to the substrate by a thick, ribbed or grooved stem-like structure made up of massed mycelial cords. Gleba (spore mass) at first whitish and firm, then marbled white and brown, finally deep olive-brown, powdery. **Spores** dark brown, decorated with spines or warts, spherical, 8-12 μm. Basidia 2-4-spored. Cystidia absent.
Odour not distinctive. **Taste** not distinctive.
Chemical tests none.
Occurrence late summer to autumn; infrequent.
■ Inedible.

Astraeus hygrometricus (Pers.) Morg.   **Astraeaceae**
= *Geastrum hygrometricum* Pers.

Pale greyish 'bulb' surmounting a star-shaped brownish base which opens flattish reflexed in damp conditions and closes up when dry; sometimes solitary but more often in small trooping groups on sandy soil in woods and on open dunes.

Dimensions unopened bulb 2 - 3.5 cm dia, fully expanded 2 - 5 cm dia from the tips of rays; spore sac 1- 3 cm dia.
Fruiting body tan or darker brown, leathery, bulb-shaped, sessile, the thick outer peridium splitting at maturity into 6 - 15 pointed starfish-like rays which reflex outwards when damp to reveal the brownish inner fleshy layer; inner peridium (spore sac) sub-spherical, greyish, thin and papery, sessile, opening by an apical pore. Gleba (spore mass) at first pallid and firm becoming dark brown, powdery. **Spores** cinnamon-brown, finely warty, spherical, 7-10 μm. Basidia 2-4-spored. Cystidia absent.
Odour not distinctive. **Taste**: not distinctive.
Chemical tests none.
Occurrence late summer to early autumn; very rare in Britain, rare in Europe. (Near Truro, Cornwall)
■ Inedible.
Note: the photograph is of dried material collected in 1966.

Mutinus caninus (Huds.: Pers.) Fr.   **Phallaceae**
Dog Stinkhorn

Phallic extrusion from a partially submerged 'egg', whitish-yellow other than an orange fertile head covered with dark olive-green foul-smelling slime; typically in small troops on soil amongst leaf litter in mixed woods.

Dimensions egg 1 - 2 cm dia; receptacle 10 - 12 cm tall x 1 - 1.5 cm dia.
Fruiting body whitish 'egg', elongated sub-spherical, consisting of a delicate, rubbery outer membrane enclosing a gelatinous matrix separated from the embryonic spore mass and stem by an inner membrane; attached to the substrate by a mycelial cord; at maturity the egg ruptures and the spore mass is carried rapidly upward on a weak stem-like receptacle, yellowish-buff to bright orange, hollow-pitted, fragile, surmounted by orange-red, narrowly conical head carrying the spore mass. Gleba (spore mass) dark olive-green.
Spores pallid yellow, smooth, ellipsoid, 4-5 x 1-2 μm. Basidia 6-spored. Cystidia absent.
Odour faint, sickly. **Taste** not distinctive.
Chemical tests none.
Occurrence summer to autumn; infrequent but locally plentiful.
■ Inedible.
Note: typically the stems collapse within an hour or so of extrusion.

Phallus impudicus Pers.   **Phallaceae**
= *Ithyphallus impudicus* (L.) Fr.   Common Stinkhorn

Phallic extrusion from a partially submerged 'egg', whitish other than a fertile head covered with olive-green foul-smelling slime; scattered, associated with rotted and often submerged wood in woodlands and parks.

Dimensions egg 3 - 6 cm dia; receptacle 10 - 25 cm tall x 2 - 4 cm dia.
Fruiting body white 'egg', pear-shaped or sub-spherical, consisting of a resistant, rubbery outer membrane enclosing a gelatinous matrix separated from the embryonic spore mass and stem by an inner membrane; attached to the substrate by a mycelial cord; at maturity the egg ruptures and the spore mass is carried rapidly upward on a stem-like receptacle, whitish, spongy, hollow-pitted, fragile, surmounted by a whitish conical head carrying the spore mass. The head (when exposed after the slime is removed) bears a reticulate mesh of raised ribs. Gleba (spore mass) dark olive-green.
Spores pallid yellow, smooth, ellipsoid, 3.5-4 x 1.5-2 µm. Basidia 6-8-spored. Cystidia absent.
Odour at maturity foul. **Taste** not distinctive.
Chemical tests none.
Occurrence early summer to autumn; infrequent but locally plentiful.
☐ Edible but poor (in the 'egg' phase only)

Clathrus archeri (Berk.) Dring   **Clathraceae**
= *Anthurus archeri* (Berk.) Fischer

Arching bright red 'arms' extruding from a partially submerged egg, the arms covered with greenish spore mass, foul-smelling; typically in small troops on soil amongst leaf litter in shrubberies and in grassy places close by trees, sometimes on wood chips.

Dimensions egg 2 - 3 cm dia; receptacle 4 - 10 cm tall x variable dia.
Fruiting body ochraceous-brown marbled 'egg', elongated sub-spherical, consisting of a delicate but leathery, outer membrane enclosing the compressed arms surrounding the olive-green gleba; attached to the substrate by mycelial threads; at maturity the egg ruptures and the spore mass is carried upward on the inside of 4 - 6 bright red arms, at first joined at the tips but then extending outwards. Gleba (spore mass) dark olive-green. **Spores** hyaline (green in the mass), smooth, elongated-ellipsoid, 5-6.5 x 2-2.5 µm. Basidia 4-8-spored. Cystidia absent.
Odour foul, of rotting meat. **Taste** not distinctive.
Chemical tests none.
Occurrence summer to autumn; rare, spreading slowly in southern counties of England, mainly in more southern parts of Europe. (Lullingstone, Kent)
■ Inedible.

Clathrus ruber Pers.   **Clathraceae**
= *Anthurus cancellatus* Tournef.

Lattice-like red sphere extruding from a partially submerged 'egg', the lattice covered (inside) with slimy olive-green foul-smelling spore mass; typically in small troops but also solitary on soil amongst leaf litter in gardens, shrubberies and in grassy places close by trees.

Dimensions egg 2 - 3 cm dia; receptacle 4 - 10 cm tall x variable dia.
Fruiting body whitish 'egg', sub-spherical, consisting of a delicate but leathery outer membrane enclosing the compressed lattice surrounding the olive-green gleba; attached to the substrate by a mycelial cord; at maturity the egg ruptures and the spore mass is carried upward on the inside of a bright red lattice. Gleba (spore mass) dark olive-green. **Spores** greenish-yellow, smooth, cylindrical, with droplets, 5-6 x 1.6-2 µm. Basidia 4-8-spored. Cystidia absent.
Odour foul, of rotting meat. **Taste** not distinctive.
Occurrence summer to autumn; rare, spreading slowly in southern counties of England, though mainly Mediterranean. (Exmouth, Devon)
Chemical tests none.
■ Inedible.

HETEROBASIDIOMYCETES

This comparatively small group characterised, in general, by a rather gelatinous structure to the fruiting bodies, which are capable of withstanding desiccation and subsequent revival, is commonly referred to as the Jelly Fungi. Microscopically the Heterobasidiomycetes differ from the rest of the Basidiomycotina in the structure of the *basidium* which is either divided by longitudinal or transverse septa or is split into prongs like a tuning fork. In one family, the Exobasidiaceae, the basidia are not divided but are long and sinuous.

In the Dacrymycetales [40], which grow on wood, fruiting bodies are typically yellow or orange in colour, smallish or microscopic, spherical, top-shaped (*Dacrymyes, Femsjonia*) or in the form of prongs or antlers (*Calocera*) and grow gregariously. The hymenial surfaces are exposed during development and all possess tuning-fork basidia.

The Tremellales [140] constitute the largest order including four families of which the largest is the Tremellaceae, again growing chiefly on wood or, less frequently, on soil, though some members parasitise other fungi. The fruiting bodies are often lobed or brain-like with colours ranging through white, yellow, pink, brown and almost black. Some, including *Pseudohydnum,*

develop spiny hymenial surfaces and there are many encrusting species (though not represented here). Again hymenial surfaces are exposed though basidia are typically divided by longitudinal septa into hypobasidia and carry long extensions or *sterigmata.*

The Exobasidiales [30] includes one family, the Exobasidiaceae, of which only a common representative species is included here, and grow parasitically as encrustations, typically white and smooth, on the leaves and twigs of green plants including *Vaccinium* and *Rhododendron*. These forms develop only rudimentary fruiting bodies on or within the host tissues and the infection causes parts of the host plant to become stunted or otherwise deformed, often producing swelling and a reddish colour change in the leaves. The basidia in these species are generally long and cylindrical, without septa.

The Auriculariales [50], including a single family, the Auriculariaceae, also grow on wood. Most are obscure but the commonest members include the Jew's Ear fungus (*Auricularia auricula-judae*). They are typically ear-shaped or lobed, often wrinkled, with colours ranging from brown to grey. In these species the hymenial surfaces are exposed and basidia possess transverse septa with small outgrowths known as *epibasidia.*

Calocera cornea (Batsch: Fr.) Fr.   **Dacrymycetaceae**

Yellow pencil-like gelatinous fruiting bodies; typically densely crowded on dead and rotting branches and twigs of broad-leaf trees.

Dimensions 0.4 - 1.3 cm tall
Fruiting body yellowish when damp, more orange-yellow when dry, awl- or spine-shaped, simple or very rarely forked, smooth. Flesh yellow, gelatinous but tough. **Spores** hyaline, smooth, allantoid or elongated-ellipsoid, non-amyloid, 1-septate at maturity, 7-10 x 3-4 µm. Basidia 2-spored shaped as tuning forks. Cystidia absent.
Odour not distinctive. **Taste** not distinctive.
Chemical tests none.
Occurrence throughout the year but mainly summer and autumn; frequent.
■ Inedible.

Calocera glossoides (Pers.: Fr.) Fr.   **Dacrymycetaceae**
= *Dacryomitra glossoides* (Pers.) Bref.

Yellow club-shaped gelatinous fruiting bodies; typically densely crowded, or in smaller trooping groups, on dead and rotting stumps, branches and twigs of broad-leaf trees, favouring oak.

Dimensions 0.3 - 1 cm tall
Fruiting body yellowish when damp, darker and more brownish when dry, clavate, simple, upper part typically compressed or wrinkled lengthwise, otherwise smooth. Flesh yellow, gelatinous but tough. **Spores** hyaline, smooth, allantoid or elongated-ellipsoid, non-amyloid, 1-3-septate at maturity, 12-17 x 3-5 µm. Basidia 2-spored, shaped as tuning forks. Cystidia absent.
Odour not distinctive. **Taste** not distinctive.
Chemical tests none.
Occurrence autumn; infrequent.
■ Inedible.

Calocera pallido-spathulata Reid   **Dacrymycetaceae**

Yellow, flatly spatulate, gelatinous fruiting bodies; typically densely crowded, or in smaller trooping groups, on dead and rotting stumps, branches and twigs of coniferous trees, favouring larch and spruce.

Dimensions 0.3 - 1 cm tall.
Fruiting body pallid yellowish, upper part flatly clavate or spatulate, simple, wrinkled lengthwise, otherwise smooth, narrowing into a distinct stem. Flesh yellow, gelatinous but tough. **Spores** hyaline, smooth, allantoid or elongated-ellipsoid, non-amyloid, 1-3-septate at maturity, 10-15 x 3.5-4 µm. Basidia 2-spored shaped as tuning forks. Cystidia absent.
Odour not distinctive. **Taste** not distinctive.
Chemical tests none.
Occurrence autumn; infrequent.
■ Inedible.

Calocera viscosa (Pers.: Fr.) Fr.   Dacrymycetaceae

Yellow, antler-like, gelatinous fruiting bodies; in tufts on stumps and roots of coniferous trees, often attached to roots by long stems which penetrate deeply into the substrate, thus appearing as if growing on soil.

Dimensions 3 - 10 cm tall.
Fruiting body golden-yellow when damp, more orange-yellow when dry, antler-shaped with repeated dichotomous branching, the forks acutely angled, compressed and grooved above, cylindrical and smooth below, somewhat slimy when wet. Flesh yellow, gelatinous but tough. **Spores** hyaline, smooth, elongated-ellipsoid, non-amyloid, 0-1-septate at maturity, 9-12 x 3.5-4.5 µm. Basidia 2-spored, shaped as tuning forks. Cystidia absent.
Odour not distinctive. **Taste** not distinctive.
Chemical tests none.
Occurrence autumn; common.
■ Inedible.

Dacrymyces stillatus Nees : Fr.   Dacrymycetaceae
= *Dacrymyces delequescens* (Mérat) Duby

Small, orange-yellow, cushion-like, gelatinous fruiting bodies; typically densely crowded, and often merging, on dead and rotting wood of broad-leaf and coniferous trees including structural timbers.

Dimensions 0.1 - 0.5 cm dia.
Fruiting body yellowish-orange when damp, deeper orange when dry, sub-spherical or more saucer-shaped, smooth, slightly glistening, somewhat wrinkled with age. Flesh orange-yellow, gelatinous, soft but firm. **Spores** orange-yellow, smooth, allantoid or elongated-ellipsoid, non-amyloid, 1-3-septate at maturity, 12-15 x 5-6 µm. Basidia 2-spored shaped as tuning forks. Cystidia absent.
Odour not distinctive. **Taste** not distinctive.
Chemical tests none.
Occurrence throughout the year but fruiting mainly in late summer and autumn; common.
■ Inedible.

Femsjonia pezizaeformis (Lév.) Karst. Dacrymycetaceae
= *Ditiola pezizaeformis* (Lév.) Karst.

Yellow, cushion- or cup-like, gelatinous fruiting bodies; in small or large groups on dead and rotting branches and twigs of broad-leaf trees, favouring oak.

Dimensions 0.5 - 1.5 cm dia.
Fruiting body pallid yellow when damp, more custard-yellow when dry, awl, or cushion-like, flattened or concave, slightly wavy margin, smooth. Flesh pallid yellow, gelatinous and soft.
Spores pallid yellow, smooth, elongated-ellipsoid or cylindrical, non-amyloid, 7-15-septate at maturity, 18-25 x 8-10.5 µm. Basidia 2-spored, shaped as tuning forks very large (up to 100 µm). Cystidia absent.
Odour not distinctive. **Taste** not distinctive.
Chemical tests none.
Occurrence summer to autumn; infrequent.
■ Inedible.

Exobasidium vaccinii (Fuckel) Woron.   Exobasidiaceae

Whitish distorted powdery 'skin', covering leaves of Ericaceae, favouring *Vaccinium* species and some cultivated azaleas.

Dimensions variable cm dia x 0.01 - 0.05 cm thick.
Fruiting body spore-bearing tissue white or tinged pink, powdery, closely applied to the upper surface of the host leaf and, less frequently, to the shoot. Flesh very thin, mould-like. **Spores** white, smooth, irregularly elongated-cylindrical, 1-6 septate, 10-20 x 2.5-5 µm. Basidia 4-5-spored. Cystidia absent.
Odour not distinctive. **Taste** not distinctive.
Chemical tests none
Occurrence spring to early autumn; infrequent.
■ Inedible.

Exidia glandulosa Fr.   Tremellaceae
= *Exidia plana* (Wigg.) Donk   Witches' Butter

Blackish, contorted disc-shape, gelatinous stalked or sessile fruiting bodies; in small or large groups on dead and rotting logs and branches of broad-leaf trees, also on wound tissue on living wood.

Dimensions 10 - 30 cm dia.
Fruiting body blackish-brown when damp, wholly black when dry, expanses of tissue contorted into brain-like folds and fused with adjacent sporophores, smooth and shiny with small glandular warts, sessile. Flesh when damp, black, gelatinous, soft; when dry, hard, shiny, thin and membraneous. **Spores** hyaline, smooth, cylindrical or allantoid, non-amyloid, 12-14 x 4.5-5 µm. Basidia 4-spored, pear-shaped and longitudinally septate. Cystidia absent.
Odour not distinctive. **Taste** not distinctive.
Chemical tests none.
Occurrence throughout the year but fruiting mainly in late summer and autumn; infrequent.
■ Inedible.

Exidia thuretiana (Lév.) Fr.   Tremellaceae
= *Exidia albida* (Huds.) Bref.

Pure white, contorted, gelatinous fruiting bodies; in small or large groups on dead and rotting branches and twigs of broad-leaf trees, favouring beech.

Dimensions 0.2 - 1 cm dia but fusing into larger masses
Fruiting body pure white, small specimens drying almost invisible, at first cushion-like becoming contorted into brain-like folds and fused with adjacent fruit bodies, smooth and shiny. Flesh white, gelatinous and tough. **Spores** hyaline, smooth, cylindrical or allantoid, non-amyloid, 13-20 x 5-7 µm. Basidia 2-4-spored, subspherical or pear-shaped and longitudinally septate. Cystidia absent.
Odour not distinctive. **Taste** not distinctive.
Chemical tests none.
Occurrence mainly in autumn; infrequent or rare. (Sullington Warren, West Sussex)
■ Inedible.

Exidia truncata Fr.   **Tremellaceae**
= *Exidia glandulosa* (Bull.) Fr.

Blackish, contorted disc-shape, gelatinous stalked or sessile fruiting
bodies; in small or large groups on dead and rotting branches and
twigs of broad-leaf trees, often on dead branches still attached to
living wood, favouring oak .

Dimensions 2 - 6 cm dia.
Fruiting body blackish-brown when damp, wholly black when dry, at
first disc- or cup-shaped, become contorted into brain-like folds and
fused with adjacent sporophores, smooth and shiny, sometimes
attached to the substrate by a stem. Flesh black, gelatinous and soft.
Spores hyaline, smooth, cylindrical or allantoid, non-amyloid, 14-19 x
4.5-5.5 µm. Basidia 4-spored, clavate and longitudinally septate.
Cystidia absent.
Odour not distinctive. **Taste** not distinctive.
Chemical tests none.
Occurrence throughout the year but fruiting mainly summer and
autumn; frequent.
■ Inedible.

Pseudohydnum gelatinosum (Scop.: Fr.) Karst.
= *Tremellodon gelatinosum* (Scop.: Fr.) Fr.   **Tremellaceae**
                                                    Jelly Hedgehog

Opalescent greyish or brownish, tongue-shaped, gelatinous fruiting
bodies; in small groups on dead and rotting stumps and logs of
coniferous trees.

Dimensions 2 - 6 cm dia.
Fruiting body opalescent greyish-white or greyish-brown when
damp, sometimes with bluish tinges, ligulate or spatulate, upper surface finely downy, under surface decorated with pallid spines, often
glistening. Flesh whitish, gelatinous and soft. **Spores** hyaline,
smooth, sub-spherical or broadly ellipsoid, non-amyloid, 5-6 x
4.5-5.5 µm. Basidia 4-spored, elongated pear-shape and longitudinally septate. Cystidia absent.
Odour not distinctive. **Taste** not distinctive.
Chemical tests none.
Occurrence autumn; infrequent.
■ Inedible.

Tremella foliacea (Pers.) Pers.   **Tremellaceae**

Brownish, contorted and lobed, gelatinous fruiting bodies; solitary or
in small groups on dead and rotting branches and logs of broad-leaf
trees, very rarely on coniferous wood.

Dimensions 3 - 10 cm dia.
Fruiting body brown when damp, blackish-brown when dry, at first
cushion-shaped, becoming contorted into wrinkled folds of leaf-like
lobes arising from a common base, smooth and shiny or matt. Flesh
brown, gelatinous and soft. **Spores** hyaline, smooth, broadly ellipsoid
or sub-spherical, non-amyloid, 9-11 x 6-8 µm. Basidia 2-4-spored,
sub-spherical and longitudinally septate. Cystidia absent.
Odour not distinctive. **Taste** not distinctive.
Chemical tests none.
Occurrence throughout the year but fruiting mainly late summer and
autumn; rare. (Edford Wood, near Midsomer Norton, Somerset)
■ Inedible.

Tremella lutescens (Pers.: Fr.) Donk   Tremellaceae

Cream or pale yellow, contorted, gelatinous fruiting bodies; solitary or in small groups on dead and rotting branches and twigs of broad-leaf trees.

Dimensions 2 - 8 cm dia.
Fruiting body sulphur- or lemon-yellow when damp, cream when dry, at first fairly regular, then contorted into brain-like or leaf-like folds, smooth and shiny. Flesh pallid yellowish, gelatinous, soft and flabby when damp, horny when dry. **Spores** hyaline, smooth, broadly ellipsoid, non-amyloid, 10-16 x 7-10 µm. Basidia 4-spored, clavate and longitudinally septate. Cystidia absent.
Odour not distinctive. **Taste** not distinctive.
Chemical tests none.
Occurrence throughout the year but fruiting mainly late summer and autumn; rare. (Stourhead, Wiltshire)
■ Inedible.
Considered to be a colour variant on *T. mesenterica* but apparently lacking conidia.
Note: the specimen photographed is immature.

Tremella mesenterica Retz.   Tremellaceae
Yellow Brain Fungus

Golden-yellow, contorted, gelatinous fruiting bodies; solitary or in small groups on dead and rotting branches and twigs of broad-leaf trees.

Dimensions 2 - 10 cm dia.
Fruiting body golden-yellow when damp, orange when dry, contorted into brain-like or leaf-like folds, smooth and shiny. Flesh yellow, gelatinous, soft and flabby when damp, horny when dry. **Spores** hyaline, smooth, broadly ellipsoid, non-amyloid, 10-16 x 7-8 µm. Basidia 4-spored, clavate and longitudinally septate. Cystidia absent.
Odour not distinctive. **Taste** not distinctive.
Chemical tests none.
Occurrence throughout the year but fruiting mainly late summer and autumn; common.
■ Inedible.

Tremiscus helvelloides (DC.: Pers.) Donk   Tremellaceae
= *Guepinia helvelloides* Fr.
= *Gyrocephalus rufus* Bref.
= *Phlogiotis helvelloides* (Fr.) Martin

Salmon pink, ear-shaped, gelatinous fruiting bodies; solitary or in small tufted (caespitose) groups on soil but associated with buried rotting wood, typically on shady path sides.

Dimensions 3 - 10 cm dia x 2 - 5 cm tall.
Fruiting body salmon or orange-pink, becoming tinged more brown with age, wavy conical or ear-shaped arising from a short pallid stem, smooth, sometimes faintly pruinose on the inner (upper) surface. Flesh concolorous, gelatinous and elastic, soft and flabby when damp, tough when dry. **Spores** hyaline, smooth, irregularly ellipsoid, truncated at one end, non-amyloid, 9.5-11 x 5.5-6 µm. Basidia 2-4-spored, ovoid and longitudinally septate. Cystidia absent.
Odour not distinctive. **Taste** not distinctive.
Chemical tests none.
Occurrence autumn; infrequent but widespread.
■ Inedible.

Auricularia auricula-judae (Bull.) Wettst.

= *Hirneola auricula-judae* (Bull.) Berk.   **Auriculariaceae**
= *Auricularia auricula* (Hook.) Underwood   Jew's Ear

Brown, ear-shaped, gelatinous fruiting bodies; in small or large groups on dying branches and trunks of living broad-leaf trees, favouring elder.

Dimensions 3 - 8 cm dia.
Fruiting body outer surface tan-brown, with purplish tinge when damp, covered with fine greyish down, inner surface grey-brown, overall darker when dry, shiny, smooth becoming wrinkled like an ear; inverted cup-shaped, often fused with adjacent fruiting bodies, narrowly attached. Flesh brownish-purple, gelatinous, tough and rubbery when damp, hard and brittle when dry. **Spores** hyaline, smooth, cylindrical or allantoid, non-amyloid, 17-19 x 6-8 µm. Basidia 3-spored, cylindrical and transversely septate. Cystidia absent.
Odour not distinctive. **Taste** not distinctive.
Chemical tests none.
Occurrence throughout the year but fruiting mainly late summer and autumn; common.
☐ Edible.

Auricularia mesenterica (Dicks.) Pers.   **Auriculariaceae**

Greyish-brown, at first disc-shaped then bracket-like, tiered, gelatinous fruiting bodies; in small or large groups on dead and rotting branches and logs of broad-leaf trees, also less commonly on living wood.

Dimensions 2 - 4 cm dia.
Fruiting body upper surface brownish-grey, zoned, more pallid at margin, hairy, margin lobed, at first disc-shaped, becoming expanded into densely tiered brackets, under-surface reddish-purple, wrinkled, white, pruinose. Flesh brownish, gelatinous, elastic when damp, hard and brittle when dry. **Spores** hyaline, smooth, cylindrical or allantoid, non-amyloid, 11-17 x 4-5 µm. Basidia 2-4-spored, cylindrical and transversely septate. Cystidia absent.
Odour not distinctive. **Taste** not distinctive.
Chemical tests none.
Occurrence throughout the year but fruiting mainly late summer and autumn; common.
■ Inedible.

MYXOMYCETES

The purpose of this volume is, primarily, to provide a field guide to the two 'higher classes' of fungi, the Ascomycotina and Basidiomycotina. The amateur field mycologist will, however, frequently encounter members of the Myxomycetes, commonly referred to as Slime Moulds. Most of these are microscopic and only a few are large enough to be distinctive to the naked eye.

The Myxomycetes comprise a specialist discipline but a limited number are included here to offer a representative 'feel' of the class.

They also constitute a comparatively primitive group of fungi most of which live on dead or rotting wood. In their vegetative state they consist of a mass of naked and somewhat amorphous protoplasm, the *plasmodium*, which matures as minute dry fruiting bodies enclosing a powdery spore mass.

Diderma hemisphaericum (Bull.) Horn.   Didymiaceae

Small, stalked, disc-like bodies, whitish, breaking down to reveal powdery brown spore mass; on dead leaves and other plant debris.

Dimensions fruiting body 0.05 - 0.15 cm dia.
Fruiting body plasmodium stage white; sporangia discoid on short stalks, smooth or slightly roughened at the centre, peridium 2-layered, outer surface consisting of a lime shell, crumbling to reveal the membraneous inner wall which then also disintegrates; capillitium of thin, pallid, branched and occasionally anastomosing threads.
Spores brown, sub-spherical, covered with small, pallid warts and isolated groups of larger ones, 7-9 µm.
Occurrence throughout the year; frequent.
■ Inedible.

Mucilago crustacea Wiggers   Didymiaceae
= *Spumaria mucilago* Pers.
= *Mucilago spongiosa* (Leyss.) Morgan

Irregular, very fragile and crumbly, creamy yellow masses; spreading over grasses and other plants as well as dead leaves.

Dimensions fruiting body up to 7cm dia.
Fruiting body plasmodium stage creamy-white or yellow; aethalium irregularly branching, creamy-white or pallid ochraceous, becoming more pallid grey; wall two-layered, the outer constructed of lime crystals which crumble to give a rough surface appearance, resting on a yellowish membrance (hypothallus); capillitium of thick tubes branching to form a network, purplish-brown, more pallid at ends.
Spores purplish-brown or black, sub-spherical, coarsely warty, 11-14 µm.
Occurrence throughout the year but more prevalent in the autumn; frequent.
■ Inedible.

Fuligo septica (L.) Wiggers   **Physaraceae**
= *Mucor septicus* L.

Fragile, irregularly cushion-like, yellow bodies, often spreading extensively over the substrate, breaking down to release greyish-brown spore mass; on dead wood generally.

Dimensions fruiting body 2 - 13 cm dia.
Fruiting body plasmodium stage yellow; peridium more or less rounded, lemon-yellow or ochraceous, spongy and fragile, crumbling readily, outer wall thin or absent in humid conditions; capillitium either abundant or sparse, thin colourless tubes. **Spores** brown with grey or lilaceous tinges in the mass, sub-spherical, finely warty, 7-9 μm.
Occurrence throughout the year but more prevalent in the summer and autumn; common.
■ Inedible.

Ceratiomyxa fructiculosa (Mull.) Macbr.
= *Byssus fructiculosa* Mull.   **Ceratiomyxaceae**

Colonies of minute, fragile, whitish, club-shaped bodies arranged in rosettes; on bark of fallen trees and branches generally.

Dimensions fruiting body up to 10 cm dia.
Fruiting body plasmodium stage watery, more or less translucent, inconspicuous; individual branches of fruiting body arising as simple or, occasionally, branched, white, narrowly clavate columns, spores arising from the 'paved' surface on hollow stalks. **Spores** more or less hyaline, smooth, sub-spherical or broadly oval, 10-13 x 6-7 μm.
Occurrence throughout the year; frequent.
■ Inedible.

Trichia botrytis (Gmelin) Pers.   **Trichaceae**
= *Stemonitis botrytis* Gmelin

Clusters of minute brown or black, rounded, stalked fruiting bodies breaking open to reveal yellowish spore mass with tangled threads; on dead wood and bark generally.

Dimensions fruiting body 0.1 - 0.4 cm tall x 0.06 - 0.08 cm dia.
Fruiting body plasmodium stage black or dark brown, inconspicuous; stalked sporangia brown with reddish or purple tinges then black but covered with a network of more pallid lines, sub-spherical, ovoid or pear-shaped, thick-walled, smooth; stem concolorous, very short or slightly taller than the sporangium; threads (elaters) spiral, smooth, unbranched. **Spores** yellowish brown, sub-spherical, minutely warty, 9-10 μm.
Occurrence throughout the year; frequent.
■ Inedible.

Lycogala terrestre (L). Fr.   Reticulariaceae

Small, spherical, puffball-like bodies, pink or orange, becoming pallid or grey at maturity, breaking down to reveal powdery pink spore mass; on dead and rotten wood generally.

Dimensions fruiting body 0.5 - 1.5 cm dia.
Fruiting body plasmodium stage orange, peach or cream; peridium more or less spherical (angular if compacted together), maturing to rosy-buff, wall-sturdy, multiple-layered, outer surface containing vesicles filled with yellowish fluid which dries into rounded scaly patches; opening by an apical pore or crack; pseudocapillitium of irregular tubes, wrinkled or minutely warty, connected to the wall of the fruiting body, elastic, 6-12 µm dia. **Spores** at first pink or salmon then yellowish, sub-spherical, covered with a fine network of thin ridges, 6-7.5 µm.
Occurrence throughout the year; frequent.
■ Inedible.
Note: this species is often mistaken for *L. epidendrum* which has a grey peridium and spore mass.

Reticularia lycoperdon Bull.   Reticulariaceae
= *Enteridium lycoperdon* (Bull.) Farr.

Comparatively large, cushion-like, whitish bodies with a somewhat rubbery feel, breaking down to reveal reddish-brown spore mass; on dead wood generally, though often still standing.

Dimensions fruiting body 0.5 - 8 cm dia.
Fruiting body plasmodium stage white; peridium more or less rounded, at first silvery-white then brown from spore deposits, outer wall thick, smooth, brittle; pseudocapillitium of thin threads on which the spores are clustered in batches of 20-80. **Spores** sub-spherical, outer members of cluster covered with a fine network of thin ridges, inner members smooth, 8-9 µm.
Occurrence throughout the year but more prevalent in the spring; frequent.
■ Inedible.

Tubifera ferruginosa (Batsch) Gmelin   Reticulariaceae
= *Stemonitis ferruginosa* Batsch

Small, orange, strawberry-like bodies, becoming dark brown at maturity; on dead and rotten wood of broad-leaf trees but also pine species.

Dimensions fruiting body 0.5 - 5 cm dia (sometimes larger); individual sporangia 0.5 cm tall x 0.03 cm dia.
Fruiting body plasmodium stage orange, vermilion or pink, maturing to beige or greyish-brown; pseudoaethalium more or less rounded, with massed sporangia, cylindrical or ovoid, the thin-walled sporangia opening either by a distinct lid or irregularly; pseudocapillitium absent or sparse, consisting of hollow tubes.
Spores umber brown, sub-spherical, covered with a fine network, 6-8 µm.
Occurrence throughout the year; frequent.
■ Inedible.

Index

When using this index, please note that the specific name of each entry precedes that of the genus. Where single names appear in bold type, they indicate the commencement of a genus section. The full name of a species, in roman, indicates current nomenclature. The full name of a species, in italics, indicates a synonym which the reader may find in use elsewhere.

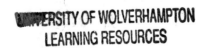